T0331707

Statistical Physics of Fields

While many scientists are familiar with fractals, fewer are cognizant of the concepts of scale-invariance and universality which underlie the ubiquity of such fascinating shapes. These inherent properties emerge from the collective behavior of simple fundamental constituents. The initial chapters smoothly connect the particulate perspective developed in the companion volume, Statistical Physics of Particles, to the coarse grained statistical fields studied in this textbook. It carefully demonstrates how such theories are constructed from basic principles such as symmetry and locality, and studied by innovative methods like the renormalization group. Perturbation theory, exact solutions, renormalization, and other tools are employed to demonstrate the emergence of scale invariance and universality. The book concludes with chapters related to the research of the author on non-equilibrium dynamics of interfaces, and directed paths in random media.

Covering the more advanced applications of statistical mechanics, this textbook is ideal for advanced graduate students in physics. It is based on lectures for a course in statistical physics taught by Professor Kardar at Massachusetts Institute of Technology (MIT). The large number of integrated problems introduce the reader to novel applications such as percolation and roughening. The selected solutions at the end of the book are ideal for self-study and honing calculation methods. Additional solutions are available to lecturers on a password protected website at www.cambridge.org/9780521873413.

MEHRAN KARDAR is Professor of Physics at MIT, where he has taught and researched in the field of Statistical Physics for the past 20 years. He received his B.A. in Cambridge, and gained his Ph.D. at MIT. Professor Kardar has held research and visiting positions as a junior fellow at Harvard, a Guggenheim fellow at Oxford, UCSB, and at Berkeley as a Miller fellow.

In this much-needed modern text, Kardar presents a remarkably clear view of statistical mechanics as a whole, revealing the relationships between different parts of this diverse subject. In two volumes, the classical beginnings of thermodynamics are connected smoothly to a thoroughly modern view of fluctuation effects, stochastic dynamics, and renormalization and scaling theory. Students will appreciate the precision and clarity in which difficult concepts are presented in generality and by example. I particularly like the wealth of interesting and instructive problems inspired by diverse phenomena throughout physics (and beyond!), which illustrate the power and broad applicability of statistical mechanics.

Statistical Physics of Particles includes a concise introduction to the mathematics of probability for physicists, an essential prerequisite to a true understanding of statistical mechanics, but which is unfortunately missing from most statistical mechanics texts. The old subject of kinetic theory of gases is given an updated treatment which emphasizes the connections to hydrodynamics.

As a graduate student at Harvard, I was one of many students making the trip to MIT from across the Boston area to attend Kardar's advanced statistical mechanics class. Finally, in *Statistical Physics of Fields* Kardar makes his fantastic course available to the physics community as a whole! The book provides an intuitive yet rigorous introduction to field-theoretic and related methods in statistical physics. The treatment of renormalization group is the best and most physical I've seen, and is extended to cover the often-neglected (or not properly explained!) but beautiful problems involving topological defects in two dimensions. The diversity of lattice models and techniques are also well-illustrated and complement these continuum approaches. The final two chapters provide revealing demonstrations of the applicability of renormalization and fluctuation concepts beyond equilibrium, one of the frontier areas of statistical mechanics.
Leon Balents, Department of Physics, University of California, Santa Barbara

Statistical Physics of Particles is the welcome result of an innovative and popular graduate course Kardar has been teaching at MIT for almost twenty years. It is a masterful account of the essentials of a subject which played a vital role in the development of twentieth century physics, not only surviving, but enriching the development of quantum mechanics. Its importance to science in the future can only increase with the rise of subjects such as quantitative biology.

Statistical Physics of Fields builds on the foundation laid by the Statistical Physics of Particles, with an account of the revolutionary developments of the past 35 years, many of which were facilitated by renormalization group ideas. Much of the subject matter is inspired by problems in condensed matter physics, with a number of pioneering contributions originally due to Kardar himself. This lucid exposition should be of particular interest to theorists with backgrounds in field theory and statistical mechanics.
David R Nelson, Arthur K Solomon Professor of Biophysics, Harvard University

If Landau and Lifshitz were to prepare a new edition of their classic Statistical Physics text they might produce a book not unlike this gem by Mehran Kardar. Indeed, Kardar is an extremely rare scientist, being both brilliant in formalism and an astoundingly careful and thorough teacher. He demonstrates both aspects of his range of talents in this pair of books, which belong on the bookshelf of every serious student of theoretical statistical physics.

Kardar does a particularly thorough job of explaining the subtleties of theoretical topics too new to have been included even in Landau and Lifshitz's most recent Third Edition (1980), such as directed paths in random media and the dynamics of growing surfaces, which are not in any text to my knowledge. He also provides careful discussion of topics that do appear in most modern texts on theoretical statistical physics, such as scaling and renormalization group.
H Eugene Stanley, Director, Center for Polymer Studies, Boston University

This is one of the most valuable textbooks I have seen in a long time. Written by a leader in the field, it provides a crystal clear, elegant and comprehensive coverage of the field of statistical physics. I'm sure this book will become "the" reference for the next generation of researchers, students and practitioners in statistical physics. I wish I had this book when I was a student but I will have the privilege to rely on it for my teaching.
Alessandro Vespignani, Center for Biocomplexity, Indiana University

Statistical Physics of Fields

Mehran Kardar

Department of Physics
Massachusetts Institute of Technology

CAMBRIDGE
UNIVERSITY PRESS

CAMBRIDGE
UNIVERSITY PRESS

University Printing House, Cambridge CB2 8BS, United Kingdom

Cambridge University Press is part of the University of Cambridge.

It furthers the University's mission by disseminating knowledge in the pursuit of education, learning and research at the highest international levels of excellence.

www.cambridge.org
Information on this title: www.cambridge.org/9780521873413

First published 2007
6th printing 2013

A catalogue record for this publication is available from the British Library

ISBN 978-0-521-87341-3 Hardback

Cambridge University Press has no responsibility for the persistence or accuracy of URLs for external or third-party internet websites referred to in this publication, and does not guarantee that any content on such websites is, or will remain, accurate or appropriate.

Contents

Preface

Many scientists and non-scientists are familiar with fractals, abstract self-similar entities which resemble the shapes of clouds or mountain landscapes. Fewer are familiar with the concepts of scale-invariance and universality which underlie the ubiquity of these shapes. Such properties may emerge from the collective behavior of simple underlying constituents, and are studied through statistical field theories constructed easily on the basis of symmetries. This book demonstrates how such theories are formulated, and studied by innovative methods such as the renormalization group.

The material covered is directly based on my lectures for the second semester of a graduate course on statistical mechanics, which I have been teaching on and off at MIT since 1988. The first semester introduces the student to the basic concepts and tools of statistical physics, and the corresponding material is presented in a companion volume. The second semester deals with more advanced applications – mostly collective phenomena, phase transitions, and the renormalization group, and familiarity with basic concepts is assumed. The primary audience is physics graduate students with a theoretical bent, but also includes postdoctoral researchers and enterprising undergraduates. Since the material is comparatively new, there are fewer textbooks available in this area, although a few have started to appear in the last few years. Starting with the problem of phase transitions, the book illustrates how appropriate statistical field theories can be constructed on the basis of symmetries. Perturbation theory, renormalization group, exact solutions, and other tools are then employed to demonstrate the emergence of scale invariance and universality. The final two chapters deal with non-equilibrium dynamics of interfaces, and directed paths in random media, closely related to the research of the author.

An essential part of learning the material is doing problems; and in teaching the course I developed a large number of problems (and solutions) that have been integrated into the text. Following each chapter there are two sets of problems: solutions to the first set are included at the end of the book, and are intended to introduce additional topics and to reinforce technical tools. There are no solutions provided for a second set of problems which can be used in assignments.

I am most grateful to my many former students for their help in formulating the material, problems, and solutions, typesetting the text and figures, and pointing out various typos and errors. The final editing of the book was accomplished during visits to the Kavli Institute for Theoretical Physics. The support of the National Science Foundation through research grants is also acknowledged.

1
Collective behavior, from particles to fields

1.1 Introduction

One of the most successful aspects of physics in the twentieth century was revealing the atomistic nature of matter, characterizing its elementary constituents, and describing the laws governing the interactions and dynamics of these particles. A continuing challenge is to find out how these underlying elements lead to the myriad of different forms of matter observed in the real world. The emergence of new collective properties in the *macroscopic* realm from the dynamics of the *microscopic* particles is the topic of statistical mechanics.

The microscopic description of matter is in terms of the many degrees of freedom: the set of positions and momenta $\{\vec{p}_i, \vec{q}_i\}$, of particles in a gas, configurations of spins $\{\vec{s}_i\}$, in a magnet, or occupation numbers $\{n_i\}$, in a grand canonical ensemble. The evolution of these degrees of freedom is governed by classical or quantum equations of motion derived from an underlying Hamiltonian $\mathcal{H}$.

The macroscopic description usually involves only a few phenomenological variables. For example, the equilibrium properties of a gas are specified by its pressure P, volume V, temperature T, internal energy E, entropy S. The laws of thermodynamics constrain these equilibrium state functions.

A step-by-step derivation of the macroscopic properties from the microscopic equations of motion is generally impossible, and largely unnecessary. Instead, statistical mechanics provides a *probabilistic connection* between the two regimes. For example, in a canonical ensemble of temperature T, each microstate, μ, of the system occurs with a probability $p(\mu) = \exp\left(-\beta\mathcal{H}(\mu)\right)/Z$, where $\beta = (k_{\mathrm{B}}T)^{-1}$. To insure that the total probability is normalized to unity, the *partition function* $Z(T)$ must equal $\sum_\mu \exp\left(-\beta\mathcal{H}(\mu)\right)$. Thermodynamic information about the *macroscopic* state of the system is then extracted from the *free energy* $F = -k_{\mathrm{B}}T \ln Z$.

While circumventing the dynamics of particles, the recipes of statistical mechanics can be fully carried out only for a small number of simple systems; mostly describing non-interacting collections of particles where the partition function can be calculated exactly. Some effects of interactions can be included by perturbative treatments around such exact solutions. However, even for the relatively simple case of an imperfect gas, the perturbative approach breaks down close to the condensation point. On the other hand, it is precisely the multitude of new phases and properties resulting from interactions that renders macroscopic physics interesting. In particular, we would like to address the following questions:

(1) In the thermodynamic limit ($N \rightarrow \infty$), strong interactions lead to new phases of matter such as solids, liquid crystals, magnets, superconductors, etc. How can we describe the emergence of such distinct macroscopic behavior from the interactions of the underlying particles? What are the thermodynamic variables that describe the macroscopic state of these phases; and what are their identifying signatures in measurements of bulk response functions (heat capacity, susceptibility, etc.)?

(2) What are the characteristic low energy excitations of the system? As in the case of phonons in solids or in superfluid helium, low energy excitations are typically *collective modes*, which involve the coordinated motions of many microscopic degrees of freedom (particles). These modes are easily excited by thermal fluctuations, and probed by scattering experiments.

The underlying microscopic Hamiltonian for the interactions of particles is usually quite complicated, making an *ab initio* particulate approach to the problem intractable. However, there are many common features in the macroscopic behavior of many such systems that can still be fruitfully studied by the methods of statistical mechanics. Although the interactions between constituents are quite specific at the microscopic scale, one may hope that averaging over sufficiently many particles leads to a simpler description. (In the same sense that the central limit theorem ensures that the sum over many random variables has a simple Gaussian probability distribution function.) This expectation is indeed justified in many cases where the collective behavior of the interacting system becomes more simple at long wavelengths and long times. (This is sometimes called the *hydrodynamic limit* by analogy to the Navier–Stokes equations for a fluid of particles.) The averaged variables appropriate to these length and time scales are no longer the discrete set of particle degrees of freedom, but slowly varying continuous *fields*. For example, the velocity field that appears in the Navier–Stokes equations is quite distinct from the velocities of the individual particles in the fluid. Hence the appropriate method for the study of collective behavior in interacting systems is the statistical mechanics of fields. Accordingly, the aims of this book are as follows:

- **Goal:** To learn to describe and classify states of matter, their collective properties, and the mechanisms for transforming from one phase to another.
- **Tools:** Methods of classical field theories; use of symmetries, treatment of nonlinearities by perturbation theory, and the renormalization group (RG) method.
- **Scope:** To provide sufficient familiarity with the material to follow the current literature on such subjects as phase transitions, growth phenomena, polymers, superconductors, etc.

1.2 Phonons and elasticity

The theory of elasticity represents one of the simplest examples of a field theory. We shall demonstrate how certain properties of an elastic medium can be obtained, either by the complicated method of starting from first principles, or by the much simpler means of appealing to symmetries of the problem. As such, it represents a prototype of how much can be learned from a phenomenological approach. The actual example has little to do with the topics that will be covered later on, but it fully illustrates the methodology that will be employed. The task of computing the low temperature heat capacity of a solid can be approached by either *ab initio* or *phenomenological* methods.

Particulate approach

Calculating the heat capacity of a solid material from first principles is rather complicated. We breifly sketch some of the steps:

- The *ab initio* starting point is the Schrödinger equation for electrons and ions which can only be treated approximately, say by a density functional formalism. Instead, we start with a many-body potential energy for the *ionic* coordinates $\mathcal{V}(\vec{q}_1, \vec{q}_2, \cdots, \vec{q}_N)$, which may itself be the outcome of such a quantum mechanical treatment.
- Ideal lattice positions at zero temperature are obtained by minimizing $\mathcal{V}$, typically forming a lattice $\vec{q}^*(\ell, m, n) = [\ell\hat{a} + m\hat{b} + n\hat{c}] \equiv \vec{q}_{\vec{r}}^*$, where $\vec{r} = \{\ell, m, n\}$ is a triplet of integers, and $\hat{a}$, $\hat{b}$, and $\hat{c}$ are unit vectors.
- Small fluctuations about the ideal positions (due to finite temperature or quantum effects) are included by setting $\vec{q}_{\vec{r}} = \vec{q}_{\vec{r}}^* + \vec{u}(\vec{r})$. The cost of deformations in the potential energy is given by

$$\mathcal{V} = \mathcal{V}^* + \frac{1}{2}\sum_{\substack{\vec{r},\vec{r}' \\ \alpha,\beta}} \frac{\partial^2 \mathcal{V}}{\partial q_{\vec{r},\alpha}\partial q_{\vec{r}',\beta}} u_\alpha(\vec{r})\, u_\beta(\vec{r}') + O(u^3), \tag{1.1}$$

where the indices α and β denote spatial components. (Note that the first derivative of $\mathcal{V}$ vanishes at the equilibrium position.) The full Hamiltonian for small deformations is obtained by adding the kinetic energy $\sum_{\vec{r},\alpha} p_\alpha(\vec{r})^2/2m$ to Eq. (1.1), where $p_\alpha(\vec{r})$ is the momentum conjugate to $u_\alpha(\vec{r})$.
- The next step is to find the normal modes of vibration (phonons) by diagonalizing the matrix of derivatives. Since the ground state configuration is a regular lattice, the elements of this matrix must satisfy various translation and rotation symmetries. For example, they can only depend on the difference between the position vectors of ions $\vec{r}$ and $\vec{r}'$, i.e.

$$\frac{\partial^2 \mathcal{V}}{\partial q_{\vec{r},\alpha}\partial q_{\vec{r}',\beta}} = K_{\alpha\beta}(\vec{r} - \vec{r}'). \tag{1.2}$$

This translational symmetry allows us to at least partially diagonalize the Hamiltonian by using the Fourier modes,

$$u_\alpha(\vec{r}) = \sum_{\vec{k}}' \frac{e^{i\vec{k}\cdot\vec{r}}}{\sqrt{N}} u_\alpha(\vec{k}). \tag{1.3}$$

(The above sum is restricted, in that only *wavevectors* $\vec{k}$ inside the first *Brillouin zone* contribute to the sum.) The Hamiltonian then reads

$$\mathcal{H} = \mathcal{V}^* + \frac{1}{2} \sum_{\vec{k},\alpha,\beta} \left[\frac{|p_\alpha(\vec{k})|^2}{m} + K_{\alpha\beta}(\vec{k})u_\alpha(\vec{k})u_\beta(\vec{k})^* \right], \qquad (1.4)$$

where $u_\beta(\vec{k})^*$ is the complex conjugate of $u_\beta(\vec{k})$. While the precise form of the Fourier transformed matrix $K_{\alpha\beta}(\vec{k})$ is determined by the microscopic interactions, it has to respect the underlying symmetries of the crystallographic point group. Let us assume that diagonalizing this 3×3 matrix yields eigenvalues $\left\{ \kappa_\alpha(\vec{k}) \right\}$. The quadratic part of the Hamiltonian is now decomposed into a set of independent (non-interacting) harmonic oscillators.

• The final step is to quantize each oscillator, leading to

$$\mathcal{H} = \mathcal{V}^* + \sum_{\vec{k},\alpha} \hbar\omega_\alpha(\vec{k}) \left(n_\alpha(\vec{k}) + \frac{1}{2} \right), \qquad (1.5)$$

where $\omega_\alpha(\vec{k}) = \sqrt{\kappa_\alpha(\vec{k})/m}$, and $\{n_\alpha(\vec{k})\}$ are the set of occupation numbers. The average energy at a temperature T is given by

$$E(T) = \mathcal{V}^* + \sum_{\vec{k},\alpha} \hbar\omega_\alpha(\vec{k}) \left(\left\langle n_\alpha(\vec{k}) \right\rangle + \frac{1}{2} \right), \qquad (1.6)$$

where we know from elementary statistical mechanics that the average occupation numbers are given by $\langle n_\alpha(\vec{k}) \rangle = 1/\left(\exp(\frac{\hbar\omega_\alpha}{k_B T}) - 1 \right)$. Clearly $E(T)$, and other macroscopic functions, have a complex behavior, dependent upon microscopic details through $\left\{ \kappa_\alpha(\vec{k}) \right\}$. Are there any features of these functions (e.g. the functional dependence as $T \to 0$) that are independent of microscopic features? The answer is positive, and illustrated with a one-dimensional example.

Fig. 1.1 Displacements $\{u_n\}$ of a one-dimensional chain of particles, and the coarse-grained field $u(x)$ of the continuous string.

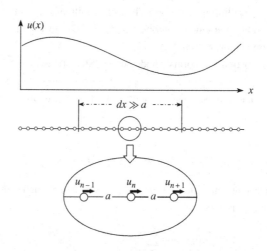

Consider a chain of particles, constrained to move in one dimension. A most general quadratic potential energy for deformations $\{u_n\}$, around an average separation of a, is

$$\mathcal{V} = \mathcal{V}^* + \frac{K_1}{2} \sum_n (u_{n+1} - u_n)^2 + \frac{K_2}{2} \sum_n (u_{n+2} - u_n)^2 + \cdots, \qquad (1.7)$$

where $\{K_i\}$ can be regarded as the Hookian constants of springs connecting particles that are i-th neighbors. The decomposition to normal modes is achieved via

$$u_n = \int_{-\pi/a}^{\pi/a} \frac{dk}{2\pi} e^{-ikna} u(k), \quad \text{where} \quad u(k) = a \sum_n e^{ikna} u_n . \qquad (1.8)$$

(Note the difference in normalizations from Eq. 1.3.) The potential energy,

$$\mathcal{V} = \mathcal{V}^* + \frac{K_1}{2} \sum_n \int_{-\pi/a}^{\pi/a} \frac{dk}{2\pi} \frac{dk'}{2\pi} (e^{ika} - 1)(e^{ik'a} - 1) e^{-i(k+k')na} u(k)u(k') + \cdots, \qquad (1.9)$$

can be simplified by using the identity $\sum_n e^{-i(k+k')na} = \delta(k+k')2\pi/a$, and noting that $u(-k) = u^*(k)$, to

$$\mathcal{V} = \mathcal{V}^* + \frac{1}{2a} \int_{-\pi/a}^{\pi/a} \frac{dk}{2\pi} [K_1(2 - 2\cos ka) + K_2(2 - 2\cos 2ka) + \cdots] |u(k)|^2. \qquad (1.10)$$

A typical frequency spectrum of normal modes, given by $\omega(k) = \sqrt{[2K_1(1 - \cos ka) + \cdots]/m}$, is depicted in Fig. 1.2. In the limit $k \to 0$, the dispersion relation becomes linear, $\omega(k) \to v|k|$, and from its slope we can identify a 'sound velocity' $v = a\sqrt{\overline{K}/m}$, where $\overline{K} = K_1 + 4K_2 + \cdots$.

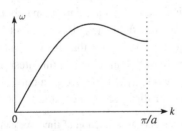

Fig. 1.2 Typical dispersion relation for phonons along a chain.

The internal energy of these excitations, for a chain of N particles, is

$$E(T) = \mathcal{V}^* + Na \int_{-\pi/a}^{\pi/a} \frac{dk}{2\pi} \frac{\hbar\omega(k)}{\exp(\hbar\omega(k)/k_B T) - 1}. \qquad (1.11)$$

As $T \to 0$, only modes with $\hbar\omega(k) < k_B T$ are excited. Hence only the $k \to 0$ part of the excitation spectrum is important and $E(T)$ simplifies to

$$E(T) \approx \mathcal{V}^* + Na \int_{-\infty}^{\infty} \frac{dk}{2\pi} \frac{\hbar v|k|}{\exp(\hbar v|k|/k_B T) - 1} = \mathcal{V}^* + Na \frac{\pi^2}{6\hbar v} (k_B T)^2. \qquad (1.12)$$

Note

(1) While the full spectrum of excitation energies can be quite complicated, as $k \to 0$,

$$\frac{K(k)}{2} = K_1(1 - \cos ka) + K_2(1 - \cos 2ka) + \cdots \to \frac{\bar{K}}{2}k^2 \quad \text{where,}$$

$$\overline{K} = K_1 + 4K_2 + \cdots .$$

(1.13)

Thus, further neighbor interactions change the speed of sound, but not the form of the dispersion relation as $k \to 0$.

(2) The heat capacity $C(T) = dE/dT$ is proportional to T. This dependence is a *universal* property, i.e. not material specific, and independent of the choice of the interatomic interactions.

(3) The T^2 dependence of energy comes from excitations with $k \to 0$ (or $\lambda \to \infty$), i.e. from collective modes involving many particles. These are precisely the modes for which statistical considerations may be meaningful.

Phenomenological (field) approach

We now outline a mesoscopic approach to the same problem, and show how it provides additional insights and is easily generalized to higher dimensions. Typical excitations at low temperatures have wavelengths $\lambda > \lambda(T) \approx (\hbar v / k_B T)$ $\gg a$, where a is the lattice spacing. We can eliminate the unimportant short wavelength modes by an averaging process known as **coarse graining**. The idea is to consider a point x, and an interval of size dx around it (Fig. 1.1). We shall choose $a \ll dx \ll \lambda(T)$, i.e. the interval is large enough to contain many lattice points, but much shorter than the characteristic wavelength of typical phonons. In this interval all the displacements u are approximately the same; and we can define an average deformation field $u(x)$. By construction, the function $u(x)$ varies slowly over dx, and despite the fact that this interval contains many lattice points, from the perspective of the function it is infinitesimal in size. We should always keep in mind that while $u(x)$ is treated as a continuous function, it does not have any variations over distances comparable to the lattice spacing a.

- By examining the displacements as a function of time, we can define a velocity field $\dot{u}(x) \equiv \partial u / \partial t$. The kinetic energy is then related to the mass density $\rho = m/a$ via $\rho \int dx \dot{u}(x)^2 / 2$.
- What is the most general potential energy functional $\mathcal{V}[u]$, for the chain? A priori, we don't know much about the form of $\mathcal{V}[u]$, but we can construct it by using the following general principles:

  Locality: In most situations, the interactions between particles are short range, allowing us to define a potential energy *density* Φ at each point x, with $\mathcal{V}[u] = \int dx \Phi\big(u(x), \partial u / \partial x, \cdots\big)$. Naturally, by including all derivatives we can also describe long-range interactions. In this context, the term *locality* implies that the higher derivative terms are less significant.

Translational symmetry: A uniform translation of the chain does not change its internal energy, and hence the energy density must satisfy the constraint $\Phi[u(x) + c] = \Phi[u(x)]$. This implies that Φ cannot depend directly on $u(x)$, but only on its derivatives $\partial u/\partial x$, $\partial^2 u/\partial x^2$, $\cdots$.

Stability: Since the fluctuations are around an *equilibrium* solution, there can be no linear terms in u or its derivatives. (Stability further requires that the quadratic part of $\mathcal{V}[u]$ must be positive definite.)

The most general potential consistent with these constraints can be expanded as a power series

$$\mathcal{V}[u] = \int dx \left[\frac{K}{2} \left(\frac{\partial u}{\partial x} \right)^2 + \frac{L}{2} \left(\frac{\partial^2 u}{\partial x^2} \right)^2 + \cdots + M \left(\frac{\partial u}{\partial x} \right)^2 \left(\frac{\partial^2 u}{\partial x^2} \right) + \cdots \right], \quad (1.14)$$

which after Fourier transformation gives

$$\mathcal{V}[u] = \int \frac{dk}{2\pi} \left[\frac{K}{2} k^2 + \frac{L}{2} k^4 + \cdots \right] |u(k)|^2$$
$$- iM \int \frac{dk_1}{2\pi} \frac{dk_2}{2\pi} k_1 k_2 (k_1 + k_2)^2 u(k_1) u(k_2) u(-k_1 - k_2) + \cdots . \quad (1.15)$$

As $k \to 0$, higher order gradient terms (such as the term proportional to L) become unimportant. Also, for small deformations we may neglect terms beyond second order in u (such as the cubic term with coefficient M). Another assumption employed in constructing Eq. (1.14) is that the mirror image deformations $u(x)$ and $u(-x)$ have the same energy. This may not be valid in more complicated lattices without inversion symmetry.

Adding the kinetic energy, we get a simple one-dimensional field theory, with a Hamiltonian

$$\mathcal{H} = \frac{\rho}{2} \int dx \left[\left(\frac{\partial u}{\partial t} \right)^2 + v^2 \left(\frac{\partial u}{\partial x} \right)^2 \right] .$$

This is a one-dimensional elastic (string) theory with material dependent constants ρ and $v = \sqrt{K/\rho}$. While the phenomenological approach cannot tell us the value of these parameters, it does show that the low energy excitations satisfy the dispersion relation $\omega = v|k|$ (obtained by examining the Fourier modes).

We can now generalize the elastic theory of the string to arbitrary dimensions d: The discrete particle deformations $\{\vec{u}_n\}$ are coarse grained into a continuous deformation field $\vec{u}(\vec{x})$. For an *isotropic* material, the potential energy $\mathcal{V}[\vec{u}]$ must be invariant under both rotations and translations (described by $u_\alpha(\vec{x}) \mapsto R_{\alpha\beta} u_\beta(\vec{x}) + c_\alpha$, where $R_{\alpha\beta}$ is a rotation matrix). A useful local quantity is the symmetric *strain field*,

$$u_{\alpha\beta}(\vec{x}) = \frac{1}{2} \left(\frac{\partial u_\alpha}{\partial x_\beta} + \frac{\partial u_\beta}{\partial x_\alpha} \right), \quad (1.16)$$

in terms of which the most general quadratic deformation Hamiltonian is

$$\mathcal{H} = \frac{1}{2} \int d^d \vec{x} \left[\sum_{\alpha} \left(\rho \frac{\partial u_\alpha}{\partial t} \frac{\partial u_\alpha}{\partial t} \right) + \sum_{\alpha,\beta} \left(2\mu\, u_{\alpha\beta} u_{\alpha\beta} + \lambda\, u_{\alpha\alpha} u_{\beta\beta} \right) \right]. \tag{1.17}$$

The elastic moduli μ and λ are known as *Lamé coefficients*. Summing over the repeated indices ensures that the result is rotationally invariant. This rotational invariance is more transparent in the Fourier basis, $\vec{u}(\vec{k}) = \int d^d\vec{x} e^{i\vec{k}.\vec{x}} \vec{u}(\vec{x})$, since the Hamiltonian

$$\mathcal{H} = \int \frac{d^d\mathbf{k}}{(2\pi)^d} \left[\frac{\rho}{2} |\dot{\vec{u}}(\vec{k})|^2 + \frac{\mu}{2} k^2 |\vec{u}(\vec{k})|^2 + \frac{\mu+\lambda}{2} \left(\vec{k}.\vec{u}(\vec{k}) \right)^2 \right], \tag{1.18}$$

manifestly includes only rotationally invariant quantities $\vec{k} \cdot \vec{k}$, $\vec{u} \cdot \vec{u}$, and $\vec{k} \cdot \vec{u}$. We can further decompose the Hamiltonian into two types of sound modes: *longitudinal modes* where $\vec{k} \parallel \vec{u}$, with $v_\ell = \sqrt{(2\mu+\lambda)/\rho}$, and *transverse modes* with $\vec{k} \perp \vec{u}$, where $v_t = \sqrt{\mu/\rho}$. The internal energy in a volume L^d is then given by

$$\begin{aligned} E(t) &= L^d \int \frac{d^d\mathbf{k}}{(2\pi)^d} \left[\frac{\hbar v_\ell k}{\exp(\hbar v_\ell k/k_B T) - 1} + \frac{(d-1)\hbar v_t k}{\exp(\hbar v_t k/k_B T) - 1} \right] \\ &\approx \mathcal{A}(v_\ell, v_t) L^d (k_B T)^{d+1}. \end{aligned} \tag{1.19}$$

The specific heat now vanishes as $C \propto T^d$, as $T \to 0$.

Note

(1) All material dependent parameters end up in the coefficient $\mathcal{A}$, while the scaling with T is *universal*.
(2) The universal exponent originates from the (hydrodynamic) modes with $\vec{k} \to 0$. The high frequency (short wavelength) modes come into play only at high temperatures.
(3) The scaling exponent depends on dimensionality and the range of interactions. (Long-range Coulomb interactions lead to a different result.)
(4) Experimental observation of a power law alerts us to the physics. For example, in superfluid helium, the observation of $C \propto T^3$ (as opposed to $C \propto T^{3/2}$ expected for an ideal Bose gas), immediately implies phonon-like excitations as noted by Landau.

There are many other well known examples demonstrating the universality and importance of power laws. For example, consider a cloud of tracers moving in some unspecified medium. The scaling of some characteristic dimension x with time t can alert us to the possible dynamics that governs the motion of the particles. Three simple possibilities are:

(1) *Diffusion,* in which case $x \propto \sqrt{Dt}$.
(2) *Dissipate transport,* where $x \propto vt$.
(3) *Free forced motion,* where $x \propto gt^2/2$, as in a gravitational field.

The Navier–Stokes equation for fluid flow is yet another example. We can use these examples to construct general guidelines for setting up and analyzing phenomenological field theories. Some of the steps in the procedure are:

(1) **Input** for construction of the coarse grained Hamiltonian comes from considerations of symmetry, range of interactions, and dimensionality.
(2) Unlike the above example, in general nonlinearities cannot be ignored in the resulting effective field theory. We shall learn how to treat such nonlinearities by the methods of perturbation theory and the renormalization group.
(3) **Output** of the analysis is expressed in terms of universal exponents, and other functional dependencies that can then be compared with observations.

1.3 Phase transitions

The most spectacular consequence of interactions among particles is the appearance of new phases of matter whose collective behavior bears little resemblance to that of a few particles. How do the particles then transform from one macroscopic state to a completely different one? From a formal perspective, all macroscopic properties can be deduced from the free energy or the partition function. Since phase transitions typically involve dramatic changes in various response functions, they must correspond to singularities in the free energy. The canonical partition function for a finite collection of particles is always an analytical function. Hence phase transitions, and their associated non-analyticities, are only obtained for infinitely many particles, i.e. in the *thermodynamic limit*, $N \to \infty$. The study of phase transitions is thus related to finding the origin of various singularities in the free energy and characterizing them.

The classical example of a phase transition is the condensation of a gas into a liquid. Two perspectives of the phase diagram for a typical system of particles is given in Fig. 1.3. Some important features of the liquid–gas condensation transition are:

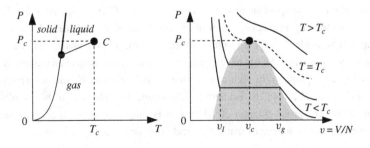

Fig. 1.3 Schematic phase diagrams of a typical system in the pressure, temperature coordinates (*left*). The isotherms in the pressure, specific volume coordinates (*right*) are flat in the coexistence region below the critical point.

(1) In the temperature/pressure plane, (T, P), the phase transition occurs along a line that terminates at a *critical point* (T_c, P_c).

(2) In the volume/pressure plane, $(P, v \equiv V/N)$, the transition appears as a *coexistence interval*, corresponding to a mixture of gas and liquids of densities $\rho_g = 1/v_g$, and $\rho_\ell = 1/v_\ell$, at temperatures $T < T_c$.

(3) Due to the termination of the coexistence line, it is possible to go from the gas phase to the liquid phase continuously (without a phase transition) by going around the critical point. Thus there are no fundamental differences between liquid and gas phases.

From a mathematical perspective, the free energy of the system is an analytical function in the (P, T) plane, except for a branch cut along the phase boundary. Observations in the vicinity of the critical point further indicate that:

(4) The difference between the densities of coexisting liquid and gas phases vanishes on approaching T_c, i.e. $\rho_{\text{liquid}} \to \rho_{\text{gas}}$, as $T \to T_c^-$.

(5) The pressure versus volume isotherms become progressively more flat on approaching T_c from the high temperature side. This implies that the isothermal compressibility, $\kappa_T = -\partial V/\partial P|_T /V$, diverges as $T \to T_c^+$.

(6) The fluid appears "milky" close to criticality. This phenomenon, known as *critical opalescence*, suggests that collective fluctuations occur in the gas at long enough wavelengths to scatter visible light. These fluctuations must necessarily involve many particles, and a coarse-graining procedure may thus be appropriate to their description.

Fig. 1.4 *Left:* Phase diagram for a typical magnet in the magnetic-field, temperature coordinates. *Right:* Magnetization versus field isotherms.

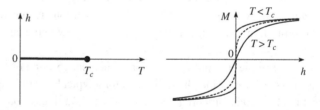

A related, but possibly less familiar, phase transition occurs between paramagnetic and ferromagnetic phases of certain substances such as iron or nickel. These materials become spontaneously magnetized below a Curie temperature T_c. There is a discontinuity in magnetization of the substance as the magnetic field h goes through zero for $T < T_c$. The phase diagram in the (h, T) plane, and the magnetization isotherms $M(h)$, have much in common with their counterparts in the condensation problem. In both cases a line of discontinuous transitions terminates at a critical point, and the isotherms exhibit singular behavior in the vicinity of this point. The phase diagram of the magnet is simpler in appearance, because the symmetry $h \mapsto -h$ ensures that the critical point occurs at $h_c = M_c = 0$.

1.4 Critical behavior

The singular behavior in the vicinity of a critical point is characterized by a set of *critical exponents*. These exponents describe the non-analyticity of various thermodynamic functions. The most commonly encountered exponents are listed below:

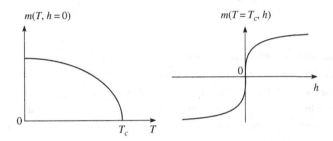

Fig. 1.5 Singular behavior for the magnetization of a magnet.

- **The order parameter:** By definition, there is more than one equilibrium phase on a coexistence line. The order parameter is a thermodynamic function that is different in each phase, and hence can be used to distinguish between them. For a magnet, the magnetization

$$m(T) = \frac{1}{V} \lim_{h \to 0} M(h, T),$$

serves as the order parameter. In zero field, m vanishes for a paramagnet and is non-zero in a ferromagnet, and close to the transition behaves as

$$m(T, h = 0) \propto \begin{cases} 0 & \text{for} \quad T > T_c, \\ |t|^\beta & \text{for} \quad T < T_c, \end{cases} \tag{1.20}$$

where $t = (T_c - T)/T_c$ is the *reduced temperature*. The singular behavior of the order parameter along the coexistence line is traditionally indicated by the critical exponent β. The phase transition disappears in a finite magnetic field, i.e. the magnetization varies continuously with T at a finite h. This non-analyticity is thus reminiscent of a *branch cut* in a complex plane. The magnetization is non-analytic for any trajectory in the (T, h) plane that passes through the critical point at $T = T_c$ and $h = 0$. In particular, it vanishes along the critical isotherm, with another exponent denoted by δ, defined through

$$m(T = T_c, h) \propto h^{1/\delta}. \tag{1.21}$$

The two phases along the liquid–gas coexistence line can be differentiated by their densities ρ_g and ρ_ℓ. Their differences from the critical point density, $\rho_\ell - \rho_c$ and $\rho_c - \rho_g$ can serve as the order parameter. In the vicinity of the critical point, both

Fig. 1.6 Divergence of the
susceptibility is described
by the critical exponents
$\gamma_{\pm}$.

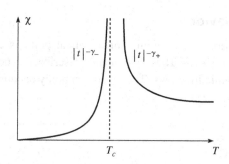

these quantities, as well as $\rho_\ell - \rho_g$ vanish with the exponent β. The exponent δ now
describes the singular form of the critical isotherm in Fig. 1.3.

- **Response functions:** The critical system is quite sensitive to external perturbations,
  as typified by the infinite compressibility at the liquid–gas critical point.

    The divergence in the response of the order parameter to a field conjugate to it is
    indicated by an exponent γ. For example, in a magnet,

$$\chi_{\pm}(T, h = 0) \propto |t|^{-\gamma_{\pm}}, \tag{1.22}$$

where in principle two exponents γ_+ and γ_- are necessary to describe the divergences
on the two sides of the phase transition. Actually in almost all cases, the same
singularity governs both sides and $\gamma_+ = \gamma_- = \gamma$.

Fig. 1.7 A negative heat
capacity exponent α may
(*left*) or may not (*right*)
describe a cusp
singularity depending on
the relative signs of
amplitudes.

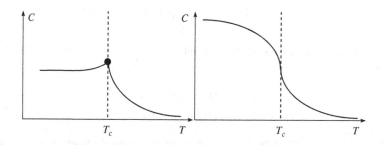

The heat capacity is the thermal response function, and its singularities at zero
field are described by the exponent α, i.e.

$$C_{\pm}(T, h = 0) \propto |t|^{-\alpha_{\pm}}. \tag{1.23}$$

A positive α corresponds to a divergence as in Fig. 1.6, while a negative α describes
a finite heat capacity, possibly with a cusp, as in Fig. 1.7. We shall also encounter
cases where α is zero.

- **Long-range correlations:** There is an intimate connection between the macroscopic
  response functions, and fluctuations of microscopic constituents of a material. In fact
  divergences of the response functions at a critical point imply, and are a consequence
  of, correlated fluctuations over large distances. Let us explore this connection in the

context of the susceptibility of a magnet. A convenient starting point is the (Gibbs) partition function in a magnetic field h, given by

$$Z(h) = \text{tr}\{\exp[-\beta\mathcal{H}_0 + \beta hM]\}, \qquad (1.24)$$

where $\mathcal{H}_0$ describes the internal energy of the magnet, while $-hM$ is the work done against the magnetic field to produce a magnetization M. The symbol tr is used to indicate the sum over all microscopic degrees of freedom. The equilibrium magnetization is computed from

$$\langle M \rangle = \frac{\partial \ln Z}{\partial(\beta h)} = \frac{1}{Z}\text{tr}\{M\exp[-\beta\mathcal{H}_0 + \beta hM]\}, \qquad (1.25)$$

and the susceptibility is then related to the variance of magnetization by

$$\chi = \frac{\partial M}{\partial h} = \beta\left\{\frac{1}{Z}\text{tr}\left[M^2\exp(-\beta\mathcal{H}_0 + \beta hM)\right] - \frac{1}{Z^2}\text{tr}\left[M\exp(-\beta\mathcal{H}_0 + \beta hM)\right]^2\right\}$$

$$= \frac{1}{k_B T}\left(\langle M^2\rangle - \langle M\rangle^2\right). \qquad (1.26)$$

The overall magnetization is obtained by adding contributions from different parts of the system as

$$M = \int d^3\vec{r}\, m(\vec{r}), \qquad (1.27)$$

where $m(\vec{r})$ is the local magnetization at position $\vec{r}$. (For the time being we treat the magnetization as a scalar quantity.) Substituting the above into Eq. (1.26) gives

$$k_B T\chi = \int d^3\vec{r}\, d^3\vec{r}'\left(\langle m(\vec{r})m(\vec{r}')\rangle - \langle m(\vec{r})\rangle\langle m(\vec{r}')\rangle\right). \qquad (1.28)$$

Translational symmetry of a homogeneous system implies that $\langle m(\vec{r})\rangle = m$ is a constant, while $\langle m(\vec{r})m(\vec{r}')\rangle = G(\vec{r}-\vec{r}')$ depends only on the separation. We can express the result in terms of the *connected correlation function*, defined as

$$\langle m(\vec{r})m(\vec{r}')\rangle_c \equiv \langle(m(\vec{r})-\langle m(\vec{r})\rangle)(m(\vec{r}')-\langle m(\vec{r}')\rangle)\rangle = G(\vec{r}-\vec{r}') - m^2. \quad (1.29)$$

Integrating over the center of mass coordinates in Eq. (1.28) results in a factor of volume V, and the susceptibility is given by

$$\chi = \beta V\int d^3\vec{r}\,\langle m(\vec{r})m(0)\rangle_c. \qquad (1.30)$$

The left-hand side of the above expression is the bulk response function, which is related to the integral of the connected correlation functions of the microscopic degrees of freedom. The connected correlation function is a measure of how the local fluctuations in one part of the system effect those of another part. Typically such influences occur over a characteristic distance ξ, called the *correlation length*. (It can be shown rigorously that this function must decay to zero at large separations; in many cases $G_c(\vec{r}) \equiv \langle m(\vec{r})m(0)\rangle_c$ decays as $\exp(-|\vec{r}|/\xi)$ at separations $|\vec{r}| > \xi$.) The microscopic correlations can be probed by scattering experiments. For example, the milky appearance (critical opalescence) of a near critical liquid–gas mixture is due to the scattering of light by density fluctuations. Since the wavelength of light is much larger than typical atomic distances, the density fluctuations must be correlated over long distances to scatter visible light.

Let g denote a typical value of the correlation function for $|\vec{r}| < \xi$. It then follows from Eq. (1.30) that $k_B T \chi / V < g \xi^3$; and $\chi \to \infty$, necessarily implies $\xi \to \infty$. This divergence of the correlation length explains the observation of critical opalescence. On approaching the critical point, the correlation length diverges as

$$\xi_\pm(T, h = 0) \propto |t|^{-\nu_\pm},\tag{1.31}$$

characterized by exponents $\nu_+ = \nu_- = \nu$.

Problems for chapter 1

1. *The binary alloy:* A binary alloy (as in β brass) consists of N_A atoms of type A, and N_B atoms of type B. The atoms form a simple cubic lattice, each interacting only with its six nearest neighbors. Assume an attractive energy of $-J$ $(J > 0)$ between like neighbors $A - A$ and $B - B$, but a repulsive energy of $+J$ for an $A - B$ pair.

   (a) What is the minimum energy configuration, or the state of the system at zero temperature?

   (b) Estimate the total interaction energy assuming that the atoms are randomly distributed among the N sites; i.e. each site is occupied independently with probabilities $p_A = N_A/N$ and $p_B = N_B/N$.

   (c) Estimate the mixing entropy of the alloy with the same approximation. Assume $N_A, N_B \gg 1$.

   (d) Using the above, obtain a free energy function $F(x)$, where $x = (N_A - N_B)/N$. Expand $F(x)$ to the fourth order in x, and show that the requirement of convexity of F breaks down below a critical temperature T_c. For the remainder of this problem use the expansion obtained in (d) in place of the full function $F(x)$.

   (e) Sketch $F(x)$ for $T > T_c$, $T = T_c$, and $T < T_c$. For $T < T_c$ there is a range of compositions $x < |x_{sp}(T)|$ where $F(x)$ is not convex and hence the composition is locally unstable. Find $x_{sp}(T)$.

   (f) The alloy globally minimizes its free energy by separating into A rich and B rich phases of compositions $\pm x_{eq}(T)$, where $x_{eq}(T)$ minimizes the function $F(x)$. Find $x_{eq}(T)$.

   (g) In the (T, x) plane sketch the phase separation boundary $\pm x_{eq}(T)$; and the so called spinodal line $\pm x_{sp}(T)$. (The spinodal line indicates onset of metastability and hysteresis effects.)

2. *The Ising model of magnetism:* The local environment of an electron in a crystal sometimes forces its spin to stay parallel or anti-parallel to a given lattice direction. As a model of magnetism in such materials we denote the direction of the spin by a

single variable $\sigma_i = \pm 1$ (an Ising spin). The energy of a configuration $\{\sigma_i\}$ of spins is then given by

$$\mathcal{H} = \frac{1}{2} \sum_{i,j=1}^{N} J_{ij}\sigma_i\sigma_j - h \sum_i \sigma_i;$$

where h is an external magnetic field, and J_{ij} is the interaction energy between spins at sites i and j.

(a) For N spins we make the drastic *approximation* that the interaction between all spins is the same, and $J_{ij} = -J/N$ (the equivalent neighbor model). Show that the energy can now be written as $E(M, h) = -N[Jm^2/2 + hm]$, with a magnetization $m = \sum_{i=1}^{N} \sigma_i/N = M/N$.

(b) Show that the partition function $Z(h, T) = \sum_{\{\sigma_i\}} \exp(-\beta\mathcal{H})$ can be rewritten as $Z = \sum_M \exp[-\beta F(m, h)]$; with $F(m, h)$ easily calculated by analogy to problem (1). For the remainder of the problem work only with $F(m, h)$ expanded to fourth order in m.

(c) By saddle point integration show that the actual free energy $F(h, T) = -k_B T \ln Z(h, T)$ is given by $F(h, T) = \min[F(m, h)]_m$. When is the saddle point method valid? Note that $F(m, h)$ is an analytic function but not convex for $T < T_c$, while the true free energy $F(h, T)$ is convex but becomes non-analytic due to the minimization.

(d) For $h = 0$ find the critical temperature T_c below which spontaneous magnetization appears; and calculate the magnetization $\overline{m}(T)$ in the low temperature phase.

(e) Calculate the singular (non-analytic) behavior of the response functions

$$C = \left.\frac{\partial E}{\partial T}\right|_{h=0}, \quad \text{and} \quad \chi = \left.\frac{\partial \overline{m}}{\partial h}\right|_{h=0}.$$

3. *The lattice-gas model:* Consider a gas of particles subject to a Hamiltonian

$$\mathcal{H} = \sum_{i=1}^{N} \frac{\vec{p}_i^{\,2}}{2m} + \frac{1}{2}\sum_{i,j} \mathcal{V}(\vec{r}_i - \vec{r}_j), \quad \text{in a volume } V.$$

(a) Show that the grand partition function Ξ can be written as

$$\Xi = \sum_{N=0}^{\infty} \frac{1}{N!}\left(\frac{e^{\beta\mu}}{\lambda^3}\right)^N \int \prod_{i=1}^{N} d^3\vec{r}_i \exp\left[-\frac{\beta}{2}\sum_{i,j} \mathcal{V}(\vec{r}_i - \vec{r}_j)\right].$$

(b) The volume V is now subdivided into $\mathcal{N} = V/a^3$ cells of volume a^3, with the spacing a chosen small enough so that each cell α is either empty or occupied by one particle; i.e. the cell occupation number n_α is restricted to 0 or 1 ($\alpha = 1, 2, \cdots, \mathcal{N}$). After approximating the integrals $\int d^3\vec{r}$ by sums $a^3 \sum_{\alpha=1}^{\mathcal{N}}$, show that

$$\Xi \approx \sum_{\{n_\alpha = 0,1\}} \left(\frac{e^{\beta\mu}a^3}{\lambda^3}\right)^{\sum_\alpha n_\alpha} \exp\left[-\frac{\beta}{2}\sum_{\alpha,\beta=1}^{\mathcal{N}} n_\alpha n_\beta \mathcal{V}(\vec{r}_\alpha - \vec{r}_\beta)\right].$$

(c) By setting $n_\alpha = (1 + \sigma_\alpha)/2$ and approximating the potential by $\mathcal{V}(\vec{r}_\alpha - \vec{r}_\beta) = -J/\mathcal{N}$, show that this model is identical to the one studied in problem (2). What does this imply about the behavior of this imperfect gas?

---

4. *Surfactant condensation:* N surfactant molecules are added to the surface of water over an area A. They are subject to a Hamiltonian

$$\mathcal{H} = \sum_{i=1}^{N} \frac{\vec{p}_i^{\,2}}{2m} + \frac{1}{2} \sum_{i,j} \mathcal{V}(\vec{r}_i - \vec{r}_j),$$

where $\vec{r}_i$ and $\vec{p}_i$ are two dimensional vectors indicating the position and momentum of particle i.

(a) Write down the expression for the partition function $Z(N, T, A)$ in terms of integrals over $\vec{r}_i$ and $\vec{p}_i$, and perform the integrals over the momenta.

The interparticle potential $\mathcal{V}(\vec{r})$ is infinite for separations $|\vec{r}| < a$, and attractive for $|\vec{r}| > a$ such that $\int_a^\infty 2\pi r dr \mathcal{V}(r) = -u_0$.

(b) Estimate the total non-excluded area available in the positional phase space of the system of N particles.

(c) Estimate the total *potential* energy of the system, *assuming a uniform density* $n = N/A$. Using this potential energy for all configurations allowed in the previous part, write down an approximation for Z.

(d) The surface tension of water without surfactants is σ_0, approximately independent of temperature. Calculate the surface tension $\sigma(n, T)$ in the presence of surfactants.

(e) Show that below a certain temperature, T_c, the expression for σ is manifestly incorrect. What do you think happens at low temperatures?

(f) Compute the heat capacities, C_A, and write down an expression for C_σ without explicit evaluation, due to the surfactants.

5. *Critical behavior of a gas:* The pressure P of a gas is related to its density $n = N/V$, and temperature T by the truncated expansion

$$P = k_B T n - \frac{b}{2} n^2 + \frac{c}{6} n^3,$$

where b and c are assumed to be positive, temperature independent constants.

(a) Locate the critical temperature T_c below which this equation must be invalid, and the corresponding density n_c and pressure P_c of the critical point. Hence find the ratio $k_B T_c n_c / P_c$.

(b) Calculate the isothermal compressibility $\kappa_T = -\dfrac{1}{V} \dfrac{\partial V}{\partial P}\bigg|_T$, and sketch its behavior as a function of T for $n = n_c$.

(c) On the critical isotherm give an expression for $(P - P_c)$ as a function of $(n - n_c)$.

(d) The instability in the isotherms for $T < T_c$ is avoided by phase separation into a liquid of density n_+ and gas of density n_-. For temperatures close to T_c, these densities behave as $n_\pm \approx n_c (1 \pm \delta)$. Using a Maxwell construction, or otherwise, find an implicit equation for $\delta(T)$, and indicate its behavior for $(T_c - T) \to 0$. (Hint: Along an isotherm, variations of chemical potential obey $d\mu = dP/n$.)

(e) Now consider a gas obeying Dieterici's equation of state:

$$P(v - b) = k_B T \exp\left(-\frac{a}{k_B T v}\right),$$

where $v = V/N$. Find the ratio $Pv/k_B T$ at its critical point.

(f) Calculate the isothermal compressibility κ_T for $v = v_c$ as a function of $T - T_c$ for the Dieterici gas.

(g) On the Dieterici critical isotherm expand the pressure to the lowest non-zero order in $(v - v_c)$.

6. *Magnetic thin films*: A crystalline film (simple cubic) is obtained by depositing a finite number of layers n. Each atom has a three component (Heisenberg) spin, and they interact through the Hamiltonian

$$-\beta\mathcal{H} = \sum_{\alpha=1}^{n} \sum_{\langle i,j \rangle} J_H \vec{s}_i^\alpha \cdot \vec{s}_j^\alpha + \sum_{\alpha=1}^{n-1} \sum_i J_V \vec{s}_i^\alpha \cdot \vec{s}_i^{\alpha+1}.$$

(The unit vector $\vec{s}_i^\alpha$ indicates the spin at site i in the αth layer.) A mean-field approximation is obtained from the variational density $\rho_0 \propto \exp(-\beta\mathcal{H}_0)$, with the trial Hamiltonian

$$-\beta\mathcal{H}_0 = \sum_{\alpha=1}^{n} \sum_i \vec{h}^\alpha \cdot \vec{s}_i^\alpha.$$

(Note that the most general single-site density matrix may include the higher order terms $L_{c_1,\cdots,c_p}^\alpha s_{c_1}^\alpha \cdots s_{c_1}^\alpha$, where s_c indicates component c of the vector $\vec{s}$.)

(a) Calculate the partition function $Z_0\left(\left\{\vec{h}^\alpha\right\}\right)$, and $\beta F_0 = -\ln Z_0$.

(b) Obtain the magnetizations $m_\alpha = |\langle \vec{s}_i^\alpha \rangle_0|$, and $\langle \beta\mathcal{H}_0 \rangle_0$, in terms of the *Langevin function* $\mathcal{L}(h) = \coth(h) - 1/h$.

(c) Calculate $\langle \beta\mathcal{H} \rangle_0$, with the (reasonable) assumption that all the variational fields $\left(\left\{\vec{h}^\alpha\right\}\right)$ are parallel.

(d) The exact free energy, $\beta F = -\ln Z$, satisfies the Gibbs inequality, $\beta F \le \beta F_0 + \langle \beta\mathcal{H} - \beta\mathcal{H}_0 \rangle_0$. Give the self-consistent equations for the magnetizations $\{m_\alpha\}$ that optimize $\beta\mathcal{H}_0$. How would you solve these equations numerically?

(e) Find the critical temperature, and the behavior of the magnetization *in the bulk* by considering the limit $n \to \infty$. (Note that $\lim_{m \to 0} \mathcal{L}^{-1}(m) = 3m + 9m^3/5 + \mathcal{O}(m^5)$.)

(f) By linearizing the self-consistent equations, show that the critical temperature of film depends on the number of layers n, as $kT_c(n \gg 1) \approx kT_c(\infty) - J_V \pi^2/(3n^2)$.

(g) Derive a continuum form of the self-consistent equations, and keep terms to cubic order in m. Show that the resulting nonlinear differential equation has a solution of the form $m(x) = m_{\text{bulk}} \tanh(kx)$. What circumstances are described by this solution?

(h) How can the above solution be modified to describe a *semi-infinite* system? Obtain the critical behaviors of the healing length $\lambda \sim 1/k$.

(i) Show that the magnetization of the surface layer vanishes as $|T - T_c|$.

The result in (f) illustrates a quite general result that the transition temperature of a finite system of size L, approaches its asymptotic (infinite-size) limit from below, as $T_c(L) = T_c(\infty) - A/L^{1/\nu}$, where ν is the exponent controlling the divergence of the correlation length. However, some liquid crystal films appeared to violate this behavior. In fact, in these films the couplings are stronger on the surface layers, which thus order before the bulk. For a discussion of the dependence of T_c on the number of layers in this case, see H. Li, M. Paczuski, M. Kardar, and K. Huang, Phys. Rev. B **44**, 8274 (1991).

2
Statistical fields

2.1 Introduction

We noted in the previous chapter that the singular behavior of thermodynamic functions at a critical point (the termination of a coexistence line) can be characterized by a set of critical exponents $\{\alpha, \beta, \gamma, \cdots\}$. Experimental observations indicate that these exponents are quite *universal*, i.e. independent of the material under investigation, and to some extent, of the nature of the phase transition. For example, the vanishing of the coexistence boundary in the condensation of CO_2 has the same singular behavior as that of the phase separation of protein solutions into dilute and dense components. This universality of behavior needs to be explained. We also noted that the divergence of the response functions, as well as direct observations of fluctuations via scattering studies, indicate that fluctuations have long wavelengths in the vicinity of the critical point, and are correlated over distances $\xi \gg a$, where a is a typical interparticle spacing. Such correlated fluctuations involve many particles and a coarse-graining approach, in the spirit of the theory of elasticity, may be appropriate to their description. Here we shall construct such a *statistical field theory*.

We shall frame the discussion in the language of a magnetic system whose symmetries are more transparent, although the results are of more general applicability. Consider a material such as iron, which is experimentally observed to be ferromagnetic below a Curie temperature T_c, as in Fig. 1.4. The microscopic origin of magnetism is quantum mechanical, involving such elements as itinerant electrons, their spins, and their interactions, described by a microscopic Hamiltonian $\mathcal{H}_{\text{mic}}$. In principle, all thermodynamic properties of the system can be extracted from a partition function obtained by summing over all degrees of freedom, written symbolically as

$$Z(T) = \text{tr}\left[e^{-\beta \mathcal{H}_{\text{mic}}}\right], \quad \text{with} \quad \beta = \frac{1}{k_B T}. \tag{2.1}$$

In practice, this formula is not of much use, as the microscopic Hamiltonian, and degrees of freedom, are too complicated to make a calculation possible.

A microscopic theory is certainly necessary to find out which elements are likely to produce ferromagnetism. However, given that there is magnetic behavior, such a theory is not necessarily useful to describe the disappearance of magnetization as a result of thermal fluctuations. For addressing the latter, the (quantum) statistical mechanics of the collection of interacting electrons is an excessively complicated starting point. Instead, we make the observation that the important degrees of freedom close to the Curie point are long wavelength collective excitations of spins (much like the long wavelength phonons that dominate the heat capacity of solids at low temperatures). It thus makes more sense to focus on the statistical properties of these fluctuations which are ultimately responsible for the phase transition. Towards this end, we change focus from the microscopic scales to intermediate *mesoscopic* scales which are much larger than the lattice spacing, but much smaller than the system size. In a manner similar to the coarse graining process depicted in Fig. 1.1, we define a magnetization field $\vec{m}(\mathbf{x})$, which represents the average of the elemental spins in the vicinity of a point $\mathbf{x}$. It is important to emphasize that while $\mathbf{x}$ is treated as a continuous variable, the function $\vec{m}(\mathbf{x})$ does not exhibit any variations at distances of the order of the lattice spacing, i.e. its Fourier transform involves only wavevectors less than some *upper cutoff* $\Lambda \sim 1/a$.

The transformation from the original degrees of freedom to the field $\vec{m}(\mathbf{x})$ is a change of variables. (This mapping is not invertible, as many microscopic details are washed out in the averaging process.) Again, it is in principle possible to obtain the corresponding probabilities for configurations of the field $\vec{m}(\mathbf{x})$, by transforming the original microscopic probabilities arising from the Boltzmann weight $e^{-\beta \mathcal{H}_{\text{mic}}}$. The partition function is preserved in the process, and can be written as

$$Z(T) = \text{tr}\left[e^{-\beta \mathcal{H}_{\text{mic}}}\right] \equiv \int \mathcal{D}\vec{m}(\mathbf{x}) \mathcal{W}\left[\vec{m}(\mathbf{x})\right]. \tag{2.2}$$

The symbol $\mathcal{D}\vec{m}(\mathbf{x})$ indicates integrating over all allowed configurations of the field, and will be discussed later. The different configurations of the field are weighted with a probability $\mathcal{W}\left[\vec{m}(\mathbf{x})\right]$, which is what we would like to find out.

While obtaining the precise form of $\mathcal{W}\left[\vec{m}(\mathbf{x})\right]$ is not easier than solving the full problem, it is in fact possible to describe it in terms of a few phenomenological parameters. (This is similar to describing the energy cost of deforming a solid in terms of a few elastic moduli.) In the context of phase transitions, this approach was first applied by Landau to describe the onset of superfluidity in helium. In fact, the method can quite generally be applied to different types of systems undergoing phase transition, with $\vec{m}(\mathbf{x})$ describing the corresponding order parameter field. We shall thus generalize the problem, and consider an n-component field, existing in d-dimensional space, i.e.

$$\mathbf{x} \equiv (x_1, x_2, ..., x_d) \in \Re^d \quad \text{(space)}, \quad \vec{m} \equiv (m_1, m_2, ..., m_n) \in \Re^n$$

(order parameter).

Some specific cases covered in this general framework are:

$n = 1$ describes liquid–gas transitions, binary mixtures, as well as uniaxial magnets;

$n = 2$ applies to superfluidity, superconductivity, and planar magnets;

$n = 3$ corresponds to classical magnets.

While most physical situations occur in three-dimensional space ($d = 3$), there are also important phenomena on surfaces ($d = 2$), and in wires ($d = 1$). Relativistic field theory is described by a similar structure, but in $d = 4$.

2.2 The Landau–Ginzburg Hamiltonian

Using the coarse-grained weight in Eq. 2.2, we can define an effective Hamiltonian

$$\beta \mathcal{H}\left[\vec{m}(\mathbf{x})\right] \equiv -\ln \mathcal{W}\left[\vec{m}(\mathbf{x})\right], \tag{2.3}$$

which gives the probabilities of the field configurations by a Boltzmann factor. Guided by the steps outlined in the last chapter for generating the elastic theory of a deformed solid, we shall construct $\beta \mathcal{H}\left[\vec{m}(\mathbf{x})\right]$ using the following principles:

Locality and uniformity: If a system consists of disconnected parts, the overall probability is obtained as a product of independent probabilities. A corresponding Hamiltonian as in Eq. (2.3) then decomposes into a sum of contributions from each location, going over to an integral in the continuum representation, i.e.

$$\beta \mathcal{H} = \int d^d \mathbf{x} \Phi\left[\vec{m}(\mathbf{x}), \mathbf{x}\right]. \tag{2.4}$$

Here, Φ is an energy density, which can in principle have a different functional form at different locations. For a material that is *uniform* in space, different positions in $\mathbf{x}$ are equivalent, and we can drop the explicit dependence of Φ on $\mathbf{x}$. This will not be the case when the system is in an external potential, or has internal impurities. We are of course interested in more complicated systems in which, as a result of interactions, there is some coupling between different parts of the system. To this end, we generalize Eq. (2.4), by including gradients of the field, to

$$\beta \mathcal{H} = \int d^d \mathbf{x} \Phi\left[\vec{m}(\mathbf{x}), \nabla \vec{m}, \nabla^2 \vec{m}, \cdots\right]. \tag{2.5}$$

Once more, general non-local interactions can be described by including many derivatives. The "local" representation is useful when a good description can be obtained by including only a few derivatives. This is the case for short-range interactions (even including van der Waals interactions in liquid gas mixtures, but fails for long-range (such as Coulomb) interactions.[1]

---

[1] The precise borderline between long- and short-range interactions will be discussed later.

Analyticity and polynomial expansions: The functional form of Φ is next written as an expansion in powers of $\vec{m}$, and its gradients. To justify such an expansion, let us again examine the simple example of a collection of independent degrees of freedom, say spins. The probability distribution at the microscopic level may be complicated; e.g. spins may be constrained to a fixed magnitude, or quantized to specific values. At the mesoscopic scale, the field $\vec{m}$ is obtained by averaging many such spins. The averaging process typically simplifies the probability distribution; in most cases the central limit theorem implies that the probability distribution for the sum approaches a Gaussian form.[2] In constructing the statistical field theory, we are in fact searching for generalizations of the central limit theorem for describing interacting degrees of freedom. The key point is that in going from the microscopic to mesoscopic scales, non-analyticities associated with the microscopic degrees of freedom are washed out, and the probability distribution for the coarse-grained field is obtained by an analytic expansion in powers of $\vec{m}$. There are of course non-analyticities associated with the phase transition at the macroscopic scale. However, such singularities involve an infinity (macroscopic number) of degrees of freedom. By focusing on the mesoscopic scale we thus avoid possible singularities at both the short and long scales!

Symmetries: One element that survives the averaging process is any underlying microscopic symmetry. Such symmetries in turn constrain possible forms and expansions of the effective Hamiltonian. For example, in the absence of an external magnetic field, all directions for magnetization are equivalent, and hence $\mathcal{H}[R_n\vec{m}(\mathbf{x})] = \mathcal{H}[\vec{m}(\mathbf{x})]$, where R_n is a rotation in the n-dimensional order parameter space. A linear term in $\vec{m}$ is not consistent with this symmetry, and the first term in an expansion is proportional to

$$m^2(\mathbf{x}) \equiv \vec{m}(\mathbf{x}) \cdot \vec{m}(\mathbf{x}) \equiv \sum_{i=1}^{n} m_i(\mathbf{x})m_i(\mathbf{x}), \qquad (2.6)$$

where $\{m_i\}$ indicate the components of the vector field. Higher order terms in the series are constructed from

$$m^4(\mathbf{x}) \equiv \left(m^2(\mathbf{x})\right)^2, \quad m^6(\mathbf{x}) \equiv \left(m^2(\mathbf{x})\right)^3, \quad \cdots$$

In constructing terms that involve gradients of the vector field, we should keep in mind the spatial symmetries of the system. In an *isotropic* system all directions in space are equivalent, and we should use combinations of derivatives that are invariant under spatial rotations. The simplest such term is

$$(\nabla\vec{m})^2 \equiv \sum_{i=1}^{n}\sum_{\alpha=1}^{d} \partial_\alpha m_i \partial_\alpha m_i, \qquad (2.7)$$

---

[2] More precisely, upon averaging over n spins, the coefficient of m^2 approaches a constant, while higher terms in the expansion decay as powers of n.

with ∂_α indicating the partial derivative along the α-th direction in space. If the different directions are not equivalent more terms are allowed. For example, in a two dimensional magnet on a rectangular lattice (unit cell aligned with the axes), the coefficients of $\partial_1 m_i \partial_1 m_i$ and $\partial_2 m_i \partial_2 m_i$ can be different. However, by appropriate rescaling of coordinates the leading gradient term can again be changed into the form of Eq. (2.7). A fourth order gradient term in an isotropic system is

$$(\nabla^2 \vec{m})^2 \equiv \sum_{i=1}^n \sum_{\alpha=1}^d \sum_{\beta=1}^d (\partial_\alpha \partial_\alpha m_i)(\partial_\beta \partial_\beta m_i),$$

and a possible quartic term in $\vec{m}$ is

$$m^2 (\nabla \vec{m})^2 \equiv \sum_{i=1}^n \sum_{j=1}^n \sum_{\alpha=1}^d m_i m_i \partial_\alpha m_j \partial_\alpha m_j.$$

Anisotropies again lead to higher order terms which in general cannot be removed by simple rescalings.

We shall demonstrate shortly that to describe magnetic systems, and in fact most transitions, it is sufficient to include only a few terms, leading to the so called *Landau–Ginzburg* Hamiltonian:

$$\beta \mathcal{H} = \beta F_0 + \int d^d \mathbf{x} \left[\frac{t}{2} m^2(\mathbf{x}) + u m^4(\mathbf{x}) + \frac{K}{2} (\nabla m)^2 + \cdots - \vec{h} \cdot \vec{m}(\mathbf{x}) \right]. \tag{2.8}$$

The integration over the magnetic and non-magnetic degrees of freedom at short scales also generates an overall constant βF_0. This contribution to the overall free energy is analytic (as discussed earlier), and will be mostly ignored. Equation (2.8) also includes the contribution from the magnetic work $\vec{B} \cdot \vec{m}$ to the Hamiltonian, where $\vec{h} \equiv \beta \vec{B}$ and $\vec{B}$ is the magnetic field.[3] The magnetic field may also generate higher order terms, such as $m^2 \vec{m} \cdot \vec{h}$, which are less important than the terms explicitly included above.

Stability: Since it originates from a well defined physical problem, the coarse-grained Boltzmann weight must not lead to any unphysical field configurations. In particular, the probability should not diverge for infinitely large values of $\vec{m}$. This condition implies that the coefficient of the highest order power of $\vec{m}$, e.g. the parameter u in Eq. (2.8), should be positive. There are related constraints on the signs of the terms involving gradients to avoid oscillatory instabilities.

The Landau–Ginzburg Hamiltonian depends on a set of *phenomenological parameters* $\{t, u, K, \cdots\}$. These parameters are non-universal functions of

---

[3] Note that with the inclusion of work against the external field we are dealing with a Gibbs canonical ensemble, and $-k_B T \ln Z$ is now the Gibbs free energy. This technical distinction is usually ignored, and we shall set $-k_B T \ln Z = F$, using the symbol usually reserved for the Helmholtz free energy.

microscopic interactions, *as well as external parameters such as temperature and pressure*. It is essential to fully appreciate the latter point which is a possible source of confusion. While the probability for a particular configuration of the field is given by the Boltzmann weight $\exp\{-\beta\mathcal{H}[\vec{m}(\mathbf{x})]\}$, this *does not imply* that all terms in the exponent are proportional to $(k_BT)^{-1}$. Such dependence holds only for the true microscopic Hamiltonian. The Landau–Ginzburg Hamiltonian is more correctly an effective free energy obtained by integrating over (coarse graining) the microscopic degrees of freedom, while constraining their average to $\vec{m}(\mathbf{x})$. It is precisely because of the difficulty of carrying out such a first principles program that we postulate the form of the resulting effective free energy on the basis of symmetries alone. The price paid is that the phenomenological parameters have unknown functional dependences on the original microscopic parameters, as well as on such external constraints as temperature (since we have to account for the entropy of the short distance fluctuations lost in the coarse-graining process). The constraints on these functional forms again arise from *symmetry*, *analyticity*, and *stability*, as discussed in the context of constructing the function form of $\Phi[\vec{m}]$. Notably, these mesoscopic coefficients will be analytic functions of the external parameters, e.g. expressible as power series in temperature T.

2.3 Saddle point approximation, and mean-field theory

By focusing only on the coarse-grained magnetization field, we have considerably simplified the original problem. Various thermodynamic functions (and their singular behavior) should now be obtained from the partition function

$$Z = \int \mathcal{D}\vec{m}(\mathbf{x}) \exp\{-\beta\mathcal{H}[\vec{m}(\mathbf{x})]\}, \qquad (2.9)$$

corresponding to the Landau–Ginzburg Hamiltonian in Eq. (2.8). The degrees of freedom appearing in the Hamiltonian are functions of $\mathbf{x}$, the symbol $\int \mathcal{D}\vec{m}(\mathbf{x})$ refers to a *functional integral*. In practice, the functional integral is obtained as a limit of discrete integrals: The continuous coordinate $\mathbf{x} \equiv (x_1, x_2, \cdots x_d)$ is first discretized into a lattice of $\mathcal{N}$ points $\mathbf{i} \equiv (i_1, i_2, \cdots i_d)$, at a lattice distance a; the various derivatives are replaced with appropriate differences, and the functional integral is obtained as

$$\int \mathcal{D}\vec{m}(\mathbf{x})\mathcal{F}\left[\vec{m}(\mathbf{x}), \frac{\partial\vec{m}}{\partial x_\alpha}, \cdots\right] \equiv \lim_{\mathcal{N}\to\infty}\prod_{i=1}^{\mathcal{N}}d\vec{m}_i\mathcal{F}\left[\vec{m}_i, \frac{\vec{m}_{i_\alpha+1}-\vec{m}_{i_\alpha}}{a}, \cdots\right].$$

There are in fact many mathematical concerns regarding the existence of functional integrals. These problems are mostly associated with having too many degrees of freedom at short distances, which allow for rather badly behaved functions. These issues need not concern us since we know that the underlying problem has a well defined lattice spacing that restricts and controls the short distance behavior.

Even after all these simplifications, it is still not easy to calculate the Landau–Ginzburg partition function in Eq. (2.9). As a first step, we perform a *saddle point approximation* in which the integral in Eq. (2.9) is replaced by the maximum value of the integrand, corresponding to the *most probable* configuration of the field $\vec{m}(\mathbf{x})$. The natural tendency of interactions in a magnet is to keep the magnetizations vectors parallel, and hence we expect the parameter K in Eq. (2.8) to be positive.[4] Any variations in magnitude or direction of $\vec{m}(\mathbf{x})$ incur an "energy penalty" from the term $K(\nabla \vec{m})^2$ in Eq. (2.8). Thus the field is uniform in its most probable configuration, and restricting the integration to this subspace gives

$$Z \approx Z_{\text{sp}} = e^{-\beta F_0} \int d\vec{m}\, \exp\left[-V\left(\frac{t}{2}m^2 + um^4 + \cdots - \vec{h}.\vec{m}\right)\right], \qquad (2.10)$$

where V is the system volume. In the limit of $V \to \infty$ the integral is governed by the saddle point $\vec{m}$, which maximizes the exponent of the integrand. The corresponding saddle point free energy is

$$\beta F_{\text{sp}} = -\ln Z_{\text{sp}} \approx \beta F_0 + V \min\{\Psi(\vec{m})\}_{\vec{m}}, \qquad (2.11)$$

where

$$\Psi(\vec{m}) \equiv \frac{t}{2}\vec{m}^2 + u\left(\vec{m}^2\right)^2 + \cdots - \vec{h}.\vec{m}. \qquad (2.12)$$

The most likely magnetization will be aligned to the external field, i.e. $\vec{m}(\mathbf{x}) = \overline{m}\hat{h}$; its magnitude is obtained from

$$\Psi'(\overline{m}) = t\overline{m} + 4u\overline{m}^3 + \cdots - h = 0. \qquad (2.13)$$

Surprisingly, this simple equation captures the qualitative behavior at a phase transition.

While the function $\Psi(m)$ is analytic, and has no singularities, the saddle point free energy is Eq. (2.11) may well be non-analytic. This is because the minimization operation is not an analytic procedure, and introduces singularities as we shall shortly demonstrate. What justifies the saddle point evaluation of Eq. (2.10) is the thermodynamic limit of $V \to \infty$, and for finite V, the integral is perfectly analytic. In the vicinity of the critical point, the magnetization is small, and it is justified to keep only the lowest powers in the expansion of $\Psi(\vec{m})$. (We can later check self-consistently that the terms left out of the expansion are indeed small corrections.) The behavior of $\Psi(m)$ depends strongly on the sign of the parameter t.

(1) For $t > 0$, the quartic term is not necessary for stability and can be ignored. The solution to Eq. (2.13) is $\overline{m} = h/t$, and describes paramagnetic behavior. The most

---

[4] This is also required by the stability condition.

Fig. 2.1 The function $\Psi(m)$ for $t > 0$, and three values of h. The most probable magnetization occurs at the minimum of this function (indicated by solid dots), and goes to zero continuously as $h \to 0$.

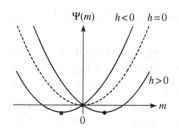

Fig. 2.2 The function $\Psi(m)$ for $t < 0$, and three values of h. The most probable magnetization occurs at the *global* minimum of this function (indicated by solid dots), and goes to zero continuously as $h \to 0$. The metastable minimum is indicated by open dots.

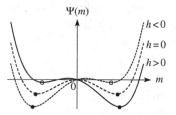

Fig. 2.3 The magnetization curves obtained from the solution of Eq. (2.13) (*left*), and the corresponding phase diagram (*right*).

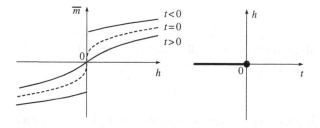

probable magnetization is aligned to the magnetic field, and vanishes continuously as $\vec{h} \to 0$. The susceptibility $\chi = 1/t$, diverges as $t \to 0$.

(2) For $t < 0$, a quartic term with a positive value of u is required to insure stability (i.e. a finite magnetization). The function $\Psi(m)$ can now have two local minima, the global minimum being aligned with the field $\vec{h}$. As $\vec{h} \to 0$, the global minimum moves towards a non-zero value, indicating a spontaneous magnetization, even at $\vec{h} = 0$, as in a ferromagnet. The direction of $\vec{m}$ at $\vec{h} = 0$ is determined by the system's history, and can be realigned by an external field $\vec{h}$.

The resulting curves for the most probable magnetization $\overline{m}(h)$ are quite similar to those of Fig. 1.4. The saddle point evaluation of the Landau–Ginzburg partition function thus results in paramagnetic behavior for $t > 0$, and ferromagnetic behavior for $t < 0$, with a line of phase transitions terminating at the point $t = h = 0$.

As noted earlier, the parameters $(t, u, K, \cdots)$ of the Landau–Ginzburg Hamiltonian are analytic functions of temperature, and can be expanded around the critical point at $T = T_c$ in a Taylor series as

$$\begin{cases} t(T, \cdots) = a_0 + a_1(T - T_c) + \mathcal{O}(T - T_c)^2, \\ u(T, \cdots) = u + u_1(T - T_c) + \mathcal{O}(T - T_c)^2, \\ K(T, \cdots) = K + K_1(T - T_c) + \mathcal{O}(T - T_c)^2. \end{cases} \qquad (2.14)$$

The expansion coefficients can be regarded as phenomenological parameters which can be determined by comparing to experiments. In particular, matching the phase diagrams in Figs. 1.4 and 2.3 requires that t should be a monotonic function of temperature which vanishes at T_c, necessitating $a_0 = 0$ and $a_1 = a > 0$. Stability of the ferromagnetic phase in the vicinity of the critical point requires that u and K should be positive. The *minimal set of conditions* for matching the experimental phase diagram of a magnet to that obtained from the saddle point is

$$t = a(T - T_c) + \mathcal{O}(T - T_c)^2, \quad \text{with} \quad (a, u, K) > 0. \qquad (2.15)$$

It is of course possible to set additional terms in the expansion, e.g. a or u, to zero, and maintain the phase diagram and stability by appropriate choice of the higher order terms. However, such choices are not *generic*, and there is no reason for imposing more constraints than absolutely required by the experiment.

Using Eq. (2.15), we can quantify the singular behaviors predicted by the saddle point evaluations of the free energy in Eqs. (2.11)–(2.13).

- **Magnetization:** In zero field, Eq. (2.13) reduces to $\partial \Psi / \partial m = t \overline{m} + 4u \overline{m}^3 = \overline{m}(t + 4u \overline{m}^2) = 0$, and we obtain

$$\overline{m}(h = 0) = \begin{cases} 0 & \text{for } t > 0, \\ \sqrt{\dfrac{-t}{4u}} = \sqrt{\dfrac{a}{4u}} (T_c - T)^{1/2} & \text{for } t < 0. \end{cases} \qquad (2.16)$$

For $t < 0$, the non-magnetized solution $\overline{m} = 0$ is a maximum of $\Psi(m)$, and there is a spontaneous magnetization that vanishes with a universal exponent of $\beta = 1/2$. The overall amplitude is non-universal and material dependent.

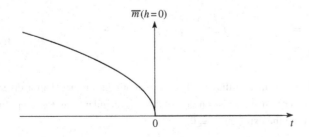

$\overline{m}(h=0)$

0

t

Fig. 2.4 The saddle point spontaneous magnetization vanishes with a square-root singularity.

Along the critical isotherm (the dashed line of Fig. 2.3), with $t = 0$, Eq. (2.13) gives

$$\overline{m}(t = 0) = \left(\frac{h}{4u}\right)^{1/3},\tag{2.17}$$

i.e. $h \propto \overline{m}^{\delta}$, with an exponent $\delta = 3$.

- **Susceptibility:** The magnetization is aligned to the external field, $\vec{m} = \overline{m}(h)\hat{h}$, its magnitude given by the solution to $t\overline{m} + 4u\overline{m}^3 = h$. The changes in the magnitude are governed by the *longitudinal susceptibility* χ_ℓ, whose inverse is easily obtained as

$$\chi_\ell^{-1} = \left.\frac{\partial h}{\partial \overline{m}}\right|_{h=0} = t + 12u\overline{m}^2 = \begin{cases} t & \text{for } t > 0, \text{ and } h = 0, \\ -2t & \text{for } t < 0, \text{ and } h = 0. \end{cases}\tag{2.18}$$

On approaching the critical point from either side, the zero field susceptibility diverges as $\chi_\pm \sim A_\pm |t|^{-\gamma_\pm}$, with $\gamma_+ = \gamma_- = 1$. Although the amplitudes $A_\pm$ are material dependent, Eq. (2.18) predicts that their ratio is universal, given by $A_+/A_- = 2$. (We shall shortly encounter the *transverse susceptibility* χ_t, which describes the change in magnetization in response to a field perpendicular to it. For $h = 0$, χ_t is always infinite in the magnetized phase.)

Fig. 2.5 The zero field longitudinal susceptibility diverges at $t = 0$.

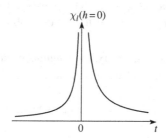

$\chi_\ell(h=0)$

0

t

- **Heat capacity:** The free energy for $h = 0$ is given by

$$\beta F = \beta F_0 + V\Psi(\overline{m}) = \beta F_0 + V \begin{cases} 0 & \text{for } t > 0, \\ -\dfrac{t^2}{16u} & \text{for } t < 0. \end{cases}\tag{2.19}$$

Since $t = a(T - T_c) + \cdots$, to leading order in $(T - T_c)$, we have $\partial/\partial T \sim a\partial/\partial t$. Using similar approximations in the vicinity of the critical point, we find the behavior of the heat capacity at zero field as

$$C(h = 0) = -T\frac{\partial^2 F}{\partial T^2} \approx -T_c a^2 \frac{\partial^2}{\partial t^2}(k_B T_c \beta F) = C_0 + Vk_B a^2 T_c^2 \times \begin{cases} 0 & \text{for } t > 0, \\ \dfrac{1}{8u} & \text{for } t < 0. \end{cases}\tag{2.20}$$

The saddle point method thus predicts a discontinuity, rather than a divergence, in the heat capacity. If we insist on describing this singularity as a power law $t^{-\alpha}$, we have to choose the exponent $\alpha = 0$.

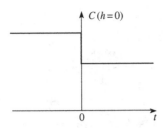

$C(h=0)$

0 t

Fig. 2.6 The saddle point approximation predicts a discontinuous heat capacity.

2.4 Continuous symmetry breaking and Goldstone modes

For zero field, although the microscopic Hamiltonian has full rotational symmetry, the low-temperature phase does not. As a specific direction in n-space is selected for the net magnetization $\vec{M}$, there is *spontaneous symmetry breaking*, and a corresponding *long-range order* is established in which the majority of the spins in the system are oriented along $\vec{M}$. The original symmetry is still present globally, in the sense that if all local spins are rotated together (i.e. the field transforms as $\vec{m}(\mathbf{x}) \mapsto \Re\vec{m}(\mathbf{x})$), there is no change in energy. Such a rotation transforms one ordered state into an equivalent one. Since a uniform rotation costs no energy, by continuity we expect a rotation that is slowly varying in space (e.g. $\vec{m}(\mathbf{x}) \mapsto \Re(\mathbf{x})\vec{m}(\mathbf{x})$, where $\Re(\mathbf{x})$ only has long wavelength variations) to cost very little energy. Such low energy excitations are called *Goldstone modes*. These collective modes appear in any system with a *broken continuous symmetry*.[5] Phonons in a solid provide a familiar example of Goldstone modes, corresponding to the breaking of translation and rotation symmetries by a crystal structure.

The Goldstone modes appearing in diverse systems share certain common characteristics. Let us explore the origin and behavior of Goldstone modes in the context of *superfluidity*. In analogy to Bose condensation, the superfluid phase has a macroscopic occupation of a single quantum ground state. The order parameter,

$$\psi(\mathbf{x}) \equiv \psi_\Re(\mathbf{x}) + i\psi_\Im(\mathbf{x}) \equiv |\psi(\mathbf{x})|e^{i\theta(\mathbf{x})}, \tag{2.21}$$

can be roughly regarded as the ground state component (overlap) of the actual wavefunction in the vicinity of $\mathbf{x}$.[6] The phase of the wavefunction is not an observable quantity and should not appear in any physically measurable

---

[5] There are no Goldstone modes when a discrete symmetry is broken, since it is impossible to produce slowly varying rotations from one state to an equivalent one.

[6] A more rigorous derivation proceeds by second quantization of the Hamiltonian for interacting bosons, and is beyond our current scope.

Fig. 2.7 The Landau–Ginzburg Hamiltonian of a superfluid for $t < 0$.

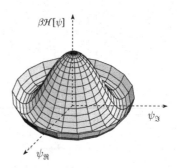

probability. This observation constrains the form of the effective coarse grained Hamiltonian, leading to an expansion

$$\beta \mathcal{H} = \beta F_0 + \int d^d \mathbf{x} \left[\frac{K}{2} |\nabla \psi|^2 + \frac{t}{2} |\psi|^2 + u |\psi|^4 + \cdots \right]. \tag{2.22}$$

Equation (2.22) is in fact equivalent to the Landau–Ginzburg Hamiltonian with $n = 2$, as can be seen by changing variables to the two component field $\vec{m}(\mathbf{x}) \equiv (\psi_\Re(\mathbf{x}), \psi_\Im(\mathbf{x}))$. The superfluid transition is signaled by the onset of a finite value of ψ for $t < 0$. The Landau–Ginzburg Hamiltonian (for a uniform ψ) has the shape of a wine bottle (or Mexican hat) for $t < 0$.

Minimizing this function sets the magnitude of ψ, but does not fix its phase θ. Now consider a state with a slowly varying phase, i.e. with $\psi(\mathbf{x}) = \bar{\psi} e^{i\theta(\mathbf{x})}$. Inserting this form into the Hamiltonian yields an energy

$$\beta \mathcal{H} = \beta \mathcal{H}_0 + \frac{\overline{K}}{2} \int d^d \mathbf{x} (\nabla \theta)^2, \tag{2.23}$$

where $\overline{K} = K \bar{\psi}^2$. As in the case of phonons we could have guessed the above form by appealing to the invariance of the energy function under a uniform rotation: Since a transformation $\theta(\mathbf{x}) \mapsto \theta(\mathbf{x}) + \theta_0$ should not change the energy, the energy density can only depend on gradients $\theta(\mathbf{x})$, and the first term in the expansion leads to Eq. (2.23). The reasoning based on symmetry does not give the value of the *stiffness parameter*. By starting from the Landau–Ginzburg form which incorporates both the normal and superfluid phases we find that $\overline{K}$ is proportional to the square of the order parameter, and vanishes (softens) at the critical point as $\overline{K} \propto \bar{\psi}^2 \propto t$.

We can decompose the variations in phase of the order parameter into independent normal modes by setting (in a region of volume V)

$$\theta(\mathbf{x}) = \frac{1}{\sqrt{V}} \sum_{\mathbf{q}} e^{i\mathbf{q} \cdot \mathbf{x}} \theta_{\mathbf{q}}. \tag{2.24}$$

Taking advantage of translational symmetry, Eq. (2.23) then gives

$$\beta \mathcal{H} = \beta \mathcal{H}_0 + \frac{\overline{K}}{2} \sum_{\mathbf{q}} q^2 |\theta(\mathbf{q})|^2. \tag{2.25}$$

We can see that, as in the case of phonons, the energy of a Goldstone mode of wavenumber $\mathbf{q}$ is proportional to q^2, and becomes very small at long wavelengths.

2.5 Discrete symmetry breaking and domain walls

For a one component (scalar) field, there are two possible values for the magnetization in the ordered phase. While the two possible states have the same energy, it is not possible to continuously deform one into the other. In this case, as well as in other systems with discrete symmetry breaking, different states in the same sample are separated by sharp domain walls. To demonstrate this, consider a scalar field with the Landau–Ginzburg Hamiltonian for $t < 0$ and $h = 0$. A domain wall can be introduced by forcing the two sides of the system to be in different states, e.g. by requiring $m(x \to -\infty) = -\overline{m}$ and $m(x \to +\infty) = +\overline{m}$. In between, the most probable field configuration is obtained by minimizing the energy, and satisfies

$$\frac{d^2 m_w(x)}{dx^2} = t m_w(x) + 4u m_w(x)^3. \tag{2.26}$$

By using the identity $d^2 \tanh(ax)/dx^2 = a^2 \tanh(ax)\left[1 - \tanh^2(ax)\right]$, it can be easily checked that the profile

$$m_w(x) = \overline{m} \tanh\left[\frac{x - x_0}{w}\right], \tag{2.27}$$

is a solution to the above nonlinear differential equation, provided

$$w = \sqrt{\frac{2K}{-t}}, \quad \text{and} \quad \overline{m} = \sqrt{\frac{-t}{4u}}. \tag{2.28}$$

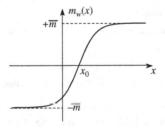

Fig. 2.8 Profile of the domain wall in a system with discrete symmetry breaking.

The above solution separates regions where the magnetization approaches its two possible bulk values of $\pm\overline{m}$. The domain wall between the two regions is centered at an arbitrary position x_0, and has a width w. On approaching the phase transition at $t = 0$ this width diverges as $(T_c - T)^{-1/2}$. The width w is in fact proportional to the correlation length of the system, which will be calculated in the next chapter.

The free energy cost of creating a domain wall in the system can be obtained by examining the energy difference to a uniformly magnetized solution, as

$$\beta F_w \equiv \beta F\left[m_w(x)\right] - \beta F\left[\overline{m}\right]$$

$$= \int d^d x \left[\frac{K}{2}\left(\frac{dm_w}{dx}\right)^2 + \frac{t}{2}\left(m_w^2 - \overline{m}^2\right) + u\left(m_w^4 - \overline{m}^4\right)\right]. \qquad (2.29)$$

Simple algebraic manipulations then give

$$\beta F_w = -\frac{t}{2}\overline{m}^2 \int d^d x \cosh^{-4}\left(\frac{x - x_0}{w}\right) = -\frac{t}{2}\overline{m}^2 w\mathcal{A} \int_{-\infty}^{\infty} \frac{dy}{\cosh^4 y} = -\frac{2}{3} t\overline{m}^2 w\mathcal{A}, \quad (2.30)$$

where $\mathcal{A}$ is the cross-sectional area of the system normal to the x direction. The free energy per unit area is proportional to the bulk energy density, multiplied by the width of the domain wall. On approaching the phase transition, the above calculation predicts that the interfacial free energy vanishes as $(T_c - T)^{3/2}$.

Problems for chapter 2

1. *Cubic invariants:* When the order parameter m, goes to zero discontinuously, the phase transition is said to be first order (discontinuous). A common example occurs in systems where symmetry considerations do not exclude a cubic term in the Landau free energy, as in

$$\beta\mathcal{H} = \int d^d x \left[\frac{K}{2}(\nabla m)^2 + \frac{t}{2}m^2 + cm^3 + um^4\right] \quad (K, c, u > 0).$$

   (a) By plotting the energy density $\Psi(m)$, for uniform m at various values of t, show that as t is reduced there is a discontinuous jump to $\overline{m} \neq 0$ for a positive $\overline{t}$ in the saddle point approximation.

   (b) By writing down the two conditions that $\overline{m}$ and $\overline{t}$ must satisfy at the transition, solve for $\overline{m}$ and $\overline{t}$.

   (c) Note that the correlation length ξ is related to the curvature of $\Psi(m)$ at its minimum by $K\xi^{-2} = \partial^2\Psi/\partial m^2|_{eq}$. Plot ξ as a function of t.

2. *Tricritical point:* By tuning an additional parameter, a second order transition can be made first order. The special point separating the two types of transitions is known as a tricritical point, and can be studied by examining the Landau–Ginzburg Hamiltonian

$$\beta\mathcal{H} = \int d^d x \left[\frac{K}{2}(\nabla m)^2 + \frac{t}{2}m^2 + um^4 + vm^6 - hm\right],$$

   where u can be positive or negative. For $u < 0$, a positive v is necessary to insure stability.

(a) By sketching the energy density $\Psi(m)$, for various t, show that in the saddle point approximation there is a first order transition for $u < 0$ and $h = 0$.

(b) Calculate $\bar{t}$ and the discontinuity $\bar{m}$ at this transition.

(c) For $h = 0$ and $v > 0$, plot the phase boundary in the (u, t) plane, identifying the phases, and order of the phase transitions.

(d) The special point $u = t = 0$, separating first and second order phase boundaries, is a *tricritical* point. For $u = 0$, calculate the tricritical exponents β, δ, γ, and α, governing the singularities in magnetization, susceptibility, and heat capacity. (Recall: $C \propto t^{-\alpha}$; $\bar{m}(h = 0) \propto t^{\beta}$; $\chi \propto t^{-\gamma}$; and $\bar{m}(t = 0) \propto h^{1/\delta}$.)

3. *Transverse susceptibility:* An n-component magnetization field $\vec{m}(\mathbf{x})$ is coupled to an external field $\vec{h}$ through a term $-\int d^d\mathbf{x}\ \vec{h} \cdot \vec{m}(\mathbf{x})$ in the Hamiltonian $\beta\mathcal{H}$. If $\beta\mathcal{H}$ for $\vec{h} = 0$ is invariant under rotations of $\vec{m}(\mathbf{x})$; then the free energy density ($f = -\ln Z/V$) only depends on the absolute value of $\vec{h}$; i.e. $f(\vec{h}) = f(h)$, where $h = |\vec{h}|$.

(a) Show that $m_\alpha = \langle \int d^d\mathbf{x}\, m_\alpha(\mathbf{x}) \rangle / V = -h_\alpha f'(h)/h$.

(b) Relate the susceptibility tensor $\chi_{\alpha\beta} = \partial m_\alpha/\partial h_\beta$, to $f''(h)$, $\vec{m}$, and $\vec{h}$.

(c) Show that the transverse and longitudinal susceptibilities are given by $\chi_t = m/h$ and $\chi_\ell = -f''(h)$; where m is the magnitude of $\vec{m}$.

(d) Conclude that χ_t diverges as $\vec{h} \to 0$, whenever there is a spontaneous magnetization. Is there any similar a priori reason for χ_ℓ to diverge?

---

4. *Superfluid He^4–He^3 mixtures:* The superfluid He^4 order parameter is a complex number $\psi(\mathbf{x})$. In the presence of a concentration $c(\mathbf{x})$ of He^3 impurities, the system has the following Landau–Ginzburg energy

$$\beta\mathcal{H}[\psi, c] = \int d^d\mathbf{x}\left[\frac{K}{2}|\nabla\psi|^2 + \frac{t}{2}|\psi|^2 + u|\psi|^4 + v|\psi|^6 + \frac{c(\mathbf{x})^2}{2\sigma^2} - \gamma c(\mathbf{x})|\psi|^2\right],$$

with positive K, u and v.

(a) Integrate out the He^3 concentrations to find the effective Hamiltonian, $\beta\mathcal{H}_{\text{eff}}[\psi]$, for the superfluid order parameter, given by

$$Z = \int \mathcal{D}\psi \exp\left(-\beta\mathcal{H}_{\text{eff}}[\psi]\right) \equiv \int \mathcal{D}\psi \mathcal{D}c \exp\left(-\beta\mathcal{H}[\psi, c]\right).$$

(b) Obtain the phase diagram for $\beta\mathcal{H}_{\text{eff}}[\psi]$ using a saddle point approximation. Find the limiting value of σ^* above which the phase transition becomes discontinuous.

(c) The discontinuous transition is accompanied by a jump in the magnitude of ψ. How does this jump vanish as $\sigma \to \sigma^*$?

(d) Show that the discontinuous transition is accompanied by a jump in He^3 concentration.

(e) Sketch the phase boundary in the (t, σ) coordinates, and indicate how its two segments join at σ^*.

(f) Going back to the original joint probability for the fields $c(\mathbf{x})$ and $\Psi(\mathbf{x})$, show that $\langle c(\mathbf{x}) - \gamma\sigma^2 |\Psi(\mathbf{x})|^2 \rangle = 0$.

(g) Show that $\langle c(\mathbf{x})c(\mathbf{y}) \rangle = \gamma^2\sigma^4 \langle |\Psi(\mathbf{x})|^2 |\Psi(\mathbf{y})|^2 \rangle$, for $\mathbf{x} \neq \mathbf{y}$.

(h) Qualitatively discuss how $\langle c(\mathbf{x})c(0) \rangle$ decays with $x = |\mathbf{x}|$ in the disordered phase.

(i) Qualitatively discuss how $\langle c(\mathbf{x})c(0) \rangle$ decays to its asymptotic value in the ordered phase.

5. *Crumpled surfaces:* The configurations of a crumpled sheet of paper can be described by a vector field $\vec{r}(\mathbf{x})$, denoting the position in three-dimensional space, $\vec{r} = (r_1, r_2, r_3)$, of the point at location $\mathbf{x} = (x_1, x_2)$ on the flat sheet. The energy of each configuration is assumed to be invariant under translations and rotations of the sheet of paper.

(a) Show that the two lowest order (in derivatives) terms in the quadratic part of a Landau–Ginzburg Hamiltonian for this system are:

$$\beta\mathcal{H}_0[\vec{r}] = \sum_{\alpha=1,2} \int d^2\mathbf{x} \left[\frac{t}{2} \partial_\alpha \vec{r} \cdot \partial_\alpha \vec{r} + \frac{K}{2} \partial_\alpha^2 \vec{r} \cdot \partial_\alpha^2 \vec{r} \right].$$

(b) Write down the lowest order terms (there are two) that appear at the quartic level.

(c) Discuss what happens when t changes sign, assuming that quartic terms provide the required stability (and $K > 0$).

3
Fluctuations

3.1 Scattering and fluctuations

In addition to bulk thermodynamic experiments, scattering measurements can be used to probe microscopic fluctuations at length scales of the order of the probe wavelength λ. In a typical set up, a beam of wavevector $\mathbf{k}_i$ is incident upon the sample and the scattered intensity is measured at wavevector $\mathbf{k}_s = \mathbf{k}_i + \mathbf{q}$. For *elastic* scattering, $|\mathbf{k}_i| = |\mathbf{k}_s| \equiv k$, and $q \equiv |\mathbf{q}| = 2k \sin \theta$, where θ is the angle between incident and scattered beams. Standard treatments of scattering start with the Fermi golden rule, and usually lead to a scattering amplitude of the form

$$A(\mathbf{q}) \propto \langle \mathbf{k}_s \otimes f | \mathcal{U} | \mathbf{k}_i \otimes i \rangle \propto \sigma(\mathbf{q}) \int d^d \mathbf{x} e^{i\mathbf{q}\cdot\mathbf{x}} \rho(\mathbf{x}). \tag{3.1}$$

In the above expression, $|i\rangle$ and $|f\rangle$ refer to the initial and final states of the sample, and $\mathcal{U}$ is the scattering potential that can be decomposed as a sum due to the various scattering elements in the sample. The amplitude has a *local* form factor $\sigma(\mathbf{q})$ describing the scattering from an individual element. For our purposes, the more interesting *global* information is contained in $\rho(\mathbf{q})$, the Fourier transform of the global density of scatterers $\rho(\mathbf{x})$. The appropriate scattering density depends on the nature of the probe. Light scattering senses the actual atomic density, electron scattering measures the charge density, while neutron scattering is usually used to probe the magnetization density. Most such probes actually do not respond to a snapshot of the system, but look at time averaged configurations. Thus the observed scattering intensity is

$$S(\mathbf{q}) \propto \langle |A(\mathbf{q})|^2 \rangle \propto \langle |\rho(\mathbf{q})|^2 \rangle. \tag{3.2}$$

Here $\langle \bullet \rangle$ indicates the thermal average of $\bullet$, which can be used in place of the time average in most cases due to ergodicity.

Equation (3.2) indicates that a uniform density only leads to forward scattering ($\mathbf{q} = \mathbf{0}$), while the long-wavelength fluctuations can be studied by working at small angles or with small k. If scattering is caused by the magnetization density, we can use the Landau–Ginzburg Hamiltonian to compute its intensity. The probability of a particular configuration is given by

$$\mathcal{P}[\vec{m}(\mathbf{x})] \propto \exp\left\{-\int d^d\mathbf{x}\left[\frac{K}{2}(\nabla m)^2 + \frac{t}{2}m^2 + um^4\right]\right\}. \tag{3.3}$$

As discussed earlier, the most probable configuration is *uniform*, with $\vec{m}(\mathbf{x}) = \overline{m}\hat{e}_1$, where $\hat{e}_1$ is a unit vector ($\overline{m}$ is zero for $t > 0$, and equal to $\sqrt{-t/4u}$ for $t < 0$). We can examine small fluctuations around such a configuration by setting

$$\vec{m}(\mathbf{x}) = [\overline{m} + \phi_\ell(\mathbf{x})]\hat{e}_1 + \sum_{\alpha=2}^{n} \phi_{t,\alpha}(\mathbf{x})\hat{e}_\alpha, \tag{3.4}$$

where ϕ_ℓ and ϕ_t refer to *longitudinal* and *transverse* fluctuations, respectively. The latter can take place along any of the $n-1$ directions perpendicular to the average magnetization.

After the substitution of Eq. (3.4), the terms appearing in the Landau–Ginzburg Hamiltonian can be expanded to second order as

$$(\nabla m)^2 = (\nabla\phi_\ell)^2 + (\nabla\phi_t)^2,$$

$$m^2 = \overline{m}^2 + 2\overline{m}\phi_\ell + \phi_\ell^2 + \phi_t^2,$$

$$m^4 = \overline{m}^4 + 4\overline{m}^3\phi_\ell + 6\overline{m}^2\phi_\ell^2 + 2\overline{m}^2\phi_t^2 + \mathcal{O}(\phi_\ell^3, \phi_t^3),$$

resulting in a quadratic energy cost

$$\beta\mathcal{H} \equiv -\ln\mathcal{P} = V\left(\frac{t}{2}\overline{m}^2 + u\overline{m}^4\right) + \int d^d\mathbf{x}\left[\frac{K}{2}(\nabla\phi_\ell)^2 + \frac{t+12u\overline{m}^2}{2}\phi_\ell^2\right]$$
$$+ \int d^d\mathbf{x}\left[\frac{K}{2}(\nabla\phi_t)^2 + \frac{t+4u\overline{m}^2}{2}\phi_t^2\right] + \mathcal{O}(\phi_\ell^3, \phi_t^3). \tag{3.5}$$

For uniform distortions, the longitudinal and transverse *restoring potentials* have "stiffness constants" given by

$$\frac{K}{\xi_\ell^2} \equiv t + 12u\overline{m}^2 = \left.\frac{\partial^2\Psi(m)}{\partial\phi_\ell^2}\right|_{\overline{m}} = \begin{cases} t & \text{for } t>0 \\ -2t & \text{for } t<0, \end{cases} \tag{3.6}$$

and

$$\frac{K}{\xi_t^2} \equiv t + 4u\overline{m}^2 = \left.\frac{\partial^2\Psi(m)}{\partial\phi_t^2}\right|_{\overline{m}} = \begin{cases} t & \text{for } t>0 \\ 0 & \text{for } t<0. \end{cases} \tag{3.7}$$

(The physical significance of the length scales ξ_ℓ and ξ_t will soon become apparent.) Note that there is no distinction between longitudinal and transverse components for the paramagnet ($t > 0$). For the ordered magnet in $t < 0$, there is no restoring force for the transverse fluctuations which correspond to the Goldstone modes discussed in the previous section.

Following the change of variables to the Fourier modes, $\phi(\mathbf{x}) = \sum_\mathbf{q} \phi_\mathbf{q} e^{i\mathbf{q} \cdot \mathbf{x}} / \sqrt{V}$, the probability of a particular fluctuation configuration is given by

$$\mathcal{P}\left[\{\phi_{\ell,\mathbf{q}};\phi_{t,\mathbf{q}}\}\right] \propto \prod_\mathbf{q} \exp\left\{-\frac{K}{2}(q^2 + \xi_\ell^{-2})|\phi_{\ell,\mathbf{q}}|^2\right\} \cdot \exp\left\{-\frac{K}{2}(q^2 + \xi_t^{-2})|\phi_{t,\mathbf{q}}|^2\right\}. \quad (3.8)$$

Clearly each mode behaves as a Gaussian random variable of zero mean, and the two-point correlation functions are

$$\langle \phi_{\alpha,\mathbf{q}} \phi_{\beta,\mathbf{q}'} \rangle = \frac{\delta_{\alpha,\beta}\delta_{\mathbf{q},-\mathbf{q}'}}{K(q^2 + \xi_\alpha^{-2})}, \quad (3.9)$$

where the indices refer to the longitudinal, or any of the transverse components. By using a spin polarized source of neutrons, the relative orientations can be adjusted to probe either the longitudinal or the transverse correlations. The *Lorentzian form*, $S(\mathbf{q}) \propto 1/(q^2 + \xi^{-2})$, usually provides an excellent fit to scattering line shapes away from the critical point. Equation (3.9) indicates that in the ordered phase, longitudinal scattering still gives a Lorentzian form (on top of a delta function at $\mathbf{q} = 0$ due to the spontaneous magnetization), while transverse scattering always grows as $1/q^2$. The same power law decay is also predicted to hold at the critical point, $t = 0$. Actual experimental fits yield a power law of the form

$$S(\mathbf{q}, T = T_c) \propto \frac{1}{q^{2-\eta}}, \quad (3.10)$$

with a small positive value of η.

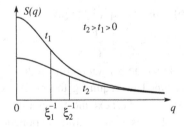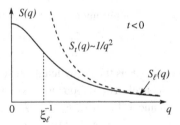

Fig. 3.1 The intensity of scattering by magnetic fluctuations for $t > 0$ (*left*), and $t < 0$ (*right*). The dashed line on the right indicates the transverse scattering intensity.

3.2 Correlation functions and susceptibilities

We can also examine the extent of fluctuations in real space. The averages $\langle \phi_\alpha(\mathbf{x}) \rangle = \langle m_\alpha(\mathbf{x}) - \overline{m}_\alpha \rangle$, are clearly zero, and the *connected correlation function* is

$$G^c_{\alpha,\beta}(\mathbf{x},\mathbf{x}') \equiv \langle (m_\alpha(\mathbf{x}) - \overline{m}_\alpha)(m_\beta(\mathbf{x}') - \overline{m}_\beta) \rangle$$

$$= \langle \phi_\alpha(\mathbf{x})\phi_\beta(\mathbf{x}') \rangle = \frac{1}{V}\sum_{\mathbf{q},\mathbf{q}'} e^{i\mathbf{q} \cdot \mathbf{x} + i\mathbf{q}' \cdot \mathbf{x}'} \langle \phi_{\alpha,\mathbf{q}}\phi_{\beta,\mathbf{q}'} \rangle. \quad (3.11)$$

Using Eq. (3.9), we obtain

$$G^c_{\alpha,\beta}(\mathbf{x},\mathbf{x}') = \frac{\delta_{\alpha,\beta}}{V} \sum_{\mathbf{q}} \frac{e^{i\mathbf{q}\cdot(\mathbf{x}-\mathbf{x}')}}{K(q^2+\xi_\alpha^{-2})} \equiv -\frac{\delta_{\alpha,\beta}}{K} I_d(\mathbf{x}-\mathbf{x}',\xi_\alpha), \qquad (3.12)$$

where in the continuum limit,

$$I_d(\mathbf{x},\xi) = -\int \frac{d^d\mathbf{q}}{(2\pi)^d} \frac{e^{i\mathbf{q}\cdot\mathbf{x}}}{q^2+\xi^{-2}}. \qquad (3.13)$$

Alternatively, I_d is the solution to the following differential equation

$$\nabla^2 I_d(x) = \int \frac{d^d\mathbf{q}}{(2\pi)^d} \frac{q^2 e^{i\mathbf{q}\cdot\mathbf{x}}}{q^2+\xi^{-2}} = \int \frac{d^d\mathbf{q}}{(2\pi)^d} \left[1 - \frac{\xi^{-2}}{q^2+\xi^{-2}}\right] e^{i\mathbf{q}\cdot\mathbf{x}} = \delta^d(\mathbf{x}) + \frac{I_d(x)}{\xi^2}. \qquad (3.14)$$

The solution is spherically symmetric, satisfying

$$\frac{d^2 I_d}{dx^2} + \frac{d-1}{x} \frac{dI_d}{dx} = \frac{I_d}{\xi^2} + \delta^d(\mathbf{x}). \qquad (3.15)$$

We can try out a solution that decays exponentially at large distances as

$$I_d(x) \propto \frac{\exp(-x/\xi)}{x^p}. \qquad (3.16)$$

(We have anticipated the presence of a subleading power law.) The derivatives of I_d are given by

$$\begin{aligned} \frac{dI_d}{dx} &= -\left(\frac{p}{x} + \frac{1}{\xi}\right) I_d, \\ \frac{d^2 I_d}{dx^2} &= \left(\frac{p(p+1)}{x^2} + \frac{2p}{x\xi} + \frac{1}{\xi^2}\right) I_d. \end{aligned} \qquad (3.17)$$

For $x \neq 0$, the requirement that Eq. (3.16) satisfies Eq. (3.15) gives

$$\frac{p(p+1)}{x^2} + \frac{2p}{x\xi} + \frac{1}{\xi^2} - \frac{p(d-1)}{x^2} - \frac{(d-1)}{x\xi} = \frac{1}{\xi^2}. \qquad (3.18)$$

The choice of ξ as the decay length ensures that the constant terms in the above equation cancel. The exponent p is determined by requiring the next largest terms to cancel. For $x \ll \xi$, the $1/x^2$ terms are the next most important; we must set $p(p+1) = p(d-1)$, and $p = d-2$. This is the familiar exponent for Coulomb interactions, and indeed at this length scale the correlations don't feel the presence of ξ. As demonstrated in the next section, the properly normalized result in this limit is

$$I_d(x) \simeq C_d(x) = \frac{x^{2-d}}{(2-d)S_d} \qquad (x \ll \xi). \qquad (3.19)$$

(Note that a constant term can always be added to the solution to satisfy the limits appropriate to the correlation function under study.) At large distances $x \gg \xi$, the $1/(x\xi)$ term dominates Eq. (3.18), and its vanishing implies $p = (d-1)/2$. Matching to Eq. (3.19) at $x \approx \xi$ yields

$$I_d(x) \simeq \frac{\xi^{(3-d)/2}}{(2-d)S_d x^{(d-1)/2}} \exp(-x/\xi) \qquad (x \gg \xi). \qquad (3.20)$$

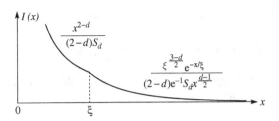

Fig. 3.2 The decay of correlations as a function of separation, away from criticality.

From Eq. (3.12), we observe that transverse and longitudinal correlations behave differently. Close to the critical point, the longitudinal correlation length (Eq. 3.6) behaves as

$$\xi_\ell = \begin{cases} t^{-1/2}/\sqrt{K} & \text{for } t > 0 \\ (-2t)^{-1/2}/\sqrt{K} & \text{for } t < 0. \end{cases} \tag{3.21}$$

The singularities can be described by $\xi_\pm \simeq \xi_0 B_\pm |t|^{-\nu_\pm}$, where $\nu_\pm = 1/2$ and $B_+/B_- = \sqrt{2}$ are universal, while $\xi_0 \propto 1/\sqrt{K}$ is not. The transverse correlation length (Eq. 3.7) equals ξ_ℓ for $t > 0$, and is infinite for all $t < 0$.

Equation (3.19) implies that right at T_c, correlations decay as $1/x^{d-2}$. Actually, the decay exponent is usually indicated by $1/x^{d-2+\eta}$, where η is the same exponent introduced in Eq. (3.10). Integrating the connected correlation functions results in bulk susceptibilities. For example, the divergence of the longitudinal susceptibility is also obtained from,

$$\chi_\ell \propto \int d^d \mathbf{x} G_\ell^c(\mathbf{x}) \propto \int_0^{\xi_\ell} \frac{d^d x}{x^{d-2}} \propto \xi_\ell^2 \simeq A_\pm t^{-1}. \tag{3.22}$$

The universal exponents and amplitude ratios are again recovered from the above equation. For $T < T_c$, there is no upper cut-off length for transverse correlations, and the divergence of the transverse susceptibility can be related to the system size L, as

$$\chi_t \propto \int d^d \mathbf{x} G_t^c(\mathbf{x}) \propto \int_0^L \frac{d^d x}{x^{d-2}} \propto L^2. \tag{3.23}$$

3.3 Lower critical dimension

In chapter 2 we discussed how the breaking of a continuous symmetry is accompanied by the appearance of low energy excitations (Goldstone modes). These modes are easily excited by thermal fluctuations, and we may inquire about the effect of such fluctuations on the ordered phase. Let us first consider the case of a superfluid in the ordered phase, with a local order parameter (see Eq. 2.21) $\psi(\mathbf{x} = |\psi(\mathbf{x})|e^{i\theta(\mathbf{x})}$. Assuming that the amplitude of the order parameter is uniform, the probability of a particular configuration is given by

$$\mathcal{P}[\theta(\mathbf{x})] \propto \exp\left[-\frac{\overline{K}}{2} \int d^d \mathbf{x} (\nabla \theta)^2 \right]. \tag{3.24}$$

Alternatively, in terms of the Fourier components,

$$\mathcal{P}[\theta(\mathbf{q})] \propto \exp\left[-\frac{\overline{K}}{2}\sum_{\mathbf{q}} q^2 |\theta(\mathbf{q})|^2\right] \propto \prod_{\mathbf{q}} p(\theta_{\mathbf{q}}). \qquad (3.25)$$

Each mode $\theta_{\mathbf{q}}$ is an *independent* random variable with a Gaussian distribution of zero mean, and with[1]

$$\langle \theta_{\mathbf{q}} \theta_{\mathbf{q}'} \rangle = \frac{\delta_{\mathbf{q},-\mathbf{q}'}}{\overline{K} q^2}. \qquad (3.26)$$

From Eq. (3.26) we can calculate the correlations in the phase $\theta(\mathbf{x})$ in real space. Clearly $\langle \theta(\mathbf{x}) \rangle = 0$ by symmetry, while

$$\langle \theta(\mathbf{x})\theta(\mathbf{x}') \rangle = \frac{1}{V}\sum_{\mathbf{q},\mathbf{q}'} e^{i\mathbf{q}\cdot\mathbf{x}+i\mathbf{q}'\cdot\mathbf{x}'} \langle \theta_{\mathbf{q}}\theta_{\mathbf{q}'} \rangle = \frac{1}{V}\sum_{\mathbf{q}} \frac{e^{i\mathbf{q}\cdot(\mathbf{x}-\mathbf{x}')}}{\overline{K} q^2}. \qquad (3.27)$$

In the continuum limit, the sum can be replaced by an integral ($\sum_{\mathbf{q}} \mapsto V\int d^d\mathbf{q}/(2\pi)^d$), and

$$\langle \theta(\mathbf{x})\theta(\mathbf{x}') \rangle = \int \frac{d^d\mathbf{q}}{(2\pi)^d} \frac{e^{i\mathbf{q}\cdot(\mathbf{x}-\mathbf{x}')}}{\overline{K} q^2} = -\frac{C_d(\mathbf{x}-\mathbf{x}')}{\overline{K}}. \qquad (3.28)$$

The function,

$$C_d(\mathbf{x}) = -\int \frac{d^d\mathbf{q}}{(2\pi)^d} \frac{e^{i\mathbf{q}\cdot\mathbf{x}}}{q^2}, \qquad (3.29)$$

is the Coulomb potential due to a unit charge at the origin in a d-dimensional space, since it is the solution to

$$\nabla^2 C_d(\mathbf{x}) = \int \frac{d^d\mathbf{q}}{(2\pi)^d} \frac{q^2}{q^2} e^{i\mathbf{q}\cdot\mathbf{x}} = \delta^d(\mathbf{x}). \qquad (3.30)$$

We can easily find a solution by using Gauss' theorem,

$$\int d^d x \nabla^2 C_d = \oint dS \cdot \nabla C_d.$$

---

[1] Note that the Fourier transform of a real field $\theta(\mathbf{x})$ is complex $\theta_{\mathbf{q}} = \theta_{\mathbf{q},\Re} + i\theta_{\mathbf{q},\Im}$. However, the number of fields is not doubled, due to the constraint of $\theta_{-\mathbf{q}} = \theta_{\mathbf{q}}^* = \theta_{\mathbf{q},\Re} - i\theta_{\mathbf{q},\Im}$. The corresponding Gaussian weight has the generic form

$$\mathcal{P}[\{\theta_{\mathbf{q}}\}] \propto \prod_{\mathbf{q}} \exp\left[-\frac{K(q)}{2}\theta_{\mathbf{q}}\theta_{-\mathbf{q}}\right] = \prod_{q>0} \exp\left[-\frac{2K(q)}{2}\left(\theta_{\mathbf{q},\Re}^2 + \theta_{\mathbf{q},\Im}^2\right)\right].$$

While the first product is over all $\mathbf{q}$, the second is restricted to half of the space. There are clearly no cross correlations for differing $\mathbf{q}$, and the Gaussian variances are

$$\langle \theta_{\mathbf{q},\Re}^2 \rangle = \langle \theta_{\mathbf{q},\Im}^2 \rangle = \frac{1}{2K(q)},$$

from which we can immediately construct

$$\langle \theta_{\mathbf{q}}\theta_{\mp\mathbf{q}} \rangle = \langle \theta_{\mathbf{q},\Re}^2 \rangle \pm \langle \theta_{\mathbf{q},\Im}^2 \rangle = \frac{1\pm 1}{2K(q)}.$$

For a spherically symmetric solution, $\nabla C_d = (dC_d/dx)\hat{x}$, and the above equation simplifies to

$$1 = S_d x^{d-1} \frac{dC_d}{dx}, \tag{3.31}$$

where

$$S_d = \frac{2\pi^{d/2}}{(d/2-1)!} \tag{3.32}$$

is the total solid angle (area of unit sphere) in d dimensions. Hence

$$\frac{dC_d}{dx} = \frac{1}{S_d x^{d-1}} \implies C_d(x) = \frac{x^{2-d}}{(2-d)S_d} + c_0, \tag{3.33}$$

where c_0 is a constant of integration.

The long distance behavior of $C_d(x)$ changes dramatically at $d = 2$, as

$$\lim_{x \to \infty} C_d(x) = \begin{cases} c_0 & d > 2 \\ \dfrac{x^{2-d}}{(2-d)S_d} & d < 2 \\ \dfrac{\ln(x)}{2\pi} & d = 2. \end{cases} \tag{3.34}$$

The constant of integration can obtained by looking at

$$\langle [\theta(\mathbf{x}) - \theta(\mathbf{x}')]^2 \rangle = 2\langle \theta(\mathbf{x})^2 \rangle - 2\langle \theta(\mathbf{x})\theta(\mathbf{x}') \rangle, \tag{3.35}$$

which goes to zero as $\mathbf{x} \to \mathbf{x}'$. Hence,

$$\langle [\theta(\mathbf{x}) - \theta(\mathbf{x}')]^2 \rangle = \frac{2\left(|\mathbf{x} - \mathbf{x}'|^{2-d} - a^{2-d}\right)}{\overline{K}(2-d)S_d}, \tag{3.36}$$

where a is of the order of the lattice spacing.

For $d > 2$, the phase fluctuations are finite, while they become asymptotically large for $d \le 2$. Since the phase is bounded by 2π, this implies that long-range order in the phase is destroyed. This result becomes more apparent by examining the effect of phase fluctuations on the two-point correlation function

$$\langle \psi(\mathbf{x})\psi^*(\mathbf{0}) \rangle = \overline{\psi}^2 \langle e^{i[\theta(\mathbf{x}) - \theta(\mathbf{0})]} \rangle. \tag{3.37}$$

(Since amplitude fluctuations are ignored, we are in fact looking at a transverse correlation function.) We shall prove later on that for any collection of Gaussian distributed variables,

$$\langle \exp(\alpha\theta) \rangle = \exp\left(\frac{\alpha^2}{2}\langle \theta^2 \rangle\right).$$

Taking this result for granted, we obtain

$$\langle \psi(\mathbf{x})\psi^*(0) \rangle = \overline{\psi}^2 \exp\left[-\frac{1}{2}\langle [\theta(\mathbf{x}) - \theta(\mathbf{0})]^2 \rangle\right] = \overline{\psi}^2 \exp\left[-\frac{x^{2-d} - a^{2-d}}{\overline{K}(2-d)S_d}\right], \tag{3.38}$$

and asymptotically

$$\lim_{x \to \infty} \langle \psi(\mathbf{x}) \psi^*(\mathbf{0}) \rangle = \begin{cases} \overline{\psi'}^2 & \text{for } d > 2 \\ 0 & \text{for } d \leq 2. \end{cases} \tag{3.39}$$

The saddle point approximation to the order parameter $\overline{\psi}$ was obtained by ignoring fluctuations. The above result indicates that inclusion of phase fluctuations leads to a reduction of order in $d > 2$, and its complete destruction in $d \leq 2$.

The above example typifies a more general result known as the *Mermin–Wagner theorem*. The theorem states that there is no spontaneous breaking of a continuous symmetry in systems with short-range interactions in dimensions $d \leq 2$. Some corollaries to this theorem are:

(1) The borderline dimensionality of two, known as the *lower critical dimension*, has to be treated carefully. As we shall demonstrate later on in the course, there is in fact a phase transition for the two-dimensional superfluid, although there is no true long-range order.
(2) There are no Goldstone modes when the broken symmetry is discrete (e.g. for $n = 1$). In such cases, long-range order is possible down to the lower critical dimension of $d_\ell = 1$.

3.4 Comparison to experiments

The true test of the validity of the theoretical results comes from comparison to experiments. A rather rough table of critical exponents is provided below for a quick check:

Transition type	Material	α	β	γ	ν
Ferromagnets ($n = 3$)	Fe, Ni	−0.1	0.4	1.3	
Superfluid ($n = 2$)	He^4	0	0.3	1.3	0.7
Liquid–gas ($n = 1$)	CO_2, Xe	0.1	0.3	1.2	0.7
Ferroelectrics and superconductors	TGS	0	1/2	1	1/2
Mean-field theory		0	1/2	1	1/2

The exponents are actually known to much better accuracy than indicated in this table. The final row (mean-field theory) refers to the results obtained from the saddle point approximation. They agree only with the experiments on ferroelectric and superconducting materials. The disagreement between the exponents for different values of n suggests that the mean-field results are too universal, and leave out some essential dependence on n (and d). How do we account for these discrepancies? The starting point of the Landau–Ginzburg Hamiltonian is sufficiently general to be trustworthy. The difficulty is in the saddle point method used in the evaluation of its partition function, as will become apparent in the following sections.

3.5 Gaussian integrals

In the previous section the energy cost of fluctuations was calculated at quadratic order. These fluctuations also modify the saddle point free energy. Before calculating this modification, we take a short (but necessary) mathematical diversion on computing Gaussian integrals.

The simplest Gaussian integral involves one variable ϕ,

$$\mathcal{I}_1 = \int_{-\infty}^{\infty} \mathrm{d}\phi \, \mathrm{e}^{-\frac{K}{2}\phi^2 + h\phi} = \sqrt{\frac{2\pi}{K}} \, \mathrm{e}^{\frac{h^2}{2K}}. \tag{3.40}$$

By taking derivatives of the above expression with respect to h, integrals involving powers of ϕ are generated; e.g.

$$
\begin{aligned}
\frac{\mathrm{d}}{\mathrm{d}h} &: \int_{-\infty}^{\infty} \mathrm{d}\phi \, \phi \, \mathrm{e}^{-\frac{K}{2}\phi^2 + h\phi} = \sqrt{\frac{2\pi}{K}} \, \mathrm{e}^{\frac{h^2}{2K}} \cdot \frac{h}{K}, \\
\frac{\mathrm{d}^2}{\mathrm{d}h^2} &: \int_{-\infty}^{\infty} \mathrm{d}\phi \, \phi^2 \mathrm{e}^{-\frac{K}{2}\phi^2 + h\phi} = \sqrt{\frac{2\pi}{K}} \, \mathrm{e}^{\frac{h^2}{2K}} \cdot \left[\frac{1}{K} + \frac{h^2}{K^2}\right].
\end{aligned}
\tag{3.41}
$$

If the integrand represents the probability density of the random variable ϕ, the above integrals imply the moments $\langle \phi \rangle = h/K$, and $\langle \phi^2 \rangle = h^2/K^2 + 1/K$. The corresponding cumulants are $\langle \phi \rangle_c = \langle \phi \rangle = h/K$, and $\langle \phi^2 \rangle_c = \langle \phi^2 \rangle - \langle \phi \rangle^2 = 1/K$. In fact all higher order cumulants of the Gaussian distribution are zero since

$$\langle \mathrm{e}^{-\mathrm{i}k\phi} \rangle \equiv \exp\left[\sum_{\ell=1}^{\infty} \frac{(-\mathrm{i}k)^\ell}{\ell!} \langle \phi^\ell \rangle_c\right] = \exp\left[-\mathrm{i}kh - \frac{k^2}{2K}\right]. \tag{3.42}$$

Now consider the following Gaussian integral involving N variables,

$$\mathcal{I}_N = \int_{-\infty}^{\infty} \prod_{i=1}^{N} \mathrm{d}\phi_i \exp\left[-\sum_{i,j} \frac{K_{i,j}}{2} \phi_i \phi_j + \sum_i h_i \phi_i\right]. \tag{3.43}$$

It can be reduced to a product of N one-dimensional integrals by diagonalizing the matrix $\mathbf{K} \equiv K_{i,j}$. Since we need only consider *symmetric matrices* ($K_{i,j} = K_{j,i}$), the eigenvalues are real, and the eigenvectors can be made orthonormal. Let us denote the eigenvectors and eigenvalues of $\mathbf{K}$ by $\hat{q}$ and K_q, respectively, i.e. $\mathbf{K}\hat{q} = K_q\hat{q}$. The vectors $\{\hat{q}\}$ form a new coordinate basis in the original N-dimensional space. Any point in this space can be represented either by coordinates $\{\phi_i\}$, or $\left\{\tilde{\phi}_q\right\}$ with $\phi_i = \sum_q \tilde{\phi}_q\hat{q}_i$. We can now change the integration variables from $\{\phi_i\}$ to $\left\{\tilde{\phi}_q\right\}$. The Jacobian associated with this unitary transformation is unity, and

$$\mathcal{I}_N = \prod_{q=1}^{N} \int_{-\infty}^{\infty} \mathrm{d}\tilde{\phi}_q \exp\left[-\frac{K_q}{2}\tilde{\phi}_q^2 + \tilde{h}_q\tilde{\phi}_q\right] = \prod_{q=1}^{N} \sqrt{\frac{2\pi}{K_q}}\exp\left[\frac{\tilde{h}_q K_q^{-1}\tilde{h}_q}{2}\right]. \qquad (3.44)$$

The final expression can be represented in terms of the original coordinates by using the *inverse* matrix $\mathbf{K}^{-1}$, such that $\mathbf{K}^{-1}\mathbf{K} = \mathbf{1}$. Since the determinant of the matrix is independent of the choice of basis, $\det\mathbf{K} = \prod_q K_q$, and

$$\mathcal{I}_N = \sqrt{\frac{(2\pi)^N}{\det\mathbf{K}}}\exp\left[\sum_{i,j}\frac{K_{i,j}^{-1}}{2}h_i h_j\right]. \qquad (3.45)$$

Regarding $\{\phi_i\}$ as Gaussian random variable distributed with a joint probability distribution function proportional to the integrand of Eq. (3.43), the *joint characteristic function* is given by

$$\langle e^{-i\sum_j k_j\phi_j}\rangle = \exp\left[-i\sum_{i,j}K_{i,j}^{-1}h_i k_j - \sum_{i,j}\frac{K_{i,j}^{-1}}{2}k_i k_j\right]. \qquad (3.46)$$

Moments of the distribution are obtained from derivatives of the characteristic function with respect to k_i, and *cumulants* from derivatives of its logarithm. Hence, Eq. (3.46) implies

$$\begin{cases} \langle\phi_i\rangle_c & = \sum_j K_{i,j}^{-1}h_j \\ \langle\phi_i\phi_j\rangle_c & = K_{i,j}^{-1}. \end{cases} \qquad (3.47)$$

Another useful form of Eq. (3.46) is

$$\langle\exp(A)\rangle = \exp\left[\langle A\rangle_c + \frac{1}{2}\langle A^2\rangle_c\right], \qquad (3.48)$$

where $A = \sum_i a_i\phi_i$ is any linear combination of Gaussian distributed variables. We used this result earlier in computing the order parameter correlations in the presence of phase fluctuations in a superfluid.

Gaussian *functional integrals* are a limiting case of the above many variable integrals. Consider the points i as the sites of a d-dimensional lattice and let the spacing go to zero. In the continuum limit, $\{\phi_i\}$ go over to a function $\phi(\mathbf{x})$, and the matrix K_{ij} is replaced by a *kernel* $K(\mathbf{x}, \mathbf{x}')$. The natural generalization of Eq. (3.45) is

$$\int_{-\infty}^{\infty}\mathcal{D}\phi(\mathbf{x})\exp\left[-\int \mathrm{d}^d x\mathrm{d}^d x'\frac{K(\mathbf{x}, \mathbf{x}')}{2}\phi(\mathbf{x})\phi(\mathbf{x}') + \int \mathrm{d}^d x h(\mathbf{x})\phi(\mathbf{x})\right]$$

$$\propto (\det\mathbf{K})^{-1/2}\exp\left[\int \mathrm{d}^d x\mathrm{d}^d x'\frac{K^{-1}(\mathbf{x}, \mathbf{x}')}{2}h(\mathbf{x})h(\mathbf{x}')\right], \qquad (3.49)$$

where the inverse kernel $K^{-1}(\mathbf{x}, \mathbf{x}')$ satisfies

$$\int d^d \mathbf{x}' K(\mathbf{x}, \mathbf{x}') K^{-1}(\mathbf{x}', \mathbf{x}'') = \delta^d(\mathbf{x} - \mathbf{x}''). \qquad (3.50)$$

The notation $\mathcal{D}\phi(\mathbf{x})$ is used to denote the functional integral. There is a constant of proportionality, $(2\pi)^{N/2}$, left out of Eq. (3.49). Although formally infinite in the continuum limit of $N \to \infty$, it does not affect the averages that are obtained as derivatives of such integrals. In particular, for Gaussian distributed functions, Eq. (3.47) generalizes to

$$\begin{cases} \langle \phi(\mathbf{x}) \rangle_c & = \int d^d \mathbf{x}' K^{-1}(\mathbf{x}, \mathbf{x}') h(\mathbf{x}') \\ \langle \phi(\mathbf{x}) \phi(\mathbf{x}') \rangle_c & = K^{-1}(\mathbf{x}, \mathbf{x}'). \end{cases} \qquad (3.51)$$

In dealing with small fluctuations to the Landau–Ginzburg Hamiltonian, we encountered the quadratic form

$$\int d^d \mathbf{x} [(\nabla \phi)^2 + \phi^2/\xi^2] \equiv \int d^d \mathbf{x} d^d \mathbf{x}' \phi(\mathbf{x}') \delta^d(\mathbf{x} - \mathbf{x}')(-\nabla^2 + \xi^{-2}) \phi(\mathbf{x}), \qquad (3.52)$$

which implies the kernel

$$K(\mathbf{x}, \mathbf{x}') = K \delta^d(\mathbf{x} - \mathbf{x}')(-\nabla^2 + \xi^{-2}). \qquad (3.53)$$

Following Eq. (3.50), the inverse kernel satisfies

$$K \int d^d \mathbf{x}'' \delta^d(\mathbf{x} - \mathbf{x}'')(-\nabla^2 + \xi^{-2}) K^{-1}(\mathbf{x}'' - \mathbf{x}') = \delta^d(\mathbf{x}' - \mathbf{x}), \qquad (3.54)$$

which implies the differential equation

$$K(-\nabla^2 + \xi^{-2}) K^{-1}(\mathbf{x}) = \delta^d(\mathbf{x}). \qquad (3.55)$$

Comparing with Eq. (3.14) indicates $K^{-1}(\mathbf{x}) = \langle \phi(\mathbf{x}) \phi(\mathbf{0}) \rangle = -I_d(\mathbf{x})/K$, as obtained before by a less direct method.

3.6 Fluctuation corrections to the saddle point

We can now examine how fluctuations around the saddle point solution modify the free energy, and other macroscopic properties. Starting with Eq. (3.5), the partition function including small fluctuations is

$$Z \approx \exp\left[-V\left(\frac{t}{2}\overline{m}^2 + u\overline{m}^4\right)\right] \int \mathcal{D}\phi_\ell(\mathbf{x}) \exp\left\{-\frac{K}{2} \int d^d \mathbf{x}\left[(\nabla \phi_\ell)^2 + \frac{\phi_\ell^2}{\xi_\ell^2}\right]\right\}$$
$$\cdot \int \mathcal{D}\phi_t(\mathbf{x}) \exp\left\{-\frac{K}{2} \int d^d \mathbf{x}\left[(\nabla \phi_t)^2 + \frac{\phi_t^2}{\xi_t^2}\right]\right\}. \qquad (3.56)$$

Each of the Gaussian kernels is diagonalized by the Fourier transforms

$$\tilde{\phi}(\mathbf{q}) = \int d^d \mathbf{x} \exp(-i\mathbf{q} \cdot \mathbf{x}) \phi(\mathbf{x})/\sqrt{V},$$

and with corresponding eigenvalues $K(\mathbf{q}) = K(q^2 + \xi^{-2})$. The resulting determinant of $\mathbf{K}$ is a product of such eigenvalues, and hence

$$\ln \det \mathbf{K} = \sum_{\mathbf{q}} \ln K(\mathbf{q}) = V \int \frac{d^d\mathbf{q}}{(2\pi)^d} \ln[K(q^2 + \xi^{-2})]. \tag{3.57}$$

The free energy resulting from Eq. (3.56) is then given by

$$\beta f - \frac{\ln Z}{V} = \frac{t\overline{m}^2}{2} + u\overline{m}^4 + \frac{1}{2} \int \frac{d^d\mathbf{q}}{(2\pi)^d} \ln[K(q^2 + \xi_\ell^{-2})]$$
$$+ \frac{n-1}{2} \int \frac{d^d\mathbf{q}}{(2\pi)^d} \ln[K(q^2 + \xi_t^{-2})]. \tag{3.58}$$

(Note that there are $n - 1$ transverse components.) Using the dependence of the correlation lengths on reduced temperature, the singular part of the heat capacity is obtained as

$$C_{\text{singular}} \propto -\frac{\partial^2(\beta f)}{\partial^2 t} = \begin{cases} 0 + \dfrac{n}{2} \int \dfrac{d^d\mathbf{q}}{(2\pi)^d} \dfrac{1}{(Kq^2 \mid t)^2} & \text{for } t > 0 \\[3mm] \dfrac{1}{8u} + 2 \int \dfrac{d^d\mathbf{q}}{(2\pi)^d} \dfrac{1}{(Kq^2 - 2t)^2} & \text{for } t < 0. \end{cases} \tag{3.59}$$

The correction terms are proportional to

$$C_F = \frac{1}{K^2} \int \frac{d^d\mathbf{q}}{(2\pi)^d} \frac{1}{(q^2 + \xi^{-2})^2}. \tag{3.60}$$

The integral has dimensions of $(\text{length})^{4-d}$, and changes behavior at $d = 4$. For $d > 4$ the integral diverges at large $\mathbf{q}$, and is dominated by the upper cutoff $\Lambda \simeq 1/a$, where a is the lattice spacing. For $d < 4$, the integral is convergent in both limits. It can be made dimensionless by rescaling $\mathbf{q}$ by ξ^{-1}, and is hence proportional to ξ^{4-d}. Therefore

$$C_F \simeq \frac{1}{K^2} \begin{cases} a^{4-d} & \text{for } d > 4 \\ \xi^{4-d} & \text{for } d < 4. \end{cases} \tag{3.61}$$

Fig. 3.3 The saddle point heat capacity (solid line), plus the corrections due to Gaussian fluctuations (dashed lines). The corrections simply modify the discontinuity in dimensions $d > 4$ (left), but are divergent in $d \leq 4$ (right).

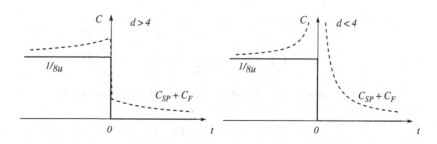

In dimensions $d > 4$, fluctuation corrections to the heat capacity add a constant term to the background on each side of the transition. However, the primary form of the singularity, a discontinuity in C, is not changed. For $d < 4$, the divergence of $\xi \propto t^{-1/2}$, at the transition leads to a correction term from Eq. (3.61) which is more important than the original discontinuity. Indeed, the correction term corresponds to an exponent $\alpha = (4 - d)/2$. However, this is only the first correction to the saddle point result. The divergence of C_F merely implies that the saddle point conclusions are no longer reliable in dimensions $d \leq 4$, below the so-called *upper critical dimension*. Although we obtained this dimension by looking at the fluctuation corrections to the heat capacity, we would have reached the same conclusion in examining the singular part of any other quantity, such as magnetization or susceptibility. The contributions due to fluctuations always modify the leading singular behavior, and hence the critical exponents, in dimensions $d \leq 4$.

3.7 The Ginzburg criterion

We have thus established the importance of fluctuations, and identified them as the probable reason for the failure of the saddle point approximation to correctly describe the observed exponents. However, as noted earlier, there are some materials, such as superconductors, in which the experimental results are well fitted to the singular forms predicted by this approximation. Can we quantify why fluctuations are less important in superconductors than in other phase transitions?

Equation (3.61) indicates that fluctuation corrections become important due to the divergence of the correlation length. Within the saddle point approximation, the correlation length diverges as $\xi \approx \xi_0 |t|^{-1/2}$, where $t = (T_c - T)/T_c$ is the reduced temperature, and $\xi_0 \approx \sqrt{K}$ is a *microscopic* length scale. In principle, ξ_0 can be measured experimentally from fitting scattering line shapes. It has to approximately equal the size of the units that undergo ordering at the phase transition. For the liquid–gas transition, ξ_0 can be estimated as $(v_c)^{1/3}$, where v_c is the critical atomic volume. In superfluids, ξ_0 is approximately the thermal wavelength $\lambda(T)$. Both these estimates are of the order of a few atomic spacings, 1–10 Å. On the other hand, the underlying unit for superconductors is a Cooper pair. The paired electrons are forced apart by their Coulomb repulsion, resulting in a relatively large separation of $\xi_0 \approx 10^3$ Å.

The importance of fluctuations can be gauged by comparing the two terms in Eq. (3.59); the saddle point discontinuity $\Delta C_{SP} \propto 1/u$, and the correction term C_F. Since $K \propto \xi_0^2$, the correction term is proportional to $\xi_0^{-d} t^{-(4-d)/2}$. Thus fluctuations are important provided,

$$\xi_0^{-d} t^{-\frac{4-d}{2}} \gg \Delta C_{SP} \quad \Longrightarrow \quad |t| \ll t_G \simeq \frac{1}{(\xi_0^d \Delta C_{SP})^{\frac{2}{4-d}}}. \tag{3.62}$$

The above requirement is known as the *Ginzburg criterion*. Naturally in $d < 4$, the inequality is satisfied sufficiently close to the critical point. However, the resolution of the experiment may not be good enough to get closer than the Ginzburg reduced temperature t_G. If so, the apparent singularities at reduced temperatures $t > t_G$ may show saddle point behavior. It is this apparent discontinuity that then appears in Eq. (3.62), and may be used to self-consistently estimate t_G. Clearly, ΔC_{SP} and ξ_0 can both be measured in dimensionless units; ξ_0 in units of atomic size a, and ΔC_{SP} in units of Nk_B. The latter is of the order of unity for most transitions, and thus $t_G \approx \xi_0^{-6}$ in $d = 3$. In cases where ξ_0 is a few atomic spacings, a resolution of $t_G \approx 10^{-1} - 10^{-2}$ will suffice. However, in superconductors with $\xi_0 \approx 10^3 a$, a resolution of $t_G < 10^{-18}$ is necessary to see any fluctuation effects. This is much beyond the ability of current apparatus. The newer ceramic high-temperature superconductors have a much smaller coherence length of $\xi_0 \approx 10a$, and they indeed show some effects of fluctuations.

Again, it is worth emphasizing that a similar criterion could have been obtained by examining any other quantity. Fluctuations corrections become important in measurement of a quantity X for $t \ll t_G(X) \simeq A(X) \xi_0^{-2d/(4-d)}$. However, the coefficient $A(X)$ may be different (by one or two orders of magnitude) for different quantities. So, it is in principle possible to observe saddle point behavior in one quantity, while fluctuations are important in another quantity measured at the same resolution. Of course, fluctuations will always become important at sufficiently high resolutions.

A summary of the results obtained so far from the Landau–Ginzburg approach is as follows:

- For dimensions d greater than an upper critical dimension of $d_u = 4$, the saddle point approximation is valid, and singular behavior at the critical point is described by exponents $\alpha = 0$, $\beta = 1/2$, $\gamma = 1$, $\nu = 1/2$, and $\eta = 0$.
- For d less than a lower critical dimension ($d_\ell = 2$ for continuous symmetry, and $d_\ell = 1$ for discrete symmetry) fluctuations are strong enough to destroy the ordered phase.
- In the intermediate dimensions, $d_\ell \le d \le d_u$, fluctuations are strong enough to change the saddle point results, but not sufficiently dominant to completely destroy order. Unfortunately, or happily, this is the case of interest to us in $d = 3$.

Problems for chapter 3

1. *Spin waves:* In the XY model of $n = 2$ magnetism, a unit vector $\vec{s} = (s_x, s_y)$ (with $s_x^2 + s_y^2 = 1$) is placed on each site of a d-dimensional lattice. There is an interaction that tends to keep nearest-neighbors parallel, i.e. a Hamiltonian

$$-\beta \mathcal{H} = K \sum_{\langle ij \rangle} \vec{s}_i \cdot \vec{s}_j.$$

The notation $\langle ij \rangle$ is conventionally used to indicate summing over all *nearest-neighbor* pairs (i, j).

(a) Rewrite the partition function $Z = \int \prod_i d\vec{s}_i \exp(-\beta \mathcal{H})$, as an integral over the set of angles $\{\theta_i\}$ between the spins $\{\vec{s}_i\}$ and some arbitrary axis.

(b) At low temperatures ($K \gg 1$), the angles $\{\theta_i\}$ vary slowly from site to site. In this case expand $-\beta \mathcal{H}$ to get a quadratic form in $\{\theta_i\}$.

(c) For $d = 1$, consider L sites with periodic boundary conditions (i.e. forming a closed chain). Find the normal modes θ_q that diagonalize the quadratic form (by Fourier transformation), and the corresponding eigenvalues $K(q)$. Pay careful attention to whether the modes are real or complex, and to the allowed values of q.

(d) Generalize the results from the previous part to a d-dimensional simple cubic lattice with periodic boundary conditions.

(e) Calculate the contribution of these modes to the free energy and heat capacity. (Evaluate the *classical* partition function, i.e. do not quantize the modes.)

(f) Find an expression for $\langle \vec{s}_0 \cdot \vec{s}_{\mathbf{x}} \rangle = \Re \langle \exp[i\theta_{\mathbf{x}} - i\theta_0] \rangle$ by adding contributions from different Fourier modes. Convince yourself that for $|\mathbf{x}| \to \infty$, only $\mathbf{q} \to \mathbf{0}$ modes contribute appreciably to this expression, and hence calculate the asymptotic limit.

(g) Calculate the transverse susceptibility from $\chi_t \propto \int d^d\mathbf{x} \langle \vec{s}_0 \cdot \vec{s}_{\mathbf{x}} \rangle_c$. How does it depend on the system size L?

(h) In $d = 2$, show that χ_t only diverges for K larger than a critical value $K_c = 1/(4\pi)$.

2. *Capillary waves:* A reasonably flat surface in d-dimensions can be described by its height h, as a function of the remaining $(d - 1)$ coordinates $\mathbf{x} = (x_1, \ldots x_{d-1})$. Convince yourself that the generalized "area" is given by $\mathcal{A} = \int d^{d-1}\mathbf{x} \sqrt{1 + (\nabla h)^2}$. With a surface tension σ, the Hamiltonian is simply $\mathcal{H} = \sigma \mathcal{A}$.

(a) At sufficiently low temperatures, there are only slow variations in h. Expand the energy to quadratic order, and write down the partition function as a functional integral.

(b) Use Fourier transformation to diagonalize the quadratic Hamiltonian into its normal modes $\{h_{\mathbf{q}}\}$ (capillary waves).

(c) What symmetry breaking is responsible for these Goldstone modes?

(d) Calculate the height–height correlations $\langle \left(h(\mathbf{x}) - h(\mathbf{x}') \right)^2 \rangle$.

(e) Comment on the form of the result (d) in dimensions $d = 4, 3, 2,$ and 1.

(f) By estimating typical values of ∇h, comment on when it is justified to ignore higher order terms in the expansion for $\mathcal{A}$.

3. *Gauge fluctuations in superconductors:* The Landau–Ginzburg model of superconductivity describes a complex superconducting order parameter $\Psi(\mathbf{x}) = \Psi_1(\mathbf{x}) +$

$i\Psi_2(\mathbf{x})$, and the electromagnetic vector potential $\vec{A}(\mathbf{x})$, which are subject to a Hamiltonian

$$\beta\mathcal{H} = \int d^3\mathbf{x} \left[\frac{t}{2}|\Psi|^2 + u|\Psi|^4 + \frac{K}{2} D_\mu \Psi D_\mu^* \Psi^* + \frac{L}{2}(\nabla \times A)^2 \right].$$

The gauge-invariant derivative $D_\mu \equiv \partial_\mu - ieA_\mu(\mathbf{x})$, introduces the coupling between the two fields. (In terms of Cooper pair parameters, $e = e^*c/\hbar$, $K = \hbar^2/2m^*$.)

(a) Show that the above Hamiltonian is invariant under the *local gauge symmetry*:

$$\Psi(\mathbf{x}) \mapsto \Psi(x)\exp(i\theta(\mathbf{x})), \quad \text{and} \quad A_\mu(\mathbf{x}) \mapsto A_\mu(\mathbf{x}) + \frac{1}{e}\partial_\mu \theta.$$

(b) Show that there is a saddle point solution of the form $\Psi(\mathbf{x}) = \overline{\Psi}$, and $\vec{A}(\mathbf{x}) = 0$, and find $\overline{\Psi}$ for $t > 0$ and $t < 0$.

(c) For $t < 0$, calculate the cost of fluctuations by setting

$$\begin{cases} \Psi(\mathbf{x}) = (\overline{\Psi} + \phi(\mathbf{x}))\exp(i\theta(\mathbf{x})), \\ A_\mu(\mathbf{x}) = a_\mu(\mathbf{x}) \quad \text{(with } \partial_\mu a_\mu = 0 \text{ in the Coulomb gauge),} \end{cases}$$

and expanding $\beta\mathcal{H}$ to quadratic order in ϕ, θ, and $\vec{a}$.

(d) Perform a Fourier transformation, and calculate the expectation values of $\langle |\phi(\mathbf{q})|^2 \rangle$, $\langle |\theta(\mathbf{q})|^2 \rangle$, and $\langle |\vec{a}(\mathbf{q})|^2 \rangle$.

4. *Fluctuations around a tricritical point:* As shown in a previous problem, the Hamiltonian

$$\beta\mathcal{H} = \int d^d\mathbf{x} \left[\frac{K}{2}(\nabla m)^2 + \frac{t}{2}m^2 + um^4 + vm^6 \right],$$

with $u = 0$ and $v > 0$ describes a tricritical point.

(a) Calculate the heat capacity singularity as $t \to 0$ by the saddle point approximation.

(b) Include both longitudinal and transverse fluctuations by setting

$$\vec{m}(\mathbf{x}) = (\overline{m} + \phi_\ell(\mathbf{x}))\hat{e}_\ell + \sum_{\alpha=2}^{n} \phi_t^\alpha(\mathbf{x})\hat{e}_\alpha,$$

and expanding $\beta\mathcal{H}$ to quadratic order in ϕ.

(c) Calculate the longitudinal and transverse correlation functions.

(d) Compute the first correction to the saddle point free energy from fluctuations.

(e) Find the fluctuation correction to the heat capacity.

(f) By comparing the results from parts (a) and (e) for $t < 0$ obtain a Ginzburg criterion, and the upper critical dimension for validity of mean-field theory at a tricritical point.

(g) A generalized multicritical point is described by replacing the term vm^6 with $u_{2n}m^{2n}$. Use simple power counting to find the upper critical dimension of this multicritical point.

---

5. *Coupling to a "massless" field:* Consider an n-component vector field $\vec{m}(\mathbf{x})$ coupled to a scalar field $A(\mathbf{x})$, through the effective Hamiltonian

$$\beta \mathcal{H} = \int d^d\mathbf{x} \left[\frac{K}{2}(\nabla\vec{m})^2 + \frac{t}{2}\vec{m}^2 + u(\vec{m}^2)^2 + e^2\vec{m}^2 A^2 + \frac{L}{2}(\nabla A)^2 \right],$$

with K, L, and u positive.

(a) Show that there is a saddle point solution of the form $\vec{m}(\mathbf{x}) = \overline{m}\hat{e}_\ell$ and $A(x) = 0$, and find $\overline{m}$ for $t > 0$ and $t < 0$.

(b) Sketch the heat capacity $C = \partial^2 \ln Z/\partial t^2$, and discuss its singularity as $t \to 0$ in the saddle point approximation.

(c) Include fluctuations by setting

$$\begin{cases} \vec{m}(\mathbf{x}) = (\overline{m} + \phi_\ell(\mathbf{x}))\hat{e}_\ell + \phi_t(\mathbf{x})\hat{e}_t, \\ A(\mathbf{x}) = a(\mathbf{x}), \end{cases}$$

and expanding $\beta\mathcal{H}$ to quadratic order in ϕ and a.

(d) Find the correlation lengths ξ_ℓ, and ξ_t, for the longitudinal and transverse components of ϕ, for $t > 0$ and $t < 0$.

(e) Find the correlation length ξ_a for the fluctuations of the scalar field a, for $t > 0$ and $t < 0$.

(f) Calculate the correlation function $\langle a(\mathbf{x})a(\mathbf{0}) \rangle$ for $t > 0$.

(g) Compute the correction to the saddle point free energy $\ln Z$, from fluctuations. (You can leave the answer in the form of integrals involving ξ_ℓ, ξ_t, and ξ_a.)

(h) Find the fluctuation corrections to the heat capacity in (b), again leaving the answer in the form of integrals.

(i) Discuss the behavior of the integrals appearing above schematically, and state their dependence on the correlation length ξ, and cutoff Λ, in different dimensions.

(j) What is the critical dimension for the validity of saddle point results, and how is it modified by the coupling to the scalar field?

6. *Random magnetic fields:* Consider the Hamiltonian

$$\beta\mathcal{H} = \int d^d\mathbf{x} \left[\frac{K}{2}(\nabla m)^2 + \frac{t}{2}m^2 + um^4 - h(\mathbf{x})m(\mathbf{x}) \right],$$

where $m(\mathbf{x})$ and $h(\mathbf{x})$ are scalar fields, and $u > 0$. The random magnetic field $h(\mathbf{x})$ results from frozen (quenched) impurities that are independently distributed in space. For simplicity $h(\mathbf{x})$ is assumed to be an independent Gaussian variable at each point $\mathbf{x}$, such that

$$\overline{h(\mathbf{x})} = 0, \quad \text{and} \quad \overline{h(\mathbf{x})h(\mathbf{x}')} = \Delta\delta^d(\mathbf{x}-\mathbf{x}'), \tag{1}$$

where the over-line indicates (*quench*) averaging over all values of the random fields. The above equation implies that the Fourier transformed random field $h(\mathbf{q})$ satisfies

$$\overline{h(\mathbf{q})} = 0, \quad \text{and} \quad \overline{h(\mathbf{q})h(\mathbf{q}')} = \Delta(2\pi)^d\delta^d(\mathbf{q}+\mathbf{q}'). \tag{2}$$

(a) Calculate the quench averaged free energy, $\overline{f_{sp}} = \overline{\min\{\Psi(m)\}_m}$, *assuming a saddle point solution with uniform magnetization* $m(\mathbf{x}) = m$. (Note that with this assumption, the random field disappears as a result of averaging and has no effect at this stage.)

(b) Include fluctuations by setting $m(\mathbf{x}) = \overline{m} + \phi(\mathbf{x})$, and expanding $\beta\mathcal{H}$ to second order in ϕ.

(c) Express the energy cost of the above fluctuations in terms of the Fourier modes $\phi(\mathbf{q})$.

(d) Calculate the mean $\langle\phi(\mathbf{q})\rangle$, and the variance $\langle|\phi(\mathbf{q})|^2\rangle_c$, where $\langle\cdots\rangle$ denotes the usual thermal expectation value *for a fixed* $h(\mathbf{q})$.

(e) Use the above results, in conjunction with Eq. (2), to calculate the quench averaged scattering line shape $S(q) = \overline{\langle|\phi(\mathbf{q})|^2\rangle}$.

(f) Perform the Gaussian integrals over $\phi(\mathbf{q})$ to calculate the fluctuation corrections, $\delta f[h(\mathbf{q})]$, to the free energy.

$$\left(\text{Reminder}: \int_{-\infty}^{\infty} d\phi d\phi^* \exp\left(-\frac{K}{2}|\phi|^2 + h^*\phi + h\phi^*\right) = \frac{2\pi}{K}\exp\left(\frac{|h|^2}{2K}\right).\right)$$

(g) Use Eq. (2) to calculate the corrections due to the fluctuations in the previous part to the quench averaged free energy $\overline{f}$. (Leave the corrections in the form of two integrals.)

(h) Estimate the singular t dependence of the integrals obtained in the fluctuation corrections to the free energy.

(i) Find the upper critical dimension, d_u, for the validity of saddle point critical behavior.

7. *Long-range interactions:* Consider a continuous spin field $\vec{s}(\mathbf{x})$, subject to a long-range ferromagnetic interaction

$$\int d^d\mathbf{x}d^d\mathbf{y}\frac{\vec{s}(\mathbf{x})\cdot\vec{s}(\mathbf{y})}{|\mathbf{x}-\mathbf{y}|^{d+\sigma}},$$

as well as short-range interactions.

(a) How is the quadratic term in the Landau–Ginzburg expansion modified by the presence of this long-range interaction? For what values of σ is the long-range interaction dominant?

(b) By estimating the magnitude of thermally excited Goldstone modes (or otherwise), obtain the lower critical dimension d_ℓ below which there is no long-range order.

(c) Find the upper critical dimension d_u, above which saddle point results provide a correct description of the phase transition.

8. *Ginzburg criterion along the magnetic field direction:* Consider the Hamiltonian

$$\beta \mathcal{H} = \int d^d \mathbf{x} \left[\frac{K}{2} (\nabla \vec{m})^2 + \frac{t}{2} \vec{m}^2 + u(\vec{m}^2)^2 - \vec{h} . \vec{m} \right],$$

describing an n-component magnetization vector $\vec{m}(\mathbf{x})$, with $u > 0$.

(a) In the saddle point approximation, the free energy is $f = \min\{\Psi(m)\}_m$. Indicate the resulting phase boundary in the (h, t) plane, and label the phases. (h denotes the magnitude of $\vec{h}$.)

(b) Sketch the form of $\Psi(m)$ for $t < 0$, on both sides of the phase boundary, and for $t > 0$ at $h = 0$.

(c) For t and h close to zero, the spontaneous magnetization can be written as $\overline{m} = t^\beta g_m(h/t^\Delta)$. Identify the exponents β and Δ in the saddle point approximation.
   For the remainder of this problem set $t = 0$.

(d) Calculate the transverse and longitudinal susceptibilities at a finite h.

(e) Include fluctuations by setting $\vec{m}(\mathbf{x}) = (\overline{m} + \phi_\ell(\mathbf{x}))\hat{e}_\ell + \vec{\phi}_t(\mathbf{x})\hat{e}_t$, and expanding $\beta\mathcal{H}$ to second order in the ϕs. ($\hat{e}_\ell$ is a unit vector parallel to the average magnetization, and $\hat{e}_t$ is perpendicular to it.)

(f) Calculate the longitudinal and transverse correlation lengths.

(g) Calculate the first correction to the free energy from these fluctuations. (The scaling form is sufficient.)

(h) Calculate the first correction to magnetization, and to longitudinal susceptibility from the fluctuations.

(i) By comparing the saddle point value with the correction due to fluctuations, find the upper critical dimension, d_u, for the validity of the saddle point result.

(j) For $d < d_u$ obtain a Ginzburg criterion by finding the field h_G below which fluctuations are important. (You may ignore the numerical coefficients in h_G, but the dependences on K and u are required.)

4

The scaling hypothesis

4.1 The homogeneity assumption

In the previous chapters the singular behavior in the vicinity of a continuous transition was characterized by a set of critical exponents $\{\alpha, \beta, \gamma, \delta, \nu, \eta, \cdots\}$. The saddle-point estimates of these exponents were found to be unreliable due to the importance of fluctuations. Since the various thermodynamic quantities are related, these exponents can not be independent of each other. The goal of this chapter is to discover the relationships between them, and to find the minimum number of independent exponents needed to describe the critical point.

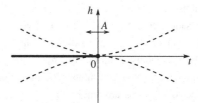

Fig. 4.1 The vicinity of the critical point in the (t, h) plane, with crossover boundaries indicated by dashed lines.

The non-analytical structure is a coexistence line for $t < 0$ and $h = 0$ that terminates at the critical point $t = h = 0$. The various exponents describe the leading singular behavior of a thermodynamic quantity $Q(t, h)$, in the vicinity of this point. A basic quantity in the canonical ensemble is the free energy, which in the saddle point approximation is given by

$$
f(t, h) = \min_m \left[\frac{t}{2} m^2 + u m^4 - h.m \right]_m = \begin{cases} -\dfrac{1}{16} \dfrac{t^2}{u} & \text{for } h = 0, \ t < 0 \\ -\dfrac{3}{4^{4/3}} \dfrac{h^{4/3}}{u^{1/3}} & \text{for } h \neq 0, \ t = 0 . \end{cases} \tag{4.1}
$$

The singularities in the free energy can in fact be described by a single *homogeneous function* in t and h, as[1]

$$f(t, h) = |t|^2 g_f \left(h/|t|^\Delta \right).$$ (4.2)

The function g_f only depends on the combination $x \equiv h/|t|^\Delta$, where Δ is known as the *gap exponent*. The asymptotic behavior of g_f is easily obtained by comparing Eqs. (4.1) and (4.2). The $h = 0$ limit is recovered if $\lim_{x \to 0} g_f(x) \sim 1/u$, while to get the proper power of h, we must set $\lim_{x \to \infty} g_f(x) \sim x^{4/3}/u^{1/3}$. The latter implies $f \sim |t|^2 h^{4/3}/(u^{1/3}|t|^{4\Delta/3})$. Since f can have no t dependence along $t = 0$, the gap exponent (corresponding to Eq. 4.1) has the value

$$\Delta = \frac{3}{2}.$$ (4.3)

The assumption of homogeneity is that, on going beyond the saddle point approximation, the singular form of the free energy (and any other thermodynamic quantity) retains the homogeneous form

$$f_{\text{sing}}(t, h) = |t|^{2-\alpha} g_f \left(h/|t|^\Delta \right).$$ (4.4)

The actual exponents α and Δ depend on the critical point being considered. The dependence on t is chosen to reproduce the heat capacity singularity at $h = 0$. The singular part of the energy is obtained from (say for $t > 0$)

$$E_{\text{sing}} \sim \frac{\partial f}{\partial t} \sim (2 - \alpha)|t|^{1-\alpha} g_f \left(h/|t|^\Delta \right) - \Delta h |t|^{1-\alpha-\Delta} g_f' \left(h/|t|^\Delta \right)$$

$$\sim |t|^{1-\alpha} \left[(2-\alpha) g_f \left(h/|t|^\Delta \right) - \frac{\Delta h}{|t|^\Delta} g_f' \left(h/|t|^\Delta \right) \right] \equiv |t|^{1-\alpha} g_E \left(h/|t|^\Delta \right).$$ (4.5)

Thus the derivative of one homogeneous function is another. Similarly, the second derivative takes the form (again for $t > 0$)

$$C_{\text{sing}} \sim -\frac{\partial^2 f}{\partial t^2} \sim |t|^{-\alpha} g_C \left(h/|t|^\Delta \right),$$ (4.6)

reproducing the scaling $C_{\text{sing}} \sim |t|^{-\alpha}$, as $h \to 0$.

It may appear that we have the freedom to postulate a more general form

$$C_\pm(t, h) = |t|^{-\alpha_\pm} g_\pm \left(h/|t|^{\Delta_\pm} \right),$$ (4.7)

with different functions and exponents for $t > 0$ and $t < 0$ that match at $t = 0$. However, this is ruled out by the condition that the free energy is analytic everywhere except on the coexistence line for $h = 0$ and $t < 0$, as shown as

---

[1] In general, a function $f(x_1, x_2, \cdots)$ is homogeneous if

$$f \left(b^{p_1} x_1, b^{p_2} x_2, \cdots \right) = b^{p_f} f(x_1, x_2, \cdots),$$

for any rescaling factor b. With the proper choice of b one of the arguments can be removed, leading to the scaling forms used in this section.

follows: Consider a point at $t = 0$ and finite h. By assumption, the function C is perfectly analytic in the vicinity of this point, expandable in a Taylor series,

$$C\left(t \ll h^{\Delta}\right) = \mathcal{A}(h) + t\mathcal{B}(h) + \mathcal{O}(t^2). \tag{4.8}$$

Furthermore, the same expansion should be obtained from both C_+ and C_-. But Eq. (4.7) leads to the expansions,

$$C_{\pm} = |t|^{-\alpha_{\pm}} \left[A_+ \left(\frac{h}{|t|^{\Delta_{\pm}}} \right)^{p_{\pm}} + B_+ \left(\frac{h}{|t|^{\Delta_{\pm}}} \right)^{q_{\pm}} + \cdots \right], \tag{4.9}$$

where $\{p_{\pm}, q_{\pm}\}$ are the leading powers in asymptotic expansions of $g_{\pm}$ for large arguments, and $\{A_{\pm}, B_{\pm}\}$ are the corresponding prefactors. Matching to the Taylor series in Eq. (4.8) requires $p_{\pm}\Delta_{\pm} = -\alpha_{\pm}$ and $q_{\pm}\Delta_{\pm} = -(1+\alpha_{\pm})$, and leads to

$$C_{\pm}\left(t \ll h^{\Delta}\right) = A_{\pm} h^{-\alpha_{\pm}/\Delta_{\pm}} + B_{\pm} h^{-(1+\alpha_{\pm})/\Delta_{\pm}} |t| + \cdots \tag{4.10}$$

Continuity at $t = 0$ now forces $\alpha_+/\Delta_+ = \alpha_-/\Delta_-$, and $(1+\alpha_+)/\Delta_+ = (1+\alpha_-)/\Delta_-$, which in turn implies

$$\begin{cases} \alpha_+ = \alpha_- \equiv \alpha \\ \Delta_+ = \Delta_- \equiv \Delta. \end{cases} \tag{4.11}$$

Despite using $|t|$ in the postulated scaling form, we can still ensure the analyticity of the function at $t = 0$ for finite h by appropriate choice of parameters, e.g. by setting $B_- = -B_+$ to match Eq. (4.10) to the analytic form in Eq. (4.8). Having established this result, we can be somewhat careless henceforth in replacing $|t|$ in the scaling equations with t. Naturally these arguments apply to any quantity $Q(t, h)$.

Starting from the free energy in Eq. (4.4), we can compute the singular parts of other quantities of interest:

- The *magnetization* is obtained from

$$m(t, h) \sim \frac{\partial f}{\partial h} \sim |t|^{2-\alpha-\Delta} g_m \left(h/|t|^{\Delta} \right). \tag{4.12}$$

In the limit $x \to 0$, $g_m(x)$ is a constant, and

$$m(t, h = 0) \sim |t|^{2-\alpha-\Delta}, \quad \implies \quad \beta = 2 - \alpha - \Delta. \tag{4.13}$$

On the other hand, if $x \to \infty$, $g_m(x) \sim x^p$, and

$$m(t = 0, h) \sim |t|^{2-\alpha-\Delta} \left(\frac{h}{|t|^{\Delta}} \right)^p. \tag{4.14}$$

Since this limit is independent of t, we must have $p\Delta = 2 - \alpha - \Delta$. Hence

$$m(t, h = 0) \sim h^{(2-\alpha-\Delta)/\Delta} \quad \implies \quad \delta = \Delta/(2 - \alpha - \Delta) = \Delta/\beta. \tag{4.15}$$

- Similarly, the *susceptibility* is computed as

$$\chi(t, h) \sim \frac{\partial m}{\partial h} \sim |t|^{2-\alpha-2\Delta} g_\chi(h/|t|^\Delta) \Rightarrow \chi(t, h=0) \sim |t|^{2-\alpha-2\Delta}$$

$$\Rightarrow \gamma = 2\Delta - 2 + \alpha. \tag{4.16}$$

Thus, the consequences of the homogeneity assumption are:

(1) The singular parts of all critical quantities $Q(t, h)$ are homogeneous, with the same exponents above and below the transition.

(2) Because of the interconnections via thermodynamic derivatives, the same gap exponent, Δ, occurs for all such quantities.

(3) All (bulk) critical exponents can be obtained from only *two* independent ones, e.g. α and Δ.

(4) As a result of the above, there are a number of *exponent identities*. For example, Eqs. (4.13), (4.15), and (4.16) imply

$$\alpha + 2\beta + \gamma = \alpha + 2(2 - \alpha - \Delta) + (2\Delta - 2 + \alpha) = 2 \quad \text{(Rushbrooke's identity)},$$

$$\delta - 1 = \frac{\Delta}{2-\alpha-\Delta} - 1 = \frac{2\Delta-2+\alpha}{2-\alpha-\Delta} = \frac{\gamma}{\beta} \quad \text{(Widom's identity)}. \tag{4.17}$$

These identities can be checked against the following table of critical exponents. The first three rows are based on a number of theoretical estimates in $d = 3$; the last row comes from an exact solution in $d = 2$. The exponent identities are completely consistent with these values, as well as with all reliable experimental data.

	α	β	γ	δ	ν	η
$n=1$	0.11	0.32	1.24	4.9	0.63	0.04
$n=2$	−0.01	0.35	1.32	4.7	0.67	0.04
$n=3$	−0.11	0.36	1.39	4.9	0.70	0.04
$n=1$	0	1/8	7/4	15	1	1/4

4.2 Divergence of the correlation length

The homogeneity assumption relates to the free energy and quantities derived from it. It says nothing about the behavior of correlation functions. An important property of a critical point is the divergence of the correlation length which is responsible for, and can be deduced from, diverging response functions. In order to obtain an identity involving the exponent ν for the divergence of the correlation length, we replace the homogeneity assumption for the free energy with the following *two* conditions:

(1) The correlation length ξ is a homogeneous function,

$$\xi(t, h) \sim |t|^{-\nu} g\left(h/|t|^{\Delta}\right). \tag{4.18}$$

(For $t = 0$, ξ diverges as $h^{-\nu_h}$ with $\nu_h = \nu/\Delta$.)

(2) Close to criticality, the correlation length ξ is the most important length in the system, and is *solely* responsible for singular contributions to thermodynamic quantities.

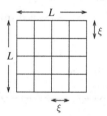

Fig. 4.2 A system of linear size L, presented as (approximately) independent components of size ξ, the correlation length.

The second condition determines the singular part of the free energy. Since $\ln Z(t, h)$ is *extensive* and *dimensionless*, it must take the form

$$\ln Z = \left(\frac{L}{\xi}\right)^d \times g_s + \cdots + \left(\frac{L}{a}\right)^d \times g_a, \tag{4.19}$$

where g_s and g_a are non-singular functions of dimensionless parameters (a is an appropriate microscopic length). The leading singular part of the free energy comes from the first term, and behaves as

$$f_{\text{sing}}(t, h) \sim \frac{\ln Z}{L^d} \sim \xi^{-d} \sim |t|^{d\nu} g_f\left(h/|t|^{\Delta}\right). \tag{4.20}$$

A simple interpretation of the above result is obtained by dividing the system into units of the size of the correlation length. Each unit is then regarded as an independent random variable, contributing a constant factor to the critical free energy. The number of units grows as $(L/\xi)^d$, leading to Eq. (4.19).

The consequences of the above assumptions are:

(1) The homogeneity of $f_{\text{sing}}(t, h)$ emerges naturally.
(2) We obtain the additional exponent relation

$$2 - \alpha = d\nu \quad \text{(Joshephson's identity).} \tag{4.21}$$

Identities obtained from the generalized homogeneity assumption involve the space dimension d, and are known as *hyperscaling relations*. The relation between α and ν is consistent with the exponents in the above table. However, it does not agree with the saddle point values, $\alpha = 0$ and $\nu = 1/2$, which are valid for $d > 4$. Any theory of critical behavior must thus account for the validity of this relation in low dimensions, and its breakdown for $d > 4$.

4.3 Critical correlation functions and self-similarity

One exponent that has not so far been accounted for is η, describing the decay of correlation functions at criticality. Exactly at the critical point, the correlation length is infinite, and there is no other length scale (except sample size) to cut off the decay of correlation functions. Thus all correlations decay as a power of the separation. As discussed in the previous chapter, the magnetization correlations fall off as

$$G^c_{m,m}(\mathbf{x}) \equiv \langle m(\mathbf{x})m(\mathbf{0})\rangle - \langle m\rangle^2 \sim 1/|\mathbf{x}|^{d-2+\eta}. \tag{4.22}$$

Similarly, we can define an exponent η' for the decay of energy–energy correlations as

$$G^c_{E,E}(\mathbf{x}) = \langle \mathcal{H}(\mathbf{x})\mathcal{H}(\mathbf{0})\rangle - \langle \mathcal{H}\rangle^2 \sim 1/|\mathbf{x}|^{d-2+\eta'}. \tag{4.23}$$

Away from criticality, the power laws are cut off for distances $|\mathbf{x}| \gg \xi$. As the response functions can be obtained from integrating the connected correlation functions, there are additional exponent identities, such as (Fisher's identity)

$$\chi \sim \int d^d\mathbf{x}\, G^c_{mm}(\mathbf{x}) \sim \int^\xi \frac{d^d x}{|x|^{d-2+\eta}} \sim \xi^{2-\eta} \sim |t|^{-\nu(2-\eta)} \implies \gamma = (2-\eta)\nu. \tag{4.24}$$

Similarly, for the heat capacity,

$$C \sim \int d^d\mathbf{x}\, G^c_{EE}(\mathbf{x}) \sim \int^\xi \frac{d^d x}{|x|^{d-2+\eta'}} \sim \xi^{2-\eta'} \sim |t|^{-\nu(2-\eta')}, \implies \alpha = (2-\eta')\nu. \tag{4.25}$$

As before, two *independent* exponents are sufficient to describe all singular critical behavior.

An important consequence of these scaling ideas is that the critical system has an additional *dilation symmetry*. Under a change of scale, the critical correlation functions behave as

$$G_{\text{critical}}(\lambda\mathbf{x}) = \lambda^p G_{\text{critical}}(\mathbf{x}). \tag{4.26}$$

This implies a *scale invariance* or *self-similarity*: if a snapshot of the critical system is blown up by a factor of λ, apart from a change of contrast (multiplication by λ^p), the resulting snapshot is statistically similar to the original one. Such statistical self-similarity is the hallmark of *fractal* geometry. As discussed by Mandelbrot, many naturally occurring forms (clouds, shore-lines, river basins, etc.) exhibit such behavior. The Landau–Ginzburg probability was constructed on the basis of *local* symmetries such as rotation invariance. If we could add to the list of constraints the requirement of *dilation symmetry*, the resulting probability would indeed describe the critical point. Unfortunately, it is not possible to directly see how such a requirement constrains the effective Hamiltonian. One notable exception is in $d = 2$, where dilation symmetry implies conformal invariance, and a lot of information can be obtained by constructing conformally invariant theories. We shall instead prescribe a less direct route of following the effects of the dilation operation on the effective energy: the *renormalization group* procedure.

4.4 The renormalization group (conceptual)

Success of the scaling theory in correctly predicting various exponent identities strongly supports the assumption that close to the critical point the correlation length, ξ, is the only important length scale, and that microscopic length scales are irrelevant. The critical behavior is dominated by fluctuations that are self-similar up to the scale ξ. The self-similarity is of course only statistical, in that a magnetization configuration is generated with a weight $W[\vec{m}(\mathbf{x})] \propto \exp\{-\beta\mathcal{H}[\vec{m}(\mathbf{x})]\}$. Kadanoff suggested taking advantage of the self-similarity of the fluctuations to gradually eliminate the correlated degrees of freedom at length scales $x \ll \xi$, until one is left with the relatively simple, uncorrelated degrees of freedom at scale ξ. This is achieved through a procedure called the *renormalization group* (RG), whose conceptual foundation is the three steps outlined in this section.

(1) **Coarse grain:** There is an implicit short distance length cutoff scale a for allowed variations of $\vec{m}(\mathbf{x})$ in the system. This is the lattice spacing for a model of spins, or the coarse graining scale that underlies the Landau–Ginzburg Hamiltonian. In a digital picture of the system, a corresponds to the pixel size. The first step of the RG is to decrease the resolution by changing this minimum scale to ba $(b > 1)$. The *coarse-grained* magnetization is then given by

$$m_i(\mathbf{x}) = \frac{1}{b^d} \int_{\text{Cell centered at } \mathbf{x}} d^d\mathbf{x}'\, m_i(\mathbf{x}'). \tag{4.27}$$

(2) **Rescale:** Due to the change in resolution, the coarse grained "picture" is grainier than the original. The original resolution of a can be restored by decreasing all length scales by a factor of b, i.e. by setting

$$\mathbf{x}_{\text{new}} = \frac{\mathbf{x}_{\text{old}}}{b}. \tag{4.28}$$

(3) **Renormalize:** The variations of fluctuations in the rescaled magnetization profile is in general different from the original, i.e. there is a difference in contrast between the pictures. This can be remedied by introducing a change of contrast by a factor ζ, through defining a *renormalized* magnetization

$$\vec{m}_{\text{new}}(\mathbf{x}_{\text{new}}) = \frac{1}{\zeta b^d} \int_{\text{Cell centered at } b\mathbf{x}_{\text{new}}} d^d\mathbf{x}'\, \vec{m}(\mathbf{x}'). \tag{4.29}$$

By following these steps, for each configuration $\vec{m}_{\rm old}(\mathbf{x})$, we generate a renormalized configuration $\vec{m}_{\rm new}(\mathbf{x})$. Equation (4.29) can be regarded as a mapping from one set of random variables to another, and can be used to construct the probability distribution, or weight $W_b[\vec{m}_{\rm new}(\mathbf{x})] \equiv \exp\{-\beta\mathcal{H}_b[\vec{m}_{\rm new}(\mathbf{x})]\}$. Kadanoff's insight was that since on length scales less than ξ, the renormalized configurations are statistically similar to the original ones, they may be distributed by a Hamiltonian $\beta\mathcal{H}_b$ that is also "close" to the original. In particular, the original Hamiltonian becomes critical by tuning the two parameters t and h to zero: at this point the original configurations are statistically similar to those of the rescaled system. The critical Hamiltonian should thus be invariant under rescaling and renormalization. In the original problem, one moves away from criticality for finite t and h. Kadanoff postulated that the corresponding renormalized Hamiltonian is similarly described by non-zero $t_{\rm new}$ and/or $h_{\rm new}$.

The assumption that the closeness of the original and renormalized Hamiltonians to criticality is described by the two parameters t and h greatly simplifies the analysis. The effect of the RG transformation on the probability of configurations is now described by the two parameter mappings $t_{\rm new} \equiv t_b(t_{\rm old}, h_{\rm old})$ and $h_{\rm new} \equiv h_b(t_{\rm old}, h_{\rm old})$. The next step is to note that since the transformation only involves changes at the shortest length scales, it cannot cause any singularities. The renormalized parameters must be *analytic* functions of the original ones, and hence expandable as

$$\begin{cases} t_b(t, h) = A(b)t + B(b)h + \cdots \\ h_b(t, h) = C(b)t + D(b)h + \cdots \end{cases} \tag{4.30}$$

Note that there are no constant terms in the above Taylor expansions. This expresses the condition that if $\beta\mathcal{H}$ is at its critical point ($t = h = 0$), then $\beta\mathcal{H}_b$ is also at criticality, and $t_{\rm new} = h_{\rm new} = 0$. Furthermore, due to rotational symmetry, under the combined transformation ($m(x) \mapsto -m(x), h \mapsto -h, t \mapsto t$) the weight of a configuration is unchanged. As this symmetry is preserved by the RG, the coefficients B and C in the above expression must be zero, leading to the further simplifications

$$\begin{cases} t_b(t, h) = A(b)t + \cdots \\ h_b(t, h) = D(b)h + \cdots \end{cases} \tag{4.31}$$

The remaining coefficients $A(b)$ and $D(b)$ depend on the (arbitrary) rescaling factor b, and trivially $A(1) = D(1) = 1$ for $b = 1$. Since the above transformations can be carried out in sequence, and the net effect of rescalings of b_1 and b_2 is a change of scale by $b_1 b_2$, the RG procedure is sometimes referred to as a *semi-group*. The term applies to the action of RG on the space of configurations: each magnetization profile is mapped uniquely to one at larger scale, but the inverse process is non-unique as some short scale information is lost in the coarse graining. (There is in fact no problem with inverting the transformation in the space of the parameters of the Hamiltonian.) The dependence of A and

D in Eqs. (4.31) on b can be deduced from this group property. Since at $b = 1$, $A = D = 1$, and $t(b_1 b_2) \approx A(b_1)A(b_2)t \approx A(b_1 b_2)t$, we must have $A(b) = b^{y_t}$, and similarly $D(b) = b^{y_h}$, yielding

$$\begin{cases} t' \equiv t_b = b^{y_t} t + \cdots \\ h' \equiv h_b = b^{y_h} h + \cdots \end{cases} \qquad (4.32)$$

If $\beta \mathcal{H}_{\text{old}}$ is slightly away from criticality, it is described by a large but finite correlation length ξ_{old}. After the RG transformation, due to the rescaling in Eq. (4.28), the new correlation length is smaller by a factor of b. Hence the renormalized Hamiltonian is less critical, and the RG procedure moves the parameters further away from the origin, i.e. the exponents y_t and y_h must be positive.

We can now explore some consequences of the assumptions leading to Eqs. (4.32).

(1) *The free energy:* The RG transformation is a many to one map of the original configurations to new ones. Since the weight of a new configuration, $W'([m'])$, is the sum of the weights $W([m])$, of old configurations, the partition function is preserved, i.e.

$$Z = \int Dm\, W([m]) = \int Dm'\, W'([m']) = Z'. \qquad (4.33)$$

Hence $\ln Z = \ln Z'$, and the corresponding free energies are related by

$$Vf(t, h) = V'f(t', h'). \qquad (4.34)$$

In d dimensions, the rescaled volume is smaller by a factor of b^d, and

$$f(t, h) = b^{-d} f(b^{y_t} t, b^{y_h} h), \qquad (4.35)$$

where we have made use of the assumption that the two free energies are obtained from the *same Hamiltonian* in which only the parameters t and h have changed according to Eqs. (4.32). Equation (4.35) describes a *homogeneous function* of t and h. This is made apparent by choosing a rescaling factor b such that b^{y_t} is a constant, say unity, i.e. $b = t^{-1/y_t}$, leading to

$$f(t, h) = t^{d/y_t} f(1, h/t^{y_h/y_t}) \equiv t^{d/y_t} g_f(h/t^{y_h/y_t}). \qquad (4.36)$$

We have thus recovered the scaling form in Eq. (4.4), and can identify the exponents as

$$2 - \alpha = d/y_t, \quad \Delta = y_h/y_t. \qquad (4.37)$$

(2) *Correlation length:* All length scales are reduced by a factor of b during the RG transformation. This is also true of the correlation length, $\xi' = \xi/b$, implying

$$\xi(t, h) = b\xi(b^{y_t} t, b^{y_h} h) = t^{-1/y_t} \xi(1, h/t^{y_h/y_t}) \sim t^{-\nu}. \qquad (4.38)$$

This identifies $\nu = 1/y_t$, and using Eq. (4.37) the hyperscaling relation, $2 - \alpha = d\nu$, is recovered.

(3) **Magnetization:** From the homogenous form of the free energy (Eq. 4.36), we can obtain other bulk quantities such as magnetization. Alternatively, from the RG results for Z, V, and h, we may directly conclude

$$m(t, h) = -\frac{1}{V}\frac{\partial \ln Z(t, h)}{\partial h} = -\frac{1}{b^d V'}\frac{\partial \ln Z'(t', h')}{b^{-y_h}\partial h'} = b^{y_h-d}m(b^{y_t}t, b^{y_h}h). \tag{4.39}$$

Choosing $b = t^{-1/y_t}$, we obtain $\beta = (y_h - d)/y_t$, and $\Delta = y_h/y_t$ as before.

It is thus apparent that quite generally, the singular part of any quantity X has a homogeneous form

$$X(t, h) = b^{y_X} X(b^{y_t}t, b^{y_h}h). \tag{4.40}$$

For any conjugate pair of variables, contributing a term $\int d^d\mathbf{x}F \cdot X$ to the Hamiltonian, the *scaling dimensions* are related by $y_X = y_F - d$, where $F' = b^{y_F}F$ under RG.

4.5 The renormalization group (formal)

In the previous section we noted that all critical properties can be obtained from the recursion relations in Eqs. (4.32). Though conceptually appealing, it is not clear how such a procedure can be formally carried out. In particular, why should the forms of the two Hamiltonians be identical, and why are two parameters t and h sufficient to describe the transformation? In this section we outline a more formal procedure for identifying the effects of the dilation operation on the Hamiltonian. The various steps of the program are as follows:

(1) Start with the most general Hamiltonian allowed by symmetries. For example, in the presence of rotational symmetry,

$$\beta\mathcal{H} = \int d^d\mathbf{x}\left[\frac{t}{2}m^2 + um^4 + vm^6 + \cdots + \frac{K}{2}(\nabla m)^2 + \frac{L}{2}(\nabla^2 m)^2 + \cdots\right]. \tag{4.41}$$

A particular system with such symmetry is therefore completely specified by a point in the (infinite-dimensional) parameter space $S \equiv (t, u, v, \cdots, K, L, \cdots)$.

(2) Apply the three steps of renormalization in configuration space: (i) coarse grain by b; (ii) rescale, $\mathbf{x}' = \mathbf{x}/b$; and (iii) renormalize, $m' = m/\zeta$. This accomplishes the change of variables,

$$m'(\mathbf{x}') = \frac{1}{\zeta b^d}\int_{\text{Cell of size } b \text{ centered at } b\mathbf{x}'} d^d\mathbf{x}m(\mathbf{x}). \tag{4.42}$$

Given the probabilities $\mathcal{P}[m(\mathbf{x})] \propto \exp(-\beta\mathcal{H}[m(\mathbf{x})])$, for the original configurations, we can use the above change of variables to construct the corresponding probabilities $\mathcal{P}'[m'(\mathbf{x}')]$, for the new configurations. Naturally this is the most difficult step in the program.

(3) Since rotational symmetry is preserved by the RG procedure, the rescaled Hamiltonian must also be described by a point in the parameter space of Eq. (4.41), i.e.

$$\beta \mathcal{H}'[m'(\mathbf{x}')] \equiv \ln \mathcal{P}'[m'(\mathbf{x}')] = f_b + \int d^d\mathbf{x}' \left[\frac{t'}{2}m'^2 + u'm'^4 \right.$$
$$\left. + v'm'^6 + \cdots + \frac{K'}{2}(\nabla m')^2 + \frac{L'}{2}(\nabla^2 m')^2 + \cdots \right]. \tag{4.43}$$

The renormalized parameters are functions of the original ones, i.e. $t' = t_b(t, u, \ldots)$; $u' = u_b(t, u, \ldots)$, etc., defining a mapping $S' = \mathfrak{R}_b S$ in parameter space.

(4) The operation $\mathfrak{R}_b$ describes the effects of dilation on the Hamiltonian of the system. Hamiltonians that describe statistically self-similar configurations must thus correspond to *fixed points* S^*, such that $\mathfrak{R}_b S^* = S^*$. Since the correlation length, a function of Hamiltonian parameters, is reduced by b under the RG operation (i.e. $\xi(S) = b\xi(\mathfrak{R}_b S)$), the correlation length at a fixed point must be zero or infinity. Fixed points with $\xi^* = 0$ describe independent fluctuations at each point and correspond to complete disorder (infinite temperature), or complete order (zero temperature). A fixed point with $\xi^* = \infty$ describes a critical point ($T = T_c$).

Fig. 4.3 The fixed point S^* has a basin of attraction spanned by the irrelevant directions of negative eigenvalues y_i, and is unstable in the relevant direction with $y > 0$.

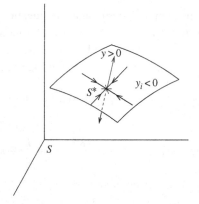

(5) Equations (4.32) represent a simplified case in which the parameter space is two dimensional. The point $t = h = 0$ is a fixed point, and the lowest order terms in these equations describe the behavior in its neighborhood. In general, we can study the stability of a fixed point by *linearizing* the recursion relations in its vicinity: Under RG, a point $S^* + \delta S$ is transformed to

$$S_\alpha^* + \delta S_\alpha' = S_\alpha^* + (\mathfrak{R}_b^L)_{\alpha\beta}\delta S_\beta + \cdots, \quad \text{where} \quad (\mathfrak{R}_b^L)_{\alpha\beta} \equiv \left. \frac{\partial S_\alpha'}{\partial S_\beta} \right|_{S^*}. \tag{4.44}$$

We now diagonalize the matrix $(\mathfrak{R}_b^L)_{\alpha\beta}$ to get the eigenvectors $\mathcal{O}_i$, and corresponding eigenvalues $\lambda(b)_i$. Because of the group property[2],

$$\mathfrak{R}_b^L \mathfrak{R}_{b'}^L \mathcal{O}_i = \lambda(b)_i \lambda(b')_i \mathcal{O}_i = \mathfrak{R}_{bb'}^L \mathcal{O}_i = \lambda(bb')_i \mathcal{O}_i. \tag{4.45}$$

Together with the condition $\lambda(1)_i = 1$, the above equation implies

$$\lambda(b)_i = b^{y_i}. \tag{4.46}$$

The vectors $\mathcal{O}_i$ are called scaling directions associated with the fixed point S^*, and y_i are the corresponding *anomalous dimensions*. Any Hamiltonian in the vicinity of the fixed point is described by a point $S = S^* + \Sigma_i g_i \mathcal{O}_i$. The renormalized Hamiltonian has interaction parameters $S' = S^* + \Sigma_i g_i b^{y_i} \mathcal{O}_i$. The following terminology is used to classify the eigenoperators:

- If $y_i > 0$, g_i *increases* under scaling, and $\mathcal{O}_i$ is a *relevant operator*.
- If $y_i < 0$, g_i *decreases* under scaling, and $\mathcal{O}_i$ is an *irrelevant operator*.
- If $y_i = 0$, g_i is called a *marginal operator*, and higher order terms are necessary to track its behavior.

The subspace spanned by the irrelevant operators is the *basin of attraction* of the fixed point S^*. Since ξ always decreases under RG, and $\xi(S^*) = \infty$, then ξ is also infinite for any point on the basin of attraction of S^*. For a general point in the vicinity of S^*, the correlation length satisfies

$$\xi(g_1, g_2, \cdots) = b\xi(b^{y_1}g_1, b^{y_2}g_2, \cdots). \tag{4.47}$$

For a sufficiently large b, all the irrelevant operators scale to zero. The leading singularities of ξ are then determined by the remaining set of *relevant* operators. In particular, if the operators are indexed in order of decreasing dimensions, we can choose b such that $b^{y_1}g_1 = 1$. In this case, Eq. (4.47) implies

$$\xi(g_1, g_2, \cdots) = g_1^{-1/y_1} f(g_2/g_1^{y_2/y_1}, \cdots). \tag{4.48}$$

We have thus obtained an exponent $\nu_1 = 1/y_1$, for the divergence of ξ, and a generalized set of gap exponents $\Delta_\alpha = y_\alpha/y_1$, associated with g_α.

Let us imagine that the fixed point S^* describes the critical point of the magnet in Eq. (4.41) at zero magnetic field. As the temperature, or some other control parameter, is changed, the coefficients of the effective Hamiltonian are altered, and the point S moves along a trajectory in parameter space. Except for a single point (at the critical temperature) the magnet has a finite correlation length. This can be achieved if the trajectory taken by the point S intersects the basis of attraction of S^* only at one point. To achieve this, the basin of attraction must have co-dimension one, i.e. the fixed point S^* must have one and only one relevant operator. This

---

[2] The group property $\mathfrak{R}_b^L \mathfrak{R}_{b'}^L = \mathfrak{R}_{bb'}^L = \mathfrak{R}_{b'}^L \mathfrak{R}_b^L$ also implies that the linearized matrices for different b commute. It is thus possible to diagonalize them simultaneously, implying that the eigenvectors $\{\mathcal{O}_i\}$ are independent of b.

provides an explanation of *universality*, in that the very many microscopic details of the system make up the huge space of irrelevant operators comprising the basin of attraction. In the presence of a magnetic field, two system parameters must be adjusted to reach the critical point ($T = T_c$ and $h = 0$). Thus the magnetic field corresponds to an additional relevant operator at S^*. Again, other "odd" interactions, such as $\{m^3, m^5, \cdots\}$, should not lead to any other relevant operators.

Although the formal procedure outlined in this section is quite rigorous, it suffers from some quite obvious shortcomings: How do we actually implement the RG transformations of step (2) analytically? There are an infinite number of interactions allowed by symmetry, and hence the space of parameters S is inconveniently large. How do we know a priori that there are fixed points for the RG transformation; that $\Re_b$ can be linearized; that relevant operators are few, etc.? Following the initial formulation of RG by Kadanoff, there was a period of uncertainty until Wilson showed how these steps can be implemented (at least perturbatively) in the Landau–Ginzburg model.

4.6 The Gaussian model (direct solution)

The RG approach will be applied to the *Gaussian model* in the next section. For the sake of later comparison, here we provide the direct solution of this problem. The Gaussian model is obtained by keeping only the quadratic terms in the Landau–Ginzburg expansion. The resulting partition function is

$$Z = \int \mathcal{D}\vec{m}(\mathbf{x}) \exp\left\{ -\int d^d\mathbf{x} \left[\frac{t}{2}m^2 + \frac{K}{2}(\nabla m)^2 + \frac{L}{2}(\nabla^2 m)^2 + \cdots - \vec{h} \cdot \vec{m} \right] \right\}. \quad (4.49)$$

Clearly the model is well defined only for $t \geq 0$, since there is no m^4 term to insure its stability for $t < 0$. The partition function still has a singularity at $t = 0$, and we can regard this as representing approaching a phase transition from the disordered side.

The quadratic form is easily evaluated by the usual rules of Gaussian integration. The kernel is first diagonalized by the Fourier modes: The allowed $\mathbf{q}$ values are discretized in a finite system of size L, with spacing of $2\pi/L$. The largest $\mathbf{q}$ are limited by the lattice spacing, and confined to a *Brillouin zone* whose shape is determined by the underlying lattice. We shall in fact use a slightly different normalization for the Fourier modes, and keep careful track of the volume factors, by setting

$$\begin{cases} \vec{m}(\mathbf{q}) = \int d^d\mathbf{x}\, e^{i\mathbf{q}\cdot\mathbf{x}} \vec{m}(\mathbf{x}) \\ \vec{m}(\mathbf{x}) = \sum_{\mathbf{q}} \frac{e^{-i\mathbf{q}\cdot\mathbf{x}}}{V} \vec{m}(\mathbf{q}) = \int \frac{d^d\mathbf{q}}{(2\pi)^d} e^{-i\mathbf{q}\cdot\mathbf{x}} \vec{m}(\mathbf{q}). \end{cases} \quad (4.50)$$

(We should really use a different symbol, such as $\tilde{m}_i(\mathbf{q})$ to indicate the Fourier modes. For the sake of brevity we use the same symbol, but explicitly include the argument $\mathbf{q}$ as the indicator of the Fourier transformed function.) The last transformation applies to the infinite size limit ($L \to \infty$), and V is the system volume.

In re-expressing the Hamiltonian in terms of Fourier modes, we encounter expressions such as

$$\int d^d\mathbf{x}\, m(\mathbf{x})^2 = \int d^d\mathbf{x} \sum_{\mathbf{q},\mathbf{q}'} \frac{e^{-i(\mathbf{q}+\mathbf{q}')\cdot\mathbf{x}}}{V^2} \vec{m}(\mathbf{q})\cdot\vec{m}(\mathbf{q}') = \sum_{\mathbf{q}} \frac{\vec{m}(\mathbf{q})\cdot\vec{m}(-\mathbf{q})}{V}. \quad (4.51)$$

The last expression follows from the vanishing of the integral over $\mathbf{x}$ unless $\mathbf{q}+\mathbf{q}'=\mathbf{0}$, in which case it equals the system volume. Similar manipulations lead to the Hamiltonian

$$\beta\mathcal{H} = \sum_{\mathbf{q}} \left(\frac{t+Kq^2+Lq^4+\cdots}{2V}\right)|m(\mathbf{q})|^2 - \vec{h}\cdot\vec{m}(\mathbf{q}=\mathbf{0}). \quad (4.52)$$

With the choice of the normalization in Eq. (4.50), the Jacobian of the transformation to Fourier modes is $1/\sqrt{V}$ per mode, and the partition function equals

$$Z = \prod_{\mathbf{q}} V^{-n/2} \int d\vec{m}(\mathbf{q}) \exp\left[-\frac{t+Kq^2+Lq^4+\cdots}{2V}|m(\mathbf{q})|^2 + \vec{h}\cdot\vec{m}(\mathbf{q}=\mathbf{0})\right]. \quad (4.53)$$

The integral for $\mathbf{q}=\mathbf{0}$ is

$$Z_0 = V^{-n/2} \int_{-\infty}^{\infty} d\vec{m}(\mathbf{0}) \exp\left[-\frac{t}{2V}|m(\mathbf{0})|^2 + \vec{h}\cdot\vec{m}(\mathbf{0})\right] = \left(\frac{2\pi}{t}\right)^{n/2} \exp\left[\frac{Vh^2}{2t}\right]. \quad (4.54)$$

After performing the integrations for $\mathbf{q}\neq\mathbf{0}$, we obtain

$$Z = \exp\left[\frac{Vh^2}{2t}\right] \prod_{\mathbf{q}} \left(\frac{2\pi}{t+Kq^2+Lq^4+\cdots}\right)^{n/2}. \quad (4.55)$$

The total number of modes, N, equals the number of original lattice points. Apart from a constant factor resulting from $(2\pi)^{nN/2}$, the free energy is

$$f(t,h) = -\frac{\ln Z}{V} = \frac{n}{2}\int_{BZ} \frac{d^dq}{(2\pi)^d} \ln\left(t+Kq^2+Lq^4+\cdots\right) - \frac{h^2}{2t}. \quad (4.56)$$

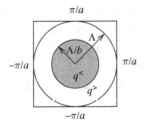

Fig. 4.4 The Brillouin zone is approximated by a hypersphere of radius Λ, which is then reduced by a factor of b in this RG scheme.

The integral in Eq. (4.56) is over the Brillouin zone, which for a hypercubic lattice of spacing a, is a cube of side $2\pi/a$ centered on the origin. However, we expect the singularities to originate from the long wavelength modes close to $\mathbf{q}=\mathbf{0}$. The contributions from the vicinity of the Brillouin zone edge are clearly analytic, since the logarithm can be simply expanded in powers of t for a finite q^2. Thus to simplify the extraction of the singular behavior as $t\to 0$, we approximate the shape of the Brillouin zone by a hypersphere of radius $\Lambda \approx \pi/a$. The spherical symmetry of the integrand then allows us to write

$$f_{\text{sing}}(t, h) = \frac{n}{2} K_d \int_0^\Lambda dq\, q^{d-1} \ln\left(t + Kq^2 + Lq^4 + \cdots\right) - \frac{h^2}{2t}, \tag{4.57}$$

where $K_d \equiv S_d/(2\pi)^d$, and S_d is the d-dimensional solid angle. The leading dependence of the integral on t can be extracted after rescaling q by a factor of $\sqrt{t/K}$, as

$$f_{\text{sing}}(t, h) = \frac{n}{2} K_d \left(\frac{t}{K}\right)^{d/2} \int_0^{\Lambda\sqrt{K}/\sqrt{t}} dx\, x^{d-1}$$
$$\times \left[\ln t + \ln\left(1 + x^2 + Ltx^4/K^2 + \cdots\right)\right] - \frac{h^2}{2t}. \tag{4.58}$$

Ignoring analytic contributions in t, the leading singular dependence of the free energy can be written as

$$f_{\text{sing}}(t, h) = -t^{d/2}\left[A + \frac{h^2}{2t^{1+d/2}}\right] \equiv t^{2-\alpha} g_f(h/t^\Delta). \tag{4.59}$$

Note that the higher order gradients, i.e. terms proportional to $L, \cdots$, do not effect the singular behavior in Eq. (4.59). On approaching the point $h = 0$ for $t = 0^+$, the singular part of the free energy is described by a homogeneous scaling form, with exponents

$$\alpha_+ = 2 - d/2, \quad \Delta = 1/2 + d/4. \tag{4.60}$$

Since the ordered phase for $t < 0$ is not stable, the exponent β is undefined. The susceptibility $\chi \propto \partial^2 f/\partial h^2 \propto 1/t$, however, diverges with the exponent $\gamma_+ = 1$.

4.7 The Gaussian model (renormalization group)

The renormalization of the Gaussian model is most conveniently performed in terms of the Fourier modes. The goal is to evaluate the partition function

$$Z \sim \int \mathcal{D}\vec{m}(\mathbf{q}) \exp\left[-\int_0^\Lambda \frac{d^d\mathbf{q}}{(2\pi)^d}\left(\frac{t + Kq^2 + Lq^4 + \cdots}{2}\right)|m(\mathbf{q})|^2 + \vec{h}\cdot\vec{m}(\mathbf{0})\right] \tag{4.61}$$

indirectly via the three steps of RG. Note that the Brillouin zone is approximated by a hypersphere of radius Λ.

(1) **Coarse grain:** Eliminating fluctuations at scales $a < x < ba$ is similar to removing Fourier modes with wavenumbers $\Lambda/b < q < \Lambda$. We thus break up the momenta into two subsets,

$$\{\vec{m}(\mathbf{q})\} = \{\vec{\sigma}(\mathbf{q}^>)\} \oplus \{\tilde{\vec{m}}(\mathbf{q}^<)\}, \tag{4.62}$$

and write

$$Z = \int \mathcal{D}\tilde{\vec{m}}(\mathbf{q}^<) \int \mathcal{D}\vec{\sigma}(\mathbf{q}^>) e^{-\beta\mathcal{H}[\tilde{\vec{m}}, \vec{\sigma}]}. \tag{4.63}$$

Since the two sets of modes are decoupled in the Gaussian model, the integration is trivial, and

$$
Z \sim \exp\left[-\frac{n}{2}V\int_{\Lambda/b}^{\Lambda}\frac{d^dq}{(2\pi)^d}\ln(t+Kq^2+Lq^4+\cdots)\right] \times \int \mathcal{D}\tilde{\tilde{m}}(\mathbf{q}^<)
$$
$$
\times \exp\left[-\int_0^{\Lambda/b}\frac{d^dq}{(2\pi)^d}\left(\frac{t+Kq^2+Lq^4+\cdots}{2}\right)|\tilde{m}(\mathbf{q})|^2+\vec{h}\cdot\tilde{\tilde{m}}(0)\right].
$$

(4.64)

(2) **Rescale:** The partition function for the modes $\tilde{\tilde{m}}(\mathbf{q}^<)$ is similar to the original, except that the upper cutoff has decreased to Λ/b, reflecting the coarsening in resolution. The rescaling $\mathbf{x}' = \mathbf{x}/b$ in real space is equivalent to $\mathbf{q}' = b\mathbf{q}$ in *momentum space*, and restores the cutoff to its original value. The rescaled partition function is

$$
Z = e^{-V\delta f_b(t)} \times \int \mathcal{D}\tilde{\tilde{m}}(\mathbf{q}') \times \exp\left[-\int_0^{\Lambda}\frac{d^dq'}{(2\pi)^d}b^{-d}\right.
$$
$$
\left.\times\left(\frac{t+Kb^{-2}q'^2+Lb^{-4}q'^4+\cdots}{2}\right)|\tilde{m}(\mathbf{q}')|^2+\vec{h}\cdot\tilde{\tilde{m}}(0)\right].
$$

(4.65)

(3) **Renormalize:** The final step of RG in real space is the renormalization of magnetization, $\vec{m}'(\mathbf{x}') = \tilde{\tilde{m}}(\mathbf{x}')/\zeta$. Alternatively, we can renormalize the Fourier modes according to $\vec{m}'(\mathbf{q}') = \tilde{\tilde{m}}(\mathbf{q}')/z$, resulting in

$$
Z = e^{-V\delta f_b(t)} \times \int \mathcal{D}\vec{m}'(\mathbf{q}') \times \exp\left[-\int_0^{\Lambda}\frac{d^dq'}{(2\pi)^d}b^{-d}z^2\right.
$$
$$
\left.\times\left(\frac{t+Kb^{-2}q'^2+Lb^{-4}q'^4+\cdots}{2}\right)|m'(\mathbf{q}')|^2+z\vec{h}\cdot\vec{m}'(0)\right].
$$

(4.66)

(Note that the factors ζ and z, for rescalings of magnetization in real and Fourier space, are different.)

Equation (4.66) indicates the renormalized modes are distributed according to a Gaussian Hamiltonian, with renormalized parameters,

$$
\begin{cases}
t' = z^2 b^{-d} t \\
h' = z h \\
K' = z^2 b^{-d-2} K \\
L' = z^2 b^{-d-4} L \\
\cdots
\end{cases}
$$

(4.67)

The singular point $t = h = 0$ is mapped onto itself as expected. To make the fluctuations scale invariant at this point, we must insure that the remaining Hamiltonian stays fixed. This is achieved by the choice of $z = b^{1+d/2}$, which sets $K' = K$, and makes the remaining parameters $L, \cdots$ scale to zero. Away from criticality, the two relevant directions now scale as

$$
\begin{cases}
t' = b^2 t \\
h' = b^{1+d/2}h
\end{cases}
\implies
\begin{cases}
y_t = 2 \\
y_h = 1+d/2.
\end{cases}
$$

(4.68)

Using the results of Section 3.4, we can identify the critical exponents,

$$\nu = 1/y_t = 1/2,$$

$$\Delta = \frac{y_h}{y_t} = \frac{1+d/2}{2} = \frac{1}{2} + \frac{d}{4},$$

$$\alpha = 2 - d\nu = 2 - d/2,$$

in agreement with the direct solution in the previous section.

The fixed point Hamiltonian ($t^* = h^* = L^* = \cdots = 0$) in *real space* is

$$\beta \mathcal{H}^* = \frac{K}{2} \int_a d^d\mathbf{x} (\nabla m)^2. \tag{4.69}$$

(The subscript a is placed on the integral as a reminder of the implicit short distance cutoff.) Under a simple rescaling $\mathbf{x} \mapsto \mathbf{x}'$, and $\vec{m}(\mathbf{x}) \mapsto \zeta \vec{m}'(\mathbf{x}'); K \mapsto K' = b^{d-2}\zeta^2 K$. Scale invariance is achieved with the choice $\zeta = b^{1-d/2}$. Forgetting the coarse-graining step, a general power of $\vec{m}(\mathbf{x})$, added as a small perturbation to $(\beta \mathcal{H})^*$, behaves as

$$\beta \mathcal{H}^* + u_n \int d^d\mathbf{x}\, m^n \quad \mapsto \quad \beta \mathcal{H}^* + u_n b^d \zeta^n \int d^d\mathbf{x}'\, m'^n, \tag{4.70}$$

suggesting that such perturbations scale as

$$u_n' = b^d b^{n\left(\frac{2-d}{2}\right)} u_n, \quad \Longrightarrow \quad y_n = n - d\left(\frac{n}{2} - 1\right). \tag{4.71}$$

The values $y_1 = 1 + d/2$, and $y_2 = 2$, reproduce the exponents for y_h and y_t in Eq. (4.68). The operators with higher powers are less relevant. The next most important operator in a system with spherical symmetry is $y_4 = 4 - d$, which is irrelevant for $d > 4$ but relevant for $d < 4$; $y_6 = 6 - 2d$ is relevant only for $d < 3$. Indeed the majority of operators are irrelevant at the Gaussian fixed point for $d > 2$.

Problems for chapter 4

1. *Scaling in fluids:* Near the liquid–gas critical point, the free energy is assumed to take the scaling form $F/N = t^{2-\alpha} g(\delta\rho/t^\beta)$, where $t = |T - T_c|/T_c$ is the reduced temperature, and $\delta\rho = \rho - \rho_c$ measures deviations from the critical point density. The leading singular behavior of any thermodynamic parameter $Q(t, \delta\rho)$ is of the form t^x on approaching the critical point along the isochore $\rho = \rho_c$; or $\delta\rho^y$ for a path along the isotherm $T = T_c$. Find the exponents x and y for the following quantities:

   (a) The internal energy per particle $\langle H \rangle/N$, and the entropy per particle $s = S/N$.
   (b) The heat capacities $C_V = T\partial s/\partial T\,|_V$, and $C_P = T\partial s/\partial T\,|_P$.
   (c) The isothermal compressibility $\kappa_T = \partial\rho/\partial P\,|_T\,/\rho$, and the thermal expansion coefficient $\alpha = \partial V/\partial T\,|_P\,/V$.

   Check that your results for parts (b) and (c) are consistent with the thermodynamic identity $C_P - C_V = TV\alpha^2/\kappa_T$.

(d) Sketch the behavior of the latent heat per particle L on the coexistence curve for $T < T_c$, and find its singularity as a function of t.

2. *The Ising model:* The differential recursion relations for temperature T, and magnetic field h, of the Ising model in $d = 1 + \epsilon$ dimensions are (for $b = e^\ell$)

$$\begin{cases} \dfrac{dT}{d\ell} = -\epsilon T + \tfrac{1}{2}T^2 \\[2mm] \dfrac{dh}{d\ell} = dh. \end{cases}$$

(a) Sketch the renormalization group flows in the (T, h) plane (for $\epsilon > 0$), marking the fixed points along the $h = 0$ axis.

(b) Calculate the eigenvalues y_t and y_h, at the critical fixed point, to order of ϵ.

(c) Starting from the relation governing the change of the correlation length ξ under renormalization, show that $\xi(t, h) = t^{-\nu} g_\xi \left(h/|t|^\Delta \right)$ (where $t = T/T_c - 1$), and find the exponents ν and Δ.

(d) Use a hyperscaling relation to find the singular part of the free energy $f_{\text{sing}}(t, h)$, and hence the heat capacity exponent α.

(e) Find the exponents β and γ for the singular behaviors of the magnetization and susceptibility, respectively.

(f) Starting with the relation between susceptibility and correlations of local magnetizations, calculate the exponent η for the critical correlations ($\langle m(\mathbf{0})m(\mathbf{x}) \rangle \sim |\mathbf{x}|^{-(d-2+\eta)}$).

(g) How does the correlation length diverge as $T \to 0$ (along $h = 0$) for $d = 1$?

---

3. *The nonlinear σ model* describes n component unit spins. As we shall demonstrate later, in $d = 2$ dimensions, the recursion relations for temperature T, and magnetic field h, are (for $b = e^\ell$)

$$\begin{cases} \dfrac{dT}{d\ell} = \dfrac{(n-2)}{2\pi} T^2 \\[2mm] \dfrac{dh}{d\ell} = 2h. \end{cases}$$

(a) How does the correlation length diverge as $T \to 0$?

(b) Write down the singular form of the free energy as $T, h \to 0$.

(c) How does the susceptibility χ diverge as $T \to 0$ for $h = 0$?

4. *Coupled scalars:* Consider the Hamiltonian

$$\beta \mathcal{H} = \int d^d\mathbf{x} \left[\frac{t}{2} m^2 + \frac{K}{2}(\nabla m)^2 - hm + \frac{L}{2}(\nabla^2 \phi)^2 + v \nabla m . \nabla \phi \right],$$

coupling two one-component fields m and ϕ.

(a) Write $\beta \mathcal{H}$ in terms of the Fourier transforms $m(\mathbf{q})$ and $\phi(\mathbf{q})$.

(b) Construct a renormalization group transformation as in the text, by rescaling distances such that $\mathbf{q}' = b\mathbf{q}$; and the fields such that $m'(\mathbf{q}') = \tilde{m}(\mathbf{q})/z$ and $\phi'(\mathbf{q}') = \tilde{\phi}(\mathbf{q})/y$. Do not evaluate the integrals that just contribute a constant additive term.

(c) There is a fixed point such that $K' = K$ and $L' = L$. Find y_t, y_h and y_v at this fixed point.

(d) The singular part of the free energy has a scaling from $f(t, h, v) = t^{2-\alpha} g(h/t^{\Delta}, v/t^{\omega})$ for t, h, v close to zero. Find α, Δ, and ω.

(e) There is another fixed point such that $t' = t$ and $L' = L$. What are the relevant operators at this fixed point, and how do they scale?

5

Perturbative renormalization group

5.1 Expectation values in the Gaussian model

Can we treat the Landau–Ginzburg Hamiltonian as a perturbation to the Gaussian model? In particular, for zero magnetic field, we shall examine

$$\beta \mathcal{H} = \beta \mathcal{H}_0 + \mathcal{U} \equiv \int d^d \mathbf{x} \left[\frac{t}{2} m^2 + \frac{K}{2} (\nabla m)^2 + \frac{L}{2} (\nabla^2 m)^2 + \cdots \right]$$

$$+ u \int d^d \mathbf{x} \, m^4 + \cdots . \tag{5.1}$$

The *unperturbed* Gaussian Hamiltonian can be decomposed into independent Fourier modes, as

$$\beta \mathcal{H}_0 = \frac{1}{V} \sum_{\mathbf{q}} \frac{t + K q^2 + L q^4 + \cdots}{2} |m(\mathbf{q})|^2$$

$$\equiv \int \frac{d^d \mathbf{q}}{(2\pi)^d} \frac{t + K q^2 + L q^4 + \cdots}{2} |m(\mathbf{q})|^2. \tag{5.2}$$

The *perturbative interaction* which mixes up the normal modes has the form

$$\mathcal{U} = u \int d^d \mathbf{x} \, m(\mathbf{x})^4 + \cdots$$

$$= u \int d^d \mathbf{x} \int \frac{d^d \mathbf{q}_1 d^d \mathbf{q}_2 d^d \mathbf{q}_3 d^d \mathbf{q}_4}{(2\pi)^{4d}} e^{-i\mathbf{x} \cdot (\mathbf{q}_1 + \mathbf{q}_2 + \mathbf{q}_3 + \mathbf{q}_4)} m_\alpha(\mathbf{q}_1) m_\alpha(\mathbf{q}_2) m_\beta(\mathbf{q}_3) m_\beta(\mathbf{q}_4)$$

$$+ \cdots , \tag{5.3}$$

where summation over α and β is implicit. The integral over $\mathbf{x}$ sets $\mathbf{q}_1 + \mathbf{q}_2 + \mathbf{q}_3 + \mathbf{q}_4 = \mathbf{0}$, and

$$\mathcal{U} = u \int \frac{d^d \mathbf{q}_1 d^d \mathbf{q}_2 d^d \mathbf{q}_3}{(2\pi)^{3d}} m_\alpha(\mathbf{q}_1) m_\alpha(\mathbf{q}_2) m_\beta(\mathbf{q}_3) m_\beta(-\mathbf{q}_1 - \mathbf{q}_2 - \mathbf{q}_3) + \cdots . \tag{5.4}$$

From the variance of the Gaussian weights, the two-point expectation values in a finite sized system with discretized modes are easily obtained as

$$\langle m_\alpha(\mathbf{q}) \, m_\beta(\mathbf{q}') \rangle_0 = \frac{\delta_{\mathbf{q}, -\mathbf{q}'} \, \delta_{\alpha, \beta} \, V}{t + K q^2 + L q^4 + \cdots}. \tag{5.5}$$

In the limit of infinite size, the spectrum becomes continuous, and Eq. (5.5) goes over to

$$\langle m_\alpha(\mathbf{q}) \, m_\beta(\mathbf{q}') \rangle_0 = \frac{\delta_{\alpha, \beta} (2\pi)^d \delta^d(\mathbf{q} + \mathbf{q}')}{t + K q^2 + L q^4 + \cdots}. \tag{5.6}$$

The subscript 0 is used to indicate that the expectation values are taken with respect to the unperturbed (Gaussian) Hamiltonian. Expectation values involving any product of m's can be obtained starting from the identity

$$\left\langle \exp\left[\sum_i a_i m_i\right]\right\rangle_0 = \exp\left[\sum_{i,j} \frac{a_i a_j}{2} \langle m_i m_j\rangle_0\right], \tag{5.7}$$

which is valid for any set of Gaussian distributed variables $\{m_i\}$. (This is easily seen by "completing the square.") Expanding both sides of the equation in powers of $\{a_i\}$ leads to

$$1 + a_i\langle m_i\rangle_0 + \frac{a_i a_j}{2}\langle m_i m_j\rangle_0 + \frac{a_i a_j a_k}{6}\langle m_i m_j m_k\rangle_0 + \frac{a_i a_j a_k a_l}{24}\langle m_i m_j m_k m_k\rangle_0 + \cdots =$$
$$1 + \frac{a_i a_j}{2}\langle m_i m_j\rangle_0 + \frac{a_i a_j a_k a_l}{24}\left(\langle m_i m_j\rangle_0\langle m_k m_l\rangle_0 + \langle m_i m_k\rangle_0\langle m_j m_l\rangle_0\right.$$
$$\left. + \langle m_i m_k\rangle_0\langle m_j m_l\rangle_0\right) + \cdots \tag{5.8}$$

Matching powers of $\{a_i\}$ on the two sides of the above equation gives

$$\left\langle \prod_{i=1}^{\ell} m_i \right\rangle_0 = \begin{cases} 0 & \text{for } \ell \text{ odd} \\ \text{sum over all pairwise contractions} & \text{for } \ell \text{ even.} \end{cases} \tag{5.9}$$

This result is known as *Wick's theorem*; and for example,

$$\langle m_i m_j m_k m_l\rangle_0 = \langle m_i m_j\rangle_0\langle m_k m_l\rangle_0 + \langle m_i m_k\rangle_0\langle m_j m_l\rangle_0 + \langle m_i m_k\rangle_0\langle m_j m_l\rangle_0.$$

5.2 Expectation values in perturbation theory

In the presence of an interaction $\mathcal{U}$, the expectation value of any operator $\mathcal{O}$ is computed perturbatively as

$$\langle \mathcal{O}\rangle = \frac{\int \mathcal{D}\vec{m}\, \mathcal{O}e^{-\beta\mathcal{H}_0 - \mathcal{U}}}{\int \mathcal{D}\vec{m}\, e^{-\beta\mathcal{H}_0 - \mathcal{U}}} = \frac{\int \mathcal{D}\vec{m}\, e^{-\beta\mathcal{H}_0}\mathcal{O}[1 - \mathcal{U} + \mathcal{U}^2/2 - \cdots]}{\int \mathcal{D}\vec{m}\, e^{-\beta\mathcal{H}_0}[1 - \mathcal{U} + \mathcal{U}^2/2 - \cdots]}$$
$$= \frac{Z_0[\langle \mathcal{O}\rangle_0 - \langle \mathcal{O}\mathcal{U}\rangle_0 + \langle \mathcal{O}\mathcal{U}^2\rangle_0/2 - \cdots]}{Z_0[1 - \langle \mathcal{U}\rangle_0 + \langle \mathcal{U}^2\rangle_0/2 - \cdots]}. \tag{5.10}$$

Inverting the denominator by an expansion in powers of $\mathcal{U}$ gives

$$\langle \mathcal{O}\rangle = \left[\langle \mathcal{O}\rangle_0 - \langle \mathcal{O}\mathcal{U}\rangle_0 + \frac{1}{2}\langle \mathcal{O}\mathcal{U}^2\rangle_0 - \cdots\right]\left[1 + \langle \mathcal{U}\rangle_0 + \langle \mathcal{U}\rangle_0^2 - \frac{1}{2}\langle \mathcal{U}^2\rangle_0 - \cdots\right]$$
$$= \langle \mathcal{O}\rangle_0 - (\langle \mathcal{O}\mathcal{U}\rangle_0 - \langle \mathcal{O}\rangle_0\langle \mathcal{U}\rangle_0) + \frac{1}{2}\left(\langle \mathcal{O}\mathcal{U}^2\rangle_0 - 2\langle \mathcal{O}\mathcal{U}\rangle_0\langle \mathcal{U}\rangle_0\right.$$
$$\left. + 2\langle \mathcal{O}\rangle_0\langle \mathcal{U}\rangle_0^2 - \langle \mathcal{O}\rangle_0\langle \mathcal{U}^2\rangle_0\right) + \cdots$$
$$\equiv \sum_{n=0}^{\infty} \frac{(-1)^n}{n!}\langle \mathcal{O}\mathcal{U}^n\rangle_0^c. \tag{5.11}$$

The *connected averages* (cumulants) are defined as the combination of unperturbed expectation values appearing at various orders in the expansion. Their significance will become apparent in diagrammatic representations, and from the following example.

Let us calculate the two-point correlation function of the Landau–Ginzburg model to first order in the parameter u. (In view of their expected irrelevance, we shall ignore higher order interactions, and also only keep the lowest order Gaussian terms.) Substituting Eq. (5.4) into Eq. (5.11) yields

$$\langle m_\alpha(\mathbf{q}) m_\beta(\mathbf{q}')\rangle = \langle m_\alpha(\mathbf{q}) m_\beta(\mathbf{q}')\rangle_0 - u \int \frac{d^d\mathbf{q}_1 d^d\mathbf{q}_2 d^d\mathbf{q}_3}{(2\pi)^{3d}}$$

$$\times [\langle m_\alpha(\mathbf{q}) m_\beta(\mathbf{q}') m_i(\mathbf{q}_1) m_i(\mathbf{q}_2) m_j(\mathbf{q}_3) m_j(-\mathbf{q}_1 - \mathbf{q}_2 - \mathbf{q}_3)\rangle_0$$

$$- \langle m_\alpha(\mathbf{q}) m_\beta(\mathbf{q}')\rangle_0 \langle m_i(\mathbf{q}_1) m_i(\mathbf{q}_2) m_j(\mathbf{q}_3) m_j(-\mathbf{q}_1 - \mathbf{q}_2 - \mathbf{q}_3)\rangle_0]$$

$$+ \mathcal{O}(u^2). \tag{5.12}$$

To calculate $\langle \mathcal{O}\mathcal{U}\rangle_0$ we need the unperturbed expectation value of the product of six m's. This can be evaluated using Eq. (5.9) as the sum of all pair-wise contractions, 15 in all. Three contractions are obtained by first pairing m_α to m_β, and then the remaining four m's in $\mathcal{U}$. Clearly these contractions cancel exactly with corresponding ones in $\langle \mathcal{O}\rangle_0 \langle \mathcal{U}\rangle_0$. The only surviving terms involve contractions that connect $\mathcal{O}$ to $\mathcal{U}$. This cancellation persists at all orders, and $\langle \mathcal{O}\mathcal{U}^n\rangle_0^c$ contains only terms in which all $n+1$ operators are connected by contractions. The remaining 12 pairings in $\langle \mathcal{O}\mathcal{U}\rangle_0$ fall into two classes:

(1) Four pairings involve contracting m_α and m_β to m's with the same index, e.g.

$$\langle m_\alpha(\mathbf{q}) m_i(\mathbf{q}_1)\rangle_0 \langle m_\beta(\mathbf{q}') m_i(\mathbf{q}_2)\rangle_0 \langle m_j(\mathbf{q}_3) m_j(-\mathbf{q}_1 - \mathbf{q}_2 - \mathbf{q}_3)\rangle_0$$

$$= \frac{\delta_{\alpha i} \delta_{\beta i} \delta_{jj} (2\pi)^{3d} \delta^d(\mathbf{q} + \mathbf{q}_1) \delta^d(\mathbf{q}' + \mathbf{q}_2) \delta^d(\mathbf{q}_1 + \mathbf{q}_2)}{(t + Kq^2)(t + Kq'^2)(t + Kq_3^2)}, \tag{5.13}$$

where we have used Eq. (5.6). After summing over i and j, and integrating over $\mathbf{q}_1$, $\mathbf{q}_2$, and $\mathbf{q}_3$, these terms make a contribution

$$-4u \frac{n\delta_{\alpha\beta} (2\pi)^d \delta^d(\mathbf{q} + \mathbf{q}')}{(t + Kq^2)^2} \int \frac{d^d\mathbf{q}_3}{(2\pi)^d} \frac{1}{t + Kq_3^2}. \tag{5.14}$$

(2) Eight pairings involve contracting m_α and m_β to m's with different indices, e.g.

$$\langle m_\alpha(\mathbf{q}) m_i(\mathbf{q}_1)\rangle_0 \langle m_\beta(\mathbf{q}') m_j(\mathbf{q}_3)\rangle_0 \langle m_i(\mathbf{q}_2) m_j(-\mathbf{q}_1 - \mathbf{q}_2 - \mathbf{q}_3)\rangle_0$$

$$= \frac{\delta_{\alpha i} \delta_{\beta j} \delta_{ij} (2\pi)^{3d} \delta^d(\mathbf{q} + \mathbf{q}_1) \delta^d(\mathbf{q}' + \mathbf{q}_3) \delta^d(\mathbf{q}_1 + \mathbf{q}_3)}{(t + Kq^2)(t + Kq'^2)(t + Kq_2^2)}. \tag{5.15}$$

Summing over all indices, and integrating over the momenta leads to an overall contribution of

$$-8u \frac{\delta_{\alpha\beta} (2\pi)^d \delta^d(\mathbf{q} + \mathbf{q}')}{(t + Kq^2)^2} \int \frac{d^d\mathbf{q}_2}{(2\pi)^d} \frac{1}{t + Kq_2^2}. \tag{5.16}$$

Adding up both contributions, we obtain

$$\langle m_\alpha(\mathbf{q})\, m_\beta(\mathbf{q'})\rangle = \frac{\delta_{\alpha\beta}\,(2\pi)^d\,\delta^d(\mathbf{q}+\mathbf{q'})}{t+Kq^2}\left[1 - \frac{4u(n+2)}{t+Kq^2}\int\frac{d^d\mathbf{k}}{(2\pi)^d}\frac{1}{t+Kk^2} + \mathcal{O}(u^2)\right].$$
(5.17)

5.3 Diagrammatic representation of perturbation theory

The calculations become more involved at higher orders in perturbation theory. A diagrammatic representation can be introduced to help keep track of all possible contractions. To calculate the ℓ-point expectation value $\langle\prod_{i=1}^{\ell} m_{\alpha_i}(\mathbf{q}_i)\rangle$, at pth order in u, proceed according to the following rules:

(1) Draw ℓ *external points* labeled by $(\mathbf{q}_i, \alpha_i)$ corresponding to the coordinates of the required correlation function. Draw p *vertices* with four legs each, labeled by *internal* momenta and indices, e.g. $\{(\mathbf{k}_1, i), (\mathbf{k}_2, i), (\mathbf{k}_3, j), (\mathbf{k}_4, j)\}$. Since the four legs are not equivalent, the four point vertex is indicated by two solid branches joined by a dotted line. (The extension to higher order interactions is straightforward.)

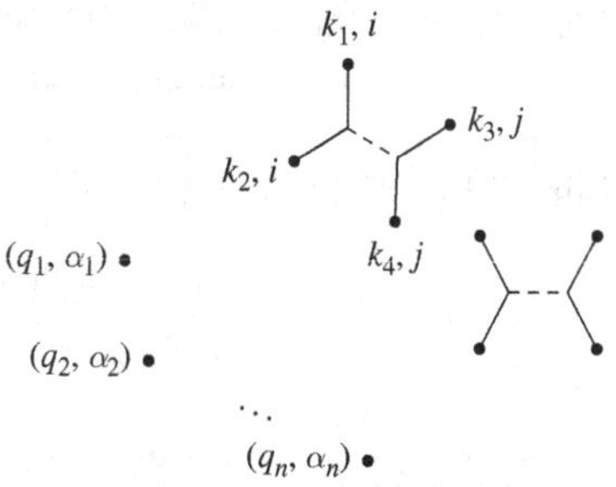

Fig. 5.1 Elements of the diagrammatic representation of perturbation theory.

(2) Each point of the graph now corresponds to one factor of $m_{\alpha_i}(\mathbf{q}_i)$, and the unperturbed average of the product is computed by Wick's theorem. This is implemented by joining all external and internal points *pair-wise*, by lines connecting one point to another, in all topologically distinct ways; see (5) below.

(3) The algebraic value of each such graph is obtained as follows: (i) A line joining a pair of points represents the two point average;[1] e.g. a connection $\overset{(q_1,\alpha_1)\ \ (q_2,\alpha_2)}{\bullet\!\!-\!\!\!-\!\!\!-\!\!\bullet}$, corresponds to $\delta_{\alpha_1\alpha_2}(2\pi)^d\delta^d(\mathbf{q}_1+\mathbf{q}_2)/(t+Kq_1^2)$; (ii) A vertex $\overset{k_1,i\quad k_3,j}{\underset{k_2,i\quad k_4,j}{\rangle\!\cdots\!\langle}}$ stands for a term $u(2\pi)^d\delta^d(\mathbf{k}_1+\mathbf{k}_2+\mathbf{k}_3+\mathbf{k}_4)$ (the delta-function insures that momentum is conserved).

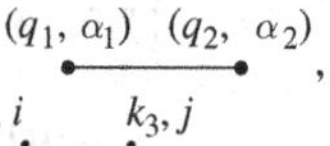

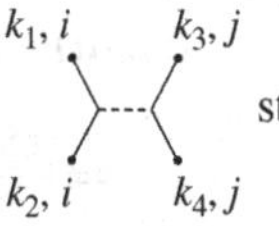

---

[1] Because of its original formulation in quantum field theory, the line joining two points is usually called a *propagator*. In this context, the line represents the world-line of a particle in time, while the perturbation $\mathcal{U}$ is an "interaction" between particles. For the same reason, the Fourier index is called a "momentum".

(4) Integrate over the $4p$ internal momenta $\{\mathbf{k}_i\}$, and sum over the $2p$ internal indices. Note that each closed loop produces a factor of $\delta_{ii} = n$ at this stage.

(5) There is a numerical factor of

$$\frac{(-1)^p}{p!} \times \text{number of different pairings leading to the same topology.}$$

The first contribution comes from the expansion of the exponential; the second merely states that graphs related by symmetry give the same result, and can be calculated once.

(6) When calculating cumulants, only fully connected diagrams (without disjoint pieces) need to be included. This is a tremendous simplification.

For example, the diagrams appearing in the expansion for the propagator

$$\langle m_\alpha(q) m_\beta(q') \rangle \equiv \underset{(q, \alpha) \quad (q', \beta)}{\rule{2cm}{0.8pt}} ,$$

to second order are

5.4 Susceptibility

It is no accident that the correction term in Eq. (5.17) is similar in form to the unperturbed value. This is because the form of the two point correlation function is constrained by symmetries, as can be seen from the identity

$$\langle m_\alpha(\mathbf{q})\, m_\beta(\mathbf{q}')\rangle = \int d^d\mathbf{x} \int d^d\mathbf{x}' e^{i\mathbf{q}\cdot\mathbf{x}+i\mathbf{q}'\cdot\mathbf{x}'} \langle m_\alpha(\mathbf{x})\, m_\beta(\mathbf{x}')\rangle. \tag{5.18}$$

The two-point correlation function in real space must satisfy translation and rotation symmetry, and (in the high temperature phase) $\langle m_\alpha(\mathbf{x})\, m_\beta(\mathbf{x}')\rangle = \delta_{\alpha\beta}\langle m_1(\mathbf{x}-\mathbf{x}')\, m_1(\mathbf{0})\rangle$. Transforming to center of mass and relative coordinates, the above integral becomes,

$$\langle m_\alpha(\mathbf{q})\, m_\beta(\mathbf{q}')\rangle$$

$$= \int d^d\left(\frac{\mathbf{x}+\mathbf{x}'}{2}\right) d^d(\mathbf{x}-\mathbf{x}')\, e^{i(\mathbf{q}+\mathbf{q}')\cdot(\mathbf{x}+\mathbf{x}')/2} e^{i(\mathbf{x}-\mathbf{x}')\cdot(\mathbf{q}-\mathbf{q}')/2} \delta_{\alpha\beta}\langle m_1(\mathbf{x}-\mathbf{x}')\, m_1(\mathbf{0})\rangle$$

$$\equiv (2\pi)^d \delta^d(\mathbf{q}+\mathbf{q}')\delta_{\alpha\beta} S(q), \tag{5.19}$$

where

$$S(q) = \langle |m_1(\mathbf{q})|^2\rangle = \int d^d\mathbf{x} e^{i\mathbf{q}\cdot\mathbf{x}} \langle m_1(\mathbf{x}-\mathbf{x}')\, m_1(\mathbf{0})\rangle \tag{5.20}$$

is the quantity observed in scattering experiments (Section 2.4).

From Eq. (5.17) we obtain

$$S(q) = \frac{1}{t+Kq^2}\left[1 - \frac{4u(n+2)}{t+Kq^2}\int \frac{d^dk}{(2\pi)^d}\frac{1}{t+Kk^2} + \mathcal{O}(u^2)\right]. \tag{5.21}$$

It is useful to examine the expansion of the inverse quantity

$$S(q)^{-1} = t+Kq^2 + 4u(n+2)\int \frac{d^dk}{(2\pi)^d}\frac{1}{t+Kk^2} + \mathcal{O}(u^2). \tag{5.22}$$

In the high-temperature phase, Eq. (5.20) indicates that the $q\to 0$ limit of $S(q)$ is just the magnetic susceptibility χ. For this reason, $S(q)$ is sometimes denoted by $\chi(q)$. From Eq. (5.22), the inverse susceptibility is given by

$$\chi^{-1}(t) = t + 4u(n+2)\int \frac{d^dk}{(2\pi)^d}\frac{1}{t+Kk^2} + \mathcal{O}(u^2). \tag{5.23}$$

The susceptibility no longer diverges at $t=0$, since

$$\chi^{-1}(0) = 4u(n+2)\int \frac{d^dk}{(2\pi)^d}\frac{1}{Kk^2} = \frac{4(n+2)u}{K}\frac{S_d}{(2\pi)^d}\int_0^\Lambda dk\, k^{d-3}$$

$$= \frac{4(n+2)u}{K}K_d\left(\frac{\Lambda^{d-2}}{d-2}\right) \tag{5.24}$$

is a finite number ($K_d \equiv S_d/(2\pi)^d$). This is because in the presence of u the critical temperature is reduced to a negative value. The modified critical point is obtained by requiring $\chi^{-1}(t_c) = 0$, and hence from Eq. (5.23), to order of u,

$$t_c = -4u(n+2)\int \frac{d^d\mathbf{k}}{(2\pi)^d}\frac{1}{t_c+Kk^2} \approx -\frac{4u(n+2)K_d\Lambda^{d-2}}{(d-2)K} < 0. \qquad (5.25)$$

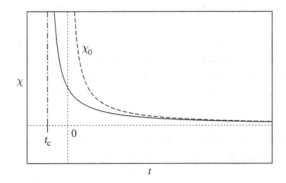

Fig. 5.2 The divergence of susceptibility occurs at a lower temperature due to the interaction u.

How does the perturbed susceptibility diverge at the shifted critical point? From Eq. (5.23),

$$\chi^{-1}(t) - \chi^{-1}(t_c) = t - t_c + 4u(n+2)\int \frac{d^d\mathbf{k}}{(2\pi)^d}\left(\frac{1}{t+Kk^2}-\frac{1}{t_c+Kk^2}\right)$$
$$(5.26)$$
$$= (t-t_c)\left[1-\frac{4u(n+2)}{K^2}\int \frac{d^d\mathbf{k}}{(2\pi)^d}\frac{1}{k^2(k^2+(t-t_c)/K)}+\mathcal{O}(u^2)\right].$$

In going from the first equation to the second, we have changed the position of t_c from one denominator to another. Since $t_c = \mathcal{O}(u)$, the corrections due to this change only appear at $\mathcal{O}(u^2)$. The final integral has dimensions of $\left[k^{d-4}\right]$. For $d > 4$ it is dominated by the largest momenta and scales as Λ^{d-4}. For $2 < d < 4$, the integral is convergent at both limits. Its magnitude is therefore set by the momentum scale $\xi^{-1} = \sqrt{(t-t_c)/K}$, which can be used to make the integrand dimensionless. Hence, in these dimensions,

$$\chi^{-1}(t) = (t-t_c)\left[1-\frac{4u(n+2)}{K^2}c\left(\frac{K}{t-t_c}\right)^{2-d/2}+\mathcal{O}(u^2)\right], \qquad (5.27)$$

where c is a constant. For $d < 4$, the correction term at the order of u diverges at the phase transition, masking the unperturbed singularity of χ with $\gamma = 1$. Thus the perturbation series is inherently inapplicable for describing the divergence of susceptibility in $d < 4$. The same conclusion arises in calculating any other quantity perturbatively. Although we start by treating u as the perturbation parameter, it is important to realize that it is not dimensionless; u/K^2 has dimensions of (length)$^{d-4}$. The perturbation series for any quantity then takes the form $X(t, u) = X_0(t)[1 + f(ua^{4-d}/K^2, u\xi^{4-d}/K^2)]$, where f is a power series. The two length scales a and ξ are available to construct dimensionless variables. Since ξ diverges close to the critical point, there is an inherent failure of the perturbation series. The effective (dimensionless) perturbation parameter diverges at t_c and is not small, making it an inherently ineffective expansion parameter.

5.5 Perturbative RG (first order)

The last section demonstrates how various expectation values associated with the Landau–Ginzburg Hamiltonian can be calculated perturbatively in powers of u. However, the perturbative series is inherently divergent close to the critical point and cannot be used to characterize critical behavior in dimensions $d \leq 4$. K.G. Wilson showed that it is possible to combine perturbative and renormalization group approaches into a systematic method for calculating critical exponents. Accordingly, we shall extend the RG calculation of Gaussian model in Section 3.7 to the Landau–Ginzburg Hamiltonian, by treating $\mathcal{U} = u \int \mathrm{d}^d x\, m^4$ as a perturbation.

(1) **Coarse grain:** This is the most difficult step of the RG procedure. As before, subdivide the fluctuations into two components as,

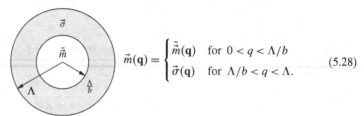

$$\vec{m}(\mathbf{q}) = \begin{cases} \tilde{\vec{m}}(\mathbf{q}) & \text{for } 0 < q < \Lambda/b \\ \vec{\sigma}(\mathbf{q}) & \text{for } \Lambda/b < q < \Lambda. \end{cases} \tag{5.28}$$

In the partition function,

$$Z = \int \mathcal{D}\tilde{\vec{m}}(\mathbf{q}) \mathcal{D}\vec{\sigma}(\mathbf{q}) \exp\left\{ -\int_0^\Lambda \frac{\mathrm{d}^d \mathbf{q}}{(2\pi)^d} \left(\frac{t + Kq^2}{2} \right) \right. \tag{5.29}$$
$$\left. \left(|\tilde{\vec{m}}(\mathbf{q})|^2 + |\vec{\sigma}(\mathbf{q})|^2 \right) - \mathcal{U}[\tilde{\vec{m}}(\mathbf{q}), \vec{\sigma}(\mathbf{q})] \right\},$$

the two sets of modes are mixed by the operator $\mathcal{U}$. Formally, the result of integrating out $\{\vec{\sigma}(\mathbf{q})\}$ can be written as

$$Z = \int \mathcal{D}\tilde{\vec{m}}(\mathbf{q}) \exp\left\{ -\int_0^{\Lambda/b} \frac{\mathrm{d}^d \mathbf{q}}{(2\pi)^d} \left(\frac{t + Kq^2}{2} \right) |\tilde{\vec{m}}(\mathbf{q})|^2 \right\}$$
$$\times \exp\left\{ -\frac{nV}{2} \int_{\Lambda/b}^\Lambda \frac{\mathrm{d}^d \mathbf{q}}{(2\pi)^d} \ln\left(t + Kq^2 \right) \right\} \left\langle \mathrm{e}^{-\mathcal{U}[\tilde{\vec{m}}, \vec{\sigma}]} \right\rangle_\sigma \tag{5.30}$$
$$\equiv \int \mathcal{D}\tilde{\vec{m}}(\mathbf{q}) \mathrm{e}^{-\beta\tilde{\mathcal{H}}[\tilde{\vec{m}}]}.$$

Here we have defined the partial averages

$$\langle \mathcal{O} \rangle_\sigma \equiv \int \frac{\mathcal{D}\vec{\sigma}(\mathbf{q})}{Z_\sigma} \mathcal{O} \exp\left[-\int_{\Lambda/b}^\Lambda \frac{\mathrm{d}^d \mathbf{q}}{(2\pi)^d} \left(\frac{t + Kq^2}{2} \right) |\sigma(\mathbf{q})|^2 \right], \tag{5.31}$$

with $Z_\sigma = \int \mathcal{D}\vec{\sigma}(\mathbf{q}) \exp\{-\beta\mathcal{H}_0[\vec{\sigma}]\}$, being the *Gaussian* partition function associated with the short wavelength fluctuations. From Eq. (5.30), we obtain

$$\beta\tilde{\mathcal{H}}[\tilde{\vec{m}}] = V\delta f_b^0 + \int_0^{\Lambda/b} \frac{\mathrm{d}^d \mathbf{q}}{(2\pi)^d} \left(\frac{t + Kq^2}{2} \right) |\tilde{\vec{m}}(\mathbf{q})|^2 - \ln\left\langle \mathrm{e}^{-\mathcal{U}[\tilde{\vec{m}}, \vec{\sigma}]} \right\rangle_\sigma. \tag{5.32}$$

The final expression can be calculated perturbatively as,

$$\ln\left\langle e^{-\mathcal{U}}\right\rangle_\sigma = -\left\langle\mathcal{U}\right\rangle_\sigma + \frac{1}{2}\left(\left\langle\mathcal{U}^2\right\rangle_\sigma - \left\langle\mathcal{U}\right\rangle_\sigma^2\right) + \cdots$$
$$+ \frac{(-1)^\ell}{\ell!} \times \ell\text{th cumulant of } \mathcal{U} + \cdots \qquad (5.33)$$

The cumulants can be computed using the rules set in the previous sections. For example, at the first order we need to compute

$$\left\langle\mathcal{U}\left[\tilde{\vec{m}},\vec{\sigma}\right]\right\rangle_\sigma = u\int \frac{d^d\mathbf{q}_1 d^d\mathbf{q}_2 d^d\mathbf{q}_3 d^d\mathbf{q}_4}{(2\pi)^{4d}}(2\pi)^d\delta^d(\mathbf{q}_1+\mathbf{q}_2+\mathbf{q}_3+\mathbf{q}_4)$$
$$\left\langle\left[\tilde{\vec{m}}(\mathbf{q}_1)+\vec{\sigma}(\mathbf{q}_1)\right]\cdot\left[\tilde{\vec{m}}(\mathbf{q}_2)+\vec{\sigma}(\mathbf{q}_2)\right]\right. \qquad (5.34)$$
$$\left.\times\left[\tilde{\vec{m}}(\mathbf{q}_3)+\vec{\sigma}(\mathbf{q}_3)\right]\cdot\left[\tilde{\vec{m}}(\mathbf{q}_4)+\vec{\sigma}(\mathbf{q}_4)\right]\right\rangle_\sigma.$$

The following types of terms result from expanding the product:

[1]　1　$\left\langle\tilde{\vec{m}}(\mathbf{q}_1)\cdot\tilde{\vec{m}}(\mathbf{q}_2)\,\tilde{\vec{m}}(\mathbf{q}_3)\cdot\tilde{\vec{m}}(\mathbf{q}_4)\right\rangle_\sigma$ 　　　$\mathcal{U}[\tilde{m}]$

[2]　4　$\left\langle\vec{\sigma}(\mathbf{q}_1)\cdot\tilde{\vec{m}}(\mathbf{q}_2)\,\tilde{\vec{m}}(\mathbf{q}_3)\cdot\tilde{\vec{m}}(\mathbf{q}_4)\right\rangle_\sigma$ 　　　0

[3]　2　$\left\langle\vec{\sigma}(\mathbf{q}_1)\cdot\vec{\sigma}(\mathbf{q}_2)\,\tilde{\vec{m}}(\mathbf{q}_3)\cdot\tilde{\vec{m}}(\mathbf{q}_4)\right\rangle_\sigma$

[4]　4　$\left\langle\vec{\sigma}(\mathbf{q}_1)\cdot\tilde{\vec{m}}(\mathbf{q}_2)\,\vec{\sigma}(\mathbf{q}_3)\cdot\tilde{\vec{m}}(\mathbf{q}_4)\right\rangle_\sigma$

[5]　4　$\left\langle\vec{\sigma}(\mathbf{q}_1)\cdot\vec{\sigma}(\mathbf{q}_2)\,\vec{\sigma}(\mathbf{q}_3)\cdot\tilde{\vec{m}}(\mathbf{q}_4)\right\rangle_\sigma$ 　　　0

[6]　1　$\left\langle\vec{\sigma}(\mathbf{q}_1)\cdot\vec{\sigma}(\mathbf{q}_2)\,\vec{\sigma}(\mathbf{q}_3)\cdot\vec{\sigma}(\mathbf{q}_4)\right\rangle_\sigma$

$$(5.35)$$

The second element in each line is the number of terms with a given "symmetry". The total of these coefficients is $2^4 = 16$. Since the averages $\langle\mathcal{O}\rangle_\sigma$ involve only the short wavelength fluctuations, only contractions with $\vec{\sigma}$ appear. The resulting internal momenta are integrated from Λ/b to Λ.

Term [1] has no $\vec{\sigma}$ factors and evaluates to $\mathcal{U}[\tilde{\vec{m}}]$. The second and fifth terms involve an odd number of $\vec{\sigma}$s and their average is zero. Term [3] has one contraction and evaluates to

$$-u \times 2 \int \frac{d^d\mathbf{q}_1 \cdots d^d\mathbf{q}_4}{(2\pi)^{4d}} (2\pi)^d \delta^d(\mathbf{q}_1 + \cdots + \mathbf{q}_4) \frac{\delta_{jj}(2\pi)^d \delta^d(\mathbf{q}_1 + \mathbf{q}_2)}{t + K q_1^2} \tilde{\vec{m}}(\mathbf{q}_3) \cdot \tilde{\vec{m}}(\mathbf{q}_4)$$

$$= -2nu \int_0^{\Lambda/b} \frac{d^d\mathbf{q}}{(2\pi)^d} |\tilde{m}(\mathbf{q})|^2 \int_{\Lambda/b}^{\Lambda} \frac{d^d\mathbf{k}}{(2\pi)^d} \frac{1}{t + Kk^2}. \tag{5.36}$$

Term [4] also has one contraction but there is no closed loop (the factor δ_{jj}) and hence no factor of n. The various contractions of 4 $\vec{\sigma}$ in term [6] lead to a number of terms with no dependence on $\tilde{m}$. We shall denote the sum of these terms by $uV\delta f_b^1$. Collecting all terms, the coarse-grained Hamiltonian at order of u is given by

$$\beta\tilde{\mathcal{H}}[\tilde{\vec{m}}] = V\left(\delta f_b^0 + u\delta f_b^1\right) + \int_0^{\Lambda/b} \frac{d^d\mathbf{q}}{(2\pi)^d} \left(\frac{\tilde{t} + Kq^2}{2}\right) |\tilde{m}(\mathbf{q})|^2$$

$$+ u \int_0^{\Lambda/b} \frac{d^d\mathbf{q}_1 d^d\mathbf{q}_2 d^d\mathbf{q}_3}{(2\pi)^{3d}} \tilde{\vec{m}}(\mathbf{q}_1) \cdot \tilde{\vec{m}}(\mathbf{q}_2) \tilde{\vec{m}}(\mathbf{q}_3) \cdot \tilde{\vec{m}}(-\mathbf{q}_1 - \mathbf{q}_2 - \mathbf{q}_3), \tag{5.37}$$

where

$$\tilde{t} = t + 4u(n+2) \int_{\Lambda/b}^{\Lambda} \frac{d^d\mathbf{k}}{(2\pi)^d} \frac{1}{t + Kk^2}. \tag{5.38}$$

The coarse-grained Hamiltonian is thus again described by three parameters $\tilde{t}$, $\tilde{K}$, and $\tilde{u}$. The last two parameters are unchanged, and

$$\tilde{K} = K, \quad \text{and} \quad \tilde{u} = u. \tag{5.39}$$

(2) **Rescale** by setting $\mathbf{q} = b^{-1}\mathbf{q}'$, and

(3) **Renormalize**, $\tilde{\vec{m}} = z\vec{m}'$, to get

$$(\beta\mathcal{H})'[m'] = V\left(\delta f_b^0 + u\delta f_b^1\right) + \int_0^{\Lambda} \frac{d^d\mathbf{q}'}{(2\pi)^d} b^{-d} z^2 \left(\frac{\tilde{t} + Kb^{-2}q'^2}{2}\right) |m'(\mathbf{q}')|^2$$

$$+ uz^4 b^{-3d} \int_0^{\Lambda} \frac{d^d\mathbf{q}'_1 d^d\mathbf{q}'_2 d^d\mathbf{q}'_3}{(2\pi)^{3d}} \vec{m}'(\mathbf{q}'_1) \cdot \vec{m}'(\mathbf{q}'_2) \vec{m}'(\mathbf{q}'_3) \cdot \vec{m}'(-\mathbf{q}'_1 - \mathbf{q}'_2 - \mathbf{q}'_3). \tag{5.40}$$

The renormalized Hamiltonian is characterized by the triplet of interactions (t', K', u'), such that

$$t' = b^{-d} z^2 \tilde{t}, \quad K' = b^{-d-2} z^2 K, \quad u' = b^{-3d} z^4 u. \tag{5.41}$$

As in the Gaussian model there is a fixed point at $t^* = u^* = 0$, provided that we set $z = b^{1 + \frac{d}{2}}$, such that $K' = K$. The recursion relations for t and u in the vicinity of this point are given by

$$\begin{cases} t'_b = b^2 \left[t + 4u(n+2) \int_{\Lambda/b}^{\Lambda} \frac{d^d\mathbf{k}}{(2\pi)^d} \frac{1}{t + Kk^2} \right] \\ u'_b = b^{4-d} u. \end{cases} \tag{5.42}$$

While the recursion relation for u at this order is identical to that obtained by dimensional analysis, the one for t is different. It is common to convert the discrete recursion relations to continuous differential flow equations by setting $b = e^{\ell}$, such that for an infinitesimal $\delta\ell$,

$$t'_b \equiv t(b) = t(1+\delta\ell) = t + \delta\ell\frac{dt}{d\ell} + \mathcal{O}(\delta\ell^2), \quad u'_b \equiv u(b) = u + \delta\ell\frac{du}{d\ell} + \mathcal{O}(\delta\ell^2).$$

Expanding Eqs. (5.42) to order of $\delta\ell$, gives

$$\begin{cases} t + \delta\ell\dfrac{dt}{d\ell} = (1+2\delta\ell)\left(t + 4u(n+2)\dfrac{S_d}{(2\pi)^d}\dfrac{1}{t+K\Lambda^2}\Lambda^d\delta\ell\right) \\ u + \delta\ell\dfrac{du}{d\ell} = (1+(4-d)\delta\ell)\,u. \end{cases} \tag{5.43}$$

The differential equations governing the evolution of t and u under rescaling are then

$$\begin{cases} \dfrac{dt}{d\ell} = 2t + \dfrac{4u(n+2)K_d\Lambda^d}{t+K\Lambda^2} \\ \dfrac{du}{d\ell} = (4-d)u. \end{cases} \tag{5.44}$$

The recursion relation for u is easily integrated to give $u(\ell) = u_0 e^{(4-d)\ell} = u_0 b^{(4-d)}$.

The recursion relations can be linearized in the vicinity of the fixed point $t^* = u^* = 0$, by setting $t = t^* + \delta t$ and $u = u^* + \delta u$, as

$$\frac{d}{d\ell}\begin{pmatrix} \delta t \\ \delta u \end{pmatrix} = \begin{pmatrix} 2 & \dfrac{4(n+2)K_d\Lambda^{d-2}}{K} \\ 0 & 4-d \end{pmatrix}\begin{pmatrix} \delta t \\ \delta u \end{pmatrix}. \tag{5.45}$$

In the differential form of the recursion relations, the eigenvalues of the matrix determine the relevance of operators. Since the above matrix has zero elements on one side, its eigenvalues are the diagonal elements, and as in the Gaussian model we can identify $y_t = 2$, and $y_u = 4-d$. The results at this order are identical to those obtained from dimensional analysis on the Gaussian model. The only difference is in the eigendirections. The exponent $y_t = 2$ is still associated with $u = 0$, while $y_u = 4-d$ is actually associated with the direction $t = -4u(n+2)K_d\Lambda^{d-2}/K$. This agrees with the shift in the transition temperature calculated to order of u from the susceptibility.

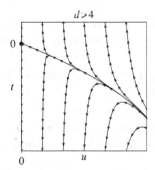
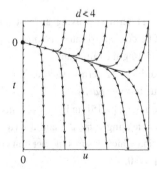

Fig. 5.3 RG flows obtained perturbatively to first order.

For $d > 4$ the Gaussian fixed point has only one unstable direction associated with y_t. It thus correctly describes the phase transition. For $d < 4$ it has two relevant directions and is unstable. Unfortunately, the recursion relations have no other fixed point at this order and it appears that we have learned little from the perturbative RG. However, since we are dealing with an alternating series we can *anticipate* that the recursion relations at the next order are modified to

$$\begin{cases} \dfrac{\mathrm{d}t}{\mathrm{d}\ell} = 2t + \dfrac{4u(n+2)K_d\Lambda^d}{t+K\Lambda^2} - Au^2 \\ \dfrac{\mathrm{d}u}{\mathrm{d}\ell} = (4-d)u - Bu^2, \end{cases} \tag{5.46}$$

with A and B positive. There is now an additional fixed point at $u^* = (4-d)/B$ for $d < 4$. For a systematic perturbation theory we need to keep the parameter u small. Thus the new fixed point can be explored systematically only for small $\epsilon = 4 - d$; we are led to consider an expansion in the dimension of space in the vicinity of $d = 4$! For a calculation valid at $\mathcal{O}(\epsilon)$ we have to keep track of terms of second order in the recursion relation for u, but only to first order in that of t. It is thus unnecessary to calculate the term A in the above recursion relation.

5.6 Perturbative RG (second order)

The coarse-grained Hamiltonian at second order in $\mathcal{U}$ is

$$\beta\tilde{\mathcal{H}}[\tilde{m}] = V\delta f_b^0 + \int_0^{\Lambda/b} \frac{\mathrm{d}^d\mathbf{q}}{(2\pi)^d} \left(\frac{t+Kq^2}{2} \right)$$
$$|\tilde{m}(\mathbf{q})|^2 + \langle \mathcal{U} \rangle_\sigma - \frac{1}{2}\left(\langle \mathcal{U}^2 \rangle_\sigma - \langle \mathcal{U} \rangle_\sigma^2 \right) + O(\mathcal{U}^3). \tag{5.47}$$

To calculate $\left(\langle \mathcal{U}^2 \rangle_\sigma - \langle \mathcal{U} \rangle_\sigma^2 \right)$ we need to consider all possible decompositions of two $\mathcal{U}$s into $\tilde{m}$ and $\vec{\sigma}$ as in Eq. (5.34). Since each $\mathcal{U}$ can be broken up into six types of terms as in Eq. (5.35), there are 36 such possibilities for two $\mathcal{U}$s which can be arranged in a 6×6 matrix, as below. Many of the elements of this matrix are either zero, or can be neglected at this stage, due to a number of considerations:

(1) All the 11 terms involving at least one factor of type [1] are zero because they cannot be contracted into a *connected* piece, and the disconnected elements cancel in calculating the cumulant.
(2) An additional 12 terms (such as [2] × [3]) involve an *odd* number of $\vec{\sigma}$s and are zero due to their *parity*.
(3) Two terms, [2] × [5] and [5] × [2], involve a vertex where two $\vec{\sigma}$s are contracted together, leaving a $\tilde{m}(\mathbf{q}^<)$ and a $\vec{\sigma}(\mathbf{q}^>)$. This configuration is not allowed by the δ-function which ensures momentum conservation for the vertex, as by construction $\mathbf{q}^> + \mathbf{q}^< \neq \mathbf{0}$.

(4) Terms $[3] \times [6]$, $[4] \times [6]$, and their partners by exchange have two factors of $\tilde{\tilde{m}}$. They involve *two-loop* integrations, and appear as corrections to the coefficient $\tilde{t}$. We shall denote their net effect by A, which as noted earlier does not need to be known precisely at this order.

(5) The term $[5] \times [5]$ also involves two factors of $\tilde{\tilde{m}}$, while $[2] \times [2]$ includes six such factors. The latter is important as it indicates that the space of parameters *is not closed* at this order. Even if initially zero, a term proportional to m^6 is generated under RG. In fact, considerations of momentum conservation indicate that both these terms are zero for $\mathbf{q} = 0$, and are thus contributions to $q^2 m^2$ and $q^2 m^6$, respectively. We shall comment on their effect later on.

(6) The contributions resulting from $[6] \times [6]$ are constants, and will be collectively denoted by $u^2 V \delta f_b^2$.

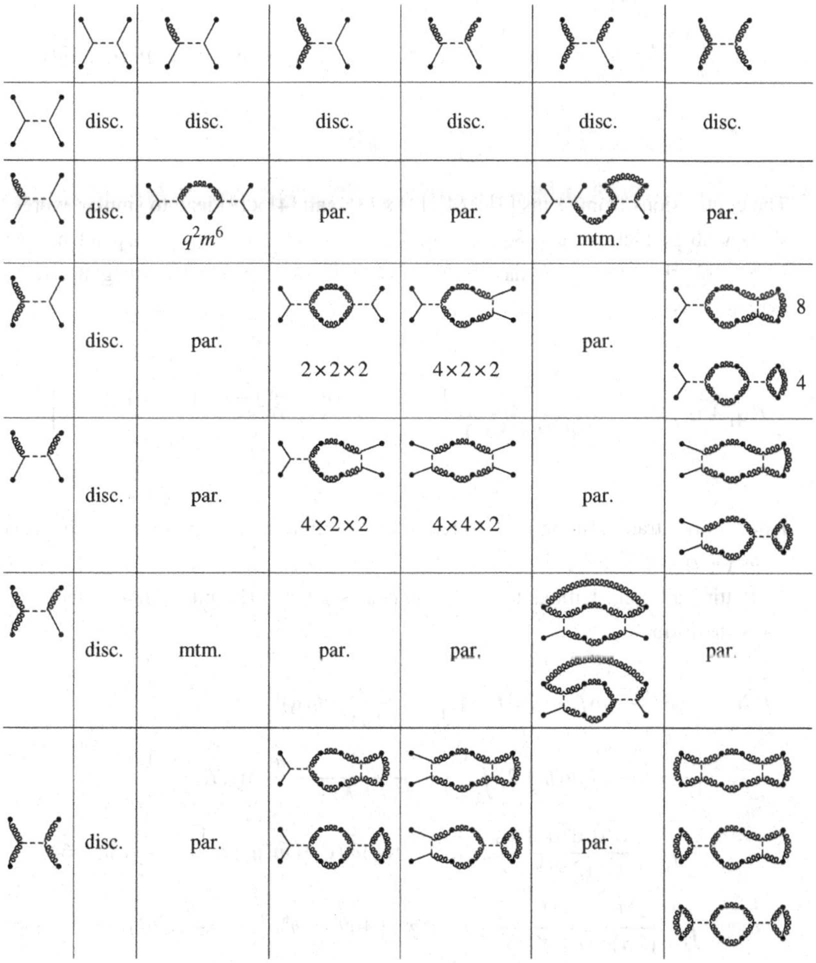

Fig. 5.4 Diagrams appearing in the second-order RG calculation (par. and disc. indicate contributions that are zero due to parity considerations, or being disconnected and mtm. is used to label diagrams that appear at higher order in q^2 due to momentum conservation).

(7) The terms $[3] \times [3]$, $[3] \times [4]$, $[4] \times [3]$, and $[4] \times [4]$ contribute to $\tilde{\tilde{m}}^4$. For example, $[3] \times [3]$ results in

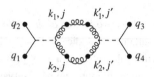

$$\frac{u^2}{2} \times 2 \times 2 \times 2 \int_0^{\Lambda/b} \frac{d^d\mathbf{q}_1 \cdots d^d\mathbf{q}_4}{(2\pi)^{4d}} \int_{\Lambda/b}^{\Lambda} \frac{d^d\mathbf{k}_1 d^d\mathbf{k}_2 d^d\mathbf{k}'_1 d^d\mathbf{k}'_2}{(2\pi)^{4d}}$$

$$\times (2\pi)^{2d} \delta^d(\mathbf{q}_1 + \mathbf{q}_2 + \mathbf{k}_1 + \mathbf{k}_2) \delta^d(\mathbf{k}_1 + \mathbf{k}_2 + \mathbf{q}_3 + \mathbf{q}_4)$$

$$\times \frac{\delta_{\alpha\alpha'}(2\pi)^d \delta^d(\mathbf{k}_1 + \mathbf{k}'_1)}{t + K k_1^{2'}} \frac{\delta_{\alpha\alpha'}(2\pi)^d \delta^d(\mathbf{k}_2 + \mathbf{k}'_2)}{t + K k_2^{2'}} \tilde{m}(\mathbf{q}_1) \cdot \tilde{m}(\mathbf{q}_2) \tilde{m}(\mathbf{q}_3) \cdot \tilde{m}(\mathbf{q}_4)$$

$$= 4nu^2 \int_0^{\Lambda/b} \frac{d^d\mathbf{q}_1 \cdots d^d\mathbf{q}_4}{(2\pi)^{4d}} (2\pi)^d \delta^d(\mathbf{q}_1 + \mathbf{q}_2 + \mathbf{q}_3 + \mathbf{q}_4) \tilde{m}(\mathbf{q}_1) \cdot \tilde{m}(\mathbf{q}_2) \tilde{m}(\mathbf{q}_3) \cdot \tilde{m}(\mathbf{q}_4)$$

$$\times \int \frac{d^d\mathbf{k}}{(2\pi)^d} \frac{1}{(t + Kk^2)(t + K(\mathbf{q}_1 + \mathbf{q}_2 - \mathbf{k})^2)}. \tag{5.48}$$

The contractions from terms $[3] \times [4]$, $[4] \times [3]$, and $[4] \times [4]$ lead to similar expressions with prefactors of 8, 8, and 16 respectively. Apart from the dependence on $\mathbf{q}_1$ and $\mathbf{q}_2$, the final result has the form of $\mathcal{U}[\tilde{m}]$. In fact the last integral can be expanded as

$$f(\mathbf{q}_1 + \mathbf{q}_2) = \int \frac{d^d\mathbf{k}}{(2\pi)^d} \frac{1}{(t + Kk^2)^2} \left[1 - \frac{2K\mathbf{k} \cdot (\mathbf{q}_1 + \mathbf{q}_2) - K(\mathbf{q}_1 + \mathbf{q}_2)^2}{(t + Kk^2)} + \cdots \right]. \tag{5.49}$$

After fourier transforming back to real space we find in addition to m^4, such terms as $m^2(\nabla m)^2$, $m^2 \nabla^2 m^2$, $\cdots$.

Putting all contributions together, the coarse grained Hamiltonian at order of u^2 takes the form

$$\beta \tilde{\mathcal{H}} = V\left(\delta f_b^0 + u\delta f_b^1 + u^2 \delta f_b^2\right) + \int_0^{\Lambda/b} \frac{d^d\mathbf{q}}{(2\pi)^d} |\tilde{m}(\mathbf{q})|^2$$

$$\left[\frac{t + Kq^2}{2} + 2u(n+2) \int_{\Lambda/b}^{\Lambda} \frac{d^d\mathbf{k}}{(2\pi)^d} \frac{1}{t + Kk^2} - \frac{u^2}{2} A(t, K, q^2) \right]$$

$$+ \int_0^{\Lambda/b} \frac{d^d\mathbf{q}_1 d^d\mathbf{q}_2 d^d\mathbf{q}_3}{(2\pi)^{3d}} \tilde{m}(\mathbf{q}_1) \cdot \tilde{m}(\mathbf{q}_2) \tilde{m}(\mathbf{q}_3) \cdot \tilde{m}(\mathbf{q}_4) \times \left[u - \frac{u^2}{2}(8n + 64) \right.$$

$$\left. \int_{\Lambda/b}^{\Lambda} \frac{d^d\mathbf{k}}{(2\pi)^d} \frac{1}{(t + Kk^2)^2} + \mathcal{O}(u^2 q^2) \right] + \mathcal{O}(u^2 \tilde{m}^6 q^2, \cdots) + \mathcal{O}(u^3). \tag{5.50}$$

5.7 The ϵ-expansion

The parameter space (K, t, u) is no longer closed at this order; several new interactions proportional to m^2, m^4, and m^6, all consistent with symmetries of the problem, appear in the coarse-grained Hamiltonian at second order in u. Ignoring these interactions for the time being, the coarse grained parameters are given by

$$\begin{cases} \tilde{K} = K - u^2 A''(0) \\ \tilde{t} = t + 4(n+2)\, u \int_{\Lambda/b}^{\Lambda} \frac{d^d\mathbf{k}}{(2\pi)^d} \frac{1}{t+Kk^2} - u^2 A(0) \\ \tilde{u} = u - 4(n+8)u^2 \int_{\Lambda/b}^{\Lambda} \frac{d^d\mathbf{k}}{(2\pi)^d} \frac{1}{(t+Kk^2)^2}, \end{cases} \tag{5.51}$$

where $A(0)$ and $A''(0)$ correspond to the first two terms in the expansion of $A(t, K, q^2)$ in Eq. (5.50) in powers of q.

After the *rescaling* $\mathbf{q} = b^{-1}\mathbf{q}'$, and *renormalization* $\tilde{\vec{m}} = z\vec{m}'$, steps of the RG procedure, we obtain

$$K' = b^{-d-2}z^2\tilde{K}, \quad t' = b^{-d}z^2\tilde{t}, \quad u' = b^{-3d}z^4\tilde{u}. \tag{5.52}$$

As before, the renormalization parameter z is chosen such that $K' = K$, leading to

$$z^2 = \frac{b^{d+2}}{(1 - u^2 A''(0)/K)} = b^{d+2}\left(1 + O(u^2)\right). \tag{5.53}$$

The value of z does depend on the fixed point position u^*. But as u^* is of the order of ϵ, $z = b^{1+\frac{d}{2}+\mathcal{O}(\epsilon^2)}$, it is not changed at the lowest order. Using this value of z, and following the previous steps for constructing differential recursion relations, we obtain

$$\begin{cases} \dfrac{dt}{d\ell} = 2t + \dfrac{4u(n+2)K_d\Lambda^d}{t+K\Lambda^2} - A(t, K, \Lambda)u^2 \\ \dfrac{du}{d\ell} = (4-d)u - \dfrac{4(n+8)K_d\Lambda^d}{(t+K\Lambda^2)^2}u^2. \end{cases} \tag{5.54}$$

The fixed points are obtained from $dt/d\ell = du/d\ell = 0$. In addition to the Gaussian fixed point at $u^* = t^* = 0$, discussed in the previous section, there is now a non-trivial fixed point located at

$$\begin{cases} u^* = \dfrac{(t^*+K\Lambda^2)^2}{4(n+8)K_d\Lambda^d}\epsilon = \dfrac{K^2}{4(n+8)K_4}\epsilon + \mathcal{O}(\epsilon^2) \\ t^* = -\dfrac{2u^*(n+2)K_d\Lambda^d}{t^*+K\Lambda^2} = -\dfrac{(n+2)}{2(n+8)}K\Lambda^2\epsilon + \mathcal{O}(\epsilon^2). \end{cases} \tag{5.55}$$

The above expressions have been further simplified by systematically keeping terms to first order in $\epsilon = 4 - d$.

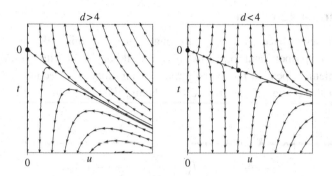

Linearizing the recursion relations in the vicinity of the fixed point results in

$$\frac{d}{d\ell}\begin{pmatrix}\delta t \\ \delta u\end{pmatrix} = \begin{pmatrix} 2 - \dfrac{4(n+2)K_d\Lambda^d}{(t^*+K\Lambda^2)^2}u^* - A'u^{*2} & \dfrac{4(n+2)K_d\Lambda^d}{t^*+K\Lambda^2} - 2Au^* \\[2mm] \dfrac{8(n+8)K_d\Lambda^d}{(t^*+K\Lambda^2)^3}u^{*2} & \epsilon - \dfrac{8(n+8)K_d\Lambda^d}{(t^*+K\Lambda^2)^2}u^* \end{pmatrix}\begin{pmatrix}\delta t \\ \delta u\end{pmatrix}. \quad (5.56)$$

At the Gaussian fixed point, $t^* = u^* = 0$, and Eq. (5.45) is reproduced. At the
new fixed point of Eqs. (5.55),

$$\frac{d}{d\ell}\begin{pmatrix}\delta t \\ \delta u\end{pmatrix} = \begin{pmatrix} 2 - \dfrac{4(n+2)K_4\Lambda^4}{K^2\Lambda^4}\dfrac{K^2\epsilon}{4(n+8)K_4} & \cdots\cdots \\[2mm] \mathcal{O}(\epsilon^2) & \epsilon - \dfrac{8(n+8)K_4\Lambda^4}{K^2\Lambda^4}\dfrac{K^2\epsilon}{4(n+8)K_4} \end{pmatrix}\begin{pmatrix}\delta t \\ \delta u\end{pmatrix}.$$
$$(5.57)$$

We have not explicitly calculated the top element of the second column as
it is not necessary for calculating the eigenvalues. This is because the lower
element of the first column is zero to order of ϵ. Hence the eigenvalues are
determined by the diagonal elements alone. The first eigenvalue is positive,
controlling the instability of the fixed point,

$$y_t = 2 - \frac{(n+2)}{(n+8)}\epsilon + \mathcal{O}(\epsilon^2). \quad (5.58)$$

The second eigenvalue,

$$y_u = -\epsilon + \mathcal{O}(\epsilon^2), \quad (5.59)$$

is negative for $d < 4$. The new fixed point thus has co-dimension of one and
can describe the phase transition in these dimensions. It is quite satisfying
that while various intermediate results, such as the position of the fixed point,
depend on such microscopic parameters as K and Λ, the final eigenvalues
are pure numbers, only depending on n and $d = 4 - \epsilon$. These eigenvalues
characterize the *universality classes* of rotational symmetry breaking in $d < 4$,
with short-range interactions. (As discussed in the problem section, long-range
interaction may lead to new universality classes.)

The divergence of the correlation length, $\xi \sim (\delta t)^{-\nu}$, is controlled by the exponent

$$\nu = \frac{1}{y_t} = \left\{ 2 \left[1 - \frac{(n+2)}{2(n+8)} \epsilon \right] \right\}^{-1} = \frac{1}{2} + \frac{1}{4} \frac{n+2}{n+8} \epsilon + \mathcal{O}(\epsilon^2). \tag{5.60}$$

The singular part of the free energy scales as $f \sim (\delta t)^{2-\alpha}$, and the heat capacity diverges with the exponent

$$\alpha = 2 - d\nu = 2 - \frac{(4-\epsilon)}{2} \left[1 + \frac{1}{2} \frac{n+2}{n+8} \epsilon \right] = \frac{4-n}{2(n+8)} \epsilon + \mathcal{O}(\epsilon^2). \tag{5.61}$$

To complete the calculation of critical exponents, we need the eigenvalue associated with the (relevant) symmetry breaking field h. This is easily found by adding a term $-\vec{h} \cdot \int d^d \mathbf{x} \vec{m}(\mathbf{x}) = -\vec{h} \cdot \vec{m}(\mathbf{q} = \mathbf{0})$ to the Hamiltonian. This term is not affected by coarse graining or rescaling, and after the renormalization step changes to $-z\vec{h} \cdot \vec{m}'(\mathbf{q}' = \mathbf{0})$, implying

$$h' = zh = b^{1+\frac{d}{2}} h, \quad \Longrightarrow \quad y_h = 1 + \frac{d}{2} + \mathcal{O}(\epsilon^2) = 3 - \frac{\epsilon}{2} + \mathcal{O}(\epsilon^2). \tag{5.62}$$

The vanishing of magnetization as $T \to T_c^-$ is controlled by the exponent

$$\beta = \frac{d - y_h}{y_t} = \left(\frac{4-\epsilon}{2} - 1 \right) \times \frac{1}{2} \left(1 + \frac{n+2}{2(n+8)} \epsilon + \mathcal{O}(\epsilon^2) \right)$$
$$= \frac{1}{2} - \frac{3}{2(n+8)} \epsilon + \mathcal{O}(\epsilon^2), \tag{5.63}$$

while the susceptibility diverges as $\chi \sim (\delta t)^{-\gamma}$, with

$$\gamma = \frac{2y_h - d}{y_t} = 2 \times \frac{1}{2} \left(1 + \frac{n+2}{2(n+8)} \epsilon \right) = 1 + \frac{n+2}{2(n+8)} \epsilon + O(\epsilon^2). \tag{5.64}$$

Using the above results, we can estimate various exponents as a function of d and n. For example, for $n = 1$, by setting $\epsilon = 1$ or 2 in Eqs. (5.60) and Eqs. (5.63) we obtain the values $\nu(1) \approx 0.58$, $\nu(2) \approx 0.67$, and $\beta(1) \approx 0.33$, $\beta(2) \approx 0.17$. The best estimates of these exponents in $d = 3$ are $\nu \approx 0.63$, and $\beta \approx 0.32$. In $d = 2$ the exact values are known to be $\nu = 1$ and $\beta = 0.125$. The estimates for β are quite good, while those for ν are less reliable. It is important to note that in all cases these estimates are an improvement over the mean field (saddle point) values. Since the expansion is around four dimensions, the results are more reliable in $d = 3$ than in $d = 2$. In any case, they correctly describe the decrease of β with lowering dimension, and the increase of ν. They also correctly describe the trends with varying n at a fixed d as indicated by the following table of exponents $\alpha(n)$.

Although the sign of α is incorrectly predicted at this order for $n = 2$ and 3, the decrease of α with increasing n is correctly described.

	$n=1$	$n=2$	$n=3$	$n=4$
$\mathcal{O}(\epsilon)$ at $\epsilon=1$	0.17	0.11	0.06	0
Experiments in $d=3$	0.11	-0.01	-0.12	–

5.8 Irrelevance of other interactions

The fixed point Hamiltonian at $\mathcal{O}(\epsilon)$ (from Eqs. 5.55) has only three terms

$$\beta\mathcal{H}^* = \frac{K}{2}\int_\Lambda \mathrm{d}^d\mathbf{x}\left[(\nabla m)^2 - \frac{(n+2)}{(n+8)}\epsilon\Lambda^2 m^2 + \frac{\epsilon\Lambda^{-\epsilon}}{2(n+8)}\frac{K}{K_4}m^4\right], \qquad (5.65)$$

and explicitly depends on the imposed cutoff $\Lambda \sim 1/a$ (unlike the exponents). However, as described in Section 3.4, the starting point for RG must be the most general Hamiltonian consistent with symmetries. We also discovered that even if some of these terms are left out of the original Hamiltonian, they are generated under coarse graining. At second order in u, terms proportional to m^6 were generated; higher powers of m will appear at higher orders in u.

Let us focus on a rotationally symmetric Hamiltonian for $\vec{h}=0$. We can incorporate all terms consistent with this symmetry in a perturbative RG by setting $\beta\mathcal{H} = \beta\mathcal{H}_0 + \mathcal{U}$, where

$$\beta\mathcal{H}_0 = \int \mathrm{d}^d\mathbf{x}\left[\frac{t}{2}m^2 + \frac{K}{2}(\nabla m)^2 + \frac{L}{2}(\nabla^2 m)^2 + \cdots\right] \qquad (5.66)$$

includes all quadratic (Gaussian terms), while the remaining higher order terms are placed in the perturbation

$$\mathcal{U} = \int \mathrm{d}^d\mathbf{x}\left[um^4 + vm^2(\nabla m)^2 + \cdots + u_6 m^6 + \cdots + u_8 m^8 + \cdots\right]. \qquad (5.67)$$

After coarse graining, and steps (ii) and (iii) of RG in real space, $\mathbf{x} = b\mathbf{x}'$ and $\tilde{\vec{m}} = \zeta\vec{m}'$, the renormalized weight depends on the parameters

$$\begin{cases}
t \mapsto & b^d\zeta^2\tilde{t} = b^2\tilde{t} \\
K \mapsto & b^{d-2}\zeta^2\tilde{K} = K \\
L \mapsto & b^{d-4}\zeta^2\tilde{L} = b^{-2}\tilde{L} \\
\quad\vdots \\
u \mapsto & b^d\zeta^4\tilde{u} = b^{4-d}\tilde{u} \\
v \mapsto & b^{d-2}\zeta^4\tilde{v} = b^{2-d}\tilde{v} \\
\quad\vdots \\
u_6 \mapsto & b^d\zeta^6\tilde{u}_6 = b^{6-2d}\tilde{u}_6 \\
u_8 \mapsto & b^d\zeta^8\tilde{u}_8 = b^{8-3d}\tilde{u}_8 \\
\quad\vdots
\end{cases} \qquad (5.68)$$

The second set of equalities are obtained by choosing $\zeta^2 = b^{2-d} K / \tilde{K} = b^{2-d}[1 + \mathcal{O}(u^2, uv, v^2, \cdots)]$, such that $K' = K$. By choosing an infinitesimal rescaling, the recursion relations take the differential forms

$$
\begin{cases}
\dfrac{dt}{d\ell} = & 2t + \mathcal{O}(u, v, u_6, u_8, \cdots) \\[2mm]
\dfrac{dK}{d\ell} = & 0 \\[2mm]
\dfrac{dL}{d\ell} = & -2L + \mathcal{O}(u^2, uv, v^2, \cdots) \\[2mm]
\quad \vdots \\[2mm]
\dfrac{du}{d\ell} = & \epsilon u - Bu^2 + \mathcal{O}(uv, v^2, \cdots) \\[2mm]
\dfrac{dv}{d\ell} = & (-2 + \epsilon)v + \mathcal{O}(u^2, uv, v^2, \cdots) \\[2mm]
\quad \vdots \\[2mm]
\dfrac{du_6}{d\ell} = & (-2 + 2\epsilon)u_6 + \mathcal{O}(u^3, u_6^2, \cdots) \\[2mm]
\dfrac{du_8}{d\ell} = & (-4 + 3\epsilon)u_8 + \mathcal{O}(u^3, u^2 u_6, \cdots) \\[2mm]
\quad \vdots
\end{cases}
\tag{5.69}
$$

These recursion relations describe two fixed points:

(1) The Gaussian fixed point, $t^* = L^* = u^* = v^* = \cdots = 0$, and $K \neq 0$, has eigenvalues

$$
y_t^0 = 2, \ y_L^0 = -2, \ \cdots, \ y_u^0 = +\epsilon, \ y_v^0 = -2 + \epsilon, \ \cdots,
$$
$$
y_6^0 = -2 + 2\epsilon, \ y_8^0 = -4 + 3\epsilon, \ \cdots.
\tag{5.70}
$$

(2) Setting Eqs. (5.69) to zero, a non-trivial fixed point is located at

$$
t^* \sim u^* \sim \mathcal{O}(\epsilon), \quad L^* \sim v^* \sim \cdots \sim \mathcal{O}(\epsilon^2), \quad u_6^* \sim \cdots \sim \mathcal{O}(\epsilon^3), \quad \cdots.
\tag{5.71}
$$

The stability of this fixed point is determined by the matrix,

$$
\frac{d}{d\ell}
\begin{pmatrix}
\delta t \\
\delta L \\
\vdots \\
\delta u \\
\delta v \\
\vdots
\end{pmatrix}
=
\begin{pmatrix}
2 - \mathcal{O}(u^*) & \mathcal{O}(\epsilon) & \cdots & \mathcal{O}(1) & \mathcal{O}(1) & \cdots \\
\mathcal{O}(\epsilon^2) & -2 + \mathcal{O}(\epsilon) & & & & \\
\vdots & \vdots & \ddots & & & \\
\mathcal{O}(\epsilon^2) & \mathcal{O}(\epsilon) & & & & \\
\mathcal{O}(\epsilon^2) & \mathcal{O}(\epsilon) & & & & \\
\vdots & \vdots & & & &
\end{pmatrix}
\begin{pmatrix}
\delta t \\
\delta L \\
\vdots \\
\delta u \\
\delta v \\
\vdots
\end{pmatrix}.
\tag{5.72}
$$

Note that as $\epsilon \to 0$, the non-trivial fixed part, its eigenvalues and eigendirections continuously go over to the Gaussian fixed points. Hence the eigenvalues can only be corrected by order of ϵ, and Eq. (5.70) is modified to

$$y_t = 2 - \frac{n+2}{n+8}\epsilon + \mathcal{O}(\epsilon^2), \qquad y_L = -2 + \mathcal{O}(\epsilon), \cdots,$$

$$y_u = -\epsilon + \mathcal{O}(\epsilon^2), y_v = -2 + \mathcal{O}(\epsilon), \cdots, y_6 = -2 + \mathcal{O}(\epsilon), y_8 = -4 + \mathcal{O}(\epsilon), \cdots$$

$$(5.73)$$

While the eigenvalues are still labeled with the coefficients of the various terms in the Landau–Ginzburg expansion, we must remember that the actual eigendirections are now rotated away from the axes of this parameter space, although their largest projection is still parallel to the corresponding axis.

Whereas the Gaussian fixed point has two relevant directions in $d < 4$, the generalized $O(n)$ fixed point has only one relevant direction corresponding to y_t. At least perturbatively, this fixed point has a basin of attraction of co-dimension one, and thus describes the phase transition. The original concept of Kadanoff scaling is thus explicitly realized and the universality of exponents is traced to the irrelevance (at least perturbatively) of the multitude of other possible interactions. The perturbative approach does not exclude the existence of other fixed points at finite values of these parameters. The uniqueness of the critical exponents observed so far for each universality class, and their proximity to the values calculated from the ϵ-expansion, suggests that postulating such *non-perturbative* fixed points is unnecessary.

5.9 Comments on the ϵ-expansion

The perturbative implementation of RG for the Landau–Ginzburg Hamiltonian was achieved by K.G. Wilson in the early 1970s; the ϵ-expansion was developed jointly with M.E. Fisher. This led to a flurry of activity in the topic which still continues. Wilson was awarded the Nobel Prize in 1982. Historical details can be found in his Nobel lecture reprinted in Rev. Mod. Phys. **55**, 583 (1983). A few comments on the ϵ-expansion are in order at this stage.

(1) *Higher orders, and convergence of the series:* Calculating the exponents to $\mathcal{O}(\epsilon)^2$ and beyond, by going to order of $\mathcal{U}^3$ and higher, is quite complicated as we have to keep track of many more interactions. It is in fact quite unappealing that the intermediate steps of the RG explicitly keep track of the cutoff scale Λ, while the final exponents must be independent of it. In fact there are a number of field theoretical RG schemes (dimensional regularization, summing leading divergences, etc.) that avoid many of these difficulties. These methods are harder to visualize and will not be described here. All higher order calculations are currently performed using one of these schemes. It is sometimes (but not always) possible to prove that these approaches are consistent with each other, and can be carried out to all orders. In principle, the problem of evaluating critical exponents in $d = 3$ is now solved: simple computations lead to approximate results, while more refined calculations should provide better answers. The situation is somewhat like finding the energy levels of a He atom, which cannot be done exactly, but which may be obtained with sufficient accuracy using various approximation methods.

To estimate how reliable the exponents are, we need some information on the convergence of the series. The ϵ expansion has been carried out to the fifth order, and the results for the exponent γ, for $n = 1$ at $d = 3$, are

$$\gamma = 1 + 0.167\epsilon + 0.077\epsilon^2 - 0.049\epsilon^3 + 0.180\epsilon^4 - 0.415\epsilon^5$$

$$1.2385 \pm 0.0025 = 1.000, 1.167, 1.244, 1.195, 1.375, 0.96.$$ (5.74)

The second line compares the values obtained at different orders by substituting $\epsilon = 1$, with the best estimate of $\gamma \approx 1.2385$ in $d = 3$. Note that the elements of the series have alternating signs. The truncated series evaluated at $\epsilon = 1$ improves up to third order, beyond which it starts to oscillate, and deviates from the left hand side. These are characteristics of an *asymptotic series*. It can be proved that for large p, the coefficients in the expansion of most quantities scale as $|f_p| \sim c p! a^{-p}$. As a result, the ϵ-expansion series is *non-convergent*, but can be evaluated by the *Borel summation* method, using the identity $\int_0^\infty \mathrm{d}x\, x^p e^{-x} = p!$, as

$$f(\epsilon) = \sum_p f_p \epsilon^p = \sum_p f_p \epsilon^p \frac{1}{p!} \int_0^\infty \mathrm{d}x\, x^p e^{-x} = \int_0^\infty \mathrm{d}x e^{-x} \sum_p \frac{f_p(\epsilon x)^p}{p!}.$$ (5.75)

The final summation (which is convergent) results in a function of x which can be integrated to give $f(\epsilon)$. Very good estimates of exponents in $d = 3$, such as the one for γ quoted above, are obtained by this summation method. There is no indication of any singularity in the exponents up to $\epsilon = 2$, corresponding to the lower critical dimension $d = 2$.

(2) **The $1/n$ expansion:** The fixed point position,

$$u^* = \frac{(t^* + K\Lambda^2)^2(4-d)}{4(n+8)K_d\Lambda^d},$$

vanishes as $n \to \infty$. This suggests that a controlled $1/n$ expansion of the critical exponents is also possible. Indeed such an expansion can be developed by a number of methods, such as a saddle point expansion that takes advantage of the exponential dependence of the Hamiltonian on n, or by an exact resummation of the perturbation series. Equation (5.58) in this limit gives,

$$y_t = \lim_{n \to \infty} \left[2 - \frac{n+2}{n-8}(4-d) \right] = d - 2 \quad \Longrightarrow \quad \nu = \frac{1}{d-2}.$$ (5.76)

This result is exact in dimensions $4 < d < 2$. Above four dimensions the mean field value of $1/2$ is recovered, while for $d < 2$ there is no order.

Problems for chapter 5

1. *Longitudinal susceptibility:* While there is no reason for the longitudinal susceptibility to diverge at the mean-field level, it in fact does so due to fluctuations in
dimensions $d < 4$. This problem is intended to show you the origin of this divergence
in perturbation theory. There are actually a number of subtleties in this calculation

which you are instructed to ignore at various steps. You may want to think about why they are justified.

Consider the Landau–Ginzburg Hamiltonian:

$$\beta \mathcal{H} = \int d^d \mathbf{x} \left[\frac{K}{2} (\nabla \vec{m})^2 + \frac{t}{2} \vec{m}^2 + u(\vec{m}^2)^2 \right],$$

describing an n-component magnetization vector $\vec{m}(\mathbf{x})$, in the ordered phase for $t < 0$.

(a) Let $\vec{m}(\mathbf{x}) = (\overline{m} + \phi_\ell(\mathbf{x})) \hat{e}_\ell + \vec{\phi}_t(\mathbf{x}) \hat{e}_t$, and expand $\beta \mathcal{H}$ *keeping all terms in the expansion.*

(b) Regard the quadratic terms in ϕ_ℓ and $\vec{\phi}_t$ as an unperturbed Hamiltonian $\beta \mathcal{H}_0$, and the lowest order term coupling ϕ_ℓ and $\vec{\phi}_t$ as a perturbation U; i.e.

$$U = 4u\overline{m} \int d^d \mathbf{x} \phi_\ell(\mathbf{x}) \vec{\phi}_t(\mathbf{x})^2.$$

Write U in Fourier space in terms of $\phi_\ell(\mathbf{q})$ and $\vec{\phi}_t(\mathbf{q})$.

(c) Calculate the Gaussian (bare) expectation values $\langle \phi_\ell(\mathbf{q})\phi_\ell(\mathbf{q}') \rangle_0$ and $\langle \phi_{t,\alpha}(\mathbf{q})\phi_{t,\beta}(\mathbf{q}') \rangle_0$, and the corresponding momentum dependent susceptibilities $\chi_\ell(\mathbf{q})_0$ and $\chi_t(\mathbf{q})_0$.

(d) Calculate $\langle \vec{\phi}_t(\mathbf{q}_1) \cdot \vec{\phi}_t(\mathbf{q}_2) \; \vec{\phi}_t(\mathbf{q}_1') \cdot \vec{\phi}_t(\mathbf{q}_2') \rangle_0$ using Wick's theorem. (Don't forget that $\vec{\phi}_t$ is an $(n-1)$ component vector.)

(e) Write down the expression for $\langle \phi_\ell(\mathbf{q})\phi_\ell(\mathbf{q}') \rangle$ to second order in the perturbation U. Note that since U is odd in ϕ_ℓ, only two terms at the second order are non-zero.

(f) Using the form of U in Fourier space, write the correction term as a product of two four-point expectation values similar to those of part (d). Note that only connected terms for the longitudinal four-point function should be included.

(g) Ignore the disconnected term obtained in (d) (i.e. the part proportional to $(n-1)^2$), and write down the expression for $\chi_\ell(\mathbf{q})$ in second order perturbation theory.

(h) Show that for $d < 4$, the correction term diverges as q^{d-4} for $q \to 0$, implying an infinite longitudinal susceptibility.

2. *Crystal anisotropy:* Consider a ferromagnet with a tetragonal crystal structure. Coupling of the spins to the underlying lattice may destroy their full rotational symmetry. The resulting anisotropies can be described by modifying the Landau–Ginzburg Hamiltonian to

$$\beta \mathcal{H} = \int d^d \mathbf{x} \left[\frac{K}{2} (\nabla \vec{m})^2 + \frac{t}{2} \vec{m}^2 + u \left(\vec{m}^2 \right)^2 + \frac{r}{2} m_1^2 + v m_1^2 \vec{m}^2 \right],$$

where $\vec{m} \equiv (m_1, \cdots, m_n)$, and $\vec{m}^2 = \sum_{i=1}^n m_i^2$ ($d = n = 3$ for magnets in three dimensions). Here $u > 0$, and *to simplify calculations we shall set $v = 0$ throughout.*

(a) For a fixed magnitude $|\vec{m}|$, what directions in the n component magnetization space are selected for $r > 0$, and for $r < 0$?

(b) Using the saddle point approximation, calculate the free energies ($\ln Z$) for phases uniformly magnetized *parallel* and *perpendicular* to direction 1.

(c) Sketch the phase diagram in the (t, r) plane, and indicate the phases (type of order), and the nature of the phase transitions (continuous or discontinuous).

(d) Are there Goldstone modes in the ordered phases?

(e) For $u = 0$, and positive t and r, calculate the unperturbed averages $\langle m_1(\mathbf{q})m_1(\mathbf{q}')\rangle_0$ and $\langle m_2(\mathbf{q})m_2(\mathbf{q}')\rangle_0$, where $m_i(\mathbf{q})$ indicates the Fourier transform of $m_i(\mathbf{x})$.

(f) Write the fourth order term $\mathcal{U} \equiv u \int d^d\mathbf{x}(\vec{m}^2)^2$, in terms of the Fourier modes $m_i(\mathbf{q})$.

(g) Treating $\mathcal{U}$ as a perturbation, calculate the *first order* correction to $\langle m_1(\mathbf{q})m_1(\mathbf{q}')\rangle$. (You can leave your answers in the form of some integrals.)

(h) Treating $\mathcal{U}$ as a perturbation, calculate the *first order* correction to $\langle m_2(\mathbf{q})m_2(\mathbf{q}')\rangle$.

(i) Using the above answer, identify the inverse susceptibility χ_{22}^{-1}, and then find the transition point, t_c, from its vanishing to first order in u.

(j) Is the critical behavior different from the isotropic $O(n)$ model in $d < 4$? In RG language, is the parameter r *relevant* at the $\mathcal{O}(n)$ fixed point? In either case indicate the universality classes expected for the transitions.

3. *Cubic anisotropy – mean-field treatment:* Consider the modified Landau–Ginzburg Hamiltonian

$$\beta\mathcal{H} = \int d^d\mathbf{x} \left[\frac{K}{2}(\nabla\vec{m})^2 + \frac{t}{2}\vec{m}^2 + u(\vec{m}^2)^2 + v\sum_{i=1}^{n} m_i^4 \right],$$

for an n-component vector $\vec{m}(\mathbf{x}) = (m_1, m_2, \cdots, m_n)$. The "cubic anisotropy" term $\sum_{i=1}^{n} m_i^4$ breaks the full rotational symmetry and selects specific directions.

(a) For a fixed magnitude $|\vec{m}|$, what directions in the n component magnetization space are selected for $v > 0$ and for $v < 0$? What is the degeneracy of easy magnetization axes in each case?

(b) What are the restrictions on u and v for $\beta\mathcal{H}$ to have finite minima? Sketch these regions of stability in the (u, v) plane.

(c) In general, higher order terms (e.g. $u_6(\vec{m}^2)^3$ with $u_6 > 0$) are present and insure stability in the regions not allowed by part (b) (as in the case of the tricritical point discussed in earlier problems). With such terms in mind, sketch the saddle point phase diagram in the (t, v) plane for $u > 0$; clearly identifying the phases, and order of the transition lines.

(d) Are there any Goldstone modes in the ordered phases?

4. *Cubic anisotropy ε-expansion:*

(a) By looking at diagrams in a second order perturbation expansion in both u and v show that the recursion relations for these couplings are

$$\begin{cases} \dfrac{du}{d\ell} = \varepsilon u - 4C\left[(n+8)u^2 + 6uv \right] \\ \dfrac{dv}{d\ell} = \varepsilon v - 4C\left[12uv + 9v^2 \right], \end{cases}$$

where $C = K_d \Lambda^d / (t + K\Lambda^2)^2 \approx K_4/K^2$ is approximately a constant.

(b) Find all fixed points in the (u, v) plane, and draw the flow patterns for $n < 4$ and $n > 4$. Discuss the relevance of the cubic anisotropy term near the stable fixed point in each case.

(c) Find the recursion relation for the reduced temperature, t, and calculate the exponent ν at the stable fixed points for $n < 4$ and $n > 4$.

(d) Is the region of stability in the (u, v) plane calculated in part (b) of the previous problem enhanced or diminished by inclusion of fluctuations? Since in reality higher order terms will be present, what does this imply about the nature of the phase transition for a small negative v and $n > 4$?

(e) Draw schematic phase diagrams in the (t, v) plane $(u > 0)$ for $n > 4$ and $n < 4$, identifying the ordered phases. Are there Goldstone modes in any of these phases close to the phase transition?

5. *Exponents:* Two critical exponents at second order are,

$$
\begin{cases}
\nu = \dfrac{1}{2} + \dfrac{(n+2)}{4(n+8)}\epsilon + \dfrac{(n+2)(n^2+23n+60)}{8(n+8)^3}\epsilon^2 \ , \\[3mm]
\eta = \dfrac{(n+2)}{2(n+8)^2}\epsilon^2 .
\end{cases}
$$

Use scaling relations to obtain ϵ-expansions for two or more of the remaining exponents α, β, γ, δ and Δ. Make a table of the results obtained by setting $\epsilon = 1$, 2 for $n = 1$, 2 and 3; and compare to the best estimates of these exponents that you can find by other sources (series, experiments, etc.).

---

6. *Anisotropic criticality:* A number of materials, such as liquid crystals, are anisotropic and behave differently along distinct directions, which shall be denoted parallel and perpendicular, respectively. Let us assume that the d spatial dimensions are grouped into n parallel directions $\mathbf{x}_{\parallel}$, and $d - n$ perpendicular directions $\mathbf{x}_{\perp}$. Consider a one-component field $m(\mathbf{x}_{\parallel}, \mathbf{x}_{\perp})$ subject to a Landau–Ginzburg Hamiltonian, $\beta\mathcal{H} = \beta\mathcal{H}_0 + U$, with

$$
\beta\mathcal{H}_0 = \int d^n\mathbf{x}_{\parallel} d^{d-n}\mathbf{x}_{\perp} \left[\frac{K}{2}(\nabla_{\parallel}m)^2 + \frac{L}{2}(\nabla_{\perp}^2 m)^2 + \frac{t}{2}m^2 - hm \right],
$$

and $\quad U = u \int d^n\mathbf{x}_{\parallel} d^{d-n}\mathbf{x}_{\perp} \ m^4$.

*(Note that $\beta\mathcal{H}$ depends on the **first** gradient in the $\mathbf{x}_{\parallel}$ directions, and on the **second** gradient in the $\mathbf{x}_{\perp}$ directions.)*

(a) Write $\beta\mathcal{H}_0$ in terms of the Fourier transforms $m(\mathbf{q}_{\parallel}, \mathbf{q}_{\perp})$.

(b) Construct a renormalization group transformation for $\beta\mathcal{H}_0$, by rescaling coordinates such that $\mathbf{q}_{\parallel}' = b\,\mathbf{q}_{\parallel}$ and $\mathbf{q}_{\perp}' = c\,\mathbf{q}_{\perp}$ and the field as $m'(\mathbf{q}') = m(\mathbf{q})/z$. Note *that parallel and perpendicular directions are scaled differently.* Write down the recursion relations for K, L, t, and h in terms of b, c, and z. (The exact shape of the Brillouin zone is immaterial at this stage, and you do not need to evaluate the integral that contributes an additive constant.)

(c) Choose $c(b)$ and $z(b)$ such that $K' = K$ and $L' = L$. At the resulting fixed point calculate the eigenvalues y_t and y_h for the rescalings of t and h.

(d) Write the relationship between the (singular parts of) free energies $f(t, h)$ and $f'(t', h')$ in the original and rescaled problems. Hence write the unperturbed free energy in the homogeneous form $f(t, h) = t^{2-\alpha} g_f(h/t^\Delta)$, and identify the exponents α and Δ.

(e) How does the unperturbed zero-field susceptibility $\chi(t, h = 0)$ diverge as $t \to 0$?

   In the remainder of this problem set $h = 0$, and treat U as a perturbation.

(f) In the unperturbed Hamiltonian calculate the expectation value $\langle m(q)m(q') \rangle_0$, and the corresponding susceptibility $\chi_0(q) = \langle |m_q|^2 \rangle_0$, where q stands for $(q_\parallel, q_\perp)$.

(g) Write the perturbation U, in terms of the normal modes $m(q)$.

(h) Using RG, or any other method, find the upper critical dimension d_u, for validity of the Gaussian exponents.

(i) Write down the expansion for $\langle m(q)m(q') \rangle$, to first order in U, and reduce the correction term to a product of two point expectation values.

(j) Write down the expression for $\chi(q)$, in first order perturbation theory, and identify the transition point t_c at order of u. (Do not evaluate any integrals explicitly.)

7. *Long-range interactions* between spins can be described by adding a term

$$\int d^d x \int d^d y J(|\mathbf{x} - \mathbf{y}|)\vec{m}(\mathbf{x}) \cdot \vec{m}(\mathbf{y}),$$

to the usual Landau–Ginzburg Hamiltonian.

(a) Show that for $J(r) \propto 1/r^{d+\sigma}$, the Hamiltonian can be written as

$$\beta\mathcal{H} = \int \frac{d^d q}{(2\pi)^d} \frac{t + K_2 q^2 + K_\sigma q^\sigma + \cdots}{2} \vec{m}(\mathbf{q}) \cdot \vec{m}(-\mathbf{q})$$

$$+ u \int \frac{d^d q_1 d^d q_2 d^d q_3}{(2\pi)^{3d}} \vec{m}(\mathbf{q}_1) \cdot \vec{m}(\mathbf{q}_2) \vec{m}(\mathbf{q}_3) \cdot \vec{m}(-\mathbf{q}_1 - \mathbf{q}_2 - \mathbf{q}_3) \ .$$

(b) For $u = 0$, construct the recursion relations for (t, K_2, K_σ) and show that K_σ is irrelevant for $\sigma > 2$. What is the fixed Hamiltonian in this case?

(c) For $\sigma < 2$ and $u = 0$, show that the spin rescaling factor must be chosen such that $K'_\sigma = K_\sigma$, in which case K_2 is irrelevant. What is the fixed Hamiltonian now?

(d) For $\sigma < 2$, calculate the generalized Gaussian exponents ν, η, and γ from the recursion relations. Show that u is irrelevant, and hence the Gaussian results are valid, for $d > 2\sigma$.

(e) For $\sigma < 2$, use a perturbation expansion in u to construct the recursion relations for (t, K_σ, u) as in the text.

(f) For $d < 2\sigma$, calculate the critical exponents ν and η to first order in $\epsilon = d - 2\sigma$.

   [See M.E. Fisher, S.-K. Ma and B.G. Nickel, Phys. Rev. Lett. **29**, 917 (1972).]

(g) What is the critical behavior if $J(r) \propto \exp(-r/a)$? Explain!

6
Lattice systems

6.1 Models and methods

While Wilson's perturbative RG provides a systematic approach to probing critical properties, carrying out the ϵ-expansion to high orders is quite cumbersome. Models defined on a discrete lattice provide a number of alternative computational routes that can complement the perturbative RG approach. Because of universality, we expect that all models with appropriate microscopic symmetries and range of interactions, no matter how simplified, lead to the same critical exponents. Lattice models are convenient for visualization, computer simulation, and series expansion purposes. We shall thus describe models in which an appropriate "spin" degree of freedom is placed on each site of a lattice, and the spins are subject to simple interaction energies. While such models are formulated in terms of explicit "microscopic" degrees of freedom, depending on their degree of complexity, they may or may not provide a more accurate description of a specific material than the Landau–Ginzburg model. The point is that universality dictates that both descriptions describe the same *macroscopic* physics, and the choice of continuum or discrete models is then a matter of computational convenience.

Fig. 6.1 Interacting "spins" $\{s_i\}$ defined on a square lattice.

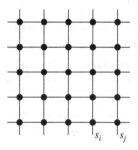

Some commonly used lattice models are described here:

(1) **The Ising model** is the simplest and most widely applied paradigm in statistical mechanics. At each site i of a lattice, there is a spin σ_i which takes the two values

of $+1$ or -1. Each state may correspond to one of two species in a binary mixture, or to empty and occupied cells in a lattice approximation to an interacting gas. The simplest possible *short-range* interaction involves only neighboring spins, and is described by a Hamiltonian

$$\mathcal{H} = \sum_{\langle i,j \rangle} \hat{B}(\sigma_i, \sigma_j), \tag{6.1}$$

where the notation $\langle i, j \rangle$ is commonly used to indicate the sum over all *nearest neighbor* pairs on the lattice. Since $\sigma_i^2 = 1$, the most general interaction between two spins is

$$\hat{B}(\sigma, \sigma') = -\hat{g} - \frac{\hat{h}}{z}(\sigma + \sigma') - J\sigma\sigma'. \tag{6.2}$$

For N spins, there are 2^N possible *microstates*, and the (Gibbs) partition function is

$$Z = \sum_{\{\sigma_i\}} e^{-\beta \mathcal{H}} = \sum_{\{\sigma_i\}} \exp\left[K \sum_{\langle i,j \rangle} \sigma_i \sigma_j + h \sum_i \sigma_i + g \right], \tag{6.3}$$

where we have set $K = \beta J$, $h = \beta \hat{h}$, and $g = z\beta\hat{g}/2$ ($\beta = 1/k_B T$, and z is the number of bonds per site, i.e. the coordination number of the lattice). For $h = 0$ at $T = 0$, the ground state has a two fold degeneracy with all spins pointing up or down ($K > 0$). This order is destroyed at a critical $K_c = J/k_B T_c$ with a phase transition to a disordered state. The field h breaks the *up–down symmetry* and removes the phase transition. The parameter g merely shifts the origin of energy, and has no effect on the relative weights of microstates, or the macroscopic properties.

All the following models can be regarded as generalizations of the Ising model.

(2) **The $O(n)$ model:** Each lattice site is now occupied by an n-component *unit* vector, i.e

$$S_i \equiv (S_i^1, S_i^2, \cdots, S_i^n), \quad \text{with} \quad \sum_{\alpha=1}^n (S_i^\alpha)^2 = 1. \tag{6.4}$$

A nearest-neighbor interaction can be written as

$$\mathcal{H} = -J \sum_{\langle i,j \rangle} \vec{S}_i \cdot \vec{S}_j - \hat{\vec{h}} \cdot \sum_i \vec{S}_i. \tag{6.5}$$

In fact, the most general interaction consistent with spherical symmetry is $f(\vec{S}_i \cdot \vec{S}_j)$ for an arbitrary function f. Similarly, the *rotational symmetry* can be broken by a number of "fields" such as $\sum_i (\hat{\vec{h}}_p \cdot \vec{S}_i)^p$. Specific cases are the Ising model ($n = 1$), the *XY model* ($n = 2$), and the *Heisenberg model* ($n = 3$).

(3) **The Potts model:** Each site of the lattice is occupied by a q-valued spin $S_i \equiv 1, 2, \cdots, q$. The interactions between the spins are described by the Hamiltonian

$$\mathcal{H} = -J \sum_{\langle i,j \rangle} \delta_{S_i, S_j} - \hat{h} \sum_i \delta_{S_i, 1}. \tag{6.6}$$

The field h now breaks the *permutation symmetry* amongst the q-states. The Ising model is recovered for $q = 2$, since $\delta_{\sigma, \sigma'} = (1 + \sigma\sigma')/2$. The three state Potts model can for example describe the distortion of a cube along one of its faces. Potts models with $q > 2$ represent new universality classes not covered by the $O(n)$ model. Actually, the transitions for $q \geq 4$ in $d = 2$, and $q > 3$ in $d = 3$ are discontinuous.

(4) **Spin s-models:** The spin at each site takes the $2s + 1$ values, $s_i = -s, -s + 1, \cdots, +s$. A general nearest-neighbor Hamiltonian is

$$\mathcal{H} = \sum_{\langle i,j \rangle} \left(J_1 s_i s_j + J_2 (s_i s_j)^2 + \cdots + J_{2s} (s_i s_j)^{2s} \right) - \hat{h} \sum_i s_i. \tag{6.7}$$

The Ising model corresponds to $s = 1/2$, while $s = 1$ is known as the Blume–Emery–Griffith (BEG) model. It describes a mixture of non-magnetic ($s = 0$) and magnetic ($s = \pm 1$) elements. This model exhibits a tricritical point separating continuous and discontinuous transitions. However, since the ordered phase breaks an up–down symmetry, the phase transition belongs to the Ising universality class for all values of s.

Some of the computational tools employed in the study of discrete models are:

(1) **Exact solutions** can be obtained for a very limited subset of lattice models. These include many one dimensional systems that can be solved by the transfer matrix method described next, and the two-dimensional Ising model discussed in the next chapter.

Fig. 6.2 A configuration of Ising spins on a square lattice.

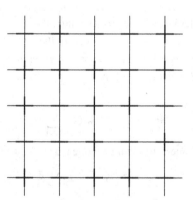

(2) **Position space renormalizations:** These are implementations of Kadanoff's RG scheme on lattice models. Some approximation is usually necessary to keep the space of interactions tractable. Most such approximations are uncontrolled; a number of them will be discussed in this chapter.

(3) **Monte Carlo simulations:** The aim of such methods is to generate configurations of spins that are distributed with the correct Boltzmann weight $\exp(-\beta \mathcal{H})$. There are a number of methods, most notably the Metropolis algorithm, for achieving this aim. Various expectation values and correlation functions are then directly computed from these configurations. With the continuing increase of computer

power, numerical simulations have become increasingly popular. Limitations of the method are due to the size of systems that can be studied, and the amount of time needed to ensure that the correctly weighted configurations are generated. There is an extensive literature on numerical simulations which will only be touched upon briefly.

(4) *Series expansions:* Low-temperature expansions start with the ordered ground state and examine the lowest energy excitations around it (see the next chapter). High temperature expansions begin with the collection of non-interacting spins at infinite temperature and include the interactions between spins perturbatively. Critical behavior is then extracted from the singularities of such series.

6.2 Transfer matrices

Consider a linear chain of N Ising spins ($\sigma_i = \pm 1$), with a nearest-neighbor coupling K, and a magnetic field h. To simplify calculations, we assume that the chain is closed upon itself such that the first and last spins are also coupled (periodic boundary conditions), resulting in the Hamiltonian

$$-\beta \mathcal{H} = K \left(\sigma_1 \sigma_2 + \sigma_2 \sigma_3 + \cdots + \sigma_{N-1}\sigma_N + \sigma_N \sigma_1 \right) + h \sum_{i=1}^{N} \sigma_i . \tag{6.8}$$

The corresponding partition function, obtained by summing over all states, can be expressed as the product of matrices, since

$$Z = \sum_{\sigma_1 = \pm 1} \sum_{\sigma_2 = \pm 1} \cdots \sum_{\sigma_N = \pm 1} \prod_{i=1}^{N} \exp \left[K \sigma_i \sigma_{i+1} + \frac{h}{2} (\sigma_i + \sigma_{i+1}) \right]$$
$$\equiv \mathrm{tr} \left[\langle \sigma_1 | T | \sigma_2 \rangle \langle \sigma_2 | T | \sigma_3 \rangle \cdots \langle \sigma_N | T | \sigma_1 \rangle \right] = \mathrm{tr} \left[T^N \right], \tag{6.9}$$

where we have introduced the 2×2 *transfer matrix* T, with elements

$$\langle \sigma_i | T | \sigma_j \rangle = \exp \left[K \sigma_i \sigma_j + \frac{h}{2} (\sigma_i + \sigma_j) \right], \quad \text{i.e.} \quad T = \begin{pmatrix} e^{K+h} & e^{-K} \\ e^{-K} & e^{K-h} \end{pmatrix}. \tag{6.10}$$

The expression for trace of the matrix can be evaluated in the basis that diagonalizes T, in which case it can be written in terms of the two eigenvalues λ_+ as

$$Z = \lambda_+^N + \lambda_-^N = \lambda_+^N \left[1 + \left(\lambda_-/\lambda_+ \right)^N \right] \approx \lambda_+^N. \tag{6.11}$$

We have assumed that $\lambda_+ > \lambda_-$, and since in the limit of $N \to \infty$ the larger eigenvalue dominates the sum, the free energy is

$$\beta f = -\ln Z / N = -\ln \lambda_+. \tag{6.12}$$

Solving the characteristic equation, we find the eigenvalues

$$\lambda_\pm = e^K \cosh h \pm \sqrt{e^{2K} \sinh^2 h + e^{-2K}}. \tag{6.13}$$

We shall leave a discussion of the singularities of the resulting free energy (at zero temperature) to the next section, and instead look at the averages and correlations in the limit of $h = 0$.

To calculate the average of the spin at site i, we need to evaluate

$$
\begin{aligned}
\langle \sigma_i \rangle &= \frac{1}{Z} \sum_{\sigma_1 = \pm 1} \sum_{\sigma_2 = \pm 1} \cdots \sum_{\sigma_N = \pm 1} \sigma_i \prod_{j=1}^{N} \exp(K\sigma_j \sigma_{j+1}) \\
&\equiv \frac{1}{Z} \mathrm{tr} \left[\langle \sigma_1 | T | \sigma_2 \rangle \cdots \langle \sigma_{i-1} | T | \sigma_i \rangle \sigma_i \langle \sigma_i | T | \sigma_{i+1} \rangle \cdots \langle \sigma_N | T | \sigma_1 \rangle \right] \\
&= \frac{1}{Z} \mathrm{tr} \left[T^{i-1} \hat{\sigma}_z T^{N-i+1} \right] \\
&= \frac{1}{Z} \mathrm{tr} \left[T^N \hat{\sigma}_z \right],
\end{aligned}
\tag{6.14}
$$

where have permuted the matrices inside the trace, and $\hat{\sigma}_z = \begin{pmatrix} 1 & 0 \\ 0 & -1 \end{pmatrix}$ is the usual 2×2 Pauli matrix. One way to evaluate the final expression in Eq. (6.14) is to rotate to a basis where the matrix T is diagonal. For $h = 0$, this is accomplished by the unitary matrix $U = \frac{1}{\sqrt{2}} \begin{pmatrix} 1 & 1 \\ 1 & -1 \end{pmatrix}$, resulting in

$$
\langle \sigma_i \rangle = \frac{1}{Z} \mathrm{tr} \left[\begin{pmatrix} \lambda_+^N & 0 \\ 0 & \lambda_-^N \end{pmatrix} \begin{pmatrix} 0 & 1 \\ 1 & 0 \end{pmatrix} \right] = \frac{1}{Z} \begin{pmatrix} 0 & \lambda_+^N \\ \lambda_-^N & 0 \end{pmatrix} = 0.
\tag{6.15}
$$

Note that under this transformation the Pauli matrix $\hat{\sigma}_z$ is rotated into $\hat{\sigma}_x = \begin{pmatrix} 0 & 1 \\ 1 & 0 \end{pmatrix}$.

The vanishing of the magnetization at zero field is of course expected by symmetry. A more interesting quantity is the two-spin correlation function

$$
\begin{aligned}
\langle \sigma_i \sigma_{i+r} \rangle &= \frac{1}{Z} \sum_{\sigma_1 = \pm 1} \sum_{\sigma_2 = \pm 1} \cdots \sum_{\sigma_N = \pm 1} \sigma_i \sigma_{i+r} \prod_{j=1}^{N} \exp(K\sigma_j \sigma_{j+1}) \\
&= \frac{1}{Z} \mathrm{tr} \left[T^{i-1} \hat{\sigma}_z T^r \hat{\sigma}_z T^{N-i-r+1} \right] = \frac{1}{Z} \mathrm{tr} \left[\hat{\sigma}_z T^r \hat{\sigma}_z T^{N-r} \right].
\end{aligned}
\tag{6.16}
$$

Once again rotating to the basis where T is diagonal simplifies the trace to

$$
\begin{aligned}
\langle \sigma_i \sigma_{i+r} \rangle &= \frac{1}{Z} \mathrm{tr} \left[\begin{pmatrix} 0 & 1 \\ 1 & 0 \end{pmatrix} \begin{pmatrix} \lambda_+^r & 0 \\ 0 & \lambda_-^r \end{pmatrix} \begin{pmatrix} 0 & 1 \\ 1 & 0 \end{pmatrix} \begin{pmatrix} \lambda_+^{N-r} & 0 \\ 0 & \lambda_-^{N-r} \end{pmatrix} \right] \\
&= \frac{1}{Z} \mathrm{tr} \begin{pmatrix} \lambda_+^{N-r} \lambda_-^r & 0 \\ 0 & \lambda_-^{N-r} \lambda_+^r \end{pmatrix} = \frac{\lambda_+^{N-r} \lambda_-^r + \lambda_-^{N-r} \lambda_+^r}{\lambda_+^N + \lambda_-^N}.
\end{aligned}
\tag{6.17}
$$

Note that because of the periodic boundary conditions, the above answer is invariant under $r \to (N - r)$. We are interested in the limit of $N \gg r$, for which

$$
\langle \sigma_i \sigma_{i+r} \rangle \approx \left(\frac{\lambda_-}{\lambda_+} \right)^r \equiv e^{-r/\xi},
\tag{6.18}
$$

with the correlation length

$$\xi = \left[\ln \left(\frac{\lambda_+}{\lambda_-} \right) \right]^{-1} = -\frac{1}{\ln \tanh K}. \tag{6.19}$$

The above transfer matrix approach can be generalized to any one dimensional chain with variables $\{s_i\}$ and nearest-neighbor interactions. The partition function can be written as

$$Z = \sum_{\{s_i\}} \exp \left[\sum_{i=1}^{N} B(s_i, s_{i+1}) \right] = \sum_{\{s_i\}} \prod_{i=1}^{N} e^{B(s_i, s_{i+1})}, \tag{6.20}$$

where we have defined a *transfer matrix* T with elements,

$$\langle s_i | T | s_j \rangle = e^{B(s_i, s_j)}. \tag{6.21}$$

In the case of *periodic boundary conditions*, we then obtain

$$Z = \mathrm{tr} \left[T^N \right] \approx \lambda_{\max}^N. \tag{6.22}$$

Note that for $N \to \infty$, the trace is dominated by the largest eigenvalue $\lambda_{\max}$. Quite generally the largest eigenvalue of the transfer matrix is related to the free energy, while the correlation lengths are obtained from ratios of eigenvalues. *Frobenius' theorem* states that for any finite matrix with finite positive elements, the largest eigenvalue is always non-degenerate. This implies that $\lambda_{\max}$ and Z are analytic functions of the parameters appearing in B, and that one dimensional models can exhibit singularities (and hence a phase transition) only at zero temperature (when some matrix elements become infinite).

While the above formulation is framed in the language of discrete variables $\{s_i\}$, the method can also be applied to continuous variables as illustrated by problems at the end of this chapter. As an example of the latter, let us consider three component *unit* spins $\vec{s}_i = \left(s_i^x, s_i^y, s_i^z \right)$, with the *Heisenberg model* Hamiltonian

$$-\beta \mathcal{H} = K \sum_{i=1}^{N} \vec{s}_i \cdot \vec{s}_{i+1}. \tag{6.23}$$

Summing over all spin configurations, the partition function can be written as

$$Z = \mathrm{tr}_{\vec{s}_i} e^{K \sum_{i=1}^{N} \vec{s}_i \cdot \vec{s}_{i+1}} = \mathrm{tr}_{\vec{s}_i} e^{K \vec{s}_1 \cdot \vec{s}_2} e^{K \vec{s}_2 \cdot \vec{s}_3} \cdots e^{K \vec{s}_N \cdot \vec{s}_1} = \mathrm{tr}\, T^N, \tag{6.24}$$

where $\langle \vec{s}_1 | T | \vec{s}_2 \rangle = e^{K \vec{s}_1 \cdot \vec{s}_2}$ is a transfer function. Quite generally we would like to bring T into the diagonal form $\sum_\alpha \lambda_\alpha | \alpha \rangle \langle \alpha |$ (in Dirac notation), such that

$$\langle \vec{s}_1 | T | \vec{s}_2 \rangle = \sum_\alpha \lambda_\alpha \langle \vec{s}_1 | \alpha \rangle \langle \alpha | \vec{s}_2 \rangle = \sum_\alpha \lambda_\alpha f_\alpha(\vec{s}_1) f_\alpha^*(\vec{s}_2). \tag{6.25}$$

From studies of plane waves in quantum mechanics you may recall that the exponential of a dot product can be decomposed in terms of the spherical harmonics $Y_{\ell m}$. In particular,

$$e^{K \vec{s}_1 \cdot \vec{s}_2} = \sum_{\ell=0}^{\infty} \sum_{m=-\ell}^{\ell} 4\pi i^\ell j_\ell(-iK) Y_{\ell m}^*(\vec{s}_1) Y_{\ell m}(\vec{s}_2) \tag{6.26}$$

is precisely in the form of Eq. (6.25), from which we can read off the eigenvalues $\lambda_{\ell m}(K) = 4\pi i^\ell j_\ell(-iK)$, which do not depend on m. The partition function is now given by

$$Z = \operatorname{tr} T^N = \sum_{\ell=0}^\infty \sum_{m=-\ell}^\ell \lambda_{\ell m}^N = \sum_{\ell=0}^\infty (2\ell+1)\lambda_\ell^N \approx \lambda_0^N, \tag{6.27}$$

with $\lambda_0 = 4\pi j_0(-iK) = 4\pi \sinh K/K$ as the largest eigenvalue. The second largest eigenvalue is three fold degenerate, and given by $\lambda_1 = 4\pi j_1(-iK) = 4\pi \left[\cosh K/K - \sinh K/K^2\right]$.

6.3 Position space RG in one dimension

An exact RG treatment can be carried out for the Ising model with nearest-neighbor interactions (Eq. 6.1) in one dimension. The basic idea is to find a transformation that reduces the number of degrees of freedom by a factor b, while preserving the partition function, i.e.

$$Z = \sum_{\{\sigma_i | i=1,\cdots,N\}} e^{-\beta \mathcal{H}[\sigma_i]} = \sum_{\{\sigma'_{i'} | i'=1,\cdots,N/b\}} e^{-\beta \mathcal{H}'[\sigma'_{i'}]}. \tag{6.28}$$

There are many mappings $\{\sigma_i\} \mapsto \{\sigma'_{i'}\}$ that satisfy this condition. The choice of the transformation is therefore guided by the simplicity of the resulting RG. With $b = 2$, for example, one possible choice is to group pairs of neighboring spins and define the renormalized spin as their average. This *majority rule*, $\sigma'_i = (\sigma_{2i-1} + \sigma_{2i})/2$, is in fact not very convenient as the new spin has three possible values $(0, \pm 1)$ while the original spins are two valued. We can remove the ambiguity by assigning one of the two spins, e.g. σ_{2i-1}, the role of tie-breaker whenever the sum is zero. In this case the transformation is simply $\sigma'_i = \sigma_{2i-1}$. Such an RG procedure effectively removes the even numbered spins, $s_i = \sigma_{2i}$ and is usually called a *decimation*.

Fig. 6.3 Renormalization treatment of a one-dimensional chain via decimation by a factor of $b = 2$.

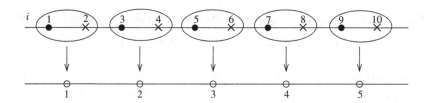

Note that since $\sigma' = \pm 1$ as in the original model, no renormalization factor ζ is necessary in this case. Since the interaction is over adjacent neighbors, the partition function can be written as

$$Z = \sum_{\{\sigma_i\}} \exp\left[\sum_{i=1}^N B(\sigma_i, \sigma_{i+1})\right] = \sum_{\{\sigma'_i\}} \sum_{\{s_i\}} \exp\left[\sum_{i=1}^{N/2} \left[B(\sigma'_i, s_i) + B(s_i, \sigma'_{i+1})\right]\right]. \tag{6.29}$$

Summing over the decimated spins, $\{s_i\}$, leads to

$$e^{-\beta\mathcal{H}'[\sigma_i']} \equiv \prod_{i=1}^{N/2}\left[\sum_{s_i=\pm 1} e^{[B(\sigma_i',s_i)+B(s_i,\sigma_{i+1}')]}\right] \equiv e^{\sum_{i=1}^{N/2} B'(\sigma_i',\sigma_{i+1}')}, \tag{6.30}$$

where following Eq. (6.2)

$$B(\sigma_1,\sigma_2) = g + \frac{h}{2}(\sigma_1+\sigma_2) + K\sigma_1\sigma_2, \tag{6.31}$$

and

$$B'(\sigma_1',\sigma_2') = g' + \frac{h'}{2}(\sigma_1'+\sigma_2') + K'\sigma_1'\sigma_2' \tag{6.32}$$

are the most general interaction forms for Ising spins.

Following Eq. (6.30), the renormalized interactions are obtained from

$$R(\sigma_1',\sigma_2') \equiv \exp\left[K'\sigma_1'\sigma_2' + \frac{h'}{2}(\sigma_1'+\sigma_2') + g'\right]$$

$$= \sum_{s_1=\pm 1} \exp\left[Ks_1(\sigma_1'+\sigma_2') + \frac{h}{2}(\sigma_1'+\sigma_2') + hs_1 + 2g\right]. \tag{6.33}$$

To solve for the renormalized interactions it is convenient to set

$$\begin{cases} x = e^K, \quad y = e^h, \quad z = e^g \\ x' = e^{K'}, \quad y' = e^{h'}, \quad z' = e^{g'}. \end{cases} \tag{6.34}$$

The four possible configurations of the bond are

$$\begin{cases} R(+,+) = x'y'z' = z^2y(x^2y+x^{-2}y^{-1}) \\ R(-,-) = x'y'^{-1}z' = z^2y^{-1}(x^{-2}y+x^2y^{-1}) \\ R(+,-) = x'^{-1}z' = z^2(y+y^{-1}) \\ R(-,+) = x'^{-1}z' = z^2(y+y^{-1}). \end{cases} \tag{6.35}$$

The last two equations are identical, resulting in three equations in three unknowns, with the solutions,

$$\begin{cases} z'^4 = z^8(x^2y+x^{-2}y^{-1})(x^{-2}y+x^2y^{-1})(y+y^{-1})^2 \\ y'^2 = y^2\dfrac{x^2y+x^{-2}y^{-1}}{x^{-2}y+x^2y^{-1}} \\ x'^4 = \dfrac{(x^2y+x^{-2}y^{-1})(x^{-2}y+x^2y^{-1})}{(y+y^{-1})^2}. \end{cases} \tag{6.36}$$

Taking the logarithms, we find recursion relations of the form

$$\begin{cases} g' = 2g + \delta g(K,h) \\ h' = h + \delta h(K,h) \\ K' = K'(K,h). \end{cases} \tag{6.37}$$

The parameter g is just an additive constant to the Hamiltonian. It does not affect the probabilities and hence does not appear in the recursion relations for K and h; $\delta g(K,h)$ is the contribution of the decimated spins to the overall free energy.

(1) **Fixed points:** The $h = 0$ subspace is closed by symmetry, and it can be checked that for $y = 1$ Eqs. (6.36) imply $y' = 1$, and

$$e^{4K'} = \left(\frac{e^{2K} + e^{-2K}}{2}\right)^2, \implies K' = \frac{1}{2}\ln\cosh 2K. \tag{6.38}$$

The recursion relation for K has the following fixed points:

(a) An infinite temperature fixed point at $K^* = 0$, which is the *sink* for the disordered phase. If K is small, $K' \approx \ln(1 + 4K^2/2)/2 \approx K^2$, is even smaller, indicating that this is a *stable* fixed point with zero correlation length.

(b) A zero temperature fixed point at $K^* \to \infty$, describing the ordered phase. For a large but finite K, the renormalized interaction $K' \approx \ln\left(e^{2K}/2\right)/2 \approx K - \ln 2/2$ is somewhat smaller. This fixed point is thus unstable with an infinite correlation length.

Fig. 6.4 Fixed points and RG flows for the coupling K in one dimension.

Clearly any finite interaction renormalizes to zero, indicating that the one-dimensional chain is always disordered at sufficiently long length scales. The absence of any other fixed point is apparent by noting that the recursion relation of Eq. (6.38) can alternatively be written as $\tanh K' = (\tanh K)^2$.

(2) **Flow diagrams** indicate that in the presence of a field h, all flows terminate on a line of fixed points with $K^* = 0$ and arbitrary h^*. These fixed points describe *independent* spins and all have zero correlation length. The flows originate from the fixed point at $h^* = 0$ and $K^* \to \infty$ which has two unstable directions in the (K, h) parameter space.

Fig. 6.5 Fixed points and RG flows in the space of coupling $(\tanh h, \tanh K)$ in one dimension.

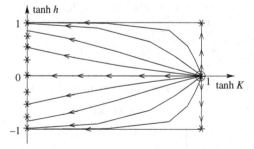

(3) **Linearizing** the recursion relations around this fixed point $(x \to \infty)$ yields

$$\begin{cases} x'^4 \approx x^4/4 \\ y'^2 \approx y^4 \end{cases} \implies \begin{cases} e^{-K'} = \sqrt{2}e^{-K} \\ h' = 2h. \end{cases} \tag{6.39}$$

We can thus regard e^{-K} and h as scaling fields. Since $\xi' = \xi/2$, the correlation length in the vicinity of the fixed point satisfies the homogeneous form ($b = 2$)

$$\xi\left(e^{-K}, h\right) = 2\xi\left(\sqrt{2}e^{-K}, 2h\right)$$
$$= 2^\ell \xi\left(2^{\ell/2}e^{-K}, 2^\ell h\right). \tag{6.40}$$

The second equation is obtained by repeating the RG procedure ℓ times. Choosing ℓ such that $2^{\ell/2}e^{-K} \approx 1$, we obtain the scaling form

$$\xi\left(e^{-K}, h\right) = e^{2K} g_\xi\left(he^{2K}\right). \tag{6.41}$$

The correlation length diverges on approaching $T = 0$ for $h = 0$. However, its divergence is not a power law of temperature. There is thus an ambiguity in identifying the exponent ν related to the choice of the measure of vicinity to $T = 0$ ($1/K$ or e^{-K}).

The hyperscaling assumption states that the singular part of the free energy in d dimensions is proportional to ξ^{-d}. Hence we expect

$$f_{\text{sing}}(K, h) \propto \xi^{-1} = e^{-2K} g_f\left(he^{2K}\right). \tag{6.42}$$

At zero field, the magnetization is always zero, while the susceptibility behaves as

$$\chi(K) \sim \left.\frac{\partial^2 f}{\partial^2 h}\right|_{h=0} \sim e^{2K}. \tag{6.43}$$

On approaching zero temperature, the divergence of the susceptibility is proportional to that of the correlation length. Using the general forms, $\langle s_i, s_{i+x}\rangle \sim e^{-x/\xi}/x^{d-2+\eta}$, and $\chi \sim \int d^d\mathbf{x}\, \langle s_0 s_\mathbf{x}\rangle_c \sim \xi^{2-\eta}$, we conclude that $\eta = 1$.

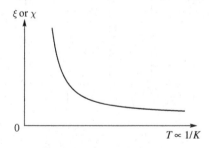

ξ or χ

0

$T \propto 1/K$

Fig. 6.6 The susceptibility (and correlation length) diverge similarly at zero temperature in the one-dimensional Ising model.

The transfer matrix method also provides an alternative RG scheme for all general one-dimensional chains with nearest-neighbor interactions. For decimation by a factor b, we can use $Z = \text{tr}\left[\left(T^b\right)^{N/b}\right]$ to construct the rescaled bond energy from

$$e^{B(s_i', s_j')} \equiv \langle s_i'|T'|s_j'\rangle = \langle s_i'|T^b|s_j'\rangle. \tag{6.44}$$

6.4 The Niemeijer–van Leeuwen cumulant approximation

Unfortunately, the decimation procedure cannot be performed exactly in higher dimensions. For example, the square lattice can be divided into two sublattices. For an RG with $b = \sqrt{2}$, we can start by decimating the spins on one sublattice. The interactions between the four spins surrounding each decimated spin are obtained by generalizing Eq. (6.33). If initially $h = g = 0$, we obtain

$$R(\sigma_1', \sigma_2', \sigma_3', \sigma_4') = \sum_{s=\pm1} e^{Ks(\sigma_1'+\sigma_2'+\sigma_3'+\sigma_4')} = 2\cosh\left[K(\sigma_1'+\sigma_2'+\sigma_3'+\sigma_4')\right]. \quad (6.45)$$

Clearly the four spins appear symmetrically in the above expression, and hence are subject to the same two-body interaction. This implies that new interactions along the diagonals of the renormalized lattice are also generated, and the nearest-neighbor form of the original Hamiltonian is not preserved. There is also a four-point interaction, and

$$R = \exp\left[g' + K'(\sigma_1'\sigma_2' + \sigma_2'\sigma_3' + \sigma_3'\sigma_4' + \sigma_4'\sigma_1' + \sigma_1'\sigma_3' + \sigma_2'\sigma_4') + K_4'\sigma_1'\sigma_2'\sigma_3'\sigma_4'\right]. \quad (6.46)$$

The number (and range) of new interactions increases with each RG step, and some truncating approximation is necessary. Two such schemes are described in the following sections.

Fig. 6.7 Removal of a spin in two (and higher) dimensions results in more than nearest-neighbor interactions.

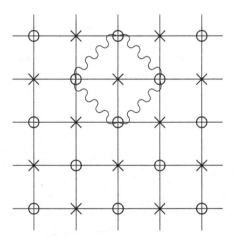

One of the earliest approaches was developed by Niemeijer and van Leeuwen (NvL) for treating the Ising model on a *triangular lattice*, subject to the usual nearest-neighbor Hamiltonian $-\beta\mathcal{H} = K\sum_{\langle ij\rangle}\sigma_i\sigma_j$. The original lattice sites are grouped into *cells* of three spins (e.g. in alternating up pointing triangles). Labeling the three spins in cell α as $\{\sigma_\alpha^1, \sigma_\alpha^2, \sigma_\alpha^3\}$, we can use a *majority rule* to define the renormalized cell spin as

$$\sigma_\alpha' = \text{sign}\left[\sigma_\alpha^1 + \sigma_\alpha^2 + \sigma_\alpha^3\right]. \quad (6.47)$$

(There is no ambiguity in the rule for any odd number of sites, and the renormalized spin is two-valued.) The renormalized interactions corresponding to the above map are obtained from the constrained sum

$$e^{-\beta\mathcal{H}'[\sigma'_\alpha]} = \sum_{\{\sigma^i_\alpha \mapsto \sigma'_\alpha\}}' e^{-\beta\mathcal{H}[\sigma^i_\alpha]}. \tag{6.48}$$

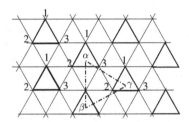

Fig. 6.8 Each cell spin is assigned from the majority of its three site spins.

To truncate the number of interactions in the renormalized Hamiltonian, NvL introduced a perturbative scheme by setting $\beta\mathcal{H} = \beta\mathcal{H}_0 + \mathcal{U}$. The unperturbed Hamiltonian

$$-\beta\mathcal{H}_0 = K \sum_\alpha \left(\sigma^1_\alpha \sigma^2_\alpha + \sigma^2_\alpha \sigma^3_\alpha + \sigma^3_\alpha \sigma^1_\alpha\right) \tag{6.49}$$

involves only *intracell interactions*. Since the cells are decoupled, this part of the Hamiltonian can be treated exactly. The remaining *intercell interactions* are treated as a perturbation

$$-\mathcal{U} = K \sum_{\langle\alpha,\beta\rangle} \left(\sigma^{(1)}_\beta \sigma^{(2)}_\alpha + \sigma^{(1)}_\beta \sigma^{(3)}_\alpha\right). \tag{6.50}$$

The sum is over all neighboring cells, each connected by two bonds. (The actual spins involved depend on the relative orientations of the cells.) Equation (6.48) is now evaluated perturbatively as

$$e^{-\beta\mathcal{H}'[\sigma'_\alpha]} = \sum_{\{\sigma^i_\alpha \mapsto \sigma'_\alpha\}}' e^{-\beta\mathcal{H}_0[\sigma^i_\alpha]} \left[1 - \mathcal{U} + \frac{\mathcal{U}^2}{2} - \cdots\right]. \tag{6.51}$$

The renormalized Hamiltonian is given by the cumulant series

$$\beta\mathcal{H}'[\sigma'_\alpha] = -\ln Z_0[\sigma'_\alpha] + \langle\mathcal{U}\rangle_0 - \frac{1}{2}\left(\langle\mathcal{U}^2\rangle_0 - \langle\mathcal{U}\rangle_0^2\right) + \mathcal{O}(\mathcal{U}^3), \tag{6.52}$$

where $\langle\ \rangle_0$ refers to the expectation values with respect to $\beta\mathcal{H}_0$, with the restriction of fixed $[\sigma'_\alpha]$, and Z_0 is the corresponding partition function.

To proceed, we construct a table of all possible configurations of spins within a cell, their renormalized value, and contribution to the cell energy:

σ'_α	σ^1_α	σ^2_α	σ^3_α	$\exp[-\beta\mathcal{H}_0]$
+	+	+	+	e^{3K}
+	−	+	+	e^{-K}
+	+	−	+	e^{-K}
+	+	+		e^{-K}
−	−	−	−	e^{3K}
−	+	−	−	e^{-K}
−	−	+	−	e^{-K}
−	−	−	+	e^{-K}

The restricted partition function is the product of contributions from the independent cells,

$$Z_0[\sigma'_\alpha] = \prod_\alpha \left[\sum_{\{\sigma^i_\alpha \mapsto \sigma'_\alpha\}}{}' e^{K(\sigma^1_\alpha \sigma^2_\alpha + \sigma^2_\alpha \sigma^3_\alpha + \sigma^3_\alpha \sigma^1_\alpha)} \right] = \left(e^{3K} + 3e^{-K}\right)^{N/3}. \tag{6.53}$$

It is *independent* of $[\sigma'_\alpha]$, thus contributing an additive constant to the Hamiltonian. The first cumulant of the interaction is

$$-\langle \mathcal{U} \rangle_0 = K \sum_{\langle \alpha,\beta \rangle} \left[\langle \sigma^1_\beta \rangle_0 \langle \sigma^2_\alpha \rangle_0 + \langle \sigma^1_\beta \rangle_0 \langle \sigma^3_\alpha \rangle_0 \right] = 2K \sum_{\langle \alpha,\beta \rangle} \langle \sigma^i_\alpha \rangle_0 \langle \sigma^j_\beta \rangle_0, \tag{6.54}$$

where we have taken advantage of the equivalence of the three spins in each cell. Using the table, we can evaluate the restricted average of site spins as

$$\langle \sigma^i_\alpha \rangle_0 = \begin{cases} \dfrac{+e^{3K} - e^{-K} + 2e^{-K}}{e^{3K} + 3e^{-K}} & \text{for } \sigma'_\alpha = +1 \\[2mm] \dfrac{-e^{3K} + e^{-K} - 2e^{-K}}{e^{3K} + 3e^{-K}} & \text{for } \sigma'_\alpha = -1 \end{cases} \equiv \frac{e^{3K} + e^{-K}}{e^{3K} + 3e^{-K}} \, \sigma'_\alpha. \tag{6.55}$$

Substituting in Eq. (6.54) leads to

$$-\beta\mathcal{H}'[\sigma'_\alpha] = \frac{N}{3} \ln\left(e^{3K} + 3e^{-K}\right) + 2K \left(\frac{e^{3K} + e^{-K}}{e^{3K} + 3e^{-K}} \right)^2 \sum_{\langle \alpha\beta \rangle} \sigma'_\alpha \sigma'_\beta + \mathcal{O}(\mathcal{U}^2). \tag{6.56}$$

At this order, the renormalized Hamiltonian involves only nearest-neighbor interactions, with the recursion relation

$$K' = 2K \left(\frac{e^{3K} + e^{-K}}{e^{3K} + 3e^{-K}} \right)^2. \tag{6.57}$$

(1) Equation (6.57) has the following *fixed points:*

(a) The high-temperature sink at $K^* = 0$. If $K \ll 1$, $K' \approx 2K(2/4)^2 = K/2 < K$, i.e. this fixed point is *stable*, and has zero correlation length.

(b) The low-temperature sink at $K^* = \infty$. If $K \gg 1$, then $K' \approx 2K > K$, i.e. unlike the one-dimensional case, this fixed point is also stable with zero correlation length.

(c) Since both of the above fixed points are stable, there must be at least one unstable fixed point at finite $K' = K = K^*$. From Eq. (6.57), the fixed point position satisfies

$$\frac{1}{\sqrt{2}} = \frac{e^{3K^*} + e^{-K^*}}{e^{3K^*} + 3e^{-K^*}}, \quad \Longrightarrow \quad \sqrt{2}e^{4K^*} + \sqrt{2} = e^{4K^*} + 3. \tag{6.58}$$

The fixed point value

$$K^* = \frac{1}{4} \ln\left(\frac{3 - \sqrt{2}}{\sqrt{2} - 1}\right) \approx 0.3356 \tag{6.59}$$

can be compared to the exactly known value of 0.2747 for the triangular lattice.

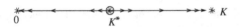

Fig. 6.9 Fixed points and RG flows for the coupling K in two dimensions.

(2) Linearizing the recursion relation around the non-trivial fixed point yields,

$$\left.\frac{\partial K'}{\partial K}\right|_{K^*} = 2\left(\frac{e^{4K^*} + 1}{e^{4K^*} + 3}\right)^2 + 32K^* e^{4K^*} \frac{(e^{4K^*} + 1)}{(e^{4K^*} + 3)^3} \approx 1.624. \tag{6.60}$$

The fixed point is indeed unstable as required by the continuity of flows. This RG scheme removes 1/3 of the degrees of freedom, and corresponds to $b = \sqrt{3}$. The thermal eigenvalue is thus obtained as

$$b^{y_t} = \left.\frac{\partial K'}{\partial K}\right|_{K^*} \quad \Longrightarrow \quad y_t \approx \frac{\ln(1.624)}{\ln(\sqrt{3})} \approx 0.883. \tag{6.61}$$

This can be compared to the exactly known value of $y_t = 1$, for the two-dimensional Ising model. It is certainly better than the mean-field (Gaussian) estimate of $y_t = 2$. From this eigenvalue we can estimate the exponents

$$\nu = 1/y_t \approx 1.13 \quad (1), \quad \text{and} \quad \alpha = 2 - 2/y_t = -0.26 \quad (0),$$

where the exact values are given in the brackets.

(3) To complete the calculation of exponents, we need the *magnetic eigenvalue* y_h, obtained after adding a magnetic field to the Hamiltonian, i.e. from

$$\beta\mathcal{H} = \beta\mathcal{H}_0 + \mathcal{U} - h\sum_i \sigma_\alpha^i. \tag{6.62}$$

Since the fixed point occurs for $h^* = 0$, the added term can also be treated perturbatively, and to the lowest order

$$\beta \mathcal{H}' = \beta \mathcal{H}_0 + \langle \mathcal{U} \rangle_0 - h \sum_\alpha \langle (\sigma_\alpha^1 + \sigma_\alpha^2 + \sigma_\alpha^3) \rangle_0 \,, \tag{6.63}$$

where the spins are grouped according to their cells. Using Eq. (6.55),

$$\beta \mathcal{H}' = \ln Z_0 + K' \sum_{\langle \alpha, \beta \rangle} \sigma_\alpha' \sigma_\beta' - 3h \sum_\alpha \left(\frac{e^{3K} + e^{-K}}{e^{3K} + 3e^{-K}} \right) \sigma_\alpha' \,, \tag{6.64}$$

thus identifying the renormalized magnetic field as

$$h' = 3h \left(\frac{e^{3K} + e^{-K}}{e^{3K} + 3e^{-K}} \right) . \tag{6.65}$$

In the vicinity of the unstable fixed point

$$b^{y_h} = \left. \frac{\partial h'}{\partial h} \right|_{K^*} = 3 \frac{e^{4K^*} + 1}{e^{4K^*} + 3} = \frac{3}{\sqrt{2}} \,, \tag{6.66}$$

and

$$y_h = \frac{\ln \left(3/\sqrt{2} \right)}{\ln \left(\sqrt{3} \right)} \approx 1.37. \tag{6.67}$$

This is lower than the exact value of $y_h = 1.875$. (The Gaussian value of $y_h = 2$ is closer to the correct result in this case.)

(4) NvL carried out the approach to the second order in $\mathcal{U}$. At this order two additional interactions over further neighbor spins are generated. The recursion relations in this three parameter space have a non-trivial fixed point with one unstable direction. The resulting eigenvalue of $y_t = 1.053$ is tantalizingly close to the exact value of 1, but this agreement is likely accidental.

6.5 The Migdal–Kadanoff bond moving approximation

Consider a $b = 2$ RG for the Ising model on a square lattice, in which every other spin along each lattice direction is decimated. As noted earlier, such decimation generates new interactions between the remaining spins. One way of overcoming this difficulty is to simply remove the bonds not connected to the retained spins. The renormalized spins are then connected to their nearest neighbors by two successive bonds. Clearly after decimation, the renormalized bond is given by the recursion relation in Eq. (6.38), characteristic of a one dimensional chain. The approximation of simply removing the unwanted bonds weakens the system to the extent that it behaves one dimensionally. This is remedied by using the unwanted bonds to strengthen those that are left behind. The spins that are retained are now connected by a pair of double bonds (of strength $2K$), and the decimation leads to

$$K' = \frac{1}{2} \ln \cosh(2 \times 2K) . \tag{6.68}$$

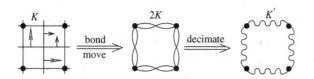

Fig. 6.10 The bond moving scheme for the Migdal–Kadanoff RG for $b = 2$ in dimension $d = 2$.

(1) Fixed points of this recursion relation are located at

 (a) $K^* = 0$: For $K \ll 1$, $K' \approx \ln(1 + 8K^2)/2 \approx 4K^2 \ll K$, i.e. this fixed point is stable.

 (b) $K^* \to \infty$: For $K \gg 1$, $K' \approx \ln(e^{4K}/2)/2 \approx 2K \gg K$, indicating that the low-temperature sink is also stable.

 (c) The domains of attractions of the above sinks are separated by a third fixed point at

$$e^{2K^*} = \frac{e^{4K^*} + e^{-4K^*}}{2} \implies K^* \approx 0.305, \tag{6.69}$$

 which can be compared with the exact value of $K_c \approx 0.441$.

(2) Linearizing Eq. (6.68) near the fixed point gives

$$b^{y_t} = \left. \frac{\partial K'}{\partial K} \right|_{K^*} = 2 \tanh 4K^* \approx 1.6786 \implies y_t \approx 0.747, \tag{6.70}$$

 compared to the exact value of $y_t = 1$.

The bond moving procedure can be extended to *higher dimensions*. For a hypercubic lattice in d-dimensions, the bond moving step strengthens each bond by a factor of 2^{d-1}. After decimation, the recursion relation is

$$K' = \frac{1}{2} \ln \cosh \left[2 \times 2^{d-1} K \right]. \tag{6.71}$$

The high and low temperature sinks at $K^* = 0$ and $K^* \to \infty$ are stable since

$$K \ll 1 \implies K' \approx \frac{1}{2} \ln(1 + 2^{2d-1} K^2) \approx 2^{2(d-1)} K^2 \ll K, \tag{6.72}$$

and

$$K \gg 1 \implies K' \approx \frac{1}{2} \ln \frac{e^{2^d k}}{2} \approx 2^{d-1} K \gg K. \tag{6.73}$$

(Note that the above result correctly identifies the lower critical dimension of the Ising model, in that the low-temperature sink is stable only for $d > 1$.) The intervening fixed point has an eigenvalue

$$2^{y_t} = \left. \frac{\partial K'}{\partial K} \right|_{K^*} = 2^{d-1} \tanh \left(2^d K^* \right). \tag{6.74}$$

The resulting values of $K^* \approx 0.065$ and $y_t \approx 0.934$ for $d = 3$ can be compared with the known values of $K_c \approx 0.222$ and $y_t \approx 1.59$ on a cubic lattice. Clearly the approximation gets worse at higher dimensions. (It fails to identify an upper critical dimension, and as $d \to \infty$, $K^* \to 2^{2(1-d)}$ and $y_t \to 1$.)

Fig. 6.11 The bond
moving scheme for the
Migdal–Kadanoff RG for
$b = 2$ in dimension $d = 3$.

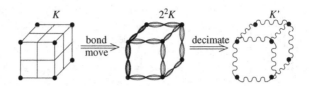

The Migdal–Kadanoff scheme can also be applied to more general spin systems. For a one-dimensional model described by the set of interactions $\{K\}$, the transfer matrix method in Eq. (6.44) allows construction of recursion relations from

$$T_b'(\{K'\}) = T(\{K\})^b.$$

For a d-dimensional lattice, the bond moving step strengthens each bond by a factor of b^{d-1}, and the generalized Migdal–Kadanoff recursion relations are

$$T_b'(\{K'\}) = T(\{b^{d-1}K\})^b. \tag{6.75}$$

The above equations can be used as a quick way of estimating phase diagrams and exponents. The procedure is exact in $d = 1$, and does progressively worse in higher dimensions. It thus complements mean-field (saddle point) approaches that are more reliable in higher dimensions. Unfortunately, it is not possible to develop a systematic scheme to improve upon its results. The RG procedure also allows evaluation of free energies, heat capacities, and other thermodynamic functions. One possible worry is that the approximations used to construct RG schemes may result in unphysical behavior, e.g. negative values of response functions C and χ. In fact most of these recursion relations (e.g. Eq. 6.75) are exact on *hierarchical* (Berker) lattices. The realizability of such lattices ensures that there are no unphysical consequences of the recursion relations.

6.6 Monte Carlo simulations

Another advantage of discrete spins on a lattice is that they are easily amenable to numerical simulations. The rapid advances of computer processing power, and the advent of clever schemes have made numerical methods quite popular and attractive, to the extent that the study of simulations may be regarded as a field onto itself. As such, the description in this section is intended only as a brief introduction to the topic, and the interested reader should consult the many specialized books devoted to this subject.

Let us consider a system with many degrees of freedom, e.g. a set of $N \gg 1$ spins denoted by $\{\underline{s}\}$, distributed according to the Boltzmann weights

$$P(\underline{s}) = \frac{1}{Z} \exp[-\beta \mathcal{H}(\underline{s})]. \tag{6.76}$$

We are interested in calculating the expectation value of some quantity $\mathcal{O}$, which is in principle obtained as

$$\langle \mathcal{O} \rangle = \text{tr}_{\underline{s}} \left[\mathcal{O}(\underline{s}) P(\underline{s}) \right], \tag{6.77}$$

where $\text{tr}_{\underline{s}}$ indicates summing over all possible values of $\{\underline{s}\}$. Unfortunately, even for a discrete spin the number of terms in the sum is prohibitively large for interesting values of N, growing as q^N if each spin can be in one of q states. Monte Carlo procedures aim at evaluating the sum approximately by summing over a representative sample of configurations $\{\underline{s}_\alpha\}$, as

$$\langle \mathcal{O} \rangle \approx \overline{\mathcal{O}} = \frac{1}{M} \sum_{\alpha=1}^{M} \mathcal{O}(\underline{s}_\alpha). \tag{6.78}$$

The number of samples M is typically much smaller than the total number of states. The difficult part is to ensure that these M samples are chosen properly.

The question is now how to generate configurations that are weighted according to the distribution in Eq. (6.76). This is achieved by stochastically changing one configuration to another. Let us denote the microstates of the system by $\alpha = 1, 2, \cdots, q^N$, and indicate by $\Pi_{\alpha\beta}$ the probability that the state α is changed to state β in the next *time step*. Note that while it is helpful to maintain the mental image of the system evolving in time, the stochastic dynamic rules are an artificial set of steps for generating the desired configurations. Since $\Pi_{\alpha\beta}$ are transition probabilities, they are constrained by

$$\Pi_{\alpha\alpha} = 1 - \sum_{\beta \neq \alpha} \Pi_{\alpha\beta}. \tag{6.79}$$

The above transition rules are an example of a *Markov process*, in which the state of a system depends on the preceding step(s). Let us assume that initially (at time $t = 0$) the probability to select a state α is $P_\alpha(0)$, e.g. $P_\alpha(0) = 1/q^N$ if all states are equally likely. These probabilities change at the next step of the Markov chain ($t = 1$), and are now given by

$$P_\alpha(1) = \sum_{\beta=1}^{q^N} P_\beta(0) \Pi_{\beta\alpha}. \tag{6.80}$$

In terms of the vector of probabilities $\vec{P}$ and the transition matrix Π this is just

$$\vec{P}(1) = \vec{P}(0) \cdot \Pi. \tag{6.81}$$

Similarly, after t steps, we have

$$\vec{P}(t) = \vec{P}(0) \cdot \Pi^t. \tag{6.82}$$

After many steps of the stochastic dynamics, the probabilities converge to steady-state values $\vec{P}^*$ which form a left eigenvector of the matrix Π,[1] since

$$\vec{P}(t) = \vec{P}(0) \cdot \Pi^t \approx \vec{P}^* \quad \Leftrightarrow \quad \vec{P}^* = \vec{P}^* \cdot \Pi. \tag{6.83}$$

Having established that irrespective of the initial choice, the Markov probabilities converge to a unique set, how can we choose the transition rates to insure that the final probabilities are the Boltzmann weights in Eq. (6.76)? The change in the probability of state α in one step is given by

$$P_\alpha(t+1) - P_\alpha(t) = \sum_{\beta \neq \alpha} \left[P_\beta(t)\Pi_{\beta\alpha} - P_\alpha(t)\Pi_{\alpha\beta} \right]. \tag{6.84}$$

For the pair of states α and β, the first term in the square brackets is the increase in probability due to transitions from β to α, the second is the decrease due to transitions in the opposite direction. The probabilities are thus unchanged if these "probability currents" are equal for every pair, i.e.

$$P_\alpha \Pi_{\alpha\beta} = P_\beta \Pi_{\beta\alpha}. \tag{6.85}$$

This is known as the condition of *detailed balance*, which in the case of Boltzmann weights implies

$$\frac{\Pi_{\alpha\beta}}{\Pi_{\beta\alpha}} = \frac{P_\beta}{P_\alpha} = \exp\left[-\beta \left(E_\beta - E_\alpha \right) \right]. \tag{6.86}$$

Note that since $\sum_\beta \Pi_{\alpha\beta} = 1$, summing Eq. (6.85) over β gives

$$\left(\vec{P} \cdot \Pi \right)_\alpha = \sum_\beta P_\beta \Pi_{\beta\alpha} = P_\alpha \sum_\beta \Pi_{\alpha\beta} = P_\alpha, \tag{6.87}$$

as expected, the detailed balance condition ensures that $\vec{P}$ is the left eigenvector of the transition matrix, corresponding to its steady state.

A popular method for implementing the detailed balance condition is the *Metropolis algorithm:* Let us attempt a transition from state α to state β. This transition is always accepted if it lowers the energy, and with a probability $\exp\left[-\beta \left(E_\beta - E_\alpha \right) \right] < 1$ if it increases the energy, i.e.

$$\Pi_{\alpha\beta} = q_{\alpha\beta} \begin{cases} 1 & \text{for } E_\alpha > E_\beta \\ \exp\left[-\beta \left(E_\beta - E_\alpha \right) \right] < 1 & \text{for } E_\alpha < E_\beta. \end{cases} \tag{6.88}$$

---

[1] Note the similarities to the transfer matrix approach described earlier in this chapter. In both cases we are interested in a matrix with non-negative elements raised to a large power. For the transition probability matrix, Eq. (6.79) implies that the vector of unit values is a right eigenvector of the matrix with eigenvalue 1. According to Frobenius' theorem, this is the non-degenerate largest eigenvalue of the matrix. The corresponding unique left eigenvector with eigenvalue one is the above steady-state probability $\vec{P}^*$. (More rigorously, the Markov process is called irreducible and converges to a unique limit, if for some finite power n, all elements of the matrix Π^n are positive.)

In the limit of zero temperature only moves that reduce the energy are accepted and the system proceeds towards a local energy minimum. At finite temperature moves that increase the energy are sometimes accepted. The factor of $q_{\alpha\beta}$ is the frequency with which the move from α to β is attempted. Since the energy change has the opposite sign for the transition in the reverse direction (from β to α) the acceptance criteria are reversed. If $q_{\alpha\beta} = q_{\beta\alpha}$, it is easy to see that Eq. (6.86) is satisfied irrespective of the energy change.

As a concrete example, let us consider simulating a system of N Ising spins, in which at each step we attempt to flip $(\sigma_i \rightarrow -\sigma_i)$ a randomly chosen spin. (In this case, the attempt frequency is $q_{\alpha\beta} = 1/N$ for states related by a single spin flip.) This change is accepted if the resulting energy change ΔE is negative, and otherwise accepted with probability $\exp(-\beta\Delta E)$. For Ising models with short-range interactions calculating ΔE is very fast, since we only need to consider the bonds emanating from the flipped spin. We can start from a randomly generated initial state, and apply the stochastic (Monte Carlo) rule many times. The average of the quantity $\mathcal{O}$ is then estimated as

$$\langle \mathcal{O} \rangle \approx \overline{\mathcal{O}} = \frac{1}{T} \sum_{t=\tau}^{T+\tau} \mathcal{O}[\sigma(t)]. \tag{6.89}$$

The first few steps are strongly influenced by the initial condition and must be discarded. The *equilibration time* τ is not known *a priori*, and will be discussed further in chapter 9. Typically it grows as some power of the correlation length, and thus as a polynomial in N at the critical point. The relaxation time τ also determines the accuracy of the estimate in Eq. (6.89): States separated by a few time steps are highly correlated, and thus the number of independent configurations appearing in the above sum is of the order of T/τ. The relaxation time also depends on the set of attempted moves for transitions, and there are schemes for accelerating equilibration by clever choices of moves. There is also an interesting way to employ Monte Carlo simulations for a position space renormalization group calculation which will not be discussed here.

Problems for chapter 6

1. *Cumulant method:* Apply the Niemeijer–van Leeuwen first order cumulant expansion to the Ising model on a *square* lattice with $-\beta\mathcal{H} = K \sum_{\langle ij \rangle} \sigma_i \sigma_j$, by following these steps:

   (a) For an RG with $b = 2$, divide the bonds into *intracell* components $\beta\mathcal{H}_0$, and *intercell* components $\mathcal{U}$.

   (b) For each cell α, define a renormalized spin $\sigma'_\alpha = \text{sign}(\sigma_\alpha^1 + \sigma_\alpha^2 + \sigma_\alpha^3 + \sigma_\alpha^4)$. This choice becomes ambiguous for configurations such that $\sum_{i=1}^{4} \sigma_\alpha^i = 0$. Distribute the weight of these configurations equally between $\sigma'_\alpha = +1$ and -1 (i.e. put a factor of 1/2 in addition to the Boltzmann weight). Make a table for all possible configurations of a cell, the internal probability $\exp(-\beta\mathcal{H}_0)$, and the weights contributing to $\sigma'_\alpha = \pm 1$.

(c) Express $\langle \mathcal{U} \rangle_0$ in terms of the cell spins σ'_α, and hence obtain the recursion relation $K'(K)$.

(d) Find the fixed point K^*, and the thermal eigenvalue y_t.

(e) In the presence of a small magnetic field $h \sum_i \sigma_i$, find the recursion relation for h, and calculate the magnetic eigenvalue y_h at the fixed point.

(f) Compare K^*, y_t, and y_h to their exact values.

2. *Migdal–Kadanoff method:* Consider Potts spins $s_i = (1, 2, \cdots, q)$, on sites i of a hypercubic lattice, interacting with their nearest neighbors via a Hamiltonian

$$-\beta \mathcal{H} = K \sum_{\langle ij \rangle} \delta_{s_i, s_j} \, .$$

(a) In $d = 1$ find the exact recursion relations by a $b = 2$ renormalization/decimation process. Indentify all fixed points and note their stability.

(b) Write down the recursion relation $K'(K)$ in d-dimensions for $b = 2$, using the Migdal–Kadanoff bond moving scheme.

(c) By considering the stability of the fixed points at zero and infinite coupling, prove the existence of a non-trivial fixed point at finite K^* for $d > 1$.

(d) For $d = 2$, obtain K^* and y_t, for $q = 3$, 1, and 0.

3. *The Potts model:* The *transfer matrix* procedure can be extended to the Potts models, where the spin s_i on each site takes q values $s_i = (1, 2, \cdots, q)$; and the Hamiltonian is $-\beta \mathcal{H} = K \sum_{i=1}^{N} \delta_{s_i, s_{i+1}} + K \delta_{s_N, s_1}$.

(a) Write down the transfer matrix and diagonalize it. Note that you do not have to solve a q_{th} order secular equation as it is easy to guess the eigenvectors from the symmetry of the matrix.

(b) Calculate the free energy per site.

(c) Give the expression for the correlation length ξ (you don't need to provide a detailed derivation), and discuss its behavior as $T = 1/K \to 0$.

---

4. *The spin-1 model:* Consider a linear chain where the spin s_i at each site takes on three values $s_i = -1$, 0, $+1$. The spins interact via a Hamiltonian

$$-\beta \mathcal{H} = \sum_i K s_i s_{i+1}.$$

(a) Write down the transfer matrix $\langle s|T|s' \rangle = e^{K s s'}$ explicitly.

(b) Use symmetry properties to find the largest eigenvalue of T and hence obtain the expression for the free energy per site ($\ln Z / N$).

(c) Obtain the expression for the correlation length ξ, and note its behavior as $K \to \infty$.

(d) If we try to perform a renormalization group by decimation on the above chain we find that additional interactions are generated. Write down the simplest generalization of $\beta \mathcal{H}$ whose parameter space is closed under such RG.

5. *Clock model:* Each site of the lattice is occupied by a q-valued spin $s_i \equiv 1, 2, \cdots, q$, with an underlying translational symmetry modulus q, i.e. the system is invariant under $s_i \to (s_i + n)_{\mathrm{mod}q}$. The most general Hamiltonian subject to this symmetry with nearest-neighbor interactions is

$$\beta \mathcal{H}_C = -\sum_{\langle i,j \rangle} J(|s_i - s_j|_{\mathrm{mod}q}),$$

where $J(n)$ is any function, e.g. $J(n) = J \cos(2\pi n/q)$. *Potts models* are a special case of clock models with full *permutation symmetry*, and the Ising model is obtained in the limit of $q = 2$.

(a) For a closed linear chain of N clock spins subject to the above Hamiltonian show that the partition function $Z = \mathrm{tr}\left[\exp(-\beta \mathcal{H})\right]$ can be written as

$$Z = \mathrm{tr}\left[\langle s_1|T|s_2 \rangle \langle s_2|T|s_3 \rangle \cdots \langle s_N|T|s_1 \rangle\right],$$

where $T \equiv \langle s_i|T|s_j \rangle = \exp\left[J(s_i - s_j)\right]$ is a $q \times q$ transfer matrix.

(b) Write down the transfer matrix explicitly and diagonalize it. Note that you do not have to solve a qth order secular equation; because of the translational symmetry, the eigenvalues are easily obtained by discrete Fourier transformation as

$$\lambda(k) = \sum_{n=1}^{q} \exp\left[J(n) + \frac{2\pi i n k}{q}\right].$$

(c) Show that $Z = \sum_{k=1}^{q} \lambda(k)^N \approx \lambda(0)^N$ for $N \to \infty$. Write down the expression for the free energy per site $\beta f = -\ln Z/N$.

(d) Show that the correlation function can be calculated from

$$\left\langle \delta_{s_i, s_{i+\ell}} \right\rangle = \frac{1}{Z} \sum_{\alpha=1}^{q} \mathrm{tr}\left[\Pi_\alpha T^\ell \Pi_\alpha T^{N-\ell}\right],$$

where Π_α is a projection matrix. Hence show that $\left\langle \delta_{s_i, s_{i+\ell}} \right\rangle_c \sim [\lambda(1)/\lambda(0)]^\ell$. (You do not have to explicitly calculate the constant of proportionality.)

6. *XY model:* Consider two component unit spins $\vec{s}_i = (\cos\theta_i, \sin\theta_i)$ in one dimension, with the nearest-neighbor interactions described by $-\beta \mathcal{H} = K \sum_{i=1}^{N} \vec{s}_i \cdot \vec{s}_{i+1}$.

(a) Write down the transfer matrix $\langle \theta|T|\theta' \rangle$, and show that it can be diagonalized with eigenvectors $f_m(\theta) \propto e^{im\theta}$ for integer m.

(b) Calulate the free energy per site, and comment on the behavior of the heat capacity as $T \propto K^{-1} \to 0$.

(c) Find the correlation length ξ, and note its behavior as $K \to \infty$.

7. *One-dimensional gas:* The transfer matrix method can also be applied to a one dimensional gas of particles with short-range interactions, as described in this problem.

(a) Show that for a potential with a hard core that screens the interactions from further neighbors, the Hamiltonian for N particles can be written as

$$\mathcal{H} = \sum_{i=1}^{N} \frac{p_i^2}{2m} + \sum_{i=2}^{N} \mathcal{V}(x_i - x_{i-1}).$$

The (indistinguishable) particles are labeled with coordinates $\{x_i\}$ such that

$$0 \le x_1 \le x_2 \le \cdots \le x_N \le L,$$

where L is the length of the box confining the particles.

(b) Write the expression for the partition function $Z(T, N, L)$. Change variables to $\delta_1 = x_1, \delta_2 = x_2 - x_1, \cdots, \delta_N = x_N - x_{N-1}$, and carefully indicate the allowed ranges of integration and the constraints.

(c) Consider the Gibbs partition function obtained from the Laplace transformation

$$\mathcal{Z}(T, N, P) = \int_0^\infty dL \exp(-\beta PL) Z(T, N, L),$$

and by extremizing the integrand find the standard formula for P in the canonical ensemble.

(d) Change variables from L to $\delta_{N+1} = L - \sum_{i=1}^{N} \delta_i$, and find the expression for $\mathcal{Z}(T, N, P)$ as a product over one-dimensional integrals over each δ_i.

(e) At a fixed pressure P, find expressions for the mean length $L(T, N, P)$, and the density $n = N/L(T, N, P)$ (involving ratios of integrals which should be easy to interpret).

Since the expression for $n(T, P)$ in part (e) is continuous and non-singular for any choice of potential, there is in fact no condensation transition for the one-dimensional gas. By contrast, the approximate van der Waals equation (or the mean-field treatment) incorrectly predicts such a transition.

(f) For a hard sphere gas, with minimum separation a between particles, calculate the equation of state $P(T, n)$.

8. *Potts chain (RG):* Consider a one-dimensional array of N Potts spins $s_i = 1, 2, \cdots, q$, subject to the Hamiltonian $-\beta \mathcal{H} = J \sum_i \delta_{s_i, s_{i+1}}$.

(a) Using the transfer matrix method (or otherwise) calculate the partition function Z, and the correlation length ξ.

(b) Is the system critical at zero temperature for antiferromagnetic couplings $J < 0$?

(c) Construct a renormalization group (RG) treatment by eliminating every other spin. Write down the recursion relations for the coupling J, and the additive constant g.

(d) Discuss the fixed points, and their stability.

(e) Write the expression for $\ln Z$ in terms of the additive constants of successive rescalings.

(f) Show that the recursion relations are simplified when written in terms of $t(J) \equiv e^{-1/\xi(J)}$.

(g) Use the result in (f) to express the series in (e) in terms of t. Show that the answer can be reduced to that obtained in part (a), upon using the result

$$\sum_{n=0}^{\infty} \frac{1}{2^{n+1}} \ln\left(\frac{1+t^{2^n}}{1-t^{2^n}}\right) = -\ln(1-t).$$

(h) Repeat the RG calculation of part (c), when a small symmetry breaking term $h \sum_i \delta_{s_i,1}$ is added to $-\beta\mathcal{H}$. You will find that an additional coupling term $K \sum_i \delta_{s_i,1}\delta_{s_{i+1},1}$ is generated under RG. Calculate the recursion relations in the three parameter space (J, K, h).

(i) Find the magnetic eigenvalues at the zero temperature fixed point where $J \to \infty$, and obtain the form of the correlation length close to zero temperature.

9. *Cluster RG:* Consider Ising spins on a *hexagonal lattice* with nearest-neighbor interactions J.

(a) Group the sites into clusters of four in preparation for a position space renormalization group with $b = 2$.

(b) How can the majority rule be modified to define the renormalized spin of each cluster?

(c) For a scheme in which the central site is chosen as the tie-breaker, make a table of all possible configurations of site-spins for a given value of the cluster-spin.

(d) Focus on a pair of neighboring clusters. Indicate the contributions of intracluster and intercluster bonds to the total energy.

(e) Show that in zero magnetic field, the Boltzmann weights of parallel and anti-parallel clusters are given by

$$R(+, +) = x^8 + 2x^6 + 7x^4 + 14x^2 + 17 + 14x^{-2} + 7x^{-4} + 2x^{-6},$$

and

$$R(+, -) = 9x^4 + 16x^2 + 13 + 16x^{-2} + 9x^{-4} + x^{-8},$$

where $x = e^J$.

(f) Find the expression for the resulting recursion relation $J'(J)$.

(g) Estimate the critical *ferromagnetic* coupling J_c, and the exponent ν obtained from this RG scheme, and compare with the exact values.

(h) What are the values of the magnetic and thermal exponents (y_h, y_t) at the zero temperature ferromagnetic fixed point?

(i) Is the above scheme also applicable for antiferromagnetic interactions? What symmetry of the original problem is not respected by this transformation?

10. *Transition probability matrix:* Consider a system of two Ising spins with a coupling K, which can thus be in one of four states.

(a) Explicitly write the 4×4 transition matrix corresponding to single spin flips for a Metropolis algorithm. Verify that the equilibrium weights are indeed a left eigenvector of this matrix.

(b) Repeat the above exercise if both single spin and double spin flips are allowed. The two types of moves are chosen randomly with probabilities p and $q = 1 - p$.

7
Series expansions

7.1 Low-temperature expansions

Lattice models can also be studied by series expansions. Such expansions start with certain exactly solvable limits, and typically represent perturbations around such limits by graphs on the lattice. High-temperature expansions are described in the next section. Here we describe the *low-temperature expansion* for the Ising model with a Hamiltonian $-\beta\mathcal{H} = K\sum_{\langle i,j\rangle}\sigma_i\sigma_j$, on a d-dimensional hypercubic lattice. The ground state with $K = \beta J > 0$ is ferromagnetic, e.g. with $\sigma_i = +1$ for all spins. A series expansion for the partition function is obtained by including low energy excitations around this state. The lowest energy excitation is a single overturned spin. Any of N sites can be chosen for this excitation, which has an energy cost of $2K \times 2d$, with respect to the ground state. The next lowest energy excitation corresponds to a dimer of negative spins with energy cost of $2K \times (4d - 2)$, and a multiplicity of $N \times d$ (there are d possible orientations for the dimer). The first few terms in the expansion give

$$Z = 2e^{NdK}\left[1 + Ne^{-4dK} + dNe^{-4(2d-1)K} + \frac{N(N-2d-1)}{2}e^{-8dK} + \cdots\right]. \quad (7.1)$$

The fourth term comes from flipping two disjoint single spins. The zeroth order term comes from the ground state, which has a two-fold degeneracy. The overall factor of 2 is insignificant in the $N \to \infty$ limit, and

$$Z \simeq e^{NdK}\sum_{\text{droplets of }\ominus\text{ spins}} e^{-2K \times \text{ boundary of droplet}}. \quad (7.2)$$

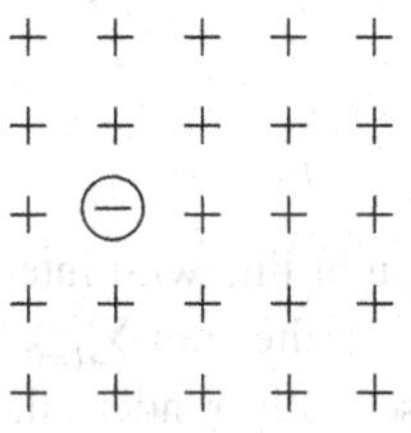

+	+	+	+	+
+	+	+	+	+
+	$\ominus$	+	+	+
+	+	+	+	+
+	+	+	+	+

excitations	energies	multiplicity
$\boxminus$	$4dK$	N
$\boxminus\!\boxminus$	$2K(4d-2)$	dN
$\boxminus$ $\boxminus$	$8dK$	$\dfrac{N(N-1-2d)}{2}$
⋮	⋮	⋮

Fig. 7.1 Low-energy excitations as islands of negative spins in the sea of positive spins. The energies and multiplicities on the right refer to a hypercubic lattice in d-dimensions.

The free energy per site is obtained from the series,

$$-\beta f = \frac{\ln Z}{N} = dK + \frac{1}{N} \ln\left[1 + Ne^{-4dK} + dNe^{-4(2d-1)K} + \frac{N(N-2d-1)}{2}e^{-8dK} + \cdots\right]$$

$$= dK + e^{-4dK} + de^{-4(2d-1)K} - \frac{(2d+1)}{2}e^{-8dK} + \cdots \qquad (7.3)$$

Note that there is an explicit cancellation of the term proportional to N^2 in the expansion for Z. Such higher N dependences result from *disconnected* diagrams with several disjoint droplets. Extensivity of the free energy insures that these terms are canceled by products of connected graphs. The energy per site is then obtained from

$$\frac{E}{N} = -\frac{\partial}{\partial\beta}\left(\frac{\ln Z}{N}\right) = -J\frac{\partial}{\partial K}\left(\frac{\ln Z}{N}\right)$$

$$= -J\left[d - 4de^{-4dK} - 4d(2d-1)e^{-4(2d-1)K} + 4d(2d+1)e^{-8dK} + \cdots\right], \qquad (7.4)$$

and the heat capacity is proportional to

$$\frac{C}{Nk_B} = \frac{1}{Nk_B}\frac{\partial E}{\partial T} = -\frac{K^2}{NJ}\frac{\partial E}{\partial K}$$

$$= K^2\left[16d^2e^{-4dK} + 16d(2d-1)^2e^{-4(2d-1)K} - 32d^2(2d+1)e^{-8dK} + \cdots\right]. \qquad (7.5)$$

Can such series be used to extract the critical coupling K_c, and more importantly, the singularities associated with the disordering transition? Suppose that we have identified a number of terms in a series $C = \sum_{\ell=0}^{\infty} a_\ell u^\ell$. From the expected divergence of the heat capacity, we expect an asymptotic expansion of the form

$$C \simeq A\left(1 - \frac{u}{u_c}\right)^{-\alpha} = A\left[1 + \frac{\alpha}{u_c}u + \frac{\alpha(\alpha+1)}{2!u_c^2}u^2 \cdots + \right.$$

$$\left. + \frac{\alpha(\alpha+1)\cdots(\alpha+\ell-1)}{\ell!u_c^\ell}u^\ell\cdots\right]. \qquad (7.6)$$

The above singular form is characterized by the three parameters A, u_c, and α. We can try to extract these parameters from the calculated coefficients in the series by requiring that they match at large ℓ, i.e. by fitting the ratio of successive terms to

$$\frac{a_\ell}{a_{\ell-1}} \simeq \left(\frac{\alpha+\ell-1}{\ell u_c}\right) = u_c^{-1}\left(1 + \frac{\alpha-1}{\ell}\right). \qquad (7.7)$$

Thus a plot of $a_\ell/a_{\ell-1}$ versus $1/\ell$ should be a straight line with intercept u_c^{-1}, and slope $u_c^{-1}(\alpha-1)$. Note, however, that adding a finite sum $\sum_{\ell=0}^{\ell_m} d_\ell u^\ell$ does not change the asymptotic singular form, but essentially renders the first ℓ_m

terms useless in determining α and u_c^{-1}. There is thus no a priori guarantee that such a fitting procedure will succeed with a finite number of coefficients. In practice, this procedure works very well and very good estimates of critical exponents in $d = 3$ (e.g. $\alpha = 0.105 \pm 0.007$) are obtained in this way by including a rather large number of terms.

The three terms calculated for the heat capacity in Eq. (7.5) have different signs, unlike those of Eq. (7.6). As this continues at higher orders in e^{-K}, the ratio fitting procedure described above cannot be used directly. The alternation of signs usually signifies singularities in the *complex* $z = e^{-K}$ plane, closer to the origin than the critical point of interest at real $z_c = e^{-K_c}$. If it is possible to construct a mapping $u(z)$ on the complex plane such that the spurious singularities are pushed further away than $u_c = u(z_c)$, then the ratio method can be used. In the case of low-temperature series, the choice of $u = \tanh K$ achieves this goal. (As we shall demonstrate shortly, $\tanh K$ is also the natural variable for the high-temperature expansion.) Quite sophisticated methods, such as *Padé approximants*, have been developed for analyzing the singular behavior of series.

Low temperature expansions can be similarly constructed for other *discrete* spin systems, such as the Potts model. For *continuous* spins, the low energy excitations are Goldstone modes, and the perturbation series cannot be represented in terms of lattice graphs. The low-temperature description in this case starts with the Gaussian treatment of the Goldstone modes, as carried out in the first problem of chapter 3 for the case of the XY model. Further terms in the series involve interactions between such modes, and a corresponding calculation will be performed in the next chapter.

7.2 High-temperature expansions

High-temperature expansions work equally well for discrete and continuous spin systems. The basic idea is to start with *independent* spins, and expand the partition function in powers of $\beta = (k_B T)^{-1}$, i.e.

$$Z = \mathrm{tr}\left(e^{-\beta \mathcal{H}}\right) = \mathrm{tr}\left[1 - \beta \mathcal{H} + \frac{\beta^2 \mathcal{H}^2}{2} - \cdots\right], \tag{7.8}$$

and

$$\frac{\ln Z}{N} = \frac{\ln Z_0}{N} - \beta \frac{\langle \mathcal{H} \rangle_0}{N} + \frac{\beta^2}{2} \frac{\langle \mathcal{H}^2 \rangle_0 - \langle \mathcal{H} \rangle_0^2}{N} - \cdots \tag{7.9}$$

The averages $\langle \ \rangle_0$ are calculated over non-interacting spins. For the Ising model, it is more convenient to organize the expansion in powers of $\tanh K$ as follows. Since $(\sigma_i \sigma_j)^2 = 1$, the Boltzmann factor for each bond can be written as

$$e^{K \sigma_i \sigma_j} = \frac{e^K + e^{-K}}{2} + \frac{e^K - e^{-K}}{2} \sigma_i \sigma_j = \cosh K \left(1 + t \sigma_i \sigma_j\right), \tag{7.10}$$

Fig. 7.2 A typical term obtained in the expansion of Eq. (7.11).

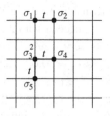

where $t \equiv \tanh K$ is a good high-temperature expansion parameter. Applying this transformation to each bond of the lattice results in

$$Z = \sum_{\{\sigma_i\}} e^{K \sum_{\langle i,j \rangle} \sigma_i \sigma_j} = (\cosh K)^{\text{number of bonds}} \sum_{\{\sigma_i\}} \prod_{\langle i,j \rangle} (1 + t\sigma_i\sigma_j). \quad (7.11)$$

For N_b bonds on the lattice, the above product generates 2^{N_b} terms, which can be represented diagrammatically by drawing a line connecting sites i and j for each factor of $t\sigma_i\sigma_j$. Note that there can at most be one such line for each lattice bond, which is either empty or occupied. This is a major simplification, and a major advantage for the use of t, rather than K, as the expansion parameter. Each site now obtains a factor of $\sigma_i^{p_i}$, where p_i is the number of occupied bonds emanating from i. Summing over the two possible values $\sigma_i = \pm 1$, gives a factor of 2 if p_i is even, and 0 if p_i is odd. Thus, the only graphs that survive the sum have an even number of lines passing through each site. The resulting graphs are collections of closed paths on the lattice, and

$$Z = 2^N \times (\cosh K)^{N_b} \sum_{\text{All closed graphs}} t^{\text{number of bonds in the graph}}. \quad (7.12)$$

Fig. 7.3 The first two terms in the expansion of Eq. (7.12) for a square lattice.

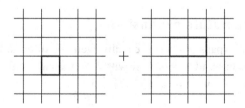

For a d-dimensional hypercubic lattice, the smallest closed graph is a square of four bonds which has $d(d-1)/2$ possible orientations. As the next graph has six bonds,

$$Z = 2^N (\cosh K)^{dN} \left[1 + \frac{d(d-1)N}{2} t^4 + d(d-1)(2d-3) t^6 + \cdots \right], \quad (7.13)$$

and

$$\frac{\ln Z}{N} = \ln 2 + d \ln \cosh K + \frac{d(d-1)}{2} t^4 + \cdots. \quad (7.14)$$

In the following sections we shall employ high-temperature expansions not as a numerical tool, but to establish the following: (a) exact solution of the Ising model in $d = 1$; (b) the duality relating models at low and high temperatures; (c) the validity of the Gaussian model in high dimensions; (d) exact solution of the Ising model in $d = 2$.

7.3 Exact solution of the one-dimensional Ising model

The graphical method provides a rapid way of solving the Ising model at zero field in $d = 1$. We can compare and contrast the solutions on chains with open and closed (periodic) boundary conditions.

(1) **An open chain** of N sites has $N_b = N - 1$ bonds. It is impossible to draw any closed graphs on such a lattice, and hence

$$Z = 2^N \cosh K^{N-1} \times 1 \quad \Longrightarrow \quad \frac{\ln Z}{N} = \ln[2 \cosh K] - \frac{\ln[\cosh K]}{N}. \tag{7.15}$$

The same method can also be used to calculate the correlation function $\langle \sigma_m \sigma_n \rangle$, since

$$\langle \sigma_m \sigma_n \rangle = \frac{1}{Z} \sum_{\{\sigma_i\}} e^{K \sum_{\langle i,j \rangle} \sigma_i \sigma_j} \sigma_m \sigma_n$$

$$= \frac{2^N (\cosh K)^{N-1}}{Z} \sum_{\{\sigma_i\}} \sigma_m \sigma_n \prod_{\langle i,j \rangle} (1 + t \sigma_i \sigma_j). \tag{7.16}$$

The terms in the numerator involve an additional factor of $\sigma_m \sigma_n$. To get a finite value after summing over $\sigma_m = \pm 1$ and $\sigma_n = \pm 1$, we have to examine graphs with an odd number of bonds emanating from these *external sites*. The only such graph for the open chain is one that directly connects these two sites, and results in

$$\langle \sigma_m \sigma_n \rangle = t^{|m-n|} = e^{-|m-n|/\xi}, \quad \text{with} \quad \xi = -\frac{1}{\ln \tanh K}. \tag{7.17}$$

These results are in agreement with the RG conclusions of Section 6.4; the correlation length diverges as e^{2K} for $K \to \infty$, and there is no power law modifying the exponential decay of correlations.

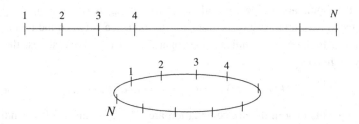

Fig. 7.4 The open chain (*top*), and the closed chain (*bottom*).

(2) *A closed chain* has the same number of sites and bonds, N. It is now possible to draw a closed graph that circles the whole chain, and

$$Z = (2\cosh K)^N \left[1 + t^N\right] = 2^N \left(\cosh K^N + \sinh K^N\right)$$

$$\implies \frac{\ln Z}{N} = \ln(2\cosh K) + \frac{\ln\left[1 + t^N\right]}{N}. \tag{7.18}$$

The difference between the free energies of closed and open chains vanishes in the thermodynamic limit. The correction to the extensive free energy, $N\ln(2\cosh K)$, is of the order of $1/N$ for the open chain, and can be regarded as the surface free energy. There is no such correction for the closed chain, which has no boundaries; there is instead an exponential term t^N. The correlation function can again be calculated from Eq. (7.16). There are two paths connecting the points m and n, and

$$\langle \sigma_m \sigma_n \rangle = \frac{t^{|m-n|} + t^{N-|m-n|}}{1 + t^N}. \tag{7.19}$$

Note that the final answer is symmetric with respect to the two ways of measuring the distance between the two sites m and n.

7.4 Self-duality in the two-dimensional Ising model

Kramers and Wannier discovered a hidden symmetry that relates the properties of the Ising model on the square lattice at low and high temperatures. One way of obtaining this symmetry is to compare the high and low temperature series expansions of the problem. The low-temperature expansion has the form

$$Z = e^{2NK} \left[1 + Ne^{-4\times 2K} + 2Ne^{-6\times 2K} + \cdots\right]$$

$$= e^{2NK} \sum_{\text{Islands of } (-) \text{ spins}} e^{-2K \times \text{ perimeter of island}}. \tag{7.20}$$

The high-temperature series is

$$Z = 2^N \cosh K^{2N} \left[1 + N\tanh K^4 + 2N\tanh K^6 + \cdots\right]$$

$$= 2^N \cosh K^{2N} \sum_{\text{graphs with 2 or 4 lines per site}} \tanh K^{\text{length of graph}}. \tag{7.21}$$

As the boundary of any island of spins serves as an acceptable graph (and vice versa), there is a one-to-one correspondence between the two series. Defining a function g to indicate the logarithm of the above series, the free energy is given by

$$\frac{\ln Z}{N} = 2K + g\left(e^{-2K}\right) = \ln 2 + 2\ln \cosh K + g\left(\tanh K\right). \tag{7.22}$$

The arguments of g in the above equation are related by the *duality* condition

$$e^{-2\tilde{K}} \leftrightarrow \tanh K, \implies \tilde{K} = D(K) \equiv -\frac{1}{2}\ln \tanh K. \tag{7.23}$$

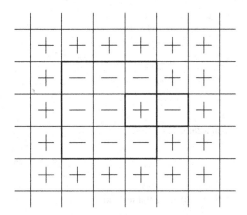

Fig. 7.5 There is a one-to-one correspondence between terms in the high and low temperature series for the Ising model on the square lattice.

The function g (which contains the singular part of the free energy) must have a special symmetry that relates its values for dual arguments. (For example the function $f(x) = x/(1+x^2)$ equals $f(x^{-1})$, establishing a duality between the arguments at x and x^{-1}.) Equation (7.23) has the following properties:

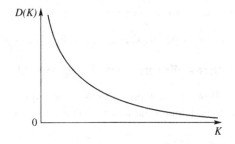

Fig. 7.6 The function $D(K)$ relating the parameter K to its dual value.

(1) Low temperatures are mapped to high temperatures, and vice versa.
(2) The mapping connects *pairs* of points since $D(D(K)) = K$. This condition is established by using trigonometric identities to show that

$$\sinh 2K = 2 \sinh K \cosh K = 2 \tanh K \cosh^2 K$$

$$= \frac{2 \tanh K}{1 - \tanh^2 K} = \frac{2e^{-2\tilde{K}}}{1 - e^{-4\tilde{K}}} = \frac{2}{e^{2\tilde{K}} - e^{-2\tilde{K}}} = \frac{1}{\sinh 2\tilde{K}}. \tag{7.24}$$

Hence, the dual interactions are symmetrically related by

$$\sinh 2K \cdot \sinh 2\tilde{K} = 1. \tag{7.25}$$

(3) If the function $g(K)$ is singular at a point K_c, it must also be singular at $\tilde{K}_c$. Since the free energy is expected to be analytic everywhere except at the transition, the critical model must be *self-dual*. At the self-dual point

$$e^{-2K_c} = \tanh K_c = \frac{1 - e^{-2K_c}}{1 + e^{-2K_c}},$$

Fig. 7.7 The duality
function maps pairs of
points from high and low
temperatures. The
self-dual point at K_c maps
onto itself.

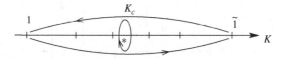

which leads to the quadratic equation,

$$e^{-4K_c} + 2e^{-K_c} - 1 = 0 \implies e^{-2K_c} = -1 \pm \sqrt{2}.$$

Only the positive solution is acceptable, and

$$K_c = -\frac{1}{2} \ln \left(\sqrt{2} - 1 \right) = \frac{1}{2} \ln \left(\sqrt{2} + 1 \right) = 0.441 \cdots \tag{7.26}$$

(4) As will be explored in the next section and problems, it is possible to obtain dual
partners for many other spin systems, such as the Potts model, the XY model, etc.
While such mappings place useful constraints on the shape of phase boundaries,
they generally provide no information on critical exponents. (The self-duality of
many two-dimensional models does restrict the ratio of critical amplitudes to unity.)

7.5 Dual of the three-dimensional Ising model

We can attempt to follow the same procedure to search for the dual of the Ising
model on a simple cubic lattice. The low-temperature series is

$$Z = e^{3NK} \left[1 + Ne^{-2K \times 6} + 3Ne^{-2K \times 10} + \cdots \right]$$
$$= e^{3NK} \sum_{\text{Islands of } (-) \text{ spins}} e^{-2K \times \text{area of island's boundary}}. \tag{7.27}$$

By contrast, the high-temperature series takes the form

$$Z = 2^N \cosh K^{3N} \left[1 + 3N \tanh K^4 + 18N \tanh K^6 + \cdots \right]$$
$$= 2^N \cosh K^{3N} \sum_{\text{graphs with 2, 4, or 6 lines per site}} \tanh K^{\text{number of lines}}. \tag{7.28}$$

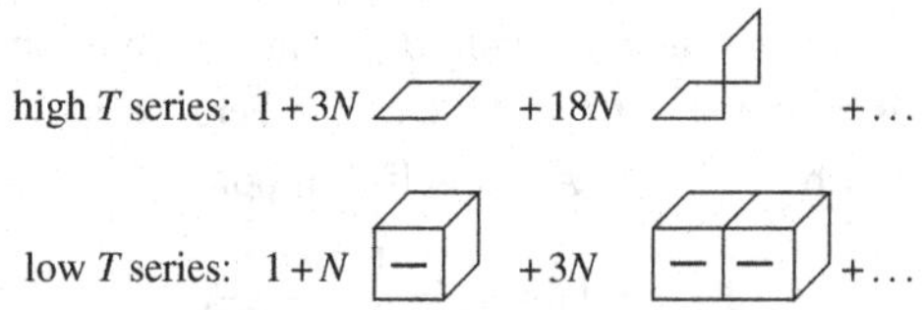

Fig. 7.8 Comparison of
the graphical expansions
of the high and low
temperature series for the
Ising model on the cubic
lattice.

high T series: $1 + 3N$ ⟋ $+ 18N$ ⟋ $+ \ldots$

low T series: $1 + N$ ⬚ $+ 3N$ ⬚ $+ \ldots$

Clearly the two sums are different, and the $3d$ Ising model is not dual to itself. Is there any other Hamiltonian whose high-temperature series reproduces the low-temperature terms for the $3d$ Ising model? To construct such a model note that:

(1) The low-temperature terms have the form $e^{-2K \times \text{area}}$, and the area is the sum of faces covered on the cubic lattice. Thus *plaquettes* must replace bonds as the building blocks of the required high temperature series.

(2) How should these plaquettes be joined together? The number of bonds next to a site can be anywhere from 3 to 12 while there are only two or four faces adjacent to each bond. This suggests "gluing" the plaquettes together by placing spins $\tilde{\sigma}_i = \pm 1$ on the *bonds* of the lattice.

(3) By analogy to the Ising model, we can construct a partition function,

$$\tilde{Z} = \sum_{\{\tilde{\sigma}_p^i = \pm 1\}} \prod_{\text{plaquettes } P} (1 + \tanh \tilde{K} \tilde{\sigma}_P^1 \tilde{\sigma}_P^2 \tilde{\sigma}_P^3 \tilde{\sigma}_P^4) \propto \sum_{\{\tilde{\sigma}_i\}} e^{\tilde{K} \sum_P \tilde{\sigma}_P^1 \tilde{\sigma}_P^2 \tilde{\sigma}_P^3 \tilde{\sigma}_P^4}, \qquad (7.29)$$

where $\tilde{\sigma}_P^i$ are used to denote the four dual spins around each plaquette, and $\tilde{K}$ is given by Eq. (7.23). This partition function describes a system dual to the original $3d$ Ising model, in the sense of reproducing its low-temperature series. (Some reflection demonstrates that the low-temperature expansion of the above partition function reproduces the high-temperature expansion of the Ising model.)

Equation (7.29) describes a Z_2 *lattice gauge theory*. The general rules for constructing such theories in all dimensions are: (i) place Ising spins $\tilde{\sigma}_i = \pm 1$ on the *bonds* of the lattice; (ii) the Hamiltonian is $-\beta \mathcal{H} = K \sum_{\text{all plaquettes}} \prod \tilde{\sigma}_P^i$. In addition to the *global* Ising symmetry, $\tilde{\sigma}_i \to -\tilde{\sigma}_i$, the Hamiltonian has a *local* or *gauge symmetry*. To observe this symmetry, select any site and change the signs of spins on all bonds emanating from it. As in each of the faces adjacent to the chosen site two bond spins change sign, their product, and hence the overall energy, is not changed.

There is a rigorous proof (*Elitzur's theorem*) that there can be no spontaneous symmetry breaking for Hamiltonians with a *local* symmetry. The essence of the proof is that even in the presence of a symmetry breaking field h, the energy cost of flipping a spin is finite ($6h$ for the gauge theory on the cubic lattice). Hence the expectation value of spin changes continuously as $h \to 0$. (By contrast, the energy cost of a spin flip in the Ising model grows as Nh.) This theorem presents us with the following paradox: Since the three dimensional Ising model undergoes a phase transition, there must be a singularity in its partition function, and also that of its dual. How can there be a singularity in the partition function of the three-dimensional gauge theory if it does not undergo a spontaneous symmetry breaking?

To resolve this contradiction, Wegner suggested the possibility of a phase transition without a local order parameter. The two phases are then distinguished by the asymptotic behavior of correlation functions. The appropriate correlation function must be invariant under the local gauge transformation. For example, the *Wilson loop* is constructed by selecting a closed path of bonds S on the lattice and examining

$$C_S = \langle \text{Product of } \tilde{\sigma} \text{ around the loop} \rangle = \left\langle \prod_{i \in S} \tilde{\sigma}_i \right\rangle. \tag{7.30}$$

As any gauge transformation changes the signs of two bonds on the loop, their product is unaffected and C_S is gauge invariant. Since the Hamiltonian encourages spins of the same sign, this expectation value is always positive. Let us examine the asymptotic dependence of C_S on the shape of the loop at high and low temperatures. In a high-temperature expansion, the correlation function is obtained as a sum of all graphs constructed from plaquettes with S as a boundary. Each plaquette contributes a factor of $\tanh \tilde{K}$, and

$$C_S = \frac{1}{Z} \sum_{\{\tilde{\sigma}_i\}} \prod_{i \in S} \tilde{\sigma}_i \ e^{K \sum_P \tilde{\sigma}_P^1 \tilde{\sigma}_P^2 \tilde{\sigma}_P^3 \tilde{\sigma}_P^4}$$

$$= \left(\tanh \tilde{K} \right)^{\text{Area of } S} \left[1 + \mathcal{O} \left(\tanh \tilde{K}^2 \right) \right] \approx \exp \left[-f \left(\tanh \tilde{K} \right) \times \text{Area of } S \right]. \tag{7.31}$$

The low-temperature expansion starts with the lowest energy configuration. There are in fact $\mathcal{N}_G = 2^N$ such ground states related by gauge transformations. The N_P plaquette interactions are satisfied in the ground states, and excitations involve creating unsatisfied plaquettes. Since C_S is gauge independent, it is sufficient to look at one of the ground states, e.g. the one with $\tilde{\sigma}_i = +1$ for all i. Flipping the sign of any of the $3N$ bonds creates an excitation of energy $8\tilde{K}$ with respect to the ground state. Denoting the number of bonds on the perimeter of the Wilson loop by P_S, we obtain

$$C_S = \frac{\mathcal{N}_G}{\mathcal{N}_G} \cdot \frac{e^{\tilde{K}N_P} \left[1 + (3N - P_S)e^{-2\tilde{K} \times 4}(+1) + (-1)P_S e^{-2\tilde{K} \times 4} + \cdots \right]}{e^{\tilde{K}N_P} \left[1 + 3Ne^{-2\tilde{K} \times 4} + \cdots \right]}$$

$$= 1 - 2P_S e^{-8\tilde{K}} + \cdots \tag{7.32}$$

$$\approx \exp \left[-2e^{-8\tilde{K}} P_S + \cdots \right].$$

The asymptotic decay of C_S is thus different at high and low temperatures. At high temperatures, the decay is controlled by the *area* of the loop, while at low temperatures it depends on its *length*. The phase transition marks the change from one type of decay to the other, and by duality has the same singularities in free energy as the Ising model.

The prototype of gauge theories in physics is quantum electrodynamics (QED), with the action

$$S = \int d^4x \left[\bar{\psi} (-i\partial + e\mathcal{A} + m) \psi + \frac{1}{4} \left(\partial_\mu A_\nu - \partial_\nu A_\mu \right)^2 \right]. \tag{7.33}$$

The spinor ψ is the Dirac field for the electron, and the 4-vector A describes the electromagnetic gauge potential. The phase of ψ is not observable, and the action is invariant under the gauge symmetry $\psi \mapsto e^{ie\phi}\psi$, and $A_\mu \mapsto A_\mu - \partial_\mu\phi$, for any $\phi(\mathbf{x})$. The Z_2 lattice gauge theory can be regarded as the Ising analog of QED with the bond spins playing the role of the electromagnetic field. We can introduce a "matter" field by placing spins $s_i = \pm 1$, on the *sites* of the lattice. The two fields are coupled by the Hamiltonian,

$$-\beta\mathcal{H} = J\sum_{\langle i,j \rangle} s_i\,\tilde{\sigma}_{ij}\,s_j + K\sum_P \tilde{\sigma}_P^1\tilde{\sigma}_P^2\tilde{\sigma}_P^3\tilde{\sigma}_P^4, \tag{7.34}$$

where $\tilde{\sigma}_{ij}$ is the spin on the bond joining i and j. The Hamiltonian has the gauge symmetry $s_i \mapsto (-1)s_i$, and $\tilde{\sigma}_{i,\mu} \mapsto -\tilde{\sigma}_{i,\mu}$, for all bonds emanating from any site i.

Regarding one of the lattice directions as time, a Wilson loop is obtained by creating two particles at a distance x, propagating them for a time t, and then removing them. The probability of such an event is roughly given by $C_S \sim e^{-U(x)t}$, where $U(x)$ is the interaction between the two particles. In the high-temperature phase, C_S decays with the area of the loop, suggesting $U(x)t = f(\tanh\tilde{K})|x|t$. The resulting potential $U(x) = f(\tanh\tilde{K})|x|$, is like a *string* that connects the particles together. This is also the potential that describes the confinement of quarks at large distances in quantum chromodynamics. The decay with the length of the loop at low temperatures implies $U(x)t \approx g(e^{-8\tilde{K}})(|x|+t)$. For $t \gg |x|$ the potential is a constant and the force vanishes. (This *asymptotic freedom* describes the behavior of quarks at short distances.) The phase transition implies a change in the nature of interactions between particles mediated by the gauge field.

7.6 Summing over phantom loops

The high-temperature series can be *approximately* summed so as to reproduce the Gaussian model. This correspondence provides a better understanding of why Gaussian behavior is applicable in high dimensions, and also prepares the way for the exact summation of the series in two dimensions in the next section. The high-temperature series for the partition function of the Ising model on a d-dimensional hypercubic lattice is obtained from

$$Z = \sum_{\{\sigma_i\}} e^{K\sum_{\langle ij \rangle} \sigma_i\sigma_j} = 2^N \cosh^{dN} K \times S, \tag{7.35}$$

where S is the sum over all allowed graphs on the lattice, each weighted by $t \equiv \tanh K$ raised to the power of the number of bonds in the graph. The allowed graphs have an even number of bonds per site. The simplest graphs have the topology of a single closed loop. There are also graphs composed of *disconnected* closed loops.

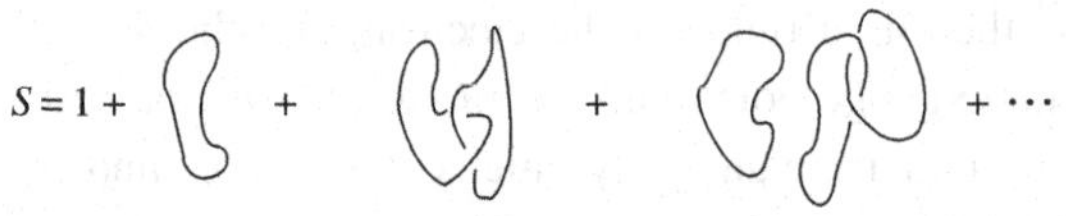

Fig. 7.9 Terms of the high-temperature series organized as sums over loops.

$$S = 1 + \quad \text{⬬} \quad + \quad \text{⬭} \quad + \quad \text{⬮} \quad + \cdots$$

Keeping in mind the cumulant expansion, we can set

$$\Xi = \text{sum over contribution of all graphs with one loop,} \tag{7.36}$$

and introduce another sum,

$$S' = \exp(\Xi) = 1 + \Xi + \frac{1}{2}(\Xi)^2 + \frac{1}{6}(\Xi)^3 + \cdots \tag{7.37}$$

$$= 1 + (\text{1 loop graphs}) + (\text{2 loop graphs}) + (\text{3 loop graphs}) + \cdots$$

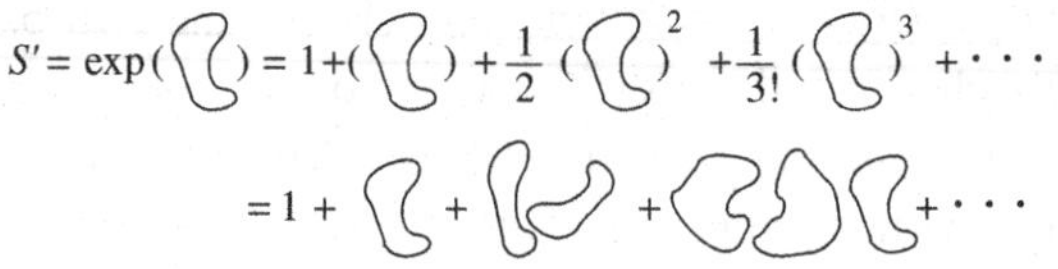

Fig. 7.10 The sum over phantom loops can be exponentiated.

$$S' = \exp\left(\text{⬬}\right) = 1 + \left(\text{⬬}\right) + \frac{1}{2}\left(\text{⬬}\right)^2 + \frac{1}{3!}\left(\text{⬬}\right)^3 + \cdots$$

$$= 1 + \quad \text{⬬} \quad + \quad \text{⬭} \quad + \quad \text{⬮} \quad + \cdots$$

Despite their similarities, the sums S and S' are not identical: There are ambiguities associated with loops that intersect at a single site, which will be discussed more fully in the next section. More importantly, S' includes additional graphs where a particular bond contributes more than once, while in the original sum S, each lattice bond contributes a factor of 1 or t. This arises because after raising Ξ to a power ℓ, a particular bond may contribute up to ℓ times for a factor of t^ℓ. In the spirit of the approximation that allows multiple appearances of a bond, we shall also include additional closed paths in Ξ, in which a particular bond is traversed more than once in completing the loop.

Fig. 7.11 Examples of graphs included in S' that are not present in S.

Qualitatively, S is the partition function of a gas of *self-avoiding polymer* loops with a monomer fugacity of t. The self-avoiding constraint is left out in the partition function S', which thus corresponds to a gas of *phantom polymer* loops, which may pass through each other with impunity. Loops of various shapes can be constructed from closed random walks on the lattice, and the corresponding free energy of phantom loops is

$\ln S' = \sum$ all closed random walks on the lattice $\times\, t^{\text{length of walk}}$

$$= N \sum_{\ell} \frac{t^{\ell}}{\ell} \, (\text{number of closed walks of } \ell \text{ steps starting and ending at } \mathbf{0}).$$

$$(7.38)$$

Note that extensivity is guaranteed since (up to boundary effects) the same loop can be started from any point on the lattice. The overall factor of $1/\ell$ accounts for the ℓ possible starting points for a loop of length ℓ.

A *transfer matrix* method can be used to count all possible (phantom) random walks on the lattice. Let us introduce a set of $N \times N$ matrices,

$$\langle \mathbf{i} | W(\ell) | \mathbf{j} \rangle \equiv \text{number of walks from } \mathbf{j} \text{ to } \mathbf{i} \text{ in } \ell \text{ steps,} \qquad (7.39)$$

in terms of which Eq. (7.38) becomes

$$\frac{\ln S'}{N} = \frac{1}{2} \sum_{\ell} \frac{t^{\ell}}{\ell} \, \langle \mathbf{0} | W(\ell) | \mathbf{0} \rangle. \qquad (7.40)$$

The additional factor of 2 arises since the same loop can be traversed by two random walks moving in opposite directions. Similarly, the spin–spin correlation function

$$\langle \sigma(\mathbf{0}) \sigma(\mathbf{r}) \rangle = \frac{1}{Z} \sum_{\{\sigma_i\}} \sigma(\mathbf{0}) \sigma(\mathbf{r}) \prod_{\langle ij \rangle} \left(1 + t \sigma_i \sigma_j\right) \qquad (7.41)$$

is related to the sum over all paths connecting the points $\mathbf{0}$ and $\mathbf{r}$ on the lattice. In addition to the simple paths that directly connect the two points, there are *disconnected* graphs that contain additional closed loops. In the same approximation of allowing all intersections between paths, the partition function S' can be factored out of the numerator and denominator of Eq. (7.41), and

$$\langle \sigma(\mathbf{0}) \sigma(\mathbf{r}) \rangle \approx \sum_{\ell} t^{\ell} \, \langle \mathbf{r} | W(\ell) | \mathbf{0} \rangle. \qquad (7.42)$$

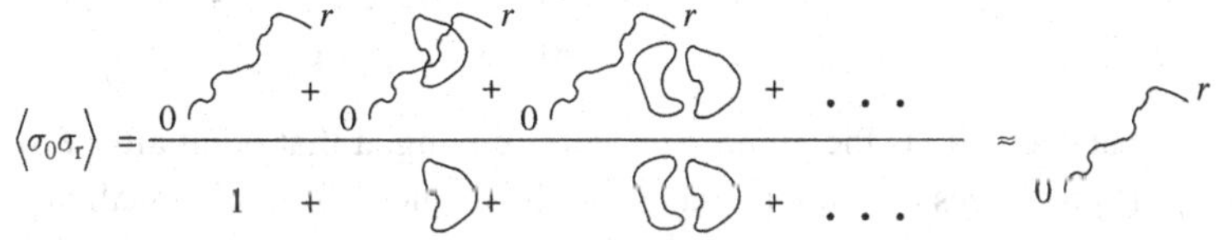

Fig. 7.12
High-temperature series for the two-point correlation function, and its approximation in the phantom path limit.

The counting of phantom paths on a lattice is easily accomplished by taking advantage of their *Markovian* property. This is the property that each step of a random walk proceeds from its last location and is independent of its previous steps. Hence, the number of walks can be calculated *recursively*. First, note that any walk from $\mathbf{0}$ to $\mathbf{r}$ in ℓ steps can be accomplished as a walk from $\mathbf{0}$ to some other point $\mathbf{r}'$ in $\ell - 1$ steps, followed by a single step from $\mathbf{r}'$ to $\mathbf{r}$. Summing over all possible locations of the intermediate point leads to

$$\langle \mathbf{r}|W(\ell)|\mathbf{0}\rangle = \sum_{\mathbf{r}'} \langle \mathbf{r}|W(1)|\mathbf{r}'\rangle \times \langle \mathbf{r}'|W(\ell-1)|\mathbf{0}\rangle$$

$$= \langle \mathbf{r}|TW(\ell-1)|\mathbf{0}\rangle, \qquad (7.43)$$

where the sum corresponds to the product of two matrices, and we have defined $T \equiv W(1)$. The recursion process can be continued and

$$W(\ell) = TW(\ell-1) = T^2 W(\ell-2)^2 = \cdots = T^\ell. \qquad (7.44)$$

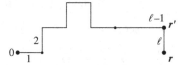

Fig. 7.13 Recursive calculation of the number of walks on a lattice.

Thus all lattice random walks are generated by the *transfer matrix* T, whose elements are

$$\langle \mathbf{r}|T|\mathbf{r}'\rangle = \begin{cases} 1 & \text{if } \mathbf{r} \text{ and } \mathbf{r}' \text{ are nearest neighbors} \\ 0 & \text{otherwise.} \end{cases} \qquad (7.45)$$

(It is also called the *adjacency*, or *connectivity* matrix.) For example in $d = 2$,

$$\langle x, y|T|x', y'\rangle = \delta_{y,y'}\left(\delta_{x,x'+1} + \delta_{x,x'-1}\right) + \delta_{x,x'}\left(\delta_{y,y'+1} + \delta_{y,y'-1}\right), \qquad (7.46)$$

and successive actions of T on a walker starting at the origin $|x, y\rangle = \delta_{x,0}\delta_{y,0}$, generate the patterns

$$
\begin{array}{ccc}
& & \begin{matrix} 0 & 0 & 1 & 0 & 0 \end{matrix} \\
\begin{matrix} 0 & 0 & 0 \\ 0 & 1 & 0 \\ 0 & 0 & 0 \end{matrix} \xrightarrow{T} & \begin{matrix} 0 & 1 & 0 \\ 1 & 0 & 1 \\ 0 & 1 & 0 \end{matrix} \xrightarrow{T} & \begin{matrix} 0 & 2 & 0 & 2 & 0 \\ 1 & 0 & 4 & 0 & 1 \\ 0 & 2 & 0 & 2 & 0 \\ 0 & 0 & 1 & 0 & 0 \end{matrix} \xrightarrow{T} \cdots
\end{array}
$$

The value at each site is the number of walks ending at that point after ℓ steps.

Various properties of random walks can be deduced from diagonalizing the matrix T. Due to the translational symmetry of the lattice, this is achieved in the Fourier basis $\langle \mathbf{r}|\mathbf{q}\rangle = e^{i\mathbf{q}\cdot\mathbf{r}}/\sqrt{N}$. For example in $d = 2$, starting from Eq. (7.46), it can be checked that

$$\langle x, y|T|q_x, q_y\rangle = \sum_{x', y'} \langle x, y|T|x', y'\rangle \langle x', y'|q_x, q_y\rangle$$

$$= \frac{1}{\sqrt{N}}\left[e^{iq_y y}\left(e^{iq_x(x+1)} + e^{iq_x(x-1)}\right) + e^{iq_x x}\left(e^{iq_y(y+1)} + e^{iq_y(y-1)}\right)\right] \qquad (7.47)$$

$$= \frac{1}{\sqrt{N}}e^{i(q_x x + q_y y)}\left[2\cos q_x + 2\cos q_y\right] = T(q_x, q_y)\langle x, y|q_x, q_y\rangle.$$

The generalized eigenvalue for a d-dimensional hypercubic lattice is

$$T(\mathbf{q}) = 2 \sum_{\alpha=1}^{d} \cos q_\alpha. \tag{7.48}$$

The correlation function in Eq. (7.42) is now evaluated as

$$\langle \sigma(\mathbf{r})\sigma(\mathbf{0}) \rangle \approx \sum_{\ell}^{\infty} t^\ell \langle \mathbf{r}|W(\ell)|\mathbf{0}\rangle = \sum_{\ell}^{\infty} \langle \mathbf{r}|(tT)^\ell|\mathbf{0}\rangle$$

$$= \left\langle \mathbf{r} \left| \frac{1}{1-tT} \right| \mathbf{0} \right\rangle = \sum_{\mathbf{q}} \langle \mathbf{r}|\mathbf{q}\rangle \frac{1}{1-tT(\mathbf{q})} \langle \mathbf{q}|\mathbf{0}\rangle \tag{7.49}$$

$$= N \int \frac{d^d\mathbf{q}}{(2\pi)^d} \frac{e^{i\mathbf{q}\cdot\mathbf{r}}}{N} \frac{1}{1-2t\sum_{\alpha=1}^{d} \cos q_\alpha} = \int \frac{d^d\mathbf{q}}{(2\pi)^d} \frac{e^{i\mathbf{q}\cdot\mathbf{r}}}{1-2t\sum_\alpha \cos q_\alpha}.$$

For $t \to 0$, the shortest path costs least energy and $\langle \sigma(\mathbf{0})\sigma(\mathbf{r}) \rangle \sim t^{|\mathbf{r}|}$. As t increases, larger paths dominate the sum because they are more numerous (i.e. entropically favored). Eventually, there is a singularity for $1 - tT(\mathbf{0}) = 0$, i.e. at $2d \times t_c = 1$, when arbitrarily long paths become important. For $t < t_c$, the partition function is dominated by small loops, and a polymer connecting two far away points is stretched by its line tension. When the fugacity exceeds t_c, the line tension vanishes and loops of arbitrary size are generated. Clearly the neglect of intersections (which stabilizes the system at a finite density) is no longer justified in this limit. This transition is the manifestation of Ising ordering in the language of paths representing the high-temperature series. On approaching the transition from the high-temperature side, the sums are dominated by very long paths. Accordingly, the denominator of Eq. (7.49) can be expanded for small $\mathbf{q}$ as

$$1 - tT(\mathbf{q}) = 1 - 2t \sum_{\alpha=1}^{d} \cos q_\alpha \simeq (1-2dt) + tq^2 + \mathcal{O}(q^4)$$

$$\approx t_c(\xi^{-2} + q^2 + \mathcal{O}(q^4)), \tag{7.50}$$

where

$$\xi = \left(\frac{1-2dt}{t_c} \right)^{-1/2}. \tag{7.51}$$

The resulting correlation functions,

$$\langle \sigma(\mathbf{0})\sigma(\mathbf{r}) \rangle \propto \int \frac{d^d\mathbf{q}}{(2\pi)^d} \frac{e^{i\mathbf{q}\cdot\mathbf{r}}}{q^2 + \xi^{-2}},$$

are identical to those obtained from the Gaussian model, and (Sec. 3.2)

$$\langle \sigma(\mathbf{0})\sigma(\mathbf{r}) \rangle \propto \begin{cases} \dfrac{1}{r^{d-2}} & \text{for } r < \xi \quad (\eta = 0) \\ \dfrac{e^{-r/\xi}}{r^{(d-1)/2}} & \text{for } r > \xi. \end{cases} \tag{7.52}$$

The correlation length in Eq. (7.51) diverges as $\xi \sim (t_c - t)^{-1/2}$, i.e. with the Gaussian exponent of $\nu = 1/2$.

We can also calculate the free energy in Eq. (7.40) as

$$
\frac{\ln S'}{N} = \frac{1}{2} \sum_{\ell}^{\infty} \frac{t^{\ell}}{\ell} \langle 0 | W(\ell) | 0 \rangle = \frac{1}{2} \left\langle 0 \left| \sum_{\ell}^{\infty} \frac{t^{\ell} T^{\ell}}{\ell} \right| 0 \right\rangle
$$

$$
= -\frac{1}{2} \langle 0 | \ln(1 - tT) | 0 \rangle = -\frac{N}{2} \int \frac{d^d \mathbf{q}}{(2\pi)^d} \langle 0 | \mathbf{q} \rangle \ln(1 - tT(\mathbf{q})) \langle \mathbf{q} | 0 \rangle
$$

$$
= -\frac{1}{2} \int \frac{d^d \mathbf{q}}{(2\pi)^d} \ln \left(1 - 2t \sum_{\alpha=1}^{d} \cos q_\alpha \right). \tag{7.53}
$$

In the vicinity of the critical point at $t_c = 1/(2d)$, the argument of the logarithm is proportional to $(q^2 + \xi^{-2})$ from Eq. (7.50). This is precisely as in the Gaussian model, and as discussed in Sec. 4.6 the singular part of the free energy scales as

$$
f_{\text{sing}} \propto \xi^{-d} \propto (t_c - t)^{d/2}. \tag{7.54}
$$

The singular part of the heat capacity, obtained after taking two derivatives, is governed by the exponent $\alpha = 2 - d/2$. Note that in evaluating the sums appearing in Eqs. (7.49) and (7.53), the lower limit for ℓ is not treated very carefully. The series in Eq. (7.49) is assumed to start from $\ell = 0$, and that of Eq. (7.53) from $\ell = 1$. In fact the first few terms of both series may be zero because the number of steps is not sufficient to reach $\mathbf{0}$ from $\mathbf{r}$, or to form a closed loop. This is not a serious omission, in that the *singular behavior* of a series is not affected by its first few terms. Treating the first few terms differently can only add analytic corrections to the singular forms calculated in Eqs. (7.49) and (7.53).

The equivalence of these results to the Gaussian model is a manifestation of field–particle duality. In a field theoretical description, (imaginary) time appears as an additional dimension, and the two-point correlations describe the probability of propagating a particle from one point in space–time to another. In a wave description, this probability is calculated by evolving the wave function using the Schrödinger equation. Alternatively, the probability can be calculated as the sum over all (Feynman) paths connecting the two points, each path weighted with the correct action. The second sum is similar to the above calculation of $\langle \sigma(\mathbf{r})\sigma(\mathbf{0}) \rangle$.

This approach provides an interesting geometrical interpretation of the phase transition. The establishment of long-range order implies that all parts of the system have selected the same state. This information is carried by the bonds connecting nearest neighbors, and can be passed from the origin to a point $\mathbf{r}$, through all paths connecting these two points. The fugacity t is a measure of the reliability of information transfer between neighboring sites. Along a one-dimensional chain, unless $t = 1$, the transferred information decays at large distances and it is impossible to establish long-range order. In higher dimensions there are many more paths, and by accumulating the information from all paths it is possible to establish order at $t_c < 1$. Since the number of paths of length ℓ grows as $(2d)^\ell$ while their information content decays as t^ℓ, the transition occurs at $t_c = 1/(2d)$. (A better approximation is obtained by including some of the constraints by noting that the random walk cannot back track. In this case the number of walks grows as $(2d-1)^\ell$.) The total information from paths of length ℓ is weighted by $(2dt)^\ell$, and decays exponentially for $t < t_c$. The characteristic path length, $\bar{\ell} = -1/\ln(2dt)$, diverges as $(t_c - t)^{-1}$ on approaching the transition. For paths of size $\ell \ll \bar{\ell}$ there is very good information transfer. Such paths execute random walks on the lattice and cover a distance $\xi \approx \bar{\ell}^{\,1/2}$. The divergence of ν with an exponent of $1/2$ is thus a consequence of the random walk nature of the paths.

Why does the classical picture fail for $d \leq 4$? Let us focus on the dominant paths close to the phase transition. Is it justified to ignore the intersections of such paths? Random walks can be regarded as geometrical entities of fractal (Hausdorf) dimension $d_f = 2$. This follows from the general definition of dimension relating the mass and extent of an object by $M \propto R^{d_f}$, and the observation that the size of a random walk ($R \propto \xi$) is the square root of its length ($M \propto \ell$). Two geometrical entities of dimensions d_1 and d_2 will generally intersect in d-dimensional space if $d_1 + d_2 \geq d$. Thus our random walkers are unlikely to intersect in $d \geq d_u = 2 + 2 = 4$, and the above (Gaussian) results obtained by neglecting the intersections are asymptotically valid. Below the upper critical dimension of 4, random walks have frequent encounters and their intersections must be treated correctly. The diagrams obtained in the perturbative calculation of the propagator with um^4 correspond precisely to taking into account the intersections of paths. (Each factor of u corresponds to one intersection.) It is now clear that the constraint of self-avoidance will swell the paths beyond their random walk size leading to an increase in the exponent ν. Below the transition the length of paths grows without bound and the self–avoiding constraint is necessary to ensure the stability of the system.

As demonstrated in the problem section, the loop expansion is easily generalized to n-component spins. The only difference is that each closed loop now contributes a factor of n. In the phantom limit, where intersections are ignored, the free energy (Eq. 7.53) is simply multiplied by n, while the correlation function is left unchanged (precisely as in the Gaussian model). Corrections due to intersections, which modify critical behavior in $d < 4$, now depend on n. For example, in correcting the two point correlation function, we have to subtract contributions from the self-intersections of the random walk, as well as from contacts with loops (which have a fugacity of n). The correspondence with the perturbative series of the propagator with a nonlinearity $u(\vec{m} \cdot \vec{m})^2$ is again apparent.

7.7 Exact free energy of the square lattice Ising model

As indicated in Eq. (7.35), the Ising partition function is related to a sum S, over collections of paths on the lattice. The allowed graphs for a square lattice have two or four bonds per site. Each bond can appear only once in each graph, contributing a factor of $t \equiv \tanh K$. While it is tempting to replace S with the exactly calculable sum S', of all phantom loops of random walks on the lattice, this leads to an overestimation of S. The differences between the two sums arise from intersections of random walks, and can be divided into two categories:

(a) There is an over-counting of graphs which intersect at a *site*, i.e. with four bonds through a point. Consider a graph composed of two loops meeting at a site. Since a walker entering the intersection has three choices, this graph can be represented by *three distinct random walks*. One choice leads to two disconnected loops; the other two are single loops with or without a self-crossing in the walker's path.

Fig. 7.14 The single high temperature graph on the left with an intersection point can be generated by the three random walks on the right.

(b) The independent random walkers in S' may go through a particular lattice *bond* more than once.

Fig. 7.15 Examples of random walks with multiple crossings of a bond, which have no counterpart in the high-temperature series.

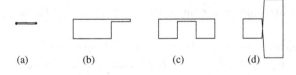

(a)   (b)   (c)   (d)

Including these constraints amounts to introducing interactions between paths. The resulting interacting random walkers are non–Markovian, as each step is no longer independent of previous ones and of other walkers. While such interacting walks are not in general amenable to exact treatment, in two dimensions an interesting topological property allows us to make the following assertion:

$$S = \sum \text{collections of loops of random walks } \textit{with no } U\textit{-turns}$$
$$\times \, t^{\text{number of bonds}} \times (-1)^{\text{number of crossings}}.$$

(7.55)

The negative signs for some terms reduce the overestimate and render the exact sum.

Justification We shall deal in turn with the two problems mentioned above.

(a) Consider a graph with many intersections and focus on a particular one. A walker must enter and leave such an intersection twice. This can be done in three ways, only one of which involves the path of the walker crossing itself (when the walker proceeds straight through the intersection twice). This configuration carries an additional factor of (-1) according to Eq. (7.55). Thus, independent of other crossings, these three configurations sum up to contribute a factor of 1. By repeating this reasoning at each intersection, we see that the over-counting problem is removed and the sum over all possible ways of tracing the graph leads to the correct factor of one.

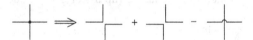

Fig. 7.16 The intersection point can be traversed in three ways, with the last carrying a relative factor of (-1).

(b) Consider a bond that is crossed by two walkers (or twice by the same walker). We can imagine the bond as an avenue with two sides. For each configuration in which the two paths enter and leave on the same side of the avenue, there is another one in which the paths go to the opposite side. The latter involves a crossing of paths and hence carries a minus sign with respect to the former. The two possibilities thus cancel out! The reasoning can be generalized to multiple passes through any bond. The only exception is when the double bond is created as a result of a U-turn. This is why such backward steps are explicitly excluded from Eq. (7.55).

Fig. 7.17 The two ways of traversing a doubly crossed bond carry opposite factors and cancel out.

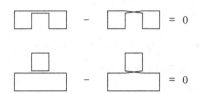

Let us label random walkers with no U-turns, and weighted by $(-1)^{\text{number of crossings}}$, as RW*s. Then as in Eq. (7.37) the terms in S can be organized as

$$S = \sum(\text{RW*s with 1 loop}) + \sum(\text{RW*s with 2 loops}) + \sum(\text{RW*s with 3 loops}) + \cdots$$

$$= \exp\left[\sum(\text{RW*s with 1 loop})\right]. \qquad (7.56)$$

The exponentiation of the sum is justified, since the only interaction between RW*s is the sign related to their crossings. As two RW* loops always cross an even number of times, this is equivalent to no interaction at all. Using Eq. (7.35), the full Ising free energy is calculated as

$$\ln Z = N \ln 2 + 2N \ln \cosh K + \sum\left(\text{RW*s with 1 loop} \times t^{\text{\# of bonds}}\right). \qquad (7.57)$$

Organizing the sum in terms of the number of bonds, and taking advantage of the translational symmetry of the lattice (up to corrections due to boundaries),

$$\frac{\ln Z}{N} = \ln\left(2\cosh^2 K\right) + \sum_{\ell}^{\infty} \frac{t^{\ell}}{\ell} \langle 0 | W^*(\ell) | 0 \rangle, \qquad (7.58)$$

where

$$\langle 0 | W^*(\ell) | 0 \rangle = \text{number of closed loops of } \ell \text{ steps, with no U-turns,}$$

$$\text{from } \mathbf{0} \text{ to } \mathbf{0} \times (-1)^{\text{\# of crossings}}. \qquad (7.59)$$

The absence of U-turns, a local constraint, does not complicate the counting of loops. On the other hand, the number of crossings is a function of the complete configuration of the loop and is a non-Markovian property. Fortunately, in two dimensions it is possible to obtain the *parity* of the number of crossings from local considerations. The first step is to construct the loops from *directed* random walks, indicated by placing an arrow along the direction that the path is traversed. Since any loop can be traversed in two directions,

$$\langle 0 | W^*(\ell) | 0 \rangle = \frac{1}{2} \sum \textit{directed } \text{RW* loops of } \ell \text{ steps, no U-turns,}$$

$$\text{from } \mathbf{0} \text{ to } \mathbf{0} \times (-1)^{n_c}, \qquad (7.60)$$

where n_c is the number of self-crossings of the loop. We can now take advantage of the following topological result:

Whitney's theorem: The number of self-crossings of a planar loop is related to the total angle Θ, through which the tangent vector turns in going around the loop by

$$(n_c)_{\text{mod } 2} = \left(1 + \frac{\Theta}{2\pi}\right)_{\text{mod } 2}. \tag{7.61}$$

This theorem can be checked by a few examples. A single loop corresponds to $\Theta = \pm 2\pi$, while a single intersection results in $\Theta = 0$.

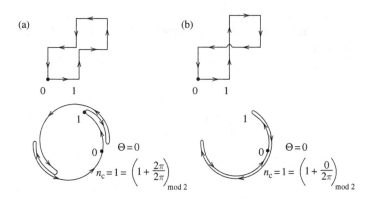

Fig. 7.19 Illustration of Whitney's theorem: The angle of the tangent vector on traversing each closed loop is plotted below it. The net angle can be related to the number of crossings.

Since the total angle Θ is the sum of the angles through which the walker turns at each step, the parity of crossings can be obtained using *local* information alone as

$$(-1)^{n_c} = e^{i\pi n_c} = \exp\left[i\pi\left(1 + \frac{\Theta}{2\pi}\right)\right] = -e^{\frac{i}{2}\sum_{j=1}^{\ell}\theta_j}, \tag{7.62}$$

where θ_j is the angle through which the walker turns on the jth step, leading to

$$\langle 0|W^*(\ell)|0\rangle = -\frac{1}{2}\sum directed\ \text{RW}^*\ \text{loops of } \ell \text{ steps, with no U-turns,}$$

$$\text{from } \mathbf{0} \ to \ \mathbf{0} \times \exp\left(\frac{i}{2}\sum \text{local change of angle by the tangent vector}\right). \tag{7.63}$$

The angle turned can be calculated at each site, if we keep track of the directions of arrival and departure of the path. To this end, we introduce a label

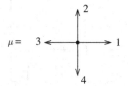

Fig. 7.20 Labeling directions out of a site.

Series expansions

μ for the four directions *going out* of each site, e.g. $\mu = 1$ for right, $\mu = 2$ for up, $\mu = 3$ for left, and $\mu = 4$ for down. We next introduce a set of $4N \times 4N$ matrices generalizing Eq. (7.39) as

$$
\begin{aligned}
\langle x_2 y_2, \mu_2 | W^*(\ell) | x_1 y_1, \mu_1 \rangle &= \sum \text{directed random walks of } \ell \text{ steps,} \\
&\text{with no } U\text{-turns, departing } (x_1, y_1) \text{ along } \mu_1, \\
&\text{proceeding along } \mu_2 \text{ after reaching } (x_2, y_2) \times e^{\frac{i}{2} \sum_{j=1}^{\ell} \theta_j}.
\end{aligned}
\tag{7.64}
$$

Thus μ_2 specifies a direction taken *after* the walker reaches its destination. It serves to exclude some paths (e.g. arriving along $-\mu_2$), and leads to an additional phase.

Fig. 7.21 Recursive construction for the weight of the directed walks.

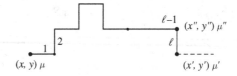

As in Eq. (7.43), due to their Markovian property, these matrices can be calculated recursively as

$$
\begin{aligned}
\langle x'y', \mu' | W^*(\ell) | xy, \mu \rangle &= \sum_{x''y'', \mu''} \langle x'y', \mu' | T^* | x''y'', \mu'' \rangle \langle x''y'', \mu'' | W^*(\ell-1) | xy, \mu \rangle \\
&= \langle x'y', \mu' | T^* W^*(\ell-1) | xy, \mu \rangle = \langle x'y', \mu' | T^{*\ell} | xy, \mu \rangle,
\end{aligned}
\tag{7.65}
$$

where $T^* \equiv W^*(1)$ describes one step of the walk. The direction of arrival uniquely determines the nearest neighbor from which the walker departed, and the angle between the two directions fixes the phase of the matrix element. We can thus generalize Eq. (7.46) to a 4×4 matrix that keeps track of both connectivity and phase between pairs of sites. The steps taken can be represented diagrammatically as

$$
T^* = \begin{bmatrix}
\overrightarrow{\rightarrow} & \rightarrow\!\!\uparrow & \rightleftarrows & \overrightarrow{\downarrow} \\
\uparrow\!\!\overrightarrow{} & \uparrow\!\!\uparrow & \leftarrow\!\!\uparrow & \uparrow\!\!\downarrow \\
\rightleftarrows & \leftarrow\!\!\uparrow & \leftarrow\!\!\leftarrow & \downarrow\!\!\leftarrow \\
\downarrow\!\!\overrightarrow{} & \uparrow\!\!\downarrow & \leftarrow\!\!\downarrow & \downarrow\!\!\downarrow
\end{bmatrix},
\tag{7.66}
$$

and correspond to the matrix

$$\langle x'y'|T^*|xy\rangle =$$

$$
\begin{bmatrix}
\langle x',y'|x+1,y\rangle & \langle x',y'|x+1,y\rangle\,e^{\frac{i\pi}{4}} & 0 & \langle x',y'|x+1,y\rangle\,e^{-\frac{i\pi}{4}} \\
\langle x',y'|x,y+1\rangle\,e^{-\frac{i\pi}{4}} & \langle x',y'|x,y+1\rangle & \langle x',y'|x,y+1\rangle\,e^{\frac{i\pi}{4}} & 0 \\
0 & \langle x',y'|x-1,y\rangle\,e^{-\frac{i\pi}{4}} & \langle x',y'|x-1,y\rangle & \langle x',y'|x-1,y\rangle\,e^{\frac{i\pi}{4}} \\
\langle x',y'|x,y-1\rangle\,e^{\frac{i\pi}{4}} & 0 & \langle x',y'|x,y-1\rangle\,e^{-\frac{i\pi}{4}} & \langle x',y'|x,y-1\rangle
\end{bmatrix},
$$

$$(7.67)$$

where $\langle x,y|x',y'\rangle \equiv \delta_{x,x'}\delta_{y,y'}$.

Because of its translational symmetry, the $4N \times 4N$ matrix takes a *block diagonal* form in the Fourier basis, $\langle xy|q_xq_y\rangle = e^{i(q_xx+q_yy)}/\sqrt{N}$, i.e.

$$\sum_{xy}\langle x'y',\mu'|T^*|xy,\mu\rangle\langle xy|q_xq_y\rangle = \langle\mu'|T^*(\mathbf{q})|\mu\rangle\langle x'y'|q_xq_y\rangle. \qquad (7.68)$$

Each 4×4 block is labeled by a wavevector $\mathbf{q} = (q_x, q_y)$, and takes the form

$$
T^*(\mathbf{q}) =
\begin{bmatrix}
e^{-iq_x} & e^{-i(q_x-\frac{\pi}{4})} & 0 & e^{-i(q_x+\frac{\pi}{4})} \\
e^{-i(q_y+\frac{\pi}{4})} & e^{-iq_y} & e^{-i(q_y-\frac{\pi}{4})} & 0 \\
0 & e^{i(q_x-\frac{\pi}{4})} & e^{iq_x} & e^{i(q_x+\frac{\pi}{4})} \\
e^{i(q_y+\frac{\pi}{4})} & 0 & e^{i(q_y-\frac{\pi}{4})} & e^{iq_y}
\end{bmatrix}.
\qquad (7.69)
$$

To ensure that a path that starts at the origin completes a loop properly, the final arrival direction at the origin must coincide with the original one. Summing over all four such directions, the total number of such loops is obtained from

$$\langle 0|W^*(\ell)|0\rangle = \sum_{\mu=1}^{4}\langle 00,\mu|T^{*\ell}|00,\mu\rangle = \frac{1}{N}\sum_{xy,\mu}\langle xy,\mu|T^{*\ell}|xy,\mu\rangle = \frac{1}{N}\,\mathrm{tr}\left(T^{*\ell}\right). \qquad (7.70)$$

Using Eq. (7.58), the free energy is calculated as

$$
\frac{\ln Z}{N} = \ln\left(2\cosh^2 K\right) - \frac{1}{2}\sum_{\ell}\frac{t^{\ell}}{\ell}\langle 0|W^*(\ell)|0\rangle = \ln\left(2\cosh^2 K\right) - \frac{1}{2N}\,\mathrm{tr}\left[\sum_{\ell}\frac{T^{*\ell}t^{\ell}}{\ell}\right]
$$

$$
= \ln\left(2\cosh^2 K\right) + \frac{1}{2N}\,\mathrm{tr}\ln\left(1 - tT^*\right) \qquad (7.71)
$$

$$
= \ln\left(2\cosh^2 K\right) + \frac{1}{2N}\sum_{\mathbf{q}}\mathrm{tr}\ln\left(1 - tT^*(\mathbf{q})\right).
$$

But for any matrix M with eigenvalues $\{\lambda_\alpha\}$,

$$\mathrm{tr}\ln M = \sum_{\alpha}\ln\lambda_\alpha = \ln\prod_{\alpha}\lambda_\alpha = \ln\det M.$$

Converting the sum over **q** in Eq. (7.71) to an integral leads to

$$\frac{\ln Z}{N} = \ln \left(2 \cosh^2 K\right) + \frac{1}{2} \int \frac{d^2 \mathbf{q}}{(2\pi)^2}$$

$$\ln \left\{ \det \begin{vmatrix} 1 - te^{-iq_x} & -te^{-i(q_x - \frac{\pi}{4})} & 0 & -te^{-i(q_x + \frac{\pi}{4})} \\ -te^{-i(q_y + \frac{\pi}{4})} & 1 - te^{-iq_y} & -te^{-i(q_y - \frac{\pi}{4})} & 0 \\ 0 & -te^{i(q_x - \frac{\pi}{4})} & 1 - te^{iq_x} & -te^{i(q_x + \frac{\pi}{4})} \\ -te^{i(q_y + \frac{\pi}{4})} & 0 & -te^{i(q_y - \frac{\pi}{4})} & 1 - te^{iq_y} \end{vmatrix} \right\}. \quad (7.72)$$

Evaluation of the above determinant is straightforward, and the final result is

$$\frac{\ln Z}{N} = \ln \left(2 \cosh^2 K\right) + \frac{1}{2} \int \frac{d^2 \mathbf{q}}{(2\pi)^2} \ln \left[\left(1 + t^2\right)^2 - 2t \left(1 - t^2\right) \left(\cos q_x + \cos q_y\right)\right]. \quad (7.73)$$

Taking advantage of trigonometric identities, the result can be simplified to

$$\frac{\ln Z}{N} = \ln 2 + \frac{1}{2} \int_{-\pi}^{\pi} \frac{dq_x dq_y}{(2\pi)^2} \ln \left[\cosh^2(2K) - \sinh(2K) \left(\cos q_x + \cos q_y\right)\right]. \quad (7.74)$$

While it is possible to obtain a closed form expression by performing the integrals exactly, the final expression involves a hypergeometric function, and is not any more illuminating.

7.8 Critical behavior of the two-dimensional Ising model

To understand the singularity in the free energy of the two-dimensional Ising model in Eq. (7.73), we start with the simpler expression obtained by the unrestricted sum over phantom loops in Section 7.6. Specializing Eq. (7.53) to $d = 2$,

$$f_G = \ln \left(2 \cosh^2 K\right) - \frac{1}{2} \int \frac{dq_x dq_y}{(2\pi)^2} \ln \left[1 - 2t \left(\cos q_x + \cos q_y\right)\right]. \quad (7.75)$$

Apart from the argument of the logarithm, this expression is similar to the exact result. Is it possible that such similar functional forms lead to distinct singular behaviors? The singularity results from the vanishing of the argument of the logarithm at $t_c = 1/4$. In the vicinity of this point we make an expansion as in Eq. (7.50),

$$A_G(t, \mathbf{q}) = (1 - 4t) + tq^2 + \mathcal{O}(q^4) \approx t_c \left(q^2 + 4\frac{\delta t}{t_c}\right), \quad (7.76)$$

where $\delta t = t_c - t$. The singular part of Eq. (7.75) can be obtained by focusing on the behavior of the integrand as $\mathbf{q} \to \mathbf{0}$, and replacing the square Brillouin zone for the range of the integral with a circle of radius $\Lambda \approx 2\pi$,

$$f_{\text{sing}} = -\frac{1}{2} \int_0^\Lambda \frac{2\pi q dq}{4\pi^2} \ln \left(q^2 + 4\frac{\delta t}{t_c}\right)$$

$$= -\frac{1}{8\pi} \left[\left(q^2 + 4\frac{\delta t}{t_c}\right) \ln \left(\frac{q^2 + 4\delta t/t_c}{e}\right)\right]_0^\Lambda. \quad (7.77)$$

Only the expression evaluated at $q = 0$ is singular, and

$$f_{\text{sing}} = \frac{1}{2\pi} \left(\frac{\delta t}{t_c} \right) \ln \left(\frac{\delta t}{t_c} \right). \qquad (7.78)$$

The resulting heat capacity, $C_G \propto \partial^2 f_G / \partial^2 t$, diverges as $1/\delta t$. Since Eq. (7.75) is not valid for $t > t_c$, we cannot obtain the behavior of heat capacity on the low-temperature side.

For the exact result of Eq. (7.73), the argument of the logarithm is

$$A^*(t, \mathbf{q}) = \left(1 + t^2 \right)^2 - 2t \left(1 - t^2 \right) \left(\cos q_x + \cos q_y \right). \qquad (7.79)$$

The minimum value of this expression, for $\mathbf{q} = \mathbf{0}$, is

$$A^*(t, \mathbf{0}) = \left(1 + t^2 \right)^2 - 4t_c \left(1 - t_c^2 \right) = \left(1 - t_c^2 \right)^2 + 4t_c^2 - 4t_c \left(1 - t_c^2 \right) = \left(1 - t_c^2 - 2t_c \right)^2. \qquad (7.80)$$

Since this expression (and hence the argument of the logarithm) is always nonnegative, the integral exists for all values of t. As required, unlike Eq. (7.75), the exact result is valid at *all* temperatures. There is a singularity when the argument vanishes for

$$t_c^2 + 2t_c - 1 = 0, \quad \Longrightarrow \quad t_c = -1 \pm \sqrt{2}. \qquad (7.81)$$

The positive solution describes a ferromagnet, and leads to a value of $K_c = \ln \left(\sqrt{2} + 1 \right) / 2$, in agreement with the duality arguments of Section 7.4. Setting $\delta t = t - t_c$, and expanding Eq. (7.79) in the vicinity of $\mathbf{q} \to 0$ gives

$$A^*(t, \mathbf{q}) \approx [(-2t_c - 2)\delta t]^2 + t_c(1 - t_c^2)q^2 + \cdots$$

$$\approx 2t_c^2 \left[q^2 + 4 \left(\frac{\delta t}{t_c} \right)^2 \right]. \qquad (7.82)$$

The important difference from Eq. (7.76) is that $(\delta t / t_c)$ appears at quadratic order. Following the steps in Eqs. (7.77) and (7.78), the singular part of the free energy is

$$\left. \frac{\ln Z}{N} \right|_{\text{sing}} = \frac{1}{2} \int_0^{\Lambda} \frac{2\pi q dq}{4\pi^2} \ln \left[q^2 + 4 \left(\frac{\delta t}{t_c} \right)^2 \right]$$

$$= \frac{1}{8\pi} \left[\left(q^2 + 4 \left(\frac{\delta t}{t_c} \right)^2 \right) \ln \left(\frac{q^2 + 4(\delta t/t_c)^2}{e} \right) \right]_0^{\Lambda} \qquad (7.83)$$

$$= -\frac{1}{\pi} \left(\frac{\delta t}{t_c} \right)^2 \ln \left| \frac{\delta t}{t_c} \right| + \text{analytic terms}.$$

The heat capacity is obtained by taking two derivatives and diverges as $C(\delta t)_{\text{sing}} = A_{\pm} \ln |\delta t|$. The logarithmic singularity corresponds to the limit $\alpha = 0$; the peak is symmetric, characterized by the amplitude ratio $A_+ / A_- = 1$.

The exact partition function of the Ising model on the square lattice was originally calculated by Lars Onsager in 1944 (Phys. Rev. **65**, 117). Onsager used a $2^L \times 2^L$ transfer matrix to study a lattice of width L. He then identified various symmetries of this matrix which allowed him to diagonalize it and obtain the largest eigenvalue as a function of L. For any finite L, this eigenvalue is non-degenerate as required by Frobenius' theorem. In the limit $L \to \infty$, the top two eigenvalues become degenerate at K_c. The result in this limit naturally coincides with Eq. (7.74). Onsager's paper is quite long and complicated, and regarded as a *tour de force* of mathematical physics. A somewhat streamlined version of this solution was developed by B. Kaufman (Phys. Rev. **76**, 1232 (1949)). The exact solution, for the first time, demonstrated that critical behavior at a phase transition can in fact be quite different from predictions of Landau (mean-field) theory. It took almost three decades to reconcile the two results by the renormalization group.

The graphical method presented in this section was originally developed by Kac and Ward (Phys. Rev. **88**, 1332 (1952)). The main ingredient of the derivation is the result that the correct accounting of the paths can be achieved by including a factor of (-1) for each intersection. (This conjecture by Feynman is proved in S. Sherman, J. Math. Phys. **1**, 202 (1960).) The change of sign is reminiscent of the exchange factor between fermions. Indeed, Schultz, Mattis, and Lieb (Rev. Mod. Phys. **36**, 856 (1964)) describe how the Onsager transfer matrix can be regarded as a Hamiltonian for the evolution of non-interacting fermions in one dimension. The Pauli exclusion principle prevents intersections of the world-lines of these fermions in two dimensional space-time.

In addition to the partition function, the correlation functions $\langle \sigma_i \sigma_j \rangle$ can also be calculated by summing over paths. Since the combination $q^2 + 4(\delta t / t_c)^2$ in Eq. (7.82) describes the behavior of these random walks, we expect a correlation length $\xi \sim |t_c / \delta t|$, i.e. diverging with an exponent $\nu = 1$ on both sides of the phase transition with an amplitude ratio of unity. The exponents α and ν are related by the hyperscaling identity $\alpha = 2 - 2\nu$. The critical correlations at t_c are more subtle and decay as $\langle \sigma_i \sigma_j \rangle_c \sim 1/|\mathbf{i} - \mathbf{j}|^\eta$, with $\eta = 1/4$. Integrating the correlation functions yields the susceptibility, which diverges as $\chi_\pm \simeq C_\pm |\delta t|^{-\gamma}$, with $\gamma = 7/4$, and $C_+/C_- = 1$. A rather simple expression for the zero field magnetization,

$$m = \left(1 - \sinh^{-4}(2K)\right)^{1/8}, \tag{7.84}$$

was presented without proof by Onsager. A rather difficult derivation of this result was finally given by C.N. Yang (Phys. Rev. **85**, 808 (1952)). The exponent $\beta = 1/8$ satisfies the required exponent identities.

Problems for chapter 7

1. *Continuous spins:* In the standard $\mathcal{O}(n)$ model, n component unit vectors are placed on the sites of a lattice. The nearest-neighbor spins are then connected by a bond

$J\vec{s}_i \cdot \vec{s}_j$. In fact, if we are only interested in universal properties, any generalized interaction $f(\vec{s}_i \cdot \vec{s}_j)$ leads to the same critical behavior. By analogy with the Ising model, a suitable choice is

$$\exp\left[f(\vec{s}_i \cdot \vec{s}_j)\right] = 1 + (nt)\vec{s}_i \cdot \vec{s}_j,$$

resulting in the so called *loop model*.

(a) Construct a high temperature expansion of the loop model (for the partition function Z) in the parameter t, on a two-dimensional *hexagonal* (honeycomb) lattice.

(b) Show that the limit $n \to 0$ describes the configurations of a single self-avoiding polymer on the lattice.

2. *Potts model I:* Consider Potts spins $s_i = (1, 2, \cdots, q)$, interacting via the Hamiltonian $-\beta\mathcal{H} = K \sum_{\langle ij \rangle} \delta_{s_i, s_j}$.

(a) To treat this problem graphically at high temperatures, the Boltzmann weight for each bond is written as

$$\exp\left(K\delta_{s_i, s_j}\right) = C(K)\left[1 + T(K)g(s_i, s_j)\right],$$

with $g(s, s') = q\delta_{s,s'} - 1$. Find $C(K)$ and $T(K)$.

(b) Show that

$$\sum_{s=1}^{q} g(s, s') = 0, \quad \sum_{s=1}^{q} g(s_1, s)g(s, s_2) = qg(s_1, s_2),$$

$$\text{and } \sum_{s,s'}^{q} g(s, s')g(s', s) = q^2(q-1).$$

(c) Use the above results to calculate the free energy, and the correlation function $\langle g(s_m, s_n) \rangle$ for a one-dimensional chain.

(d) Calculate the partition function on the square lattice to order of T^4. Also calculate the first term in the low-temperature expansion of this problem.

(e) By comparing the first terms in low- and high-temperature series, find a duality rule for Potts models. Don't worry about higher order graphs, they will work out! Assuming a single transition temperature, find the value of $K_c(q)$.

(f) How do the higher order terms in the high-temperature series for the Potts model differ from those of the Ising model? What is the fundamental difference that sets apart the graphs for $q = 2$? (This is ultimately the reason why only the Ising model is solvable.)

3. *Potts model II:* An alternative expansion is obtained by starting with

$$\exp\left[K\delta(s_i, s_j)\right] = 1 + v(K)\delta(s_i, s_j),$$

where $v(K) = e^K - 1$. In this case, the sum over spins *does not* remove any graphs, and all choices of distributing bonds at random on the lattice are acceptable.

(a) Including a magnetic field $h \sum_i \delta_{s_i,1}$, show that the partition function takes the form

$$Z(q, K, h) = \sum_{\text{all graphs}} \prod_{\text{clusters } c \text{ in graph}} \left[v^{n_b^c} \times (q - 1 + e^{h n_s^c}) \right],$$

where n_b^c and n_s^c are the numbers of bonds and sites in cluster c. This is known as the *random cluster expansion*.

(b) Show that the limit $q \to 1$ describes a *percolation* problem, in which bonds are randomly distributed on the lattice with probability $p = v/(v+1)$. What is the percolation threshold on the square lattice?

(c) Show that in the limit $q \to 0$, only a single connected cluster contributes to leading order. The enumeration of all such clusters is known as listing *branched lattice animals*.

4. *Ising model in a field:* Consider the partition function for the Ising model ($\sigma_i = \pm 1$) on a square lattice, *in a magnetic field h*; i.e.

$$Z = \sum_{\{\sigma_i\}} \exp \left[K \sum_{\langle ij \rangle} \sigma_i \sigma_j + h \sum_i \sigma_i \right].$$

(a) Find the general behavior of the terms in a low-temperature expansion for Z.

(b) Think of a model whose high-temperature series reproduces the generic behavior found in (a); and hence obtain the Hamiltonian, and interactions of the dual model.

5. *Potts duality:* Consider Potts spins, $s_i = (1, 2, \cdots, q)$, placed on the sites of a *square lattice* of N sites, interacting with their nearest-neighbors through a Hamiltonian

$$-\beta \mathcal{H} = K \sum_{\langle ij \rangle} \delta_{s_i, s_j}.$$

(a) By comparing the first terms of high and low temperature series, or by any other method, show that the partition function has the property

$$Z(K) = q e^{2NK} \Xi \left[e^{-K} \right] = q^{-N} \left[e^K + q - 1 \right]^{2N} \Xi \left[\frac{e^K - 1}{e^K + (q - 1)} \right],$$

for some function Ξ, and hence locate the critical point $K_c(q)$.

(b) Starting from the duality expression for $Z(K)$, derive a similar relation for the internal energy $U(K) = \langle \beta \mathcal{H} \rangle = -\partial \ln Z / \partial \ln K$. Use this to calculate the exact value of U at the critical point.

6. *Anisotropic random walks:* Consider the ensemble of all random walks on a square lattice starting at the origin (0,0). Each walk has a weight of $t_x^{\ell_x} \times t_y^{\ell_y}$, where ℓ_x and ℓ_y are the number of steps taken along the x and y directions, respectively.

(a) Calculate the total weight $W(x, y)$, of all walks terminating at (x, y). Show that W is well defined only for $\bar{t} = (t_x + t_y)/2 < t_c = 1/4$.

(b) What is the shape of a curve $W(x, y) = $ constant, for large x and y, and close to the transition?

(c) How does the average number of steps, $\langle \ell \rangle = \langle \ell_x + \ell_y \rangle$, diverge as $\bar{t}$ approaches t_c?

7. *Anisotropic ising model:* Consider the anisotropic Ising model on a square lattice with a Hamiltonian

$$-\beta\mathcal{H} = \sum_{x,y}\left(K_x \sigma_{x,y}\sigma_{x+1,y} + K_y \sigma_{x,y}\sigma_{x,y+1}\right);$$

i.e. with bonds of different strengths along the x and y directions.

(a) By following the method presented in the text, calculate the free energy for this model. You do not have to write down every step of the derivation. Just sketch the steps that need to be modified due to anisotropy; and calculate the final answer for $\ln Z/N$.

(b) Find the critical boundary in the (K_x, K_y) plane from the singularity of the free energy. Show that it coincides with the condition $K_x = \tilde{K}_y$, where $\tilde{K}$ indicates the standard dual interaction to K.

(c) Find the singular part of $\ln Z/N$, and comment on how anisotropy affects critical behavior in the exponent and amplitude ratios.

8. *Müller–Hartmann and Zittartz estimate* of the interfacial energy of the $d = 2$ Ising model on a square lattice:

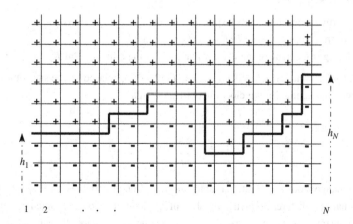

(a) Consider an interface on the square lattice with periodic boundary conditions in one direction. Ignoring islands and overhangs, the configurations can be labelled by heights h_n for $1 \leq n \leq L$. Show that for an ansiotropic Ising model of interactions (K_x, K_y), the energy of an interface along the x-direction is

$$-\beta \mathcal{H} = -2K_y L - 2K_x \sum_n |h_{n+1} - h_n|.$$

(b) Write down a column-to-column transfer matrix $\langle h|T|h' \rangle$, and diagonalize it.
(c) Obtain the interface free energy using the result in (b), or by any other method.
(d) Find the condition between K_x and K_y for which the interfacial free energy vanishes. Does this correspond to the critical boundary of the original 2d Ising model as computed in the previous problem?

9. *Anisotropic Landau theory:* Consider an n-component magnetization field $\vec{m}(\mathbf{x})$ in d–dimensions.

(a) Using the previous problems on anisotropy as a guide, generalize the standard Landau–Ginzburg Hamiltonian to include the effects of spacial anisotropy.
(b) Are such anisotropies "relevant"?
(c) In La_2CuO_4, the Cu atoms are arranged on the sites of a square lattice in planes, and the planes are then stacked together. Each Cu atom carries a "spin", which we assume to be classical, and can point along any direction in space. There is a very strong antiferromagnetic interaction in each plane. There is also a very weak interplane interaction that prefers to align successive layers. Sketch the low-temperature magnetic phase, and indicate to what universality class the order–disorder transition belongs.

---

10. *Energy by duality:* Consider the Ising model $(\sigma_i = \pm 1)$ on a square lattice with $-\beta \mathcal{H} = K \sum_{\langle ij \rangle} \sigma_i \sigma_j$.

(a) Starting from the duality expression for the free energy, derive a similar relation for the internal energy $U(K) = \langle H \rangle = -\partial \ln Z / \partial \ln K$.
(b) Using (a), calculate the exact value of U at the critical point K_c.

11. *Clock model duality:* Consider spins $s_i = (1, 2, \cdots, q)$ placed on the sites of a square lattice, interacting via the clock model Hamiltonian

$$\beta \mathcal{H}_C = -\sum_{\langle i,j \rangle} J(|s_i - s_j|_{\mathrm{mod}q}).$$

(a) Change from the N site variables to the $2N$ bond variables $b_{ij} = s_i - s_j$. Show that the difference in the number of variables can be accounted for by the constraint that around each plaquette (elementary square) the sum of the four bond variables must be zero modulus q.

(b) The constraints can be implemented by adding "delta-functions"

$$\delta\left[S_p\right]_{\mathrm{mod}q} = \frac{1}{q}\sum_{n_p=1}^{q}\exp\left[\frac{2\pi i n_p S_p}{q}\right],$$

for each plaquette. Show that after summing over the bond variables, the partition function can be written in terms of the dual variables, as

$$Z = q^{-N}\sum_{\{n_p\}}\prod_{\langle p,p'\rangle}\lambda\left(n_p-n_{p'}\right) \equiv \sum_{\{n_p\}}\exp\left[\sum_{\langle p,p'\rangle}\tilde{J}\left(n_p-n_{p'}\right)\right],$$

where $\lambda(k)$ is the discrete Fourier transform of $e^{J(n)}$.

(c) Calculate the dual interaction parameter of a Potts model, and hence locate the critical point $J_c(q)$.

(d) Construct the dual of the anisotropic Potts model, with

$$-\beta\mathcal{H} = \sum_{x,y}\left(J_x\delta_{s_{x,y},s_{x+1,y}} + J_y\delta_{s_{x,y},s_{x,y+1}}\right);$$

i.e. with bonds of different strengths along the x and y directions. Find the line of self-dual interactions in the plane $\left(J_x, J_y\right)$.

12. *Triangular/hexagonal lattices:* For any *planar* network of bonds, one can define a geometrical dual by connecting the centers of neighboring plaquettes. Each bond of the dual lattice crosses a bond of the original lattice, allowing for a *local* mapping. Clearly, the dual of a triangular lattice is a hexagonal (or honeycomb) lattice, and vice versa.

(a) Starting from a Potts model with nearest-neighbor interaction J on the triangular lattice, find the interaction parameter $J_h(J)$ of the dual hexagonal lattice.

(b) Note that the hexagonal lattice is *bipartite*, i.e. can be separated into two sublattices. In the partition function, do a partial sum over all spins in one sublattice. Show that the remaining spins form a triangular lattice with nearest-neighbor interactions $\tilde{J}(J)$, *and* a three spin interaction $\tilde{J}_3(J)$. (This is called the *star–triangle transformation.*) Why is $\tilde{J}_3$ absent in the Ising model?

(c) Clearly, the model is not self-dual due to the additional interaction. Nonetheless, obtain the critical value such that $\tilde{J}(J_c) = J_c$.

(d) Check that $\tilde{J}_3(J_c) = 0$, i.e. while the model in general is not self-dual, it is self-dual right at criticality, leading to the exact value of $J_c(q)$!

13. *Cubic lattice:* The geometric concept of duality can be extended to general dimensions d. However, the dual of a geometric element of dimension D is an entity of dimension $d-D$. For example, the dual of a bond $(D=1)$ in $d=3$ is a plaquette $(D=2)$, as demonstrated in this problem.

(a) Consider a clock model on a cubic lattice of N points. Change to the $3N$ bond variables $b_{ij} = s_i - s_j$. (Note that one must make a convention about the positive directions on the three axes.) Show that there are now $2N$ constraints associated with the plaquettes of this lattice.

(b) Implement the constraints through discrete delta-functions by associating an auxiliary variable n_p with each plaquette. It is useful to imagine n_p as defined on a bond of the dual lattice, perpendicular to the plaquette p.

(c) By summing over the bond variables in Z, obtain the dual Hamiltonian

$$\beta \tilde{\mathcal{H}} = \sum_p \tilde{J} \left(n^p_{12} - n^p_{23} + n^p_{34} - n^p_{41} \right),$$

where the sum is over all plaquettes p of the dual lattice, with $\left\{ n^p_{ij} \right\}$ indicating the four bonds around plaquette p.

(d) Note that $\beta \tilde{\mathcal{H}}$ is left invariant if all the six bonds going out of any site are simultaneously increased by the same integer. Thus unlike the original model which only had a *global translation symmetry*, the dual model has a *local*, i.e. *gauge symmetry*.

(e) Consider a Potts gauge theory defined on the plaquettes of a four-dimensional hypercubic lattice. Find its critical coupling $J_c(q)$.

Percolation

Fluids do not pass through a solid with a small concentration of holes. However, beyond a threshold concentration, the holes overlap, and the fluid can *percolate* through a connected channel in the material. *Percolation* is a classical *geometric* phase transition, and has been used as a model of many breakdown or failure processes. The loss of rigidity in an elastic network, conductivity in resistor nets, and magnetization in diluted magnets are but a few examples.

In simple models of percolation, elements of a lattice (sites or bonds) are independently occupied with a probability p. A cluster is defined as a connected (by neighboring bonds) set of these occupied elements. At small p, only small clusters exist, and the probability that two sites, separated by a distance r, are connected to each other decays as $\exp(-r/\xi)$. The correlation length $\xi(p)$ grows with increasing p, diverging at the *percolation threshold* p_c as $\xi(p) \sim |p_c - p|^{-\nu}$. An infinite cluster first appears at this threshold, and percolates through the (infinite) system for all $p > p_c$. The analog of the order parameter is the probability $P(p)$ that a site belongs to this infinite cluster. On approaching p_c from above, it vanishes as $P(p) \sim |p_c - p|^\beta$. While the value of p_c depends on the details of the model, the exponents β and ν are *universal*, varying only with the spatial dimension d.

In the following problems we shall focus on *bond* percolation, i.e. p denotes the probability that a bond on the lattice is occupied.

14. *Cayley trees:* Consider a hierarchical lattice in which each site at one level is connected to z sites at the level below. Thus the n-th level of the tree has z^n sites.

   (a) For $z = 2$, obtain a recursion relation for the probability $P_n(p)$ that the top site of a tree of n levels is connected to some site at the bottom level.

   (b) Find the limiting behavior of $P(p)$ for infinitely many levels, and give the exponent β for this tree.

   (c) In a *mean-field* approximation, $P(p)$ for a lattice of coordination number z is calculated self-consistently, in a manner similar to the above. Give the mean-field estimates for p_c and the exponent β.

   (d) The one-dimensional chain corresponds to $z = 1$. Find the probability that end-points of an open chain of $N + 1$ sites are connected, and hence deduce the correlation length ξ.

15. *Duality* has a very natural interpretation in percolation: If a bond is occupied, its dual is empty, and vice versa. Thus the occupation probability for dual bonds is $\tilde{p} = 1 - p \equiv q$. Since, by construction, the original and dual elements do not intersect, one or the other percolates through the system.

   (a) The dual of a chain in which N bonds are connected in *series* has N bonds connected in *parallel*. What is the corresponding (non-) percolation probability?

   (b) The bond percolation problem on a square lattice is self-dual. What is its threshold p_c?

   (c) Bond percolation in three dimensions is dual to plaquette percolation. This is why in $d = 3$ it is possible to percolate while maintaining solid integrity.

   (d) The triangular and hexagonal lattices are dual to each other. Combined with the star–triangle transformation that also maps the two lattices, this leads to exact values of p_c for these lattices. However, this calculation is not trivial, and it is best to follow the steps for calculations of the critical couplings of q-state Potts models in an earlier problem.

16. *Position–space renormalization group (PSRG):* In the Migdal–Kadanoff approximation, a decimation is performed after bonds are moved to connect the remaining sites in a one-dimensional geometry. For a PSRG rescaling factor b on a d-dimensional hypercubic lattice, the retained sites are connected by b^{d-1} strands in parallel, each strand having b bonds in series. This scheme is naturally exact for $d = 1$, and becomes progressively worse in higher dimensions.

   (a) Apply the above scheme to $b = d = 2$, and compare the resulting estimates of p_c and $y_t = 1/\nu$ to the exact values for the square lattice ($p_c = 1/2$ and $y_t = 3/4$).

   (b) Find the limiting values of p_c and y_t for large d. (It can be shown that $\nu = 1/2$ *exactly*, above an upper critical dimension $d_u = 6$.)

   (c) In an alternative PSRG scheme, first b^{d-1} bonds are connected in parallel, the resulting groups are then joined in series. Repeat the above calculations with this scheme.

8

Beyond spin waves

8.1 The nonlinear σ model

Previously we considered low-temperature expansions for *discrete* spins (Ising, Potts, etc.), in which the low energy excitations are droplets of incorrect spin in a uniform background selected by broken symmetry. These excitations occur at small scales, and are easily described by graphs on the lattice. By contrast, for *continuous* spins, the lowest energy excitations are long-wavelength Goldstone modes, as discussed in Section 2.4. The thermal excitation of these modes destroys the long-range order in dimensions $d \leq 2$. For d close to 2, the critical temperature must be small, making low-temperature expansions a potential tool for the study of critical phenomena. As we shall demonstrate next, such an approach requires keeping track of the interactions between Goldstone modes.

Consider unit n-component spins on the sites of a lattice, i.e.

$$\vec{s}(\mathbf{i}) = (s_1, s_2, \cdots, s_n), \quad \text{with} \quad |\vec{s}(\mathbf{i})|^2 = s_1^2 + \cdots + s_n^2 = 1. \tag{8.1}$$

The usual nearest-neighbor Hamiltonian can be written as

$$-\beta \mathcal{H} = K \sum_{\langle \mathbf{ij} \rangle} \vec{s}(\mathbf{i}) \cdot \vec{s}(\mathbf{j}) = K \sum_{\langle \mathbf{ij} \rangle} \left(1 - \frac{(\vec{s}(\mathbf{i}) - \vec{s}(\mathbf{j}))^2}{2} \right). \tag{8.2}$$

At low temperatures, the fluctuations between neighboring spins are small and the difference in Eq. (8.2) can be replaced by a gradient. Assuming a unit lattice spacing,

$$-\beta \mathcal{H} = -\beta E_0 - \frac{K}{2} \int d^d \mathbf{x} \left(\nabla \vec{s}(\mathbf{x}) \right)^2, \tag{8.3}$$

where the discrete index $\mathbf{i}$ has been replaced by a continuous vector $\mathbf{x} \in \Re^d$. A cutoff of $\Lambda \approx \pi$ is thus implicit in Eq. (8.3). Ignoring the ground state energy, the partition function is

$$Z = \int \mathcal{D}\left[\vec{s}(\mathbf{x})\delta\left(s(\mathbf{x})^2 - 1\right)\right] e^{-\frac{K}{2}\int d^d \mathbf{x}(\nabla \vec{s})^2}. \tag{8.4}$$

A possible ground state configuration is $\vec{s}(\mathbf{x}) = (0, \cdots, 1)$. There are $n-1$ Goldstone modes describing the transverse fluctuations. To examine the effects of these fluctuations close to zero temperature, set

$$\vec{s}(\mathbf{x}) = (\pi_1(\mathbf{x}), \cdots, \pi_{n-1}(\mathbf{x}), \sigma(\mathbf{x})) \equiv (\vec{\pi}(\mathbf{x}), \sigma(\mathbf{x})), \tag{8.5}$$

where $\vec{\pi}(\mathbf{x})$ is an $n-1$ component vector. The unit length of the spin fixes $\sigma(\mathbf{x})$ in terms of $\vec{\pi}(\mathbf{x})$. For each degree of freedom

$$
\int d\vec{s}\, \delta(s^2-1) = \int_{-\infty}^{\infty} d\vec{\pi} d\sigma \delta\left(\pi^2+\sigma^2-1\right)
$$
$$
= \int_{-\infty}^{\infty} d\vec{\pi} d\sigma \delta\left[\left(\sigma-\sqrt{1-\pi^2}\right)\left(\sigma+\sqrt{1-\pi^2}\right)\right] \qquad (8.6)
$$
$$
= \int_{-\infty}^{\infty} \frac{d\vec{\pi}}{2\sqrt{1-\pi^2}} \,,
$$

where we have used the identity $\delta(ax) = \delta(x)/|a|$. Using this result, the partition function in Eq. (8.4) can be written as

$$
Z \propto \int \frac{\mathcal{D}\vec{\pi}(\mathbf{x})}{\sqrt{1-\pi(\mathbf{x})^2}} e^{-\frac{K}{2}\int d^dx\left[(\nabla\vec{\pi})^2+\left(\nabla\sqrt{1-\pi^2}\right)^2\right]}
$$
$$
= \int \mathcal{D}\vec{\pi}(\mathbf{x}) \exp\left\{-\int d^dx\left[\frac{K}{2}(\nabla\vec{\pi})^2+\frac{K}{2}\left(\nabla\sqrt{1-\pi^2}\right)^2+\frac{\rho}{2}\ln(1-\pi^2)\right]\right\}.
$$
$$
(8.7)
$$

In going from the lattice to the continuum, we have introduced a density $\rho = N/V = 1/a^d$ of lattice points. For unit lattice spacing $\rho = 1$, but for the purpose of renormalization we shall keep an arbitrary ρ. Whereas the original Hamiltonian was quite simple, the one describing the Goldstone modes $\vec{\pi}(\mathbf{x})$ is rather complicated. In selecting a particular ground state, the rotational symmetry was broken. The nonlinear terms in Eq. (8.7) ensure that this symmetry is properly reflected when considering only $\vec{\pi}$.

We can expand the nonlinear terms for the effective Hamiltonian in powers of $\vec{\pi}(\mathbf{x})$, resulting in a series

$$
\beta\mathcal{H}[\vec{\pi}(\mathbf{x})] = \beta\mathcal{H}_0 + \mathcal{U}_1 + \mathcal{U}_2 + \cdots, \qquad (8.8)
$$

where

$$
\beta\mathcal{H}_0 = \frac{K}{2}\int d^dx(\nabla\vec{\pi})^2 \qquad (8.9)
$$

describes independent Goldstone modes, while

$$
\mathcal{U}_1 = \int d^dx\left[\frac{K}{2}(\vec{\pi}\cdot\nabla\vec{\pi})^2-\frac{\rho}{2}\pi^2\right] \qquad (8.10)
$$

is the first order perturbation when the terms in the series are organized according to powers of $T = 1/K$. Since we expect fluctuations $\langle\pi^2\rangle \propto T$, $\beta\mathcal{H}_0$ is

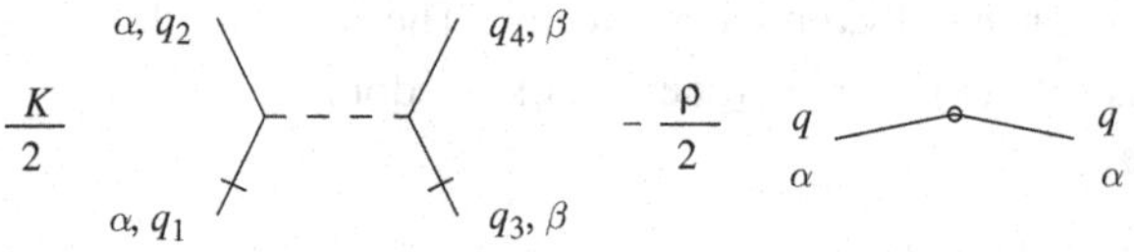

Fig. 8.1 Diagrammatic
representation of the two
terms in the perturbation
$\mathcal{U}_1$.

Fig. 8.1 Diagrammatic representation of the two terms in the perturbation $\mathcal{U}_1$.

order of one, the two terms in $\mathcal{U}_1$ are order of T; remaining terms are order of T^2 and higher. In the language of Fourier modes,

$$\beta\mathcal{H}_0 = \frac{K}{2} \int \frac{d^d\mathbf{q}}{(2\pi)^d} \, q^2 \, |\vec{\pi}(\mathbf{q})|^2,$$

$$\mathcal{U}_1 = -\frac{K}{2} \int \frac{d^d\mathbf{q}_1 \, d^d\mathbf{q}_2 \, d^d\mathbf{q}_3}{(2\pi)^{3d}} \, \pi_\alpha(\mathbf{q}_1) \, \pi_\alpha(\mathbf{q}_2) \, \pi_\beta(\mathbf{q}_3) \, \pi_\beta(-\mathbf{q}_1-\mathbf{q}_2-\mathbf{q}_3) \, (\mathbf{q}_1 \cdot \mathbf{q}_3)$$

$$-\frac{\rho}{2} \int \frac{d^d\mathbf{q}}{(2\pi)^d} \, |\vec{\pi}(\mathbf{q})|^2. \tag{8.11}$$

For the non-interacting (quadratic) theory, the correlation functions of the Goldstone modes are

$$\langle \pi_\alpha(\mathbf{q})\pi_\beta(\mathbf{q}') \rangle_0 = \frac{\delta_{\alpha\beta}(2\pi)^d\delta^d(\mathbf{q}+\mathbf{q}')}{Kq^2}. \tag{8.12}$$

The resulting fluctuations in real space behave as

$$\langle \pi(\mathbf{x})^2 \rangle_0 = \int \frac{d^d\mathbf{q}}{(2\pi)^d} \langle |\vec{\pi}(\mathbf{q})|^2 \rangle_0 = \frac{(n-1)}{K} \int_{1/L}^{1/a} \frac{d^d\mathbf{q}}{(2\pi)^d} \frac{1}{q^2}$$

$$= \frac{(n-1)}{K} \frac{K_d \left(a^{2-d} - L^{2-d}\right)}{(d-2)}. \tag{8.13}$$

For $d > 2$ the fluctuations are indeed proportional to T. However, for $d \leq 2$ they diverge as $L \to \infty$. This is a consequence of the Mermin–Wagner theorem on the absence of long range order in $d \leq 2$. Polyakov [Phys. Lett. **59B**, 79 (1975)] reasoned that this implies a critical temperature $T_c \sim \mathcal{O}(d-2)$, and that an RG expansion in powers of T may provide a systematic way to explore critical behavior close to two dimensions.

To construct a perturbative RG, consider a spherical Brillouin zone of radius Λ, and divide the modes as $\vec{\pi}(\mathbf{q}) = \vec{\pi}^<(\mathbf{q}) + \vec{\pi}^>(\mathbf{q})$. The modes $\vec{\pi}^<$ involve momenta $0 < |\mathbf{q}| < \Lambda/b$, while we shall integrate over the short wavelength fluctuations $\vec{\pi}^>$ with momenta in the shell $\Lambda/b < |\mathbf{q}| < \Lambda$. To order of T, the coarse-grained Hamiltonian is given by

$$\beta\tilde{\mathcal{H}}\left[\vec{\pi}^<\right] = V\delta f_b^0 + \beta\mathcal{H}_0\left[\vec{\pi}^<\right] + \langle \mathcal{U}_1\left[\vec{\pi}^< + \vec{\pi}^>\right] \rangle_0^> + \mathcal{O}(T^2), \tag{8.14}$$

where $\langle \ \rangle_0^>$ indicates averaging over $\vec{\pi}^>$.

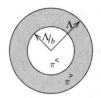

Fig. 8.2 Decomposition of Fourier modes to short and long range components for RG.

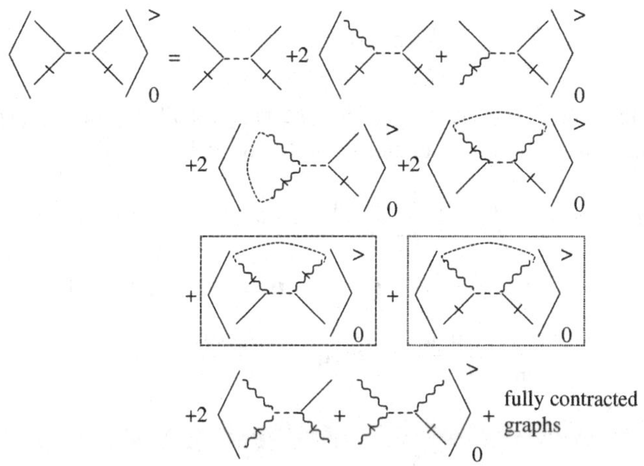

Fig. 8.3 Terms arising from the quartic interaction in first order perturbation theory.

The term proportional to ρ in Eq. (8.11) results in two contributions; one is a constant addition to the free energy (from $\langle(\pi^>)^2\rangle$), and the other is simply $\rho(\pi^<)^2$. (The cross terms proportional to $\vec{\pi}^< \cdot \vec{\pi}^>$ vanish by symmetry.) The quartic part of $\mathcal{U}_1$ generates 16 terms as indicated in Fig. 8.3. The contributions on the first and last lines involve either an odd number of $\vec{\pi}^>$, and hence vanish by symmetry, or come with either 4 or 0 factors of $\vec{\pi}^<$. The former reproduces the quartic interaction, while the latter adds an overall constant. The more interesting contributions arise from products of two $\vec{\pi}^<$ and two $\vec{\pi}^>$ as depicted in the second and third lines, and are of three types. The first type (appearing on the second line) has the form

$$\langle \mathcal{U}_1^a \rangle_0^> = 2 \times \frac{-K}{2} \int \frac{d^d\mathbf{q}_1\, d^d\mathbf{q}_2\, d^d\mathbf{q}_3}{(2\pi)^{3d}} (\mathbf{q}_1 \cdot \mathbf{q}_3)$$
$$\times \langle \pi_\alpha^>(\mathbf{q}_1)\pi_\alpha^>(\mathbf{q}_2) \rangle_0^> \pi_\beta^<(\mathbf{q}_3)\pi_\beta^<(-\mathbf{q}_1 - \mathbf{q}_2 - \mathbf{q}_3). \tag{8.15}$$

The integral over the shell momentum $\mathbf{q}_1$ is odd and this contribution is zero. (Two similar vanishing terms arise from contractions with different indices

α and β.) The second term (in the dashed rectangle on the third line) is a renormalization of ρ, arising from

$$
\begin{aligned}
\langle \mathcal{U}_1^b \rangle_0^> &= -\frac{K}{2} \int \frac{d^d\mathbf{q}_1\, d^d\mathbf{q}_2\, d^d\mathbf{q}_3}{(2\pi)^{3d}} (\mathbf{q}_1 \cdot \mathbf{q}_3) \\
&\quad \times \langle \pi_\alpha^>(\mathbf{q}_1) \pi_\beta^>(\mathbf{q}_3) \rangle_0^> \, \pi_\alpha^<(\mathbf{q}_2) \pi_\beta^<(-\mathbf{q}_1 - \mathbf{q}_2 - \mathbf{q}_3) \\
&= \frac{K}{2} \int_0^{\Lambda/b} \frac{d^d\mathbf{q}}{(2\pi)^d} |\vec{\pi}^<(\mathbf{q})|^2 \times \int_{\Lambda/b}^{\Lambda} \frac{d^d\mathbf{k}}{(2\pi)^d} \frac{k^2}{Kk^2} \\
&= \frac{\rho}{2} \int_0^{\Lambda/b} \frac{d^d\mathbf{q}}{(2\pi)^d} |\vec{\pi}^<(\mathbf{q})|^2 \times \left(1 - b^{-d}\right).
\end{aligned}
\tag{8.16}
$$

(Note that in general $\rho = N/V = \int_0^\Lambda d^d\mathbf{q}/(2\pi)^d$.) Finally, a renormalization of K is obtained from the graph enclosed by the dotted rectangle, as

$$
\begin{aligned}
\langle \mathcal{U}_1^c \rangle_0^> &= -\frac{K}{2} \int \frac{d^d\mathbf{q}_1\, d^d\mathbf{q}_2\, d^d\mathbf{q}_3}{(2\pi)^{3d}} (\mathbf{q}_1 \cdot \mathbf{q}_3) \\
&\quad \times \langle \pi_\alpha^>(\mathbf{q}_2) \pi_\beta^>(-\mathbf{q}_1 - \mathbf{q}_2 - \mathbf{q}_3) \rangle_0^> \, \pi_\alpha^<(\mathbf{q}_1) \pi_\beta^<(\mathbf{q}_3) \\
&= \frac{K}{2} \int_0^{\Lambda/b} \frac{d^d\mathbf{q}}{(2\pi)^d} q^2 |\vec{\pi}^<(\mathbf{q})|^2 \times \frac{I_d(b)}{K},
\end{aligned}
\tag{8.17}
$$

where

$$
I_d(b) \equiv \int_{\Lambda/b}^{\Lambda} \frac{d^d\mathbf{k}}{(2\pi)^d} \frac{1}{k^2} = \frac{K_d \Lambda^{d-2} \left(1 - b^{2-d}\right)}{(d-2)}.
\tag{8.18}
$$

The coarse-grained Hamiltonian in Eq. (8.14) now equals

$$
\begin{aligned}
\beta \tilde{\mathcal{H}}[\vec{\pi}^<] &= V\delta f_b^0 + V\delta f_b^1 + \frac{K}{2}\left(1 + \frac{I_d(b)}{K}\right) \int_0^{\Lambda/b} \frac{d^d\mathbf{q}}{(2\pi)^d} q^2 |\vec{\pi}^<(\mathbf{q})|^2 \\
&\quad + \frac{K}{2} \int \frac{d^d\mathbf{q}_1\, d^d\mathbf{q}_2\, d^d\mathbf{q}_3}{(2\pi)^{3d}} \pi_\alpha^<(\mathbf{q}_1)\, \pi_\alpha^<(\mathbf{q}_2)\, \pi_\beta^<(\mathbf{q}_3)\, \pi_\beta^<(-\mathbf{q}_1 - \mathbf{q}_2 - \mathbf{q}_3)(\mathbf{q}_1 \cdot \mathbf{q}_3) \\
&\quad - \frac{\rho}{2} \int_0^{\Lambda/b} \frac{d^d\mathbf{q}}{(2\pi)^d} |\vec{\pi}^<(\mathbf{q})|^2 \times \left[1 - \left(1 - b^{-d}\right)\right] + \mathcal{O}(T^2).
\end{aligned}
\tag{8.19}
$$

The most important consequence of coarse graining is the change of the stiffness coefficient K to

$$
\tilde{K} = K\left(1 + \frac{I_d(b)}{K}\right).
\tag{8.20}
$$

After rescaling, $\mathbf{x}' = \mathbf{x}/b$, and renormalizing, $\vec{\pi}'(\mathbf{x}) = \vec{\pi}^<(\mathbf{x})/\zeta$, we obtain the renormalized Hamiltonian in real space as

$$
\begin{aligned}
-\beta \mathcal{H}' &= -V\delta f_b^0 - V\delta f_b^1 - \frac{\tilde{K} b^{d-2} \zeta^2}{2} \int d^d\mathbf{x}' (\nabla' \vec{\pi}')^2 \\
&\quad - \frac{K b^{d-2} \zeta^4}{2} \int d^d\mathbf{x}' \left(\vec{\pi}'(\mathbf{x}') \nabla \vec{\pi}'(\mathbf{x}')\right)^2 + \frac{\rho \zeta^2}{2} \int d^d\mathbf{x}' \pi'(\mathbf{x}')^2 + \mathcal{O}(T^2).
\end{aligned}
\tag{8.21}
$$

The easiest method for obtaining the rescaling factor ζ is to take advantage of the rotational symmetry of spins. After averaging over the short wavelength modes, the spin is

$$
\begin{aligned}
\left\langle \vec{s} \right\rangle_0^> &= \left\langle \left(\left(\pi_1^< + \pi_1^>, \cdots, \sqrt{1 - (\vec{\pi}^< + \vec{\pi}^>)^2} \right) \right) \right\rangle_0^> \\
&= \left(\pi_1^<, \cdots, 1 - \frac{(\vec{\pi}^<)^2}{2} - \left\langle \frac{(\vec{\pi}^>)^2}{2} \right\rangle_0^> + \cdots \right) \\
&= \left(1 - \left\langle \frac{(\vec{\pi}^>)^2}{2} \right\rangle_0^> + \mathcal{O}(T^2) \right) \left(\pi_1^<, \cdots, \sqrt{1 - (\vec{\pi}^<)^2} \right).
\end{aligned}
\tag{8.22}
$$

We thus identify

$$
\zeta = 1 - \left\langle \frac{(\vec{\pi}^>)^2}{2} \right\rangle_0^> + \mathcal{O}(T^2) = 1 - \frac{(n-1)}{2} \frac{I_d(b)}{K} + \mathcal{O}(T^2)
\tag{8.23}
$$

as the length of the coarse-grained spin. The renormalized coupling constant in Eq. (8.21) is now obtained from

$$
\begin{aligned}
K' &= b^{d-2} \zeta^2 \tilde{K} \\
&= b^{d-2} \left[1 - \frac{n-1}{2K} I_d(b) \right]^2 K \left[1 + \frac{1}{K} I_d(b) \right] \\
&= b^{d-2} K \left[1 - \frac{n-2}{K} I_d(b) + \mathcal{O}\left(\frac{1}{K^2} \right) \right].
\end{aligned}
\tag{8.24}
$$

For infinitesimal rescaling, $b = (1 + \delta\ell)$, the shell integral results in

$$
I_d(b) = K_d \Lambda^{d-2} \delta\ell.
\tag{8.25}
$$

The differential recursion relation corresponding to Eq. (8.24) is thus

$$
\frac{dK}{d\ell} = (d-2)K - (n-2)K_d \Lambda^{d-2}.
\tag{8.26}
$$

Alternatively, the scaling of temperature $T = K^{-1}$ is

$$
\frac{dT}{d\ell} = -\frac{1}{K^2} \frac{dK}{d\ell} = -(d-2)T + (n-2)K_d \Lambda^{d-2} T^2.
\tag{8.27}
$$

It may appear that we should also keep track of the evolution of the coefficients of the two terms in $\mathcal{U}_1$ under RG. In fact, spherical symmetry ensures that the coefficient of the quartic term is precisely the same as K at all orders. The apparent difference between the two in Eq. (8.21) is of order of $\mathcal{O}(T^2)$, and will vanish when all terms at this order are included. The coefficient of the second order term in $\mathcal{U}_1$ merely tracks the density of points and also has trivial renormalization.

The behavior of temperature under RG changes drastically at $d = 2$. For $d < 2$, the linear flow is away from zero, indicating that the ordered phase is unstable and there is no broken symmetry. For $d > 2$, small T flows back to zero, indicating that the ordered phase is stable. The flows for $d = 2$ are controlled by the second ordered term which changes sign at $n = 2$. For $n > 2$ the flow is towards high temperatures, indicating that Heisenberg and higher spin models are disordered. The situation for $n = 2$ is ambiguous, and it can in fact be shown that $dT/d\ell$ is zero to all orders. This special case will be discussed in more detail in the next section. For $d > 2$ and $n > 2$, there is a phase transition at the fixed point,

$$T^* = \frac{\epsilon}{(n-2)K_d\Lambda^{d-2}} = \frac{2\pi\epsilon}{(n-2)} + \mathcal{O}(\epsilon^2), \tag{8.28}$$

where $\epsilon = d - 2$ is used as a small parameter. The recursion relation at order of ϵ is

$$\frac{dT}{d\ell} = -\epsilon T + \frac{(n-2)}{2\pi}T^2. \tag{8.29}$$

Stability of the fixed point is determined by the linearized recursion relation

$$\left.\frac{d\delta T}{d\ell}\right|_{T^*} = \left[-\epsilon + \frac{(n-2)}{\pi}T^*\right]\delta T = [-\epsilon + 2\epsilon]\,\delta T = \epsilon\delta T, \implies y_t = \epsilon. \tag{8.30}$$

The thermal eigenvalue, and the resulting exponents $\nu = 1/\epsilon$, and $\alpha = 2 - (2 + \epsilon)/\epsilon \approx -2/\epsilon$, are independent of n at this order.

The magnetic eigenvalue can be obtained by adding a term $-\vec{h} \cdot \int d^d\mathbf{x}\,\vec{s}(\mathbf{x})$, to the Hamiltonian. Under the action of RG, $h' = b^d\zeta h \equiv b^{y_h}h$, with

$$b^{y_h} = b^d\left[1 - \frac{n-1}{2K}I_d(b)\right]. \tag{8.31}$$

For an infinitesimal rescaling

$$1 + y_h\delta\ell = (1 + d\delta\ell)\left(1 - \frac{n-1}{2}T^*K_d\Lambda^{d-2}\delta\ell\right), \tag{8.32}$$

leading to

$$y_h = d - \frac{n-1}{2(n-2)}\epsilon = 1 + \frac{n-3}{2(n-2)}\epsilon + \mathcal{O}(\epsilon^2), \tag{8.33}$$

which does depend on n. Using exponent identities, we find

$$\eta = 2 + d - 2y_h = \frac{\epsilon}{n-2}. \tag{8.34}$$

The exponent η is zero at the lowest order in a $4 - d$ expansion, but appears at first order in the vicinity of two dimensions. The actual values of the exponents calculated at this order are not very satisfactory.

8.2 Topological defects in the XY model

As stated in the previous section, thermal excitation of Goldstone modes destroys spontaneous order in two-dimensional models with a broken continuous symmetry. The RG study of the nonlinear σ model confirms that the transition temperature of n-component spins vanishes as $T^* = 2\pi\epsilon/(n-2)$ for $\epsilon = (d-2) \to 0$. However, the same RG procedure appears to suggest a different behavior for $n = 2$. The first indication of unusual behavior for the two-dimensional spin models appeared in an analysis of high-temperature series [H.E. Stanley and T.A. Kaplan, Phys. Rev. Lett. **17**, 913 (1966)]. The series results strongly suggested the divergence of susceptibility at a finite temperature, seemingly in contradiction with the absence of symmetry breaking. It was indeed this contradiction that led Wegner to explore the possibility of a phase transition without symmetry breaking [F.J. Wegner, J. Math. Phys. **12**, 2259 (1971)]. The Z_2 lattice gauge theory, discussed in Section 7.5 as the dual of the three-dimensional Ising model, realizes such a possibility. The two phases of the Z_2 gauge theory are characterized by different functional forms for the decay of an appropriate correlation function (the Wilson loop). We can similarly examine the asymptotic behavior of the spin–spin correlation functions of the XY model at high and low temperatures.

A high-temperature expansion for the correlation function for the XY model on a lattice is constructed from

$$\langle \vec{s}_0 \cdot \vec{s}_r \rangle = \langle \cos(\theta_0 - \theta_r) \rangle = \frac{1}{Z} \prod_{i=1}^{N} \left(\int_0^{2\pi} \frac{d\theta_i}{2\pi} \right) \cos(\theta_0 - \theta_r) e^{K \sum_{\langle i,j \rangle} \cos(\theta_i - \theta_j)}$$

$$= \frac{1}{Z} \prod_{i=1}^{N} \left(\int_0^{2\pi} \frac{d\theta_i}{2\pi} \right) \cos(\theta_0 - \theta_r) \prod_{\langle i,j \rangle} \left[1 + K \cos(\theta_i - \theta_j) + \mathcal{O}(K^2) \right].$$

(8.35)

The expansion for the partition function is similar, except for the absence of the factor $\cos(\theta_0 - \theta_r)$. To the lowest order in K, each bond on the lattice contributes either a factor of one, or $K \cos(\theta_i - \theta_j)$. Since,

$$\int_0^{2\pi} \frac{d\theta_1}{2\pi} \cos(\theta_1 - \theta_2) = 0,$$

(8.36)

any graph with a single bond emanating from an *internal* site vanishes. For the numerator of Eq. (8.35) to be non-zero, there must be bonds originating from the *external* points at 0 and r. Integrating over an *internal* point with two bonds leads to

$$\int_0^{2\pi} \frac{d\theta_2}{2\pi} \cos(\theta_1 - \theta_2) \cos(\theta_2 - \theta_3) = \frac{1}{2} \cos(\theta_1 - \theta_3).$$

(8.37)

The leading graph that contributes to Eq. (8.35) is the shortest path (of length r) connecting the points 0 and r. Since integrating over the end points gives

$$\int_0^{2\pi} \frac{d\theta_0 d\theta_r}{(2\pi)^2} \cos(\theta_0 - \theta_r)^2 = \frac{1}{2},$$

(8.38)

each bond along the path contributes $(K/2)$. (In constructing graphs for the partition function, there is an additional factor of 2 for every loop.) Thus to lowest order

$$\langle \vec{s}_0 \cdot \vec{s}_r \rangle \approx \left(\frac{K}{2} \right)^r = e^{-r/\xi}, \quad \text{with} \quad \xi \approx \frac{1}{\ln(2/K)}, \tag{8.39}$$

and the disordered high-temperature phase is characterized by an exponential decay of correlations. (We have assumed that the path of shortest length is unique. Along a particular direction, the multiplicity of shortest lattice paths typically grows exponentially with r, thus modifying the correlation length, but not the exponential decay which is generic to all spin systems.)

At low temperatures, the cost of small fluctuations around the ground state is obtained by a quadratic expansion, which gives $K \int d^d\mathbf{x} (\nabla\theta)^2/2$ in the continuum limit. The standard rules of Gaussian integration yield,

$$\langle \vec{s}_0 \cdot \vec{s}_r \rangle = \langle \Re e^{i(\theta_0 - \theta_r)} \rangle = \Re \exp\left[-\frac{1}{2} \left\langle (\theta_0 - \theta_r)^2 \right\rangle \right]. \tag{8.40}$$

In two dimensions, the Gaussian fluctuations grow as

$$\frac{1}{2} \left\langle (\theta_0 - \theta_r)^2 \right\rangle = \frac{1}{2\pi K} \ln\left(\frac{r}{a} \right), \tag{8.41}$$

where a is a short distance cutoff (of the order of the lattice spacing). Hence, at low temperatures,

$$\langle \vec{s}_0 \cdot \vec{s}_r \rangle \approx \left(\frac{a}{r} \right)^{\frac{1}{2\pi K}}, \tag{8.42}$$

i.e. the decay of correlations is *algebraic* rather than *exponential*. A power law decay of correlations implies self-similarity (no correlation length), and is usually associated with a critical point. Here it arises from the logarithmic growth of angular fluctuations, which is specific to two dimensions.

The distinct asymptotic decays of correlations at high and low temperatures allows for the possibility of a finite temperature phase transition separating the two regimes. However, the arguments put forward above are not specific to the XY model. Any continuous spin model will exhibit exponential decay of correlations at high temperature, and a power law decay in a low-temperature Gaussian approximation. To show that the Gaussian behavior persists at finite temperatures, we must prove that it is not modified by the additional terms in the gradient expansion. Quartic terms, such as $\int d^d\mathbf{x} (\nabla\theta)^4$, generate interactions between the Goldstone modes. The *relevance* of these interactions was probed in the previous section using the nonlinear σ model. The zero temperature fixed point in $d = 2$ is unstable for all $n > 2$, but apparently stable for $n = 2$. (There is only one branch of Goldstone modes for $n = 2$. It is the interactions between different branches of these modes for $n > 2$ that leads to instability towards high-temperature behavior.) The low-temperature phase of the XY model is said to possess *quasi-long range order*, as opposed to *true long range order* that accompanies a finite magnetization.

What is the mechanism for the disordering of the quasi-long range ordered phase? As the RG study suggests that higher order terms in the gradient expansion are not relevant, we must search for other operators. The gradient expansion describes the energy cost of *small* deformations around the ground state, and applies to configurations that can be continuously deformed to the uniformly ordered state. Kosterlitz and Thouless [J. Phys. C **6**, 1181 (1973)] suggested that the disordering is caused by *topological defects* that can not be regarded as simple deformations of the ground state. Since the angle describing the orientation of a spin is undefined up to an integer multiple of 2π, it is possible to construct spin configurations for which in going around a closed path the angle rotates by $2\pi n$. The integer n is the *topological charge* enclosed by the path. Because of the discrete nature of this charge, it is impossible to continuously deform to the uniformly ordered state in which the charge is zero. (More generally, topological defects arise in any model with a compact group describing the order parameter.)

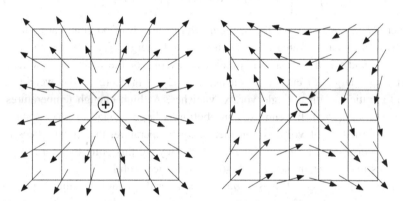

Fig. 8.4 The two elementary vortices in the XY model.

The elementary defect, or *vortex*, has unit charge. In completing a circle centered on the defect the orientation of the spin changes by $\pm 2\pi$. If the radius r, of the circle is sufficiently large, the variations in angle will be small and the lattice structure can be ignored. By symmetry, $\nabla\theta$ has a uniform magnitude, and points along the circle (i.e. perpendicular to the radial vector). The magnitude of the distortion is obtained from

$$\oint \nabla\theta \cdot \mathrm{d}\vec{s} = \frac{\mathrm{d}\theta}{\mathrm{d}r}(2\pi r) = 2\pi n, \quad \implies \quad \frac{\mathrm{d}\theta}{\mathrm{d}r} = \frac{n}{r}. \tag{8.43}$$

Since $\nabla\theta$ is a radial vector, it can be written as

$$\nabla\theta = n\left(-\frac{y}{r^2}, +\frac{x}{r^2}, 0\right) = -\frac{n}{r}\hat{r} \times \hat{z} = -n\nabla \times (\hat{z}\ln r). \tag{8.44}$$

Here, $\hat{r}$ and $\hat{z}$ are unit vectors, respectively, in the plane and perpendicular to it, and $\vec{a} \times \vec{b}$ indicates the cross product of the two vectors. This (continuum) approximation fails close to the center (core) of the vortex, where the lattice structure is important.

The energy cost of a single vortex of charge n has contributions from the core region, as well as from the relatively uniform distortions away from the center. The distinction between regions inside and outside the core is arbitrary, and, for simplicity, we shall use a circle of radius a to distinguish the two, i.e.

$$\beta \mathcal{E}_n = \beta \mathcal{E}_n^0(a) + \frac{K}{2} \int_a d^2\mathbf{x}(\nabla\theta)^2$$
$$= \beta \mathcal{E}_n^0(a) + \frac{K}{2} \int_a^L (2\pi r \mathrm{d}r) \left(\frac{n}{r}\right)^2 = \beta \mathcal{E}_n^0(a) + \pi K n^2 \ln\left(\frac{L}{a}\right). \tag{8.45}$$

The dominant part of the energy comes from the region outside the core, and diverges with the size of the system, L. The large energy cost of defects prevents their spontaneous formation close to zero temperature. The partition function for a configuration with a single vortex is

$$Z_1(n) \approx \left(\frac{L}{a}\right)^2 \exp\left[-\beta\mathcal{E}_n^0(a) - \pi K n^2 \ln\left(\frac{L}{a}\right)\right] = y_n^0(a) \left(\frac{L}{a}\right)^{2-\pi K n^2}, \tag{8.46}$$

where $(L/a)^2$ results from the *configurational entropy* of possible vortex locations in a domain of area L^2. The entropy and energy of a vortex both grow as $\ln L$, and the free energy is dominated by one or the other. At low temperatures (large K) energy dominates and Z_1, a measure of the weight of configurations with a single vortex, vanishes. At high enough temperatures, $K < K_n = 2/(\pi n^2)$, the entropy contribution is large enough to favor spontaneous formation of vortices. On increasing temperature, the first vortices to appear correspond to $n = \pm 1$ at $K_c = 2/\pi$. Beyond this point there are many vortices in the system, and Eq. (8.46) is no longer useful.

The coupling $K_c = 2/\pi$ is a *lower bound* for the stability of the system to topological defects. This is because pairs (dipoles) of defects may appear at larger couplings. Consider a pair of charges ± 1 at a separation d. The distortions far away from the dipole center, $r \gg d$, can be obtained by superposing those of the individual vortices, and

$$\nabla\theta = \nabla\theta_+ + \nabla\theta_- \approx \vec{d} \cdot \nabla\left(\frac{\hat{r} \times \hat{z}}{r}\right), \tag{8.47}$$

decays as d/r^2. Integrating this distortion leads to a *finite* energy, and hence dipoles appear with the appropriate Boltzmann weight at any temperature. The low-temperature phase should thus be visualized as a gas of tightly bound dipoles. The number (and size) of dipoles increases with temperature, and the high-temperature state is a plasma of unbound vortices. The distinction between the two regimes can be studied by examining a typical net topological charge, $Q(\ell)$, in a large area of dimension $\ell \gg a$. The average charge is always zero, while fluctuations in the low-temperature phase are due to the dipoles straddling the perimeter, i.e. $\langle Q(\ell)^2 \rangle \propto \ell$. In the high-temperature state, charges of either sign can appear without restriction and $\langle Q(\ell)^2 \rangle \propto \ell^2$. (Note the similarity to the distinct behaviors of the Wilson loop in the high and low temperature phases of the Z_2 gauge theory.)

To describe the transition between the two regimes, we need to properly account for the interactions between vortices. The distortion field $\vec{u} \equiv \nabla\theta$, in the presence of a collection of vortices, is similar to the velocity of a fluid. In the absence of vorticity, the flow is *potential*, i.e. $\vec{u} = \vec{u}_0 = \nabla\phi$, and $\nabla \times \vec{u}_0 = 0$. The topological charge can be related to $\nabla \times \vec{u}$ by noting that for any closed path,

$$\oint \vec{u} \cdot d\vec{s} = \int (d^2\mathbf{x}\,\hat{z}) \cdot \nabla \times \vec{u}, \tag{8.48}$$

where the second integral is over the area enclosed by the path. Since the left-hand side is an integer multiple of 2π, we can set

$$\nabla \times \vec{u} = 2\pi\hat{z}\sum_i n_i\,\delta^2(\mathbf{x} - \mathbf{x}_i), \tag{8.49}$$

describing a collection of vortices of charge $\{n_i\}$ at locations $\{\mathbf{x}_i\}$. The solution to Eq. (8.49) can be obtained by setting $\vec{u} = \vec{u}_0 - \nabla \times (\hat{z}\psi)$, leading to

$$\nabla \times \vec{u} = \hat{z}\nabla^2\psi, \quad \Longrightarrow \quad \nabla^2\psi = 2\pi\sum_i n_i\,\delta^2(\mathbf{x} - \mathbf{x}_i). \tag{8.50}$$

Thus ψ behaves like the potential due to a set of charges $\{2\pi n_i\}$. The solution,

$$\psi(\mathbf{x}) = \sum_i n_i \ln\left(|\mathbf{x} - \mathbf{x}_i|\right), \tag{8.51}$$

is simply a superposition of the potentials as in Eq. (8.44).

Any two dimensional distortion can thus be written as

$$\vec{u} = \vec{u}_0 + \vec{u}_1 = \nabla\phi - \nabla \times (\hat{z}\psi), \tag{8.52}$$

and the corresponding "kinetic energy", $\beta\mathcal{H} = K\int d^2\mathbf{x}\,|\vec{u}|^2/2$, decomposed as

$$\beta\mathcal{H} = \int d^2\mathbf{x}\left[(\nabla\phi)^2 - 2\nabla\phi \cdot \nabla \times (\hat{z}\psi) + (\nabla \times \hat{z}\psi)^2\right]. \tag{8.53}$$

The second term vanishes, since following an integration by parts,

$$-\int d^2\mathbf{x}\,\nabla\phi \cdot \nabla \times (\hat{z}\psi) = \int d^2\mathbf{x}\,\phi\nabla \cdot (\nabla \times \hat{z}\psi), \tag{8.54}$$

and $\nabla \cdot \nabla \times \vec{u} = 0$ for any vector. The third term in Eq. (8.53) can be simplified by noting that $\nabla\psi = (\partial_x\psi, \partial_y\psi, 0)$ and $\nabla \times (\hat{z}\psi) = (-\partial_y\psi, \partial_x\psi, 0)$ are orthogonal vectors of equal length. Hence

$$\beta\mathcal{H}_1 \equiv \frac{K}{2}\int d^2\mathbf{x}\,(\nabla \times \hat{z}\psi)^2 = \frac{K}{2}\int d^2\mathbf{x}\,(\nabla\psi)^2 = -\frac{K}{2}\int d^2\mathbf{x}\,\psi\nabla^2\psi, \tag{8.55}$$

where the second identity follows an integration by parts. Equations (8.50) and (8.51) now result in

$$\beta\mathcal{H}_1 = -\frac{K}{2}\int d^2\mathbf{x}\left(\sum_i n_i \ln\left(|\mathbf{x} - \mathbf{x}_i|\right)\right)\left(2\pi\sum_j n_j\,\delta^2(\mathbf{x} - \mathbf{x}_j)\right)$$

$$= -2\pi^2 K\sum_{i,j} n_i n_j\,C(\mathbf{x}_i - \mathbf{x}_j), \tag{8.56}$$

where $C(\mathbf{x}) = \ln(|\mathbf{x}|)/2\pi$ is the two-dimensional Coulomb potential. There is a difficulty with the above result for $i = j$ due to the divergence of the logarithm at small arguments. This is a consequence of the breakdown of the continuum treatment at short distances. The self-interaction of a vortex is simply its core energy $\beta \mathcal{E}_n^0$, and

$$\beta \mathcal{H}_1 = \sum_i \beta \mathcal{E}_{n_i}^0 - 4\pi^2 K \sum_{i<j} n_i n_j C(\mathbf{x}_i - \mathbf{x}_j). \tag{8.57}$$

The configuration space of the XY model close to zero temperature can thus be partitioned into different topological segments. The degrees of freedom in each segment are the charges $\{n_i\}$, and locations $\{\mathbf{x}_i\}$, of the *vortices*, in addition to the field $\phi(\mathbf{x})$ describing *spin waves*. The partition function of the model can thus be approximated as

$$Z = \prod_i \int_0^{2\pi} \frac{\mathrm{d}\theta_i}{2\pi} \exp\left[K \sum_{\langle i,j \rangle} \cos(\theta_i - \theta_j) \right]$$

$$\propto \int \mathcal{D}\phi(\mathbf{x}) \, \mathrm{e}^{-\frac{K}{2} \int \mathrm{d}^2\mathbf{x} (\nabla\phi)^2} \sum_{\{n_i\}} \int \mathrm{d}^2\mathbf{x}_i \, \mathrm{e}^{-\sum_i \beta \mathcal{E}_{n_i}^0 + 4\pi^2 K \sum_{i<j} n_i n_j C(\mathbf{x}_i - \mathbf{x}_j)} \tag{8.58}$$

$$\equiv Z_{\text{s.w.}} Z_Q,$$

where $Z_{\text{s.w.}}$ is the Gaussian partition function of spin waves, and Z_Q is the contribution of vortices. The latter describes a grand canonical gas of charged particles, interacting via the two dimensional Coulomb interaction. In calculating the Hamiltonian $\beta \mathcal{H}_1$ in Eq. (8.55), we performed an integration by parts. The surface integral that was ignored in the process in fact grows with system size as $(\sum_i n_i) \ln L$. Thus only configurations that are overall neutral are included in calculating Z_Q. We further simplify the problem by considering only the elementary excitations with $n_i = \pm 1$, which are most likely at low temperatures due to their lower energy. Setting $y_0 \equiv \exp\left[-\beta \mathcal{E}_{\pm 1}^0\right]$,

$$Z_Q = \sum_{N=0}^{\infty} y_0^N \int \prod_{i=1}^{N} \mathrm{d}^2\mathbf{x}_i \exp\left[4\pi^2 K \sum_{i<j} q_i q_j C(\mathbf{x}_i - \mathbf{x}_j) \right], \tag{8.59}$$

where $q_i = \pm 1$, and $\sum_i q_i = 0$.

8.3 Renormalization group for the Coulomb gas

The two partition functions in Eq. (8.58) are independent and can be calculated separately. As the Gaussian partition function is analytic, any phase transitions of the XY model must originate in the Coulomb gas. As briefly discussed earlier, in the low temperature phase the charges appear only in a small population of tightly bound dipole pairs. The dipoles dissociate in the high temperature phase, forming a plasma. The two phases can be distinguished by examining the interaction between two *external* test charges at a large separation X. In the absence of any *internal* charges (for $y_0 = 0$) in the medium, the two particles interact by the *bare* Coulomb interaction $C(X)$. A finite density of internal charges for small y_0 partially screens the external charges, and reduces the interaction between the test charges to $C(X)/\varepsilon$, where ε is an effective *dielectric* constant. There is an *insulator to metal* transition at sufficiently large y_0. In the metallic (plasma) phase, the external charges are completely screened and their effective interaction decays exponentially.

To quantify this picture, we shall compute the effective interaction between two *external* charges at $\mathbf{x}$ and $\mathbf{x}'$, perturbatively in the fugacity y_0. To lowest order, we need to include configurations with two *internal* charges (at $\mathbf{y}$ and $\mathbf{y}'$), and

$$e^{-\beta \mathcal{V}(\mathbf{x}-\mathbf{x}')} = e^{-4\pi^2 K C(\mathbf{x}-\mathbf{x}')}$$

$$\times \frac{\left[1 + y_0^2 \int d^2\mathbf{y} d^2\mathbf{y}' \, e^{-4\pi^2 K C(\mathbf{y}-\mathbf{y}') + 4\pi^2 K[C(\mathbf{x}-\mathbf{y}) - C(\mathbf{x}-\mathbf{y}') - C(\mathbf{y}'-\mathbf{x}) + C(\mathbf{x}'-\mathbf{y}')]} + \mathcal{O}(y_0^4) \right]}{\left[1 + y_0^2 \int d^2\mathbf{y} d^2\mathbf{y}' \, e^{-4\pi^2 K C(\mathbf{y}-\mathbf{y}')} + \mathcal{O}(y_0^4) \right]}$$

$$\text{(8.60)}$$

$$= e^{-4\pi^2 K C(\mathbf{x}-\mathbf{x}')} \left[1 + y_0^2 \int d^2\mathbf{y} d^2\mathbf{y}' \, e^{-4\pi^2 K C(\mathbf{y}-\mathbf{y}')} \left(e^{4\pi^2 K D(\mathbf{x},\mathbf{x}';\mathbf{y},\mathbf{y}')} - 1 \right) + \mathcal{O}(y_0^4) \right],$$

where $D(\mathbf{x}, \mathbf{x}'; \mathbf{y}, \mathbf{y}')$ is the interaction *between* the internal and external dipoles. The direct interaction between internal charges tends to keep the separation $\mathbf{r} = \mathbf{y}' - \mathbf{y}$ small. Using the center of mass $\mathbf{R} = (\mathbf{y} + \mathbf{y}')/2$, we can change variables to $\mathbf{y} = \mathbf{R} - \mathbf{r}/2$ and $\mathbf{y}' = \mathbf{R} + \mathbf{r}/2$, and expand the dipole–dipole interaction for small $\mathbf{r}$ as

$$D(\mathbf{x}, \mathbf{x}'; \mathbf{y}, \mathbf{y}') = C\left(\mathbf{x} - \mathbf{R} + \frac{\mathbf{r}}{2} \right) - C\left(\mathbf{x} - \mathbf{R} - \frac{\mathbf{r}}{2} \right) - C\left(\mathbf{x}' - \mathbf{R} + \frac{\mathbf{r}}{2} \right) + C\left(\mathbf{x}' - \mathbf{R} - \frac{\mathbf{r}}{2} \right)$$

$$= -\mathbf{r} \cdot \nabla_{\mathbf{R}} C(\mathbf{x} - \mathbf{R}) + \mathbf{r} \cdot \nabla_{\mathbf{R}} C(\mathbf{x}' - \mathbf{R}) + \mathcal{O}(r^3). \qquad \text{(8.61)}$$

To the same order

$$e^{4\pi^2 K D(\mathbf{x},\mathbf{x}';\mathbf{y},\mathbf{y}')} - 1 = -4\pi^2 K \mathbf{r} \cdot \nabla_{\mathbf{R}} \left(C(\mathbf{x} - \mathbf{R}) - C(\mathbf{x}' - \mathbf{R}) \right)$$

$$+ 8\pi^4 K^2 \left[\mathbf{r} \cdot \nabla_{\mathbf{R}} \left(C(\mathbf{x} - \mathbf{R}) - C(\mathbf{x}' - \mathbf{R}) \right) \right]^2 + \mathcal{O}(r^3).$$

$$\text{(8.62)}$$

After the change of variables $\int d^2\mathbf{y} d^2\mathbf{y}' \to \int d^2\mathbf{r} d^2\mathbf{R}$, the effective interaction becomes

$$e^{-\beta \mathcal{V}(\mathbf{x}-\mathbf{x}')} = e^{-4\pi^2 K C(\mathbf{x}-\mathbf{x}')} \left\{ \left[1 + y_0^2 \int d^2\mathbf{r} d^2\mathbf{R} \, e^{-4\pi^2 K C(\mathbf{r})} \right. \right.$$

$$\times \left(-4\pi^2 K \mathbf{r} \cdot \nabla_{\mathbf{R}} \left(C(\mathbf{x} - \mathbf{R}) - C(\mathbf{x}' - \mathbf{R}) \right) \right. \qquad \text{(8.63)}$$

$$\left. \left. \left. + 8\pi^4 K^2 \left[\mathbf{r} \cdot \nabla_{\mathbf{R}} \left(C(\mathbf{x} - \mathbf{R}) - C(\mathbf{x}' - \mathbf{R}) \right) \right]^2 + \mathcal{O}(r^3) \right) + \mathcal{O}(y_0^4) \right] \right\}.$$

Following the angular integrations in $d^2\mathbf{r}$, the term linear in $\mathbf{r}$ vanishes, while the angular average of $(\mathbf{r} \cdot \nabla_\mathbf{R} C)^2$ is $r^2 |\nabla_\mathbf{R} C|^2/2$. Hence Eq. (8.63) simplifies to

$$e^{-\beta\mathcal{V}(\mathbf{x}-\mathbf{x}')} = e^{-4\pi^2 KC(\mathbf{x}-\mathbf{x}')}$$
$$\times \left[1 + y_0^2 \int (2\pi r dr) e^{-4\pi^2 KC(r)} 8\pi^4 K^2 \frac{r^2}{2} \right. \qquad (8.64)$$
$$\left. \int d^2\mathbf{R} \left(\nabla_\mathbf{R} \left(C(\mathbf{x}-\mathbf{R}) - C(\mathbf{x}'-\mathbf{R}) \right) \right)^2 + \mathcal{O}(r^4) \right].$$

The remaining integral can be evaluated by parts,

$$\int d^2\mathbf{R} \left[\nabla_\mathbf{R} \left(C(\mathbf{x}-\mathbf{R}) - C(\mathbf{x}'-\mathbf{R}) \right) \right]^2$$
$$= -\int d^2\mathbf{R} \left(C(\mathbf{x}-\mathbf{R}) - C(\mathbf{x}'-\mathbf{R}) \right) \left(\nabla^2 C(\mathbf{x}-\mathbf{R}) - \nabla^2 C(\mathbf{x}'-\mathbf{R}) \right)$$
$$(8.65)$$
$$= -\int d^2\mathbf{R} \left(C(\mathbf{x}-\mathbf{R}) - C(\mathbf{x}'-\mathbf{R}) \right) \left(\delta^2(\mathbf{x}-\mathbf{R}) - \delta^2(\mathbf{x}'-\mathbf{R}) \right)$$
$$= 2C(\mathbf{x}-\mathbf{x}') - 2C(0).$$

The short distance divergence can again be absorbed into a proper cutoff with $C(x) \to \ln(x/a)/2\pi$, and

$$e^{-\beta\mathcal{V}(\mathbf{x}-\mathbf{x}')} = e^{-4\pi^2 KC(\mathbf{x}-\mathbf{x}')} \left[1 + 16\pi^5 K^2 y_0^2 C(\mathbf{x}-\mathbf{x}') \int drr^3 e^{-2\pi K \ln(r/a)} + \mathcal{O}(y_0^4) \right]. \quad (8.66)$$

The second order term can be exponentiated to give an effective interaction $\beta\mathcal{V}(\mathbf{x}-\mathbf{x}') \equiv 4\pi^2 K_{\text{eff}} C(\mathbf{x}-\mathbf{x}')$, with

$$K_{\text{eff}} = K - 4\pi^3 K^2 y_0^2 a^{2\pi K} \int_a^\infty drr^{3-2\pi K} + \mathcal{O}(y_0^4). \qquad (8.67)$$

We have thus evaluated the dielectric constant of the medium, $\varepsilon = K/K_{\text{eff}}$, perturbatively to order of y_0^2. However, the perturbative correction is small only as long as the integral in r converges at large r. The breakdown of the perturbation theory for $K < K_c = 2/\pi$, occurs precisely at the point where the free energy of an isolated vortex changes sign. This breakdown of perturbation theory is reminiscent of that encountered in the Landau–Ginzburg model for $d < 4$. Using the experience gained from that problem, we shall reorganize the perturbation series into a renormalization group for the parameters K and y_0.

To construct an RG for the Coulomb gas, note that the partition function for the system in Eq. (8.59) involves two parameters (K, y_0), and has an implicit cutoff a, related to the minimum separation between vortices. As discussed earlier, the distinction between regions inside and outside the core of a vortex is arbitrary. Increasing the core size to ba modifies not only the core energy, hence y_0, but also the interaction parameter K. The latter is a consequence of the change in the dielectric properties of the medium due to dipoles of separations between a and ba. The change in fugacity is obtained from Eq. (8.46) by changing a to ba as

$$\tilde{y}_0(ba) = b^{2-\pi K} y_0(a). \tag{8.68}$$

The modified Coulomb interaction due to dipoles of all sizes is given in Eq. (8.67). (The perturbative calculation at order of y_0^2 incorporates only dipoles.) From dipoles in the size range a to ba, we obtain a contribution

$$\tilde{K} = K\left[1 - (2\pi^2 K)\int_a^{ba} (2\pi r \mathrm{d}r)\left(y_0^2 e^{-4\pi^2 KC(r)}\right) r^2\right], \tag{8.69}$$

where the terms are grouped so as to make the similarity to standard computations of the dielectric constant apparent. (The probability of creating a dipole of size r is multiplied by its polarizability; the role of $\beta = (k_B T)^{-1}$ is played by $2\pi^2 K$.)

By choosing an infinitesimal $b = e^\ell \approx 1 + \ell$, Eq. (8.69) is converted to

$$\frac{\mathrm{d}K}{\mathrm{d}\ell} = -4\pi^3 K^2 a^4 y_0^2 + \mathcal{O}(y_0^4). \tag{8.70}$$

Including the fugacity, the recursion relations are

$$\begin{cases} \dfrac{\mathrm{d}K^{-1}}{\mathrm{d}\ell} = 4\pi^3 a^4 y_0^2 + \mathcal{O}(y_0^4) \\ \dfrac{\mathrm{d}y_0}{\mathrm{d}\ell} = (2 - \pi K)y_0 + \mathcal{O}(y_0^3), \end{cases} \tag{8.71}$$

originally obtained by Kosterlitz [J. Phys. C **7**, 1046 (1974)]. While $\mathrm{d}K^{-1}/\mathrm{d}\ell \geq 0$, the recursion relation for y_0 changes sign at $K_c^{-1} = \pi/2$. At smaller values of K^{-1} (high temperatures) y_0 is relevant, while at lower temperatures it is irrelevant. Thus the RG flows separate the parameter space into two regions. At low temperatures and small y_0, flows terminate on a *fixed line* at $y_0 = 0$ and $K_{\text{eff}} \geq 2/\pi$. This is the insulating phase in which only dipoles of finite size occur. (Hence the vanishing of y_0 under coarse graining.) The strength of the effective Coulomb interaction is given by the point on the fixed line that the flows terminate on. Flows not terminating on the fixed line asymptote to larger values of K^{-1} and y_0, where perturbation theory breaks down. This is the signal of the high-temperature phase with an abundance of vortices.

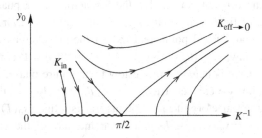

Fig. 8.5 Renormalization group flows of the Coulomb gas system. The line of fixed points at $y_0 = 0$ terminates at $K^{-1} = \pi/2$.

The critical trajectory that separates the two regions of the phase diagram flows to a fixed point at $(K_c^{-1} = \pi/2, y_0 = 0)$. To find the critical behavior at the transition, expand the recursion relations in the vicinity of this point by setting $x = K^{-1} - \pi/2$, and $y = y_0 a^2$. To lowest order, Eqs. (8.71) simplify to

$$\begin{cases} \dfrac{dx}{d\ell} = 4\pi^3 y^2 + \mathcal{O}(xy^2, y^4) \\[2mm] \dfrac{dy}{d\ell} = \dfrac{4}{\pi} xy + \mathcal{O}(x^2 y, y^3). \end{cases} \tag{8.72}$$

The recursion relations are inherently *nonlinear* in the vicinity of the fixed point. This is quite different from the linear recursion relations that we have encountered so far, and the resulting critical behavior is non-standard. First note that Eqs. (8.72) imply

$$\frac{dx^2}{d\ell} = 8\pi^3 y^2 x = \pi^4 \frac{dy^2}{d\ell}, \quad \Rightarrow \quad \frac{d}{d\ell}\left(x^2 - \pi^4 y^2\right) = 0, \quad \Rightarrow \quad x^2 - \pi^4 y^2 = c. \tag{8.73}$$

The RG flows thus proceed along hyperbolas characterized by different values of c. For $c > 0$, the focus of the hyperbola is along the y-axis, and the flows proceed to $(x, y) \to \infty$. The hyperbolas with $c < 0$ have foci along the x-axis, and have two branches in the half plane $y \geq 0$: the trajectories for $x < 0$ flow to the fixed line, while those in the $x > 0$ quadrant flow to infinity. The critical trajectory separating flows to zero and infinite y corresponds to $c = 0$, i.e. $x_c = -\pi^2 y_c$. Therefore, a small but finite fugacity y_0 reduces the critical temperature to $K_c^{-1} = \pi/2 - \pi^2 y_0 a^2 + \mathcal{O}(y_0^2)$.

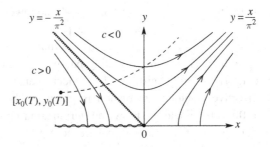

Fig. 8.6 Detailed views of the RG flows in the vicinity of the termination point of the line of low-temperature fixed points. The dashed line indicates a trajectory of the initial Hamiltonian as the temperature is varied. The phase transition occurs when this trajectory intersects the dotted line.

In terms of the original XY model, the low-temperature phase is characterized by a line of fixed points with $K_{\text{eff}} = \lim_{\ell \to \infty} K(\ell) \geq 2/\pi$. There is no correlation length at a fixed point, and indeed the correlations in this phase decay as a power law, $\langle \cos(\theta_{\mathbf{r}} - \cos\theta_{\mathbf{0}}) \rangle \sim 1/r^\eta$, with $\eta = 1/(2\pi K_{\text{eff}}) \leq 1/4$. Since the parameter c is negative in the low temperature phase, and vanishes at the critical point, we can set it to $c = -b^2(T_c - T)$ close to the transition. In other words, the trajectory of initial points tracks a line $(x_0(T), y_0(T))$. The resulting $c = x_0^2 - \pi^4 y_0^2 \propto (T_c - T)$ is a linear measure of the vicinity to the phase transition. Under renormalization, such trajectories flow to a fixed point at $y = 0$, and $x = -b\sqrt{T_c - T}$. Thus in the vicinity of transition, the effective interaction parameter

$$K_{\text{eff}} = \frac{2}{\pi} - \frac{4}{\pi^2} \lim_{\ell \to \infty} x(\ell) = \frac{2}{\pi} + \frac{4b}{\pi^2} \sqrt{T_c - T} \qquad (8.74)$$

has a square root singularity.

The *stiffness* K_{eff}, can be measured in experiments on superfluid films. In the superfluid phase, the order parameter is the condensate wavefunction $\psi(\mathbf{x}) = \Psi e^{i\theta}$. Variations in the phase θ lead to a superfluid kinetic energy,

$$\mathcal{H} = \int d^d\mathbf{x}\, \psi^* \left(-\frac{\hbar^2 \nabla^2}{2m} \right) \psi = \frac{\hbar^2 \Psi^2}{2m} \int d^d\mathbf{x}(\nabla\theta)^2, \qquad (8.75)$$

where m is the particle (helium 4) mass. The corresponding XY model has a stiffness $K = \hbar^2 \rho_s/(2m^2 k_B T)$, where $\rho_s = m\Psi^2$ is the superfluid mass density. The density ρ_s is measured by examining the changes in the inertia of a torsional oscillator; the superfluid fraction, $m\Psi^2$, experiences no friction and does not oscillate. Bishop and Reppy [Phys. Rev. Lett. **40**, 1727 (1978)] examined ρ_s for a variety of superfluid films of different thickness wrapped around a torsional cylinder. They constructed the effective stiffness K as a function of temperature, and found that for all films it undergoes a *universal jump* of $2/\pi$ at the transition. The behavior of K for $T < T_c$ was consistent with a square root singularity.

Correlations decay exponentially in the high-temperature phase. How does the correlation length ξ diverge at T_c? The parameter $c = x^2 - \pi^4 y^2 = b^2(T - T_c)$ is now positive all along the hyperbolic trajectory. The recursion relation for x,

$$\frac{dx}{d\ell} = 4\pi^3 y^2 = \frac{4}{\pi} \left(x^2 + b^2(T - T_c) \right), \qquad (8.76)$$

can be integrated to give

$$\frac{dx}{x^2 + b^2(T - T_c)} = \frac{4}{\pi} d\ell, \implies \frac{1}{b\sqrt{T - T_c}} \arctan\left(\frac{x}{b\sqrt{T - T_c}} \right) = \frac{4}{\pi} \ell. \quad (8.77)$$

The contribution of the initial point to the left-hand side of the above equation can be left out if $x_0 \propto (T - T_c) \ll 1$. The integration has to be stopped when $x(\ell) \sim y(\ell) \sim 1$, since the perturbative calculation is no longer valid beyond this point. This occurs for a value of

$$\ell^* \approx \frac{\pi}{4b\sqrt{T - T_c}} \frac{\pi}{2}, \qquad (8.78)$$

where we have used $\arctan(1/b\sqrt{T - T_c}) \approx \arctan(\infty) = \pi/2$. The resulting correlation length is

$$\xi \approx a e^{\ell^*} \approx a \exp\left(\frac{\pi^2}{8b\sqrt{T - T_c}} \right). \qquad (8.79)$$

Unlike any of the transitions encountered so far, the divergence of the correlation length is not through a power law. This is a consequence of the nonlinear nature of the recursion relations in the vicinity of the fixed point.

Vortices occur in bound pairs for distances smaller than ξ, while there can be an excess of vortices of one sign or the other at larger separations. The interactions between vortices at large distances can be obtained from the Debye–Hückel theory of polyelectrolytes. According to this theory the free charges screen each other leading to a screened Coulomb interaction, $\exp(-r/\xi)C(r)$. On approaching the transition from the high-temperature side, the singular part of the free energy,

$$f_{\text{sing}} \propto \xi^{-2} \propto \exp\left(-\frac{\pi^2}{4b\sqrt{T-T_c}}\right),\tag{8.80}$$

has only an *essential singularity*. All derivatives of this function are finite at T_c. Thus the predicted heat capacity is quite smooth at the transition. Numerical results based on the RG equations by Berker and Nelson [Phys. Rev. B **19**, 2488 (1979)] indicate a smooth maximum in the heat capacity at a temperature higher than T_c, corresponding to the point at which the majority of dipoles unbind.

The Kosterlitz–Thouless picture of vortex unbinding has found numerous applications in two-dimensional systems such as superconducting and superfluid films, thin liquid crystals, Josephson junction arrays, electrons on the surface of helium films, etc. Perhaps more importantly, the general idea of topological defects has had much impact in understanding the behavior of many systems. The theory of two-dimensional melting developed in the next sections is one such example.

8.4 Two-dimensional solids

The transition from a liquid to solid involves the breaking of translational and orientational symmetry. Here we shall examine the types of order that are present in a two-dimensional solid. The destruction of such order upon melting is discussed next.

(a) A low-temperature treatment of the problem starts with the perfect solid at $T = 0$. The equilibrium configuration of atoms forms a lattice, $\mathbf{r}_0(m, n) = m\mathbf{e}_1 + n\mathbf{e}_2$, where $\mathbf{e}_1$ and $\mathbf{e}_2$ are basis vectors, and $\{m, n\}$ are integers. At finite temperatures, the atoms fluctuate away from their equilibrium position, moving to

$$\mathbf{r}(m, n) = \mathbf{r}_0(m, n) + \mathbf{u}(m, n).\tag{8.81}$$

The low-temperature distortions do not vary substantially over nearby atoms, enabling us to define a coarse-grained *distortion field* $\mathbf{u}(\mathbf{x})$, where $\mathbf{x} \equiv (x_1, x_2)$ is treated as continuous, with an implicit short distance cutoff of the order of the lattice spacing a. The distortion $\mathbf{u}$ is the analog of the angle θ in the XY model.

(b) Due to translational symmetry, the energy depends only on the *strain matrix*,

$$u_{ij}(\mathbf{x}) = \frac{1}{2}\left(\partial_i u_j + \partial_j u_i\right), \quad \Longrightarrow \quad u_{ij}(\mathbf{q}) = \frac{i}{2}\left(q_i u_j + q_j u_i\right). \tag{8.82}$$

The elastic energy must respect the symmetries of the underlying lattice. For simplicity, we shall consider the triangular lattice whose elastic energy is fully isotropic, i.e. invariant under all rotations (see Landau and Lifshitz, *Theory of Elasticity*, Pergamon Press). In terms of the *Lamé coefficients* λ and μ,

$$\begin{aligned}\beta\mathcal{H} &= \frac{1}{2}\int d^2\mathbf{x}\left[2\mu u_{ij}u_{ij} + \lambda u_{ii}u_{jj}\right]\\ &= \frac{1}{2}\int \frac{d^2\mathbf{q}}{(2\pi)^2}\left[\mu q^2 u^2 + (\mu+\lambda)(\mathbf{q}\cdot\mathbf{u})^2\right].\end{aligned} \tag{8.83}$$

The rotational invariance of energy is insured by the implicit sum over the indices (i,j) in the above expression. In the Fourier representation, the energy depends on the quantities q^2, $|\mathbf{u}|^2$, and $(\mathbf{q}\cdot\mathbf{u})^2$ which are clearly independent of rotations. For other lattices, there are more elastic coefficients, since the energy need only be invariant under the subset of lattice rotations. For example, the symmetry of a square lattice permits a term proportional to $\left(q_x^2 u_x^2 + q_y^2 u_y^2\right)$.

(c) The Goldstone modes associated with the broken translational symmetry are *phonons*, the normal modes of vibrations. Equation (8.83) supports two types of normal modes:

(i) *Transverse modes* with $\mathbf{q}\perp\mathbf{u}$ have energy

$$\beta\mathcal{H}_T = \frac{\mu}{2}\int \frac{d^2\mathbf{q}}{(2\pi)^2}\, q^2 u_T^2, \quad \Longrightarrow \quad \left\langle|u_T(\mathbf{q})|^2\right\rangle = \frac{1}{\mu q^2}. \tag{8.84}$$

(ii) *Longitudinal modes* with $\mathbf{q}\parallel\mathbf{u}$ have energy

$$\beta\mathcal{H}_L = \frac{2\mu+\lambda}{2}\int \frac{d^2\mathbf{q}}{(2\pi)^2}\, q^2 u_L^2, \quad \Longrightarrow \quad \left\langle|u_L(\mathbf{q})|^2\right\rangle = \frac{1}{(2\mu+\lambda)q^2}. \tag{8.85}$$

Combining the above results, we find that a general correlation function is given by

$$\left\langle u_i(\mathbf{q})u_j(\mathbf{q}')\right\rangle = \frac{(2\pi)^2\delta^2(\mathbf{q}+\mathbf{q}')}{\mu q^2}\left[\delta_{ij} - \frac{\mu+\lambda}{2\mu+\lambda}\frac{q_i q_j}{q^2}\right]. \tag{8.86}$$

The extent of fluctuations in position space is given by

$$\begin{aligned}\left\langle[\mathbf{u}(\mathbf{x}) - \mathbf{u}(\mathbf{0})]^2\right\rangle &= \int \frac{d^2\mathbf{q}}{(2\pi)^2}\frac{2 - 2\cos(\mathbf{q}\cdot\mathbf{x})}{\mu q^2}\left[\delta_{ii} - \frac{\mu+\lambda}{2\mu+\lambda}\frac{q_i q_i}{q^2}\right]\\ &= \frac{3\mu+\lambda}{\mu(2\mu+\lambda)}\frac{\ln(|\mathbf{x}|/a)}{\pi}.\end{aligned} \tag{8.87}$$

The unbounded growth of fluctuations with $|\mathbf{x}|$ indicates the destruction of long-range translational order in two dimensions. A similar calculation in higher dimensions results in a *finite* maximum extent of fluctuations proportional to temperature. The *Lindemann criterion* identifies the melting point heuristically as the temperature for which these thermal fluctuations reach a fraction (roughly 0.1) of the perfect lattice spacing. According to this criterion, the two-dimensional solid melts at any finite temperature. This is of coarse another manifestation of the general absence of true long-range order in two dimensions. Is some form of quasi-long range order, similar to the XY model, possible in this case?

(d) The *order parameter* describing broken translational symmetry is $\rho_{\mathbf{G}}(\mathbf{x}) = e^{i\mathbf{G}\cdot\mathbf{r}(\mathbf{x})}$, where $\mathbf{G}$ is any *reciprocal lattice* vector. Since $\mathbf{G}\cdot\mathbf{r}_0$ is an integer multiple of 2π by definition, $\rho_{\mathbf{G}} = 1$ at zero temperature. Due to fluctuations $\langle\rho_{\mathbf{G}}(\mathbf{x})\rangle = \langle e^{i\mathbf{G}\cdot\mathbf{u}(\mathbf{x})}\rangle$ decreases at finite temperatures, and its correlations decay as

$$
\begin{aligned}
\langle\rho_{\mathbf{G}}(\mathbf{x})\rho_{\mathbf{G}}^*(\mathbf{0})\rangle &= \langle e^{i\mathbf{G}\cdot(\mathbf{u}(\mathbf{x})-\mathbf{u}(\mathbf{0}))}\rangle \\
&= \exp\left(-\frac{G_\alpha G_\beta}{2}\langle(u_\alpha(\mathbf{x})-u_\alpha(\mathbf{0}))(u_\beta(\mathbf{x})-u_\beta(\mathbf{0}))\rangle\right) \\
&= \exp\left\{-\frac{G_\alpha G_\beta}{2}\int\frac{d^2\mathbf{q}\,d^2\mathbf{q}'}{(2\pi)^2}\left(e^{i\mathbf{q}\cdot\mathbf{x}}-1\right)\left(e^{i\mathbf{q}'\cdot\mathbf{x}}-1\right)\langle u_\alpha(\mathbf{q})u_\beta(\mathbf{q}')\rangle\right\} \\
&= \exp\left\{-\frac{1}{2}\int\frac{d^2\mathbf{q}}{(2\pi)^2}\frac{2-2\cos\mathbf{q}\cdot\mathbf{x}}{q^2}\left(\frac{G^2}{\mu}-\frac{\mu+\lambda}{\mu(2\mu+\lambda)}\frac{(\mathbf{G}\cdot\mathbf{q})^2}{q^2}\right)\right\} \\
&\approx \exp\left\{-\frac{G^2(3\mu+\lambda)}{2\mu(2\mu+\lambda)}\frac{\ln(|\mathbf{x}|/a)}{2\pi}\right\} = \left(\frac{a}{|\mathbf{x}|}\right)^{\eta_G}.
\end{aligned}
\tag{8.88}
$$

The integration over $\mathbf{q}$ is performed after replacing $(\mathbf{G}\cdot\mathbf{q})^2$ with the angular average of $G^2 q^2/2$. This is not strictly correct because of the presence of the term $\cos\mathbf{q}\cdot\mathbf{x}$ which breaks the rotational symmetry. Nevertheless, this approximation captures the correct asymptotic growth of the integral. Correlations thus decay algebraically with an exponent

$$
\eta_G = \frac{G^2(3\mu+\lambda)}{4\pi\mu(2\mu+\lambda)}.
\tag{8.89}
$$

The translational correlations are measured in diffraction experiments. The scattering amplitude is the Fourier transform of $\rho_{\mathbf{q}}$, and the scattered intensity at a wavevector $\mathbf{q}$ is proportional to the structure factor

$$
\begin{aligned}
S(\mathbf{q}) &= \langle|A(\mathbf{q})|^2\rangle = \left\langle\left|\sum_{m,n}e^{i\mathbf{q}\cdot\mathbf{r}(m,n)}\right|^2\right\rangle = N\sum_{m,n}\langle e^{i\mathbf{q}\cdot(\mathbf{r}(m,n)-\mathbf{r}(0,0))}\rangle \\
&= N\sum_{m,n}e^{i\mathbf{q}\cdot(\mathbf{r}_0(m,n)-\mathbf{r}_0(0,0))}\langle e^{i\mathbf{q}\cdot(\mathbf{u}(\mathbf{x})-\mathbf{u}(\mathbf{0}))}\rangle,
\end{aligned}
\tag{8.90}
$$

where N is the total number of particles. At zero temperature, the structure factor is a set of delta-functions (Bragg peaks) at the reciprocal lattice vectors. Even at finite temperature the sum is zero due to varying phases, unless $\mathbf{q}$ is in the vicinity of a lattice vector $\mathbf{G}$. We can then replace the sum with an integral, and

$$S(\mathbf{q} \approx \mathbf{G}) \approx N \int d^2x e^{i(\mathbf{q}-\mathbf{G})\cdot\mathbf{x}} \left(\frac{a}{|\mathbf{x}|}\right)^{\eta_G} \approx \frac{N}{|\mathbf{q}-\mathbf{G}|^{2-\eta_G}}. \tag{8.91}$$

The Bragg peaks are now replaced by power law singularities. The strength of the divergence decreases with temperature and increasing $|\mathbf{G}|$. The peaks corresponding to sufficiently large $|\mathbf{G}|$ are no longer visible, and gradually more of them disappear on increasing temperature. In three dimensions, the structure factor is still a set of delta-functions, but with magnitudes diminished by the so called *Debye–Waller factor* of $\exp\left(-\frac{G^2}{12\pi a}\frac{5\mu+2\lambda}{\mu(2\mu+\lambda)}\right)$.

(e) The crystal phase is also characterized by a broken rotational symmetry. We can define an *orientational order parameter* $\Psi(\mathbf{x}) = e^{6i\theta(\mathbf{x})}$, where $\theta(\mathbf{x})$ is the angle between local lattice bonds and a reference axis. (The factor of 6 results from the equivalence of the six possible directions on the triangular lattice. The appropriate choice for a square lattice is $e^{4i\theta(\mathbf{x})}$.) The order parameter has unit magnitude at $T = 0$, and is expected to decrease due to fluctuations at finite temperature. The distortion $\mathbf{u}(\mathbf{x})$ leads to a change in bond angle given by

$$\theta(\mathbf{x}) = -\frac{1}{2}\left(\frac{\partial u_y}{\partial x} - \frac{\partial u_x}{\partial y}\right) = -\frac{1}{2}\hat{z}\cdot\nabla\times\mathbf{u}. \tag{8.92}$$

The decay of orientational fluctuations is now calculated from

$$\langle\Psi(\mathbf{x})\Psi^*(0)\rangle = \left\langle e^{i6(\theta(\mathbf{x})-\theta(0))}\right\rangle$$

$$= \exp\left[-\frac{6^2}{2}\frac{1}{4}\left\langle[\nabla\times\mathbf{u}(\mathbf{x})-\nabla\times\mathbf{u}(0)]^2\right\rangle\right]$$

$$= \exp\left\{-\frac{9}{2}\int\frac{d^2q\,d^2q'}{(2\pi)^4}(e^{i\mathbf{q}\cdot\mathbf{x}}-1)(e^{i\mathbf{q}'\cdot\mathbf{x}}-1)\,\varepsilon_{ijk}\varepsilon_{i'j'k'}q_j q'_{j'}\,\langle u_k(\mathbf{q})u_{k'}(\mathbf{q}')\rangle\right\}$$

$$= \exp\left\{-\frac{9}{2}\int\frac{d^2q}{(2\pi)^2}(2-2\cos\mathbf{q}\cdot\mathbf{x})\left(q^2\langle|u(\mathbf{q})|^2\rangle-\langle(\mathbf{q}\cdot\mathbf{u})^2\rangle\right)\right\}$$

$$= \exp\left\{-\frac{9}{2\mu}\int\frac{d^2q}{(2\pi)^2}(2-2\cos\mathbf{q}\cdot\mathbf{x})\right\} \approx \exp\left(-\frac{9}{a^2\mu}\right). \tag{8.93}$$

(Note that $\int d^2q/(2\pi)^2$ is the density $n = N/L^2 = 1/a^2$.) The final result is independent of $\mathbf{x}$ at large distances, and asymptotes to a constant that decays exponentially with temperature (since $\mu \propto 1/T$). The two-dimensional solid is thus characterized by quasi-long range translational order, and true long range orientational order.

8.5 Two-dimensional melting

There are qualitative similarities between the melting of a two-dimensional solid and the disordering of the XY spin system. Expanding on the work of Kosterlitz and Thouless [J. Phys. C **5**, L124 (1972)], Halperin and Nelson [Phys. Rev. Lett. **41**, 121 (1978)], and independently, Young [Phys. Rev. B **19**,

1855 (1978)], developed a theory of two-dimensional melting. The resulting, so-called KTHNY theory is briefly sketched here.

(a) The topological defects of a solid are *dislocations*. A single dislocation corresponds to an extra half lattice plane, and is characterized by performing a *Burger's circuit:* The circuit is a series of steps from site to site that on a perfect lattice returns to the starting point (e.g. five steps to the right, four down, followed by five left, and four up on a square lattice.) If the circuit encloses a region with dislocations, it will fail to close. The difference between the initial and final sites of the circuit defines **b**, the *Burger's vector* of the dislocation. This closure failure is possible since the distortion field $\mathbf{u}(\mathbf{x})$ is defined only up to a lattice vector. In terms of the continuous distortion field this closure failure is described by

$$\oint \nabla u_\alpha \, \mathrm{d}s = b^\alpha. \tag{8.94}$$

Fig. 8.7 Examples of dislocations, and the corresponding Burger's vectors, on the triangular and square lattices.

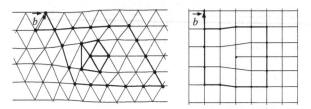

Thus each component of the distortion field behaves like the angle $\theta(\mathbf{x})$ in the XY model. Using this analogy, a possible distortion field due to a collection of dislocations $\{\mathbf{b}_i\}$, at locations $\{\mathbf{x}_i\}$, has a singular part

$$\nabla \tilde{u}_\alpha^* = -\nabla \times \left(\hat{z} \sum_i \frac{b_i^\alpha}{2\pi} \ln |\mathbf{x} - \mathbf{x}_i| \right). \tag{8.95}$$

While the above $\tilde{u}_\alpha^*$ correctly describes the singularities due to the presence of dislocations, it is not a local equilibrium configuration, i.e. not a solution to $2\mu \tilde{u}_{\alpha\beta} + \lambda \delta_{\alpha\beta} \tilde{u}_{\gamma\gamma} = 0$. Allowing the particles to equilibrate in the presence of the dislocations leads to a solution $\tilde{u}_\alpha$, which differs from $\tilde{u}_\alpha^*$ only by a regular part. The total strain field can now decomposed as $u_{\alpha\beta} = \phi_{\alpha\beta} + \tilde{u}_{\alpha\beta}$, where $\tilde{u}_{\alpha\beta}$ results from dislocations, and $\phi_{\alpha\beta}$ is the contribution of phonons. After substituting this form in the elastic energy $\beta\mathcal{H} = \int \mathrm{d}^2\mathbf{x} \left[2\mu u_{\alpha\beta} u_{\alpha\beta} + \lambda u_{\alpha\alpha} u_{\beta\beta} \right]/2$, manipulations similar to the XY model lead to $\beta\mathcal{H} = \beta\mathcal{H}_0 + \beta\mathcal{H}_1$, where

$$\beta\mathcal{H}_0 = \frac{1}{2} \int \mathrm{d}^2\mathbf{x} \left[2\mu \phi_{\alpha\beta} \phi_{\alpha\beta} + \lambda \phi_{\alpha\alpha} \phi_{\beta\beta} \right], \tag{8.96}$$

and

$$\beta\mathcal{H}_1 = -\bar{K} \sum_{i<j} b_i^\alpha b_j^\beta C_{\alpha\beta}(\mathbf{x}_i - \mathbf{x}_j) - \sum_i \ln[y_0(\mathbf{b}_i)]. \tag{8.97}$$

The dislocations behave as a grand canonical gas of particles with vector charges $\{\mathbf{b}_i\}$. The fugacity y_0 comes from the core energy of each dislocation. These charges interact via a vectorial generalization of the Coulomb interaction,

$$C_{\alpha\beta}(\mathbf{x}) = \frac{1}{2\pi}\left[\delta_{\alpha\beta}\ln\left(\frac{|\mathbf{x}|}{a}\right) - \frac{x_\alpha x_\beta}{x^2}\right]. \tag{8.98}$$

The strength of the interaction in Eq. (8.97) is $\bar{K}a^2 = 2\mu(\mu + \lambda)/(2\mu + \lambda)$. Implicit in the above derivation is an overall neutrality condition, $\sum_i \mathbf{b}_i = \mathbf{0}$.

The bare interaction between dislocations is screened by other dislocations in the medium. As in the case of vortices, an effective strength can be calculated perturbatively in y_0. The perturbative correction, however, diverges for $\pi\bar{K}a^2 < 2$, signaling the unbinding of paired neutral dislocations. Under coarse graining, the parameters of the theory evolve as

$$\begin{cases} \dfrac{d\bar{K}^{-1}}{d\ell} = Ay^2 + By^3, \\ \dfrac{dy_0}{d\ell} = (2 - \pi\bar{K})\,y + Dy^2, \end{cases} \tag{8.99}$$

where A, B, and D are constants. In contrast to the XY model, additional terms appear at order of y_0^3. This is because in a scalar Coulomb gas the only neutral configurations have an even number of particles. For dislocations on a triangular lattice, it is possible to construct a neutral configuration from three dislocations at $120°$ angles.

The low-temperature phase maps onto a line of fixed points characterized by renormalized Lamé coefficients, μ_R and λ_R, and $y_0^* = 0$. At high temperatures y_0 and $\bar{K}$ both diverge, indicating the vanishing of the shear modulus μ_R, and a finite correlation length ξ. On approaching the transition from the low-temperature side, the effective shear modulus undergoes a discontinuous jump, with a singular behavior, $\mu_R \simeq \mu_c + c(T_c - T)^{\bar{\nu}}$. The correlation length diverges from the high-temperature side as $\xi \simeq a\exp\left(c'/(T - T_c)^{\bar{\nu}}\right)$. Due to the cubic terms appearing in Eqs. (8.99), the value of $\bar{\nu} = 0.36963\cdots$ is different from the $1/2$ that appears in the XY model, indicating a difference between transitions in vector and scalar Coulomb universality classes [Nelson and Halperin, Phys. Rev. B **19**, 2457 (1979)].

(b) Does the vanishing of the shear modulus imply that the high temperature phase with unbound dislocations is a liquid? As discussed earlier, the crystalline phase has both translational and orientational order. A distortion $\mathbf{u}(\mathbf{x})$ results in a rotation in bond angles according to Eq. (8.92). The net rotation due to a collection of dislocations is

$$\tilde{\theta}(\mathbf{x}) = -\frac{1}{2}\hat{z}\cdot\nabla\times\tilde{u} = \frac{1}{2\pi}\sum_i\frac{\mathbf{b}_i\cdot(\mathbf{x}-\mathbf{x}_i)}{|\mathbf{x}-\mathbf{x}_i|^2}. \tag{8.100}$$

In terms of a continuum dislocation density $\mathbf{b}(\mathbf{x}) = \sum_i \mathbf{b}_i\delta^2(\mathbf{x}-\mathbf{x}_i)$,

$$\tilde{\theta}(\mathbf{x}) = \frac{1}{2\pi}\int d^2\mathbf{x}'\,\frac{\mathbf{b}(\mathbf{x}')\cdot(\mathbf{x}-\mathbf{x}')}{|\mathbf{x}-\mathbf{x}'|^2}. \tag{8.101}$$

Alternatively, in Fourier space,

$$\tilde{\theta}(\mathbf{q}) = \int d^2\mathbf{x}d^2\mathbf{x}'e^{i\mathbf{q}\cdot\mathbf{x}}\frac{\mathbf{b}(\mathbf{x}')\cdot(\mathbf{x}-\mathbf{x}')}{2\pi|\mathbf{x}-\mathbf{x}'|^2} = i\frac{\mathbf{b}(\mathbf{q})\cdot\mathbf{q}}{q^2}. \tag{8.102}$$

The angular fluctuations are thus related to correlations in dislocation density via

$$\left\langle |\tilde{\theta}(\mathbf{q})|^2 \right\rangle = \frac{q_\alpha q_\beta}{q^4} \left\langle b^\alpha(\mathbf{q}) b^\beta(\mathbf{q}) \right\rangle, \tag{8.103}$$

where

$$\left\langle b^\alpha(\mathbf{q}) b^\beta(\mathbf{q}) \right\rangle = \int d^2\mathbf{x}\, e^{i\mathbf{q}\cdot\mathbf{x}} \left\langle b^\alpha(\mathbf{x}) b^\beta(\mathbf{0}) \right\rangle. \tag{8.104}$$

After the dislocations are unbound for $T > T_c$, they interact via a screened Coulomb interaction, and

$$\lim_{|\mathbf{x}|\to\infty} \left\langle b^\alpha(\mathbf{x}) b^\beta(\mathbf{0}) \right\rangle \propto \delta_{\alpha\beta} e^{-|\mathbf{x}|/\xi}, \quad \implies \quad \lim_{\mathbf{q}\to 0} \left\langle b^\alpha(\mathbf{q}) b^\beta(\mathbf{q}) \right\rangle \propto \delta_{\alpha\beta} \xi^2. \tag{8.105}$$

Substituting into Eq. (8.103) leads to

$$\lim_{\mathbf{q}\to 0} \left\langle |\tilde{\theta}(\mathbf{q})|^2 \right\rangle \propto \frac{\xi^2}{q^2}. \tag{8.106}$$

(In the low-temperature phase, the neutrality of the charges at a large scale $|\mathbf{q}|^{-1}$ requires the vanishing of $\mathbf{b}(\mathbf{q})$ as $\mathbf{q} \to \mathbf{0}$, and $\lim_{\mathbf{q}\to 0} \left\langle b^\alpha(\mathbf{q}) b^\beta(\mathbf{q}) \right\rangle \propto q_\alpha q_\beta$, leading to a finite $\langle |\tilde{\theta}(\mathbf{q})|^2 \rangle$.)

Equation (8.106) implies that after the unbinding of dislocations the orientational fluctuations are still correlated. In fact such correlations would result from a Hamiltonian

$$\beta\mathcal{H} = \frac{K_A}{2} \int d^2\mathbf{x} \, (\nabla\theta)^2, \quad \text{with} \quad K_A \propto \xi^2. \tag{8.107}$$

The angular stiffness K_A is known as the Frank constant. The bond angle order correlations now decay as

$$\langle \Psi(\mathbf{x})\Psi^*(\mathbf{0}) \rangle = e^{-\frac{36}{2}\langle [\tilde{\theta}(\mathbf{x})-\tilde{\theta}(\mathbf{0})]^2 \rangle} = \left(\frac{a}{|\mathbf{x}|}\right)^{-\eta_\Psi}, \tag{8.108}$$

with $\eta_\Psi = 18/(\pi K_A)$. The quasi-long range decay of orientational fluctuations leads to the appearance of a six-fold intensity modulation in the diffraction pattern. The dislocations are thus not effective in completely destroying order, and their unbinding leads to the appearance of an orientationally ordered phase known as a *hexatic*. The stiffness of the hexatic phase diverges at the transition to a solid according to Eq. (8.107).

(c) Orientational order disappears at a higher temperature due to the unbinding of a new set of topological defects known as *disclinations*. These are very similar to the vortices in the XY model, except that since the bond angle is defined only up to $2\pi/6$, they satisfy

$$\oint \nabla\theta \cdot d\vec{s} = \frac{2\pi}{6}. \tag{8.109}$$

The energy cost of a disclination grows with system size as $\mathcal{E}_1 = \pi K_A \ln(L/a)/36$. Considering the entropy of $2\ln(L/a)$, we find a disclination unbinding transition for $K_A < 72/\pi$. This transition is in the universality class of the scalar Coulomb gas. The resulting high-temperature phase has neither orientational nor translational order and is a conventional liquid.

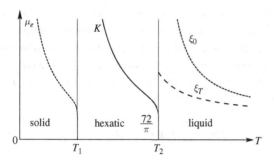

Fig. 8.8 Two-stage melting of a two-dimensional triangular lattice: The orientational correlation length ξ_o diverges at the liquid to hexatic transition at T_1. The hexatic stiffness K diverges at the hexatic to solid transition at T_2. The elastic shear coefficient has a jump at T_1.

This scenario predicts that the melting of a two-dimensional solid proceeds through an intermediate hexatic phase. In fact, computer simulations suggest that simple two-dimensional solids, e.g. particles interacting via Lennard-Jones potentials, undergo a direct first order melting transition as in three dimensions. More complicated molecular systems in three dimensions, e.g. long polymers, are known to have intermediate liquid crystal phases. Liquid crystals have order intermediate between solid and fluid and are three-dimensional analogs of the hexatic phase. Thin films made up of a few monolayers of liquid crystals are good candidates for examining the KTHNY melting scenario.

Problems for chapter 8

1. *Anisotropic nonlinear σ model:* Consider unit n-component spins, $\vec{s}(\mathbf{x}) = (s_1, \cdots, s_n)$ with $\sum_\alpha s_\alpha^2 = 1$, subject to a Hamiltonian

$$\beta\mathcal{H} = \int \mathrm{d}^d\mathbf{x} \left[\frac{1}{2T} \left(\nabla\vec{s} \right)^2 + gs_1^2 \right].$$

For $g = 0$, renormalization group equations are obtained through rescaling distances by a factor $b = e^\ell$, and spins by a factor $\zeta = b^{y_s}$ with $y_s = -\frac{(n-1)}{4\pi}T$, and lead to the flow equation

$$\frac{\mathrm{d}T}{\mathrm{d}\ell} = -\epsilon T + \frac{(n-2)}{2\pi}T^2 + \mathcal{O}(T^3),$$

where $\epsilon = d - 2$.

(a) Find the fixed point, and the thermal eigenvalue y_T.

(b) Write the renormalization group equation for g in the vicinity of the above fixed point, and obtain the corresponding eigenvalue y_g.

(c) Sketch the phase diagram as a function of T and g, indicating the phases, and paying careful attention to the shape of the phase boundary as $g \to 0$.

2. *Matrix models:* In some situations, the order parameter is a matrix rather than a vector. For example, in triangular (Heisenberg) antiferromagnets each triplet of spins aligns at 120°, locally defining a plane. The variations of this plane across the system are described by a 3×3 rotation matrix. We can construct a nonlinear σ model to describe a generalization of this problem as follows. Consider the Hamiltonian

$$\beta \mathcal{H} = \frac{K}{4} \int d^d \mathbf{x} \, \mathrm{tr} \left[\nabla M(\mathbf{x}) \cdot \nabla M^T (\mathbf{x}) \right],$$

where M is a *real, $N \times N$ orthogonal* matrix, and "tr" denotes the trace operation. The condition of orthogonality is that $MM^T = M^T M = I$, where I is the $N \times N$ identity matrix, and M^T is the transposed matrix, $M_{ij}^T = M_{ji}$. The partition function is obtained by summing over all matrix functionals, as

$$Z = \int \mathcal{D}M(\mathbf{x}) \delta \left(M(\mathbf{x}) M^T (\mathbf{x}) - I \right) e^{-\beta \mathcal{H}[M(\mathbf{x})]}.$$

(a) Rewrite the Hamiltonian and the orthogonality constraint in terms of the matrix elements M_{ij} ($i, j = 1, \cdots, N$). Describe the ground state of the system.

(b) Define the symmetric and anti-symmetric matrices

$$\begin{cases} \sigma = \dfrac{1}{2} \left(M + M^T \right) = \sigma^T \\ \pi = \dfrac{1}{2} \left(M - M^T \right) = -\pi^T. \end{cases}$$

Express $\beta \mathcal{H}$ and the orthogonality constraint in terms of the matrices σ and π.

(c) Consider *small fluctuations* about the ordered state $M(\mathbf{x}) = I$. Show that σ can be expanded in powers of π as

$$\sigma = I - \frac{1}{2} \pi \pi^T + \cdots.$$

Use the orthogonality constraint to integrate out σ, and obtain an expression for βH to fourth order in π. Note that there are two distinct types of fourth order terms. *Do not include* terms generated by the argument of the delta function. As shown for the nonlinear σ model in the text, these terms do not affect the results at lowest order.

(d) For an N-vector order parameter there are $N - 1$ Goldstone modes. Show that an orthogonal $N \times N$ order parameter leads to $N(N-1)/2$ such modes.

(e) Consider the quadratic piece of $\beta \mathcal{H}$. Show that the two point correlation function in Fourier space is

$$\langle \pi_{ij}(\mathbf{q}) \pi_{kl}(\mathbf{q}') \rangle = \frac{(2\pi)^d \delta^d (\mathbf{q} + \mathbf{q}')}{K q^2} \left[\delta_{ik} \delta_{jl} - \delta_{il} \delta_{jk} \right].$$

We shall now construct a renormalization group by removing Fourier modes $M^>(\mathbf{q})$, with $\mathbf{q}$ in the shell $\Lambda/b < |\mathbf{q}| < \Lambda$.

(f) Calculate the coarse grained expectation value for $\langle \text{tr}(\sigma) \rangle_0^>$ at low temperatures after removing these modes. Identify the scaling factor, $M'(\mathbf{x}') = M^<(\mathbf{x})/\zeta$, that restores $\text{tr}(M') = \text{tr}(\sigma') = N$.

(g) Use perturbation theory to calculate the coarse grained coupling constant $\tilde{K}$. Evaluate only the two diagrams that directly renormalize the $(\nabla \pi_{ij})^2$ term in $\beta \mathcal{H}$, and show that

$$\tilde{K} = K + \frac{N}{2} \int_{\Lambda/b}^{\Lambda} \frac{d^d \mathbf{q}}{(2\pi)^d} \frac{1}{q^2}.$$

(h) Using the result from part (f), show that after matrix rescaling, the RG equation for K' is given by:

$$K' = b^{d-2} \left[K - \frac{N-2}{2} \int_{\Lambda/b}^{\Lambda} \frac{d^d \mathbf{q}}{(2\pi)^d} \frac{1}{q^2} \right].$$

(i) Obtain the *differential* RG equation for $T = 1/K$, by considering $b = 1 + \delta \ell$. Sketch the flows for $d < 2$ and $d = 2$. For $d = 2 + \epsilon$, compute T_c and the critical exponent ν.

(j) Consider a small symmetry breaking term $-h \int d^d \mathbf{x} \, \text{tr}(M)$, added to the Hamiltonian. Find the renormalization of h, and identify the corresponding exponent y_h.

   Combining RG and symmetry arguments, it can be shown that the 3×3 matrix model is perturbatively equivalent to the $N = 4$ vector model at all orders. This would suggest that stacked triangular antiferromagnets provide a realization of the $\mathcal{O}(4)$ universality class; see P. Azaria, B. Delamotte, and T. Jolicoeur, J. Appl. Phys. **69**, 6170 (1991). However, non-perturbative (topological) aspects appear to remove this equivalence as discussed in S.V. Isakov, T. Senthil, Y.B. Kim, Phys. Rev. B **72**, 174417 (2005).

3. *The roughening transition:* In an earlier problem we examined a continuum interface model which in $d = 3$ is described by the Hamiltonian

$$\beta \mathcal{H}_0 = -\frac{K}{2} \int d^2 \mathbf{x} \, (\nabla h)^2,$$

where $h(\mathbf{x})$ is the interface height at location $\mathbf{x}$. For a crystalline facet, the allowed values of h are multiples of the lattice spacing. In the continuum, this tendency for integer h can be mimicked by adding a term

$$-\beta U = y_0 \int d^2 \mathbf{x} \cos(2\pi h),$$

to the Hamiltonian. Treat $-\beta U$ as a perturbation, and proceed to construct a renormalization group as follows:

(a) Show that

$$\left\langle \exp\left[i\sum_{\alpha} q_{\alpha} h(\mathbf{x}_{\alpha})\right]\right\rangle_{0} = \exp\left[\frac{1}{K}\sum_{\alpha<\beta} q_{\alpha} q_{\beta} C(\mathbf{x}_{\alpha}-\mathbf{x}_{\beta})\right]$$

for $\sum_{\alpha} q_{\alpha} = 0$, and zero otherwise. ($C(\mathbf{x}) = \ln|\mathbf{x}|/2\pi$ is the Coulomb interaction in two dimensions.)

(b) Prove that

$$\left\langle |h(\mathbf{x}) - h(\mathbf{y})|^2\right\rangle = -\frac{\mathrm{d}^2}{\mathrm{d}k^2} G_k(\mathbf{x}-\mathbf{y})|_{k=0},$$

where $G_k(\mathbf{x}-\mathbf{y}) = \left\langle \exp\left[ik(h(\mathbf{x}) - h(\mathbf{y}))\right]\right\rangle$.

(c) Use the results in (a) to calculate $G_k(\mathbf{x}-\mathbf{y})$ in perturbation theory to order of y_0^2. (Hint: Set $\cos(2\pi h) = \left(e^{2i\pi h} + e^{-2i\pi h}\right)/2$. The first order terms vanish according to the result in (a), while the second order contribution is identical in structure to that of the Coulomb gas described in this chapter.)

(d) Write the perturbation result in terms of an effective interaction K, and show that perturbation theory fails for K larger than a critical K_c.

(e) Recast the perturbation result in part (d) into renormalization group equations for K and y_0, by changing the "lattice spacing" from a to ae^{ℓ}.

(f) Using the recursion relations, discuss the phase diagram and phases of this model.

(g) For large separations $|\mathbf{x}-\mathbf{y}|$, find the magnitude of the discontinuous jump in $\left\langle |h(\mathbf{x}) - h(\mathbf{y})|^2\right\rangle$ at the transition.

4. *Roughening and duality:* Consider a discretized version of the Hamiltonian in the previous problem, in which for each site i of a square lattice there is an integer valued height h_i. The Hamiltonian is

$$\beta\mathcal{H} = \frac{K}{2}\sum_{\langle i,j\rangle} |h_i - h_j|^{\infty},$$

where the "∞" power means that there is no energy cost for $\Delta h = 0$; an energy cost of $K/2$ for $\Delta h = \pm 1$; and $\Delta h = \pm 2$ or higher *are not allowed* for neighboring sites. (This is known as the restricted solid on solid (RSOS) model.)

(a) Construct the dual model either diagrammatically, or by following these steps:

(i) Change from the N site variables h_i, to the $2N$ bond variables $n_{ij} = h_i - h_j$. Show that the sum of n_{ij} around any plaquette is constrained to be zero.

(ii) Impose the constraints by using the identity $\int_0^{2\pi} \mathrm{d}\theta e^{i\theta n}/2\pi = \delta_{n,0}$, for integer n.

(iii) After imposing the constraints, you can sum freely over the bond variables n_{ij} to obtain a dual interaction $\tilde{v}(\theta_i - \theta_j)$ between dual variables θ_i on neighboring plaquettes.

(b) Show that for large K, the dual problem is just the XY model. Is this conclusion consistent with the renormalization group results of the previous problem? (Also note the connection with the loop model considered in the problems of the previous chapter.)

(c) Does the one dimensional version of this Hamiltonian, i.e. a 2d interface with

$$-\beta \mathcal{H} = -\frac{K}{2}\sum_i |h_i - h_{i+1}|^\infty,$$

have a roughening transition?

---

5. *Nonlinear σ model with long-range interactions:* Consider unit n-component spins, $\vec{s}(\mathbf{x}) = (s_1, s_2, \cdots, s_n)$ with $|\vec{s}(\mathbf{x})|^2 = \sum_i s_i(\mathbf{x})^2 = 1$, interacting via a Hamiltonian

$$\beta \mathcal{H} = \int d^d\mathbf{x} \int d^d\mathbf{y}\, K(|\mathbf{x} - \mathbf{y}|)\, \vec{s}(\mathbf{x}) \cdot \vec{s}(\mathbf{y}).$$

(a) The long-range interaction, $K(x)$, is the Fourier transform of $Kq^\omega/2$ with $\omega < 2$. What kind of asymptotic decay of interactions at long distances is consistent with such decay? (Dimensional analysis is sufficient for the answer, and no explicit integrations are required.)

(b) Close to zero temperature we can set $\vec{s}(\mathbf{x}) = (\vec{\pi}(\mathbf{x}), \sigma(\mathbf{x}))$, where $\vec{\pi}(\mathbf{x})$ is an $n-1$ component vector representing *small fluctuations* around the ground state. Find the effective Hamiltonian for $\vec{\pi}(\mathbf{x})$ after integrating out $\{\sigma(\mathbf{x})\}$.

(c) Fourier transform the *quadratic part* of the above Hamiltonian focusing only on terms proportional to K, and hence calculate the expectation value $\langle \pi_i(\mathbf{q})\pi_j(\mathbf{q}')\rangle_0$.

We shall now construct a renormalization group by removing Fourier modes, $\vec{\pi}^>(\mathbf{q})$, with $\mathbf{q}$ in the shell $\Lambda/b < |\mathbf{q}| < \Lambda$.

(d) Calculate the coarse grained expectation value for $\langle \sigma \rangle_0^>$ to order of π^2 after removing these modes. Identify the scaling factor, $\vec{s}'(\mathbf{x}') = \vec{s}^<(\mathbf{x})/\zeta$, that restores $\vec{s}'$ to unit length.

(e) A simplifying feature of long-range interactions is that the coarse grained coupling constant is not modified by the perturbation, i.e. $\tilde{K} = K$ to all orders in a perturbative calculation. Use this information, along with simple dimensional analysis, to express the renormalized interaction, $K'(b)$, in terms of K, b, and ζ.

(f) Obtain the *differential* RG equation for $T = 1/K$ by considering $b = 1 + \delta \ell$.

(g) For $d = \omega + \epsilon$, compute T_c and the critical exponent ν to lowest order in ϵ.

(h) Add a small symmetry breaking term, $-\vec{h} \cdot \int d^d\mathbf{x}\, \vec{s}(\mathbf{x})$, to the Hamiltonian. Find the renormalization of h and identify the corresponding exponent y_h.

6. *The XY model in $2 + \epsilon$ dimensions:* The recursion relations of the XY model in two dimensions can be generalized to $d = 2 + \epsilon$ dimensions, and take the form:

$$\begin{cases} \dfrac{dT}{d\ell} = -\epsilon T + 4\pi^3 y^2 \\ \dfrac{dy}{d\ell} = \left(2 - \dfrac{\pi}{T}\right)y. \end{cases}$$

(a) Calculate the position of the fixed point for the finite temperature phase transition.

(b) Obtain the eigenvalues at this fixed point to *lowest* contributing order in ϵ.

(c) Estimate the exponents ν and α for the superfluid transition in $d = 3$ from these results. [Be careful in keeping track of only the lowest nontrivial power of ϵ in your expressions.]

7. *Symmetry breaking fields:* Let us investigate adding a term

$$-\beta \mathcal{H}_p = h_p \int d^2\mathbf{x} \cos(p\theta(\mathbf{x}))$$

to the XY model. There are a number of possible causes for such a symmetry breaking field: $p = 1$ is the usual "magnetic field", $p = 2, 3, 4$, and 6 could be due to couplings to an underlying lattice of rectangular, hexagonal, square, or triangular symmetry respectively. As $h_p \to \infty$, the spin becomes discrete, taking one of p possible values, and the model becomes equivalent to the previously discussed clock models.

(a) Assume that we are in the low-temperature phase so that vortices are absent, i.e. the vortex fugacity y is zero (in the RG sense). In this case, we can ignore the angular nature of θ and replace it with a scalar field ϕ, leading to the partition function

$$Z = \int D\phi(x) \exp\left\{-\int d^2\mathbf{x}\left[\frac{K}{2}(\nabla\phi)^2 + h_p \cos(p\phi)\right]\right\}.$$

This is known as the sine–Gordon model, and is equivalent to the roughening transition of a previous problem. Use similar methods to obtain the recursion relations for h_p and K.

(b) Show that once vortices are included, the recursion relations are

$$\frac{dh_p}{d\ell} = \left(2 - \frac{p^2}{4\pi K}\right)h_p$$

$$\frac{dK^{-1}}{d\ell} = -\frac{\pi p^2 h_p^2}{2}K^{-2} + 4\pi^3 y^2,$$

$$\frac{dy}{d\ell} = (2 - \pi K)y.$$

(c) Show that the above RG equations are only valid for $\frac{8\pi}{p^2} < K^{-1} < \frac{\pi}{2}$, and thus only apply for $p > 4$. Sketch possible phase diagrams for $p > 4$ and $p < 4$. In fact $p = 4$ is rather special as there is a marginal operator h_4, and the transition to the four-fold phase (cubic anisotropy) has continuously varying critical exponents!

8. *Inverse-square interactions:* Consider a scalar field $s(x)$ in one-dimension, subject to an energy

$$-\beta \mathcal{H}_s = \frac{K}{2}\int dxdy \frac{s(x)s(y)}{|x-y|^2} + \int dx\Phi[s(x)].$$

The local energy $\Phi[s]$ strongly favors $s(x) = \pm 1$ (e.g. $\Phi[s] = g(s^2 - 1)^2$, with $g \gg 1$).

(a) With $K > 0$, the ground state is ferromagnetic. Estimate the energy cost of a single domain wall in a chain of length L. You may assume that the transition from $s = +1$ to $s = -1$ occurs over a short distance cutoff a.

(b) From the probability of the formation of a single kink, obtain a lower bound for the critical coupling K_c, separating ordered and disordered phases.

(c) Show that the energy of a dilute set of domain walls located at positions $\{x_i\}$ is given by

$$-\beta \mathcal{H}_Q = 4K \sum_{i<j} q_i q_j \ln \left(\frac{|x_i - x_j|}{a} \right) + \ln y_0 \sum_i |q_i|,$$

where $q_i = \pm 1$ depending on whether $s(x)$ increases or decreases at the domain wall. (Hints: Perform integrations by part, and coarse-grain to size a. The function $\Phi[s]$ only contributes to the core energy of the domain wall, which results in the fugacity y_0.)

(d) The logarithmic interaction between two opposite domain walls at a large distance L is reduced due to screening by other domain walls in between. This interaction can be calculated perturbatively in y_0, and to lowest order is described by an effective coupling (see later)

$$K \to K_{\text{eff}} = K - 2Ky_0^2 \int_a^\infty dr \left(\frac{a}{r} \right)^{4K} + \mathcal{O}(y_0^4). \tag{1}$$

By changing the cutoff from a to $ba = (1 + \delta\ell)a$, construct differential recursion relations for the parameters K and y_0.

(e) Sketch the renormalization group flows as a function of $T = K^{-1}$ and y_0, and discuss the phases of the model.

(f) Derive the effective interaction given above as Eq. (1). (Hint: This is somewhat easier than the corresponding calculation for the two-dimensional Coulomb gas, as the charges along the chain must alternate.)

9. *Melting:* The elastic energy cost of a deformation $u_i(\mathbf{x})$, of an isotropic lattice is

$$-\beta \mathcal{H} = \frac{1}{2} \int d^d\mathbf{x} \left[2\mu u_{ij} u_{ij} + \lambda u_{ii} u_{jj} \right],$$

where $u_{ij}(\mathbf{x}) = \left(\partial_i u_j + \partial_j u_i \right)/2$ is the strain tensor.

(a) Express the energy in terms of the Fourier transforms $u_i(\mathbf{q})$, and find the normal modes of vibrations.

(b) Calculate the expectation value $\langle u_i(\mathbf{q}) u_j(\mathbf{q}') \rangle$.

(c) Assuming a short-distance cutoff of the order of the lattice spacing a, calculate $U^2(\mathbf{x}) \equiv \langle \left(\vec{u}(\mathbf{x}) - \vec{u}(0) \right)^2 \rangle$.

(d) The (heuristic) *Lindemann criterion* states that the lattice melts when deformations grow to a fraction of the lattice spacing, i.e. for $\lim_{|\mathbf{x}| \to \infty} U(\mathbf{x}) = c_L a$. Assuming that $\mu = \hat{\mu}/(k_B T)$ and $\lambda = \hat{\lambda}/(k_B T)$, use the above criterion to calculate the melting temperature T_m. Comment on the behavior of T_m as a function of dimension d.

9

Dissipative dynamics

9.1 Brownian motion of a particle

Observations under a microscope indicate that a dust particle in a liquid drop undergoes a random jittery motion. This is because of the random impacts of the much smaller fluid particles. The theory of such (*Brownian*) motion was developed by Einstein in 1905 and starts with the equation of motion for the particle [A. Einstein, Ann. d. Physik **17**, 549 (1905)]. The displacement $\vec{x}(t)$, of a particle of mass m is governed by

$$m\ddot{\vec{x}} = -\frac{\dot{\vec{x}}}{\mu} - \frac{\partial \mathcal{V}}{\partial \vec{x}} + \vec{f}_{\text{random}}(t). \tag{9.1}$$

The three forces acting on the particle are:

(1) A friction force due to the viscosity of the fluid. For a spherical particle of radius R, the mobility in the low Reynolds number limit is given by $\mu = (6\pi\bar{\eta}R)^{-1}$, where $\bar{\eta}$ is the specific viscosity.
(2) The force due to the external potential $\mathcal{V}(\vec{x})$, e.g. gravity.
(3) A random force of zero mean due to the impacts of fluid particles.

The viscous term usually dominates the inertial one (i.e. the motion is overdamped), and we shall henceforth ignore the acceleration term. Equation (9.1) now reduces to the *Langevin equation*,

$$\dot{\vec{x}} = \vec{v}(\vec{x}) + \vec{\eta}(t), \tag{9.2}$$

where $\vec{v}(\vec{x}) = -\mu \partial \mathcal{V}/\partial \vec{x}$ is the *deterministic* velocity. The *stochastic* velocity, $\vec{\eta}(t) = \mu \vec{f}_{\text{random}}(t)$, has zero mean,

$$\langle \vec{\eta}(t) \rangle = 0. \tag{9.3}$$

It is usually assumed that the probability distribution for the noise in velocity is Gaussian, i.e.

$$\mathcal{P}\left[\vec{\eta}(t)\right] \propto \exp\left[-\int d\tau \, \frac{\eta(\tau)^2}{4D}\right]. \tag{9.4}$$

Note that different components of the noise, and at different times, are independent, and the covariance is

$$\langle \eta_\alpha(t)\eta_\beta(t') \rangle = 2D\delta_{\alpha\beta}\delta(t-t').$$ (9.5)

The parameter D is related to *diffusion* of particles in the fluid. In the absence of any potential, $\mathcal{V}(\vec{x}) = 0$, the position of a particle at time t is given by

$$\vec{x}(t) = \vec{x}(0) + \int_0^t d\tau \, \vec{\eta}(\tau).$$

Clearly the separation $\vec{x}(t) - \vec{x}(0)$ which is the sum of random Gaussian variables is itself Gaussian distributed with mean zero, and a variance

$$\left\langle \left(\vec{x}(t) - \vec{x}(0) \right)^2 \right\rangle = \int_0^t d\tau_1 d\tau_2 \langle \vec{\eta}(\tau_1) \cdot \vec{\eta}(\tau_2) \rangle = 3 \times 2Dt.$$

For an ensemble of particles released at $\vec{x}(t) = 0$, i.e. with $\mathcal{P}\left(\vec{x}, t = 0\right) = \delta^3(\vec{x})$, the particles at time t are distributed according to

$$\mathcal{P}\left(\vec{x}, t\right) = \left(\frac{1}{\sqrt{4\pi Dt}} \right)^{3/2} \exp\left[-\frac{x^2}{4Dt} \right],$$

which is the solution to the diffusion equation

$$\frac{\partial \mathcal{P}}{\partial t} = D\nabla^2 \mathcal{P}.$$

A simple example is provided by a particle connected to a Hookian spring, with $\mathcal{V}(\vec{x}) = Kx^2/2$. The deterministic velocity is now $\vec{v}(\vec{x}) = -\mu K\vec{x}$, and the Langevin equation, $\dot{\vec{x}} = -\mu K\vec{x} + \vec{\eta}(t)$, can be rearranged as

$$\frac{d}{dt}\left[e^{\mu Kt}\vec{x}(t) \right] = e^{\mu Kt}\vec{\eta}(t).$$ (9.6)

Integrating the equation from 0 to t yields

$$e^{\mu Kt}\vec{x}(t) - \vec{x}(0) = \int_0^t d\tau e^{\mu K\tau}\vec{\eta}(\tau),$$ (9.7)

and

$$\vec{x}(t) = \vec{x}(0)e^{-\mu Kt} + \int_0^t d\tau e^{-\mu K(t-\tau)}\vec{\eta}(\tau).$$ (9.8)

Averaging over the noise indicates that the mean position,

$$\langle \vec{x}(t) \rangle = \vec{x}(0)e^{-\mu Kt},$$ (9.9)

decays with a characteristic *relaxation time*, $\tau = 1/(\mu K)$. Fluctuations around the mean behave as

$$\left\langle \left(\vec{x}(t) - \langle\vec{x}(t)\rangle\right)^2\right\rangle = \int_0^t d\tau_1 d\tau_2 e^{-\mu K(2t-\tau_1-\tau_2)} \overbrace{\langle\vec{\eta}(\tau_1)\cdot\vec{\eta}(\tau_2)\rangle}^{2D\delta(\tau_1-\tau_2)\times 3}$$

$$= 6D\int_0^t d\tau e^{-2\mu K(t-\tau)} \qquad (9.10)$$

$$= \frac{3D}{\mu K}\left[1 - e^{-2\mu Kt}\right] \xrightarrow{t\to\infty} \frac{3D}{\mu K}.$$

However, once the dust particle reaches equilibrium with the fluid at a temperature T, its probability distribution must satisfy the normalized Boltzmann weight

$$\mathcal{P}_{\text{eq.}}(\vec{x}) = \left(\frac{K}{2\pi k_B T}\right)^{3/2} \exp\left[-\frac{Kx^2}{2k_B T}\right], \qquad (9.11)$$

yielding $\langle x^2\rangle = 3k_B T/K$. Since the dynamics is expected to bring the particle to equilibrium with the fluid at temperature T, Eq. (9.1) implies the condition

$$D = k_B T\mu. \qquad (9.12)$$

This is the Einstein relation connecting the *fluctuations* of noise to the *dissipation* in the medium.

Clearly the Langevin equation at long times reproduces the correct mean and variance for a particle in equilibrium at a temperature T in the potential $\mathcal{V}(\vec{x}) = Kx^2/2$, provided that Eq. (9.12) is satisfied. Can we show that the whole probability distribution evolves to the Boltzmann weight for any potential? Let $\mathcal{P}(\vec{x}, t) \equiv \langle\vec{x}|\mathcal{P}(t)|0\rangle$ denote the probability density of finding the particle at $\vec{x}$ at time t, given that it was at 0 at $t = 0$. This probability can be constructed recursively by noting that a particle found at $\vec{x}$ at time $t + \epsilon$ must have arrived from some other point $\vec{x}'$ at t. Adding up all such probabilities yields

$$\mathcal{P}(\vec{x}, t+\epsilon) = \int d^3\vec{x}' \, \mathcal{P}(\vec{x}', t)\langle\vec{x}|T_\epsilon|\vec{x}'\rangle, \qquad (9.13)$$

where $\langle\vec{x}|T_\epsilon|\vec{x}'\rangle \equiv \langle\vec{x}|\mathcal{P}(\epsilon)|\vec{x}'\rangle$ is the transition probability. For $\epsilon \ll 1$,

$$\vec{x} = \vec{x}' + \vec{v}(\vec{x}')\epsilon + \vec{\eta}_\epsilon, \qquad (9.14)$$

where $\vec{\eta}_\epsilon = \int_t^{t+\epsilon} d\tau\vec{\eta}(\tau)$. Clearly, $\langle\vec{\eta}_\epsilon\rangle = 0$, and $\langle\eta_\epsilon^2\rangle = 2D\epsilon \times 3$, and following Eq. (9.4),

$$p(\vec{\eta}_\epsilon) = \left(\frac{1}{4\pi D\epsilon}\right)^{3/2} \exp\left[-\frac{\eta_\epsilon^2}{4D\epsilon}\right]. \qquad (9.15)$$

The transition rate is simply the probability of finding a noise of the right magnitude according to Eq. (9.14), and

$$\langle\vec{x}|T(\epsilon)|\vec{x}'\rangle = p(\eta_\epsilon) = \left(\frac{1}{4\pi D\epsilon}\right)^{3/2} \exp\left[-\frac{\left(\vec{x} - \vec{x}' - \epsilon\vec{v}(\vec{x}')\right)^2}{4D\epsilon}\right]$$

$$= \left(\frac{1}{4\pi D\epsilon}\right)^{3/2} \exp\left[-\epsilon\frac{\left(\dot{\vec{x}} - \vec{v}(\vec{x})\right)^2}{4D}\right]. \qquad (9.16)$$

By subdividing the time interval t into infinitesimal segments of size ϵ, repeated application of the above evolution operator yields

$$\mathcal{P}(\vec{x}, t) = \langle \vec{x} \left| T(\epsilon)^{t/\epsilon} \right| 0 \rangle$$

$$= \int_{(0,0)}^{(\vec{x},t)} \frac{\mathcal{D}\vec{x}(\tau)}{\mathcal{N}} \exp\left[-\int_0^t d\tau \frac{\left(\dot{\vec{x}} - \vec{v}(\vec{x})\right)^2}{4D}\right]. \qquad (9.17)$$

The integral is over all paths connecting the initial and final points; each path's weight is related to its deviation from the classical trajectory, $\dot{\vec{x}} = \vec{v}(\vec{x})$. The recursion relation in Eq. (9.13) can now be written as

$$\mathcal{P}(\vec{x}, t) = \int d^3\vec{x}' \left(\frac{1}{4\pi D\epsilon}\right)^{3/2} \exp\left[-\frac{\left(\vec{x} - \vec{x}' - \epsilon\vec{v}(\vec{x}')\right)^2}{4D\epsilon}\right] \mathcal{P}(\vec{x}', t-\epsilon), \qquad (9.18)$$

and simplified by the change of variables,

$$\vec{y} = \vec{x}' + \epsilon\vec{v}(\vec{x}') - \vec{x} \implies$$
$$d^3\vec{y} = d^3\vec{x}' \left(1 + \epsilon\nabla \cdot \vec{v}(\vec{x}')\right) = d^3\vec{x}' \left(1 + \epsilon\nabla \cdot \vec{v}(\vec{x}) + \mathcal{O}(\epsilon^2)\right). \qquad (9.19)$$

Keeping only terms at order of ϵ, we obtain

$$\mathcal{P}(\vec{x}, t) = \left[1 - \epsilon\nabla \cdot \vec{v}(\vec{x})\right] \int d^3\vec{y} \left(\frac{1}{4\pi D\epsilon}\right)^{3/2} e^{-\frac{y^2}{4D\epsilon}} \mathcal{P}(\vec{x} + \vec{y} - \epsilon\vec{v}(\vec{x}), t-\epsilon)$$

$$= \left[1 - \epsilon\nabla \cdot \vec{v}(\vec{x})\right] \int d^3\vec{y} \left(\frac{1}{4\pi D\epsilon}\right)^{3/2} e^{-\frac{y^2}{4D\epsilon}}$$

$$\times \left[\mathcal{P}(\vec{x}, t) + (\vec{y} - \epsilon\vec{v}(\vec{x})) \cdot \nabla\mathcal{P} + \frac{y_i y_j - 2\epsilon y_i v_j + \epsilon^2 v_i v_j}{2} \nabla_i \nabla_j \mathcal{P} - \epsilon\frac{\partial\mathcal{P}}{\partial t} + \mathcal{O}(\epsilon^2)\right]$$

$$= \left[1 - \epsilon\nabla \cdot \vec{v}(\vec{x})\right] \left[\mathcal{P} - \epsilon\vec{v} \cdot \nabla + \epsilon D\nabla^2\mathcal{P} - \epsilon\frac{\partial\mathcal{P}}{\partial t} + \mathcal{O}(\epsilon^2)\right]. \qquad (9.20)$$

Equating terms at order of ϵ leads to the *Fokker–Planck equation*,

$$\frac{\partial\mathcal{P}}{\partial t} + \nabla \cdot \vec{J} = 0, \quad \text{with} \quad \vec{J} = \vec{v}\mathcal{P} - D\nabla\mathcal{P}. \qquad (9.21)$$

The Fokker–Planck equation is simply the statement of conservation of probability. The probability current has a deterministic component $\vec{v}\mathcal{P}$, and a stochastic part $-D\nabla\mathcal{P}$. A *stationary distribution*, $\partial\mathcal{P}/\partial t = 0$, is obtained if the net current vanishes. It is now easy to check that the Boltzmann weight, $\mathcal{P}_{eq.}(\vec{x}) \propto \exp[-\mathcal{V}(\vec{x})/k_B T]$, with $\nabla\mathcal{P}_{eq.} = \vec{v}\mathcal{P}_{eq.}/(\mu k_B T)$, leads to a stationary state as long as the fluctuation–dissipation condition in Eq. (9.12) is satisfied.

9.2 Equilibrium dynamics of a field

The next step is to generalize the Langevin formalism to a collection of degrees of freedom, most conveniently described by a continuous field. Let us consider the order parameter field $\vec{m}(\mathbf{x}, t)$ of a magnet. In equilibrium, the probability to find a coarse-grained configuration of the magnetization field is governed by the Boltzmann weight of the Landau–Ginzburg Hamiltonian

$$\mathcal{H}\left[\vec{m}\right] = \int d^d\mathbf{x} \left[\frac{r}{2} m^2 + u m^4 + \frac{K}{2} (\nabla m)^2 + \cdots \right]. \tag{9.22}$$

(To avoid confusion with time, the coefficient of the quadratic term is changed from t to r.) Clearly the above energy functional contains no kinetic terms, and should be regarded as the analog of the potential energy $\mathcal{V}(\vec{x})$ employed in the previous section. To construct a Langevin equation governing the dynamics of the field $\vec{m}(\mathbf{x})$, we first calculate the analogous *force* on each field element from the variations of this potential energy. The *functional derivative* of Eq. (9.22) yields

$$F_i(\mathbf{x}) = -\frac{\delta\mathcal{H}[\vec{m}]}{\delta m_i(\mathbf{x})} = -r m_i - 4 u m_i |\vec{m}|^2 + K \nabla^2 m_i. \tag{9.23}$$

The straightforward analog of Eq. (9.2) is

$$\frac{\partial m_i(\mathbf{x}, t)}{\partial t} = \mu F_i(\mathbf{x}) + \eta_i(\mathbf{x}, t), \tag{9.24}$$

with a random "velocity", $\vec{\eta}$, such that

$$\langle \eta_i(\mathbf{x}, t) \rangle = 0, \quad \text{and} \quad \langle \eta_i(\mathbf{x}, t)\eta_j(\mathbf{x}', t') \rangle = 2D\delta_{ij}\delta(\mathbf{x} - \mathbf{x}')\delta(t - t'). \tag{9.25}$$

The resulting Langevin equation,

$$\frac{\partial \vec{m}(\mathbf{x}, t)}{\partial t} = -\mu r \vec{m} - 4\mu u m^2 \vec{m} + \mu K \nabla^2 \vec{m} + \vec{\eta}(\mathbf{x}, t), \tag{9.26}$$

is known as the *time-dependent Landau–Ginzburg equation*. Because of the nonlinear term $m^2\vec{m}$, it is not possible to integrate this equation exactly. To gain some insight into its behavior we start with the disordered phase of the model which is well described by the Gaussian weight with $u = 0$. The resulting linear equation is then easily solved by examining the Fourier components,

$$\vec{m}(\mathbf{q}, t) = \int d^d\mathbf{x}\, e^{i\mathbf{q}\cdot\mathbf{x}} \vec{m}(\mathbf{x}, t), \tag{9.27}$$

which evolve according to

$$\frac{\partial \vec{m}(\mathbf{q}, t)}{\partial t} = -\mu(r + Kq^2)\, \vec{m}(\mathbf{q}, t) + \vec{\eta}(\mathbf{q}, t). \tag{9.28}$$

The Fourier transformed noise,

$$\vec{\eta}(\mathbf{q}, t) = \int d^d\mathbf{x}\, e^{i\mathbf{q}\cdot\mathbf{x}} \vec{\eta}(\mathbf{x}, t), \tag{9.29}$$

has zero mean, $\langle \eta_i(\mathbf{q}, t) \rangle = 0$, and correlations

$$
\langle \eta_i(\mathbf{q}, t) \eta_j(\mathbf{q}', t') \rangle = \int d^d\mathbf{x} d^d\mathbf{x}' \ e^{i\mathbf{q}\cdot\mathbf{x} + i\mathbf{q}'\cdot\mathbf{x}'} \overbrace{\langle \eta_i(\mathbf{x}, t) \eta_j(\mathbf{x}', t') \rangle}^{2D\delta_{ij}\delta^d(\mathbf{x}-\mathbf{x}')\,\delta(t-t')}
$$

$$
= 2D\delta_{ij}\delta(t - t') \int d^d\mathbf{x} \, e^{i\mathbf{x}\cdot(\mathbf{q}+\mathbf{q}')}
$$

$$
= 2D\delta_{ij}\delta(t - t')(2\pi)^d \delta^d(\mathbf{q}+\mathbf{q}'). \tag{9.30}
$$

Each Fourier mode in Eq. (9.28) now behaves as an independent particle connected to a spring as in Eq. (9.6). Introducing a decay rate

$$
\gamma(\mathbf{q}) \equiv \frac{1}{\tau(\mathbf{q})} = \mu(r + Kq^2), \tag{9.31}
$$

the evolution of each mode is similar to Eq. (9.8), and follows

$$
\vec{m}(\mathbf{q}, t) = \vec{m}(\mathbf{q}, 0)e^{-\gamma(\mathbf{q})t} + \int_0^t d\tau \, e^{-\gamma(\mathbf{q})(t-\tau)} \vec{\eta}(\mathbf{q}, t). \tag{9.32}
$$

Fluctuations in each mode decay with a different *relaxation time* $\tau(\mathbf{q})$; $\langle \vec{m}(\mathbf{q}, t) \rangle = \vec{m}(\mathbf{q}, 0) \exp[-t/\tau(\mathbf{q})]$. When in equilibrium, the order parameter in the Gaussain model is correlated over the length scale $\xi = \sqrt{K/r}$. In considering relaxation to equilibrium, we find that at length scales larger than ξ (or $q \ll 1/\xi$), the relaxation time saturates to $\tau_{max} = 1/(\mu r)$. On approaching the singular point of the Gaussian model at $r = 0$, the time required to reach equilibrium diverges. This phenomenon is known as *critical slowing down*, and is also present for the nonlinear equation, albeit with modified exponents. The critical point is thus characterized by diverging *length* and *time* scales. For the critical fluctuations at distances shorter than the correlation length ξ, the characteristic time scale grows with wavelength as $\tau(q) \approx (\mu Kq^2)^{-1}$. The scaling relation between the critical length and time scales is described by a *dynamic exponent* z, as $\tau \propto \lambda^z$. The value of $z = 2$ for the critical Gaussian model is reminiscent of diffusion processes.

Time-dependent correlation functions are obtained from

$$
\langle m_i(\mathbf{q}, t) m_j(\mathbf{q}', t) \rangle_c = \int_0^t d\tau_1 d\tau_2 e^{-\gamma(\mathbf{q})(t-\tau_1) - \gamma(\mathbf{q}')(t-\tau_2)} \overbrace{\langle \eta_i(\mathbf{q}, \tau_1) \eta_j(\mathbf{q}', \tau_2) \rangle}^{2D\delta_{ij}\delta(\tau_1-\tau_2)(2\pi)^d\delta^d(\mathbf{q}+\mathbf{q}')}
$$

$$
= (2\pi)^d \delta^d(\mathbf{q}+\mathbf{q}') \, 2D\delta_{ij} \int_0^t d\tau e^{-2\gamma(\mathbf{q})(t-\tau)}
$$

$$
= (2\pi)^d \delta^d(\mathbf{q}+\mathbf{q}')\delta_{ij} \frac{D}{\gamma(\mathbf{q})} \left(1 - e^{-2\gamma(\mathbf{q})t}\right) \tag{9.33}
$$

$$
\xrightarrow{t\to\infty} (2\pi)^d \delta^d(\mathbf{q}+\mathbf{q}')\delta_{ij} \frac{D}{\mu(r + Kq^2)}.
$$

However, direct diagonalization of the Hamiltonian in Eq. (9.22) with $u = 0$ gives

$$\mathcal{H} = \int \frac{d^d \mathbf{q}}{(2\pi)^d} \frac{(r + Kq^2)}{2} |\vec{m}(\mathbf{q})|^2, \tag{9.34}$$

leading to the equilibrium correlation functions

$$\langle m_i(\mathbf{q}) m_j(\mathbf{q}') \rangle = (2\pi)^d \delta^d(\mathbf{q} + \mathbf{q}') \delta_{ij} \frac{k_B T}{r + Kq^2}. \tag{9.35}$$

Comparing Eqs. (9.33) and (9.35) indicates that the long-time dynamics reproduce the correct equilibrium behavior if the fluctuation–dissipation condition, $D = k_B T \mu$, is satisfied.

In fact, quite generally, the single-particle Fokker–Planck equation (9.21) can be generalized to describe the evolution of the whole probability functional, $\mathcal{P}([\vec{m}(\mathbf{x})], t)$, as

$$\frac{\partial \mathcal{P}([\vec{m}(\mathbf{x})], t)}{\partial t} = -\int d^d \mathbf{x} \frac{\delta}{\delta m_i(\mathbf{x})} \left[-\mu \frac{\delta \mathcal{H}}{\delta m_i(\mathbf{x})} \mathcal{P} - D \frac{\delta \mathcal{P}}{\delta m_i(\mathbf{x})} \right]. \tag{9.36}$$

For the equilibrium Boltzmann weight

$$\mathcal{P}_{\text{eq.}}[\vec{m}(\mathbf{x})] \propto \exp\left[-\frac{\mathcal{H}[\vec{m}(\mathbf{x})]}{k_B T} \right], \tag{9.37}$$

the functional derivative results in

$$\frac{\delta \mathcal{P}_{\text{eq.}}}{\delta m_i(\mathbf{x})} = -\frac{1}{k_B T} \frac{\delta \mathcal{H}}{\delta m_i(\mathbf{x})} \mathcal{P}_{\text{eq.}}. \tag{9.38}$$

The total probability current,

$$J[h(\mathbf{x})] = \left[-\mu \frac{\delta \mathcal{H}}{\delta m_i(\mathbf{x})} + \frac{D}{k_B T} \frac{\delta \mathcal{H}}{\delta m_i(\mathbf{x})} \right] \mathcal{P}_{\text{eq.}}, \tag{9.39}$$

vanishes if the fluctuation–dissipation condition, $D = \mu k_B T$, is satisfied. Once again, the Einstein equation ensures that the equilibrium weight indeed describes a steady state.

9.3 Dynamics of a conserved field

In fact it is possible to obtain the correct equilibrium weight with $\mathbf{q}$ dependent mobility and noise, as long as the generalized fluctuation–dissipation condition,

$$D(\mathbf{q}) = k_B T \mu(\mathbf{q}), \tag{9.40}$$

holds. This generalized condition is useful in considering the dissipative dynamics of a *conserved* field. The prescription that leads to the Langevin equations (9.23)–(9.25) does not conserve the field in the sense that $\int d^d \mathbf{x} \, \vec{m}(\mathbf{x}, t)$ can change with time. (Although this quantity is on average zero for $r > 0$, it undergoes stochastic fluctuations.) If we are dealing with a binary mixture ($n = 1$), the order parameter which measures the difference between densities of the two components is conserved. Any concentration that is removed from some part of the system must go to a neighboring region in any realistic dynamics. Let us then consider a local dynamical process constrained such that

$$\frac{d}{dt} \int d^d\mathbf{x}\, \vec{m}(\mathbf{x}, t) = \int d^d\mathbf{x}\, \frac{\partial \vec{m}(\mathbf{x}, t)}{\partial t} = \vec{0}. \tag{9.41}$$

How can we construct a dynamical equation that satisfies Eq. (9.41)? The integral clearly vanishes if the integrand is a total divergence, i.e.

$$\frac{\partial m_i(\mathbf{x}, t)}{\partial t} = -\nabla \cdot \mathbf{j}_i + \eta_i(\mathbf{x}, t). \tag{9.42}$$

The noise itself must be a total divergence, $\eta_i = -\nabla \cdot \sigma_i$, and hence in Fourier space,

$$\langle \eta_i(\mathbf{q}, t) \rangle = 0, \quad \text{and} \quad \langle \eta_i(\mathbf{q}, t)\eta_j(\mathbf{q}', t') \rangle = 2D\delta_{ij}q^2\delta(t - t')(2\pi)^d\delta^d(\mathbf{q} + \mathbf{q}'). \tag{9.43}$$

We can now take advantage of the generalized Einstein relation in Eq. (9.40) to ensure the correct equilibrium distribution by setting,

$$\mathbf{j}_i = \mu \nabla \cdot \left(-\frac{\delta \mathcal{H}}{\delta m_i(\mathbf{x})} \right). \tag{9.44}$$

The standard terminology for such dynamical equations is provided by Hohenberg and Halperin [P.C. Hohenberg and B.I. Halperin, Rev. Mod. Phys. **49**, 435 (1977)]. In **model A** dynamics the field $\vec{m}$ is *not conserved*, and the mobility and diffusion coefficients are constants. In **model B** dynamics the field $\vec{m}$ is *conserved*, and $\hat{\mu} = -\mu\nabla^2$ and $\hat{D} = -D\nabla^2$.

Let us now reconsider the Gaussian model ($u = 0$), this time with a conserved order parameter, with model B dynamics

$$\frac{\partial \vec{m}(\mathbf{x}, t)}{\partial t} = \mu r \nabla^2 \vec{m} - \mu K \nabla^4 \vec{m} + \vec{\eta}(\mathbf{x}, t). \tag{9.45}$$

The evolution of each Fourier mode is given by

$$\frac{\partial \vec{m}(\mathbf{q}, t)}{\partial t} = -\mu q^2(r + Kq^2)\vec{m}(\mathbf{q}, t) + \vec{\eta}(\mathbf{q}, t) \equiv -\frac{\vec{m}(\mathbf{q}, t)}{\tau(\mathbf{q})} + \vec{\eta}(\mathbf{q}, t). \tag{9.46}$$

Because of the constraints imposed by the conservation law, the relaxation of the field is more difficult, and slower. The relaxation times diverge even away from criticality. Depending on wavelength, we find scaling between length and time scales with dynamic exponents z, according to

$$\tau(\mathbf{q}) = \frac{1}{\mu q^2(r + Kq^2)} \approx \begin{cases} q^{-2} & \text{for } q \ll \xi^{-1} \quad (z = 2) \\ q^{-4} & \text{for } q \gg \xi^{-1} \quad (z = 4). \end{cases} \tag{9.47}$$

The equilibrium behavior is unchanged, and

$$\lim_{t \to \infty} \langle |\vec{m}(\mathbf{q}, t)|^2 \rangle = n\frac{Dq^2}{\mu q^2(r + Kq^2)} = \frac{nD}{\mu(r + Kq^2)}, \tag{9.48}$$

as before. Thus the same static behavior can be achieved by different dynamics. The static exponents (e.g. ν) are determined by the equilibrium (stationary) state and are unchanged, while the dynamic exponents may be different. As a result, dynamical critical phenomena involve more universality classes than the corresponding static ones. We shall not elaborate on dynamic critical phenomena any further. In addition to the standard review article by Hohenberg and Halperin [P.C. Hohenberg and B.I. Halperin, Rev. Mod. Phys. **49**, 435 (1977)], more details can be found in references [S.-K. Ma, *Modern Theory of Critical Phenomena* (Benjamin-Cummings, Reading, MA, 1976)] and [D. Forster, *Hydrodynamic Fluctuations, Broken Symmetries, and Correlation Functions* (Benjamin-Cummings, Reading, MA, 1975)].

9.4 Generic scale invariance in equilibrium systems

We live in a world full of complex spatial patterns and structures such as coastlines and river networks. There are similarly diverse temporal processes generically exhibiting "$1/f$" noise, as in resistance fluctuations, sand flowing through an hour glass, and even in traffic and stock market movements. These phenomena lack natural length and time scales and exhibit scale invariance and self-similarity. The spacial aspects of scale invariant systems can be characterized using *fractal* geometry [B.B. Mandelbrot, *The Fractal Geometry of Nature* (Freeman, San Francisco, 1982)]. In this section we explore dynamical processes that can naturally result in such scale invariant patterns.

Let us assume that the system of interest is described by a scalar field $m(\mathbf{x})$, distributed with a probability $\mathcal{P}[m]$. Scale invariance can be probed by examining the correlation functions of $m(\mathbf{x})$, such as the two point correlator, $C(|\mathbf{x} - \mathbf{y}|) \equiv \langle m(\mathbf{x})m(\mathbf{y})\rangle - \langle m(\mathbf{x})\rangle \langle m(\mathbf{y})\rangle$. (It is assumed that the system has rotational and translational symmetry.) In a system with a characteristic length scale, correlations decay to zero for separations $z = |\mathbf{x} - \mathbf{y}| \gg \xi$. By contrast, if the system possesses scale invariance, correlations are homogeneous at long distances, and $\lim_{z \to \infty} C(z) \sim z^{2\chi}$.

As we have seen, in equilibrium statistical mechanics the probability is given by $\mathcal{P}_{eq} \propto \exp(-\beta \mathcal{H}[m])$ with $\beta = (k_B T)^{-1}$. Clearly at infinite temperature there are no correlations for a finite Hamiltonian. As long as the interactions in $\mathcal{H}[m]$ are *short ranged*, it can be shown by high temperature expansions that correlations at small but finite β decay as $C(z) \propto \exp(-z/\xi)$, indicating a characteristic length scale[1]. The correlation length usually increases upon reducing temperature, and may diverge if the system undergoes a continuous (critical) phase transition. At a critical transition the system is scale invariant and $C(z) \propto z^{2-d-\eta}$. However, such scale invariance is *non-generic* in the sense that it can be obtained only by precise tuning of the system to the critical temperature. Most scale invariant processes in nature do not require such

---

[1] It is of course possible to generate long-range correlations with *long-ranged* interactions. However, it is most interesting to find out how long-range correlations are generated from local, *short-ranged* interactions.

precise tuning, and therefore the analogy to the critical point is not particularly instructive [T. Hwa and M. Kardar, Phys. Rev. A **45**, 7002 (1992). Similar motivations underlie the development of the concept of self-organized criticality by P. Bak, C. Tang and K. Wiesenfeld, Phys. Rev. Lett. **59**, 381 (1987). The distinctions and similarities to generic scale invariance are discussed in Ref. [T. Hwa and M. Kardar, Phys. Rev. A **45**, 7002 (1992)].

We shall frame our discussion of scale invariance by considering the dynamics of a surface, described by its height $h(\mathbf{x}, t)$. Specific examples are the distortions of a soap film or the fluctuations on the surface of water in a container. In both cases the minimum energy configuration is a flat surface (ignoring the small effects of gravity on the soap film). The energy cost of small fluctuations for a soap film comes from the increased area and *surface tension* σ. Expanding the area in powers of the slope results in

$$\mathcal{H}_\sigma = \sigma \int d^d\mathbf{x} \left[\sqrt{1 + (\nabla h)^2} - 1 \right] \approx \frac{\sigma}{2} \int d^d\mathbf{x} \, (\nabla h)^2 . \tag{9.49}$$

For the surface of water there is an additional gravitational potential energy, obtained by adding the contributions from all columns of water as

$$\mathcal{H}_g = \int d^d\mathbf{x} \int_0^{h(\mathbf{x})} \rho g h(\mathbf{x}) = \frac{\rho g}{2} \int d^d\mathbf{x} h(\mathbf{x})^2 . \tag{9.50}$$

The total (potential) energy of small fluctuations is thus given by

$$\mathcal{H} = \int d^d\mathbf{x} \left[\frac{\sigma}{2} (\nabla h)^2 + \frac{\rho g}{2} h^2 \right], \tag{9.51}$$

with the second term absent for the soap film.

The corresponding Langevin equation,

$$\frac{\partial h(\mathbf{x}, t)}{\partial t} = -\mu \rho g h + \mu \sigma \nabla^2 h + \eta(\mathbf{x}, t), \tag{9.52}$$

is similar to the linearized version of Eq. (9.26), and can be solved by Fourier transforms. Starting with a flat interface, $h(\mathbf{x}, t = 0) = h(\mathbf{q}, t = 0) = 0$, the profile at time t is

$$h(\mathbf{x}, t) = \int \frac{d^d\mathbf{q}}{(2\pi)^d} e^{-i\mathbf{q}\cdot\mathbf{x}} \int_0^t d\tau e^{-\mu(\rho g + \sigma q^2)(t-\tau)} \eta(\mathbf{q}, t). \tag{9.53}$$

The average height of the surface, $\bar{H} = \int d^d\mathbf{x} \langle h(\mathbf{x}, t) \rangle / L^d$, is zero, while its overall width is defined by

$$w^2(t, L) \equiv \frac{1}{L^d} \int d^d\mathbf{x} \langle h(\mathbf{x}, t)^2 \rangle = \frac{1}{L^d} \int \frac{d^d\mathbf{q}}{(2\pi)^d} \langle |h(\mathbf{q}, t)|^2 \rangle, \tag{9.54}$$

where L is the linear size of the surface. Similar to Eq. (9.33), we find that the width grows as

$$w^2(t, L) = \int \frac{d^d\mathbf{q}}{(2\pi)^d} \frac{D}{\gamma(\mathbf{q})} \left(1 - e^{-2\gamma(\mathbf{q})t} \right). \tag{9.55}$$

There are a range of time scales in the problem, related to characteristic length scales as in Eq. (9.31). The shortest time scale, $t_{\min} \propto a^2/(\mu\sigma)$, is set by an atomic size a. The longest time scale is set by either the capillary length ($\lambda_c \equiv \sqrt{\sigma/\rho g}$) or the system size ($L$). For simplicity we shall focus on the soap film where the effects of gravity are negligible and $t_{\max} \propto L^2/(\mu\sigma)$. We can now identify three different ranges of behavior in Eq. (9.55):

(a) For $t \ll t_{\min}$, none of the modes has relaxed since $\gamma(\mathbf{q})t \ll 1$ for all $\mathbf{q}$. Each mode grows diffusively, and

$$w^2(t, L) = \int \frac{d^d\mathbf{q}}{(2\pi)^d} \frac{D}{\gamma(\mathbf{q})} 2\gamma(\mathbf{q})t = \frac{2Dt}{a^d}. \tag{9.56}$$

(b) For $t \gg t_{\max}$, all modes have relaxed to their equilibrium values since $\gamma(\mathbf{q})t \gg 1$ for all $\mathbf{q}$. The height fluctuations now saturate to a maximum value given by

$$w^2(t, L) = \int \frac{d^d\mathbf{q}}{(2\pi)^d} \frac{D}{\mu\sigma q^2}. \tag{9.57}$$

The saturated value depends on the dimensionality of the surface, and in a general dimension d behaves as

$$w^2(t, L) \propto \frac{D}{\mu\sigma} \begin{cases} a^{2-d} & \text{for } d > 2, \quad (\chi = 0) \\ \ln(L/a) & \text{for } d = 2, \quad (\chi = 0^+) \\ L^{2-d} & \text{for } d < 2, \quad (\chi = \frac{2-d}{2}), \end{cases} \tag{9.58}$$

where we have defined a *roughness exponent* χ that governs the divergence of the width with system size via $\lim_{t\to\infty} w(t, L) \propto L^\chi$. (The symbol 0^+ is used to indicate a logarithmic divergence.) The exponent of $\chi = 1/2$ in $d = 1$ indicates that the one dimensional interface fluctuates like a random walk.

(c) For $t_{\min} \ll t \ll t_{\max}$ only a fraction of the shorter length scale modes are saturated. The integrand in Eq. (9.55) (for $g = 0$) is made dimensionless by setting $y = \mu\sigma q^2 t$, and

$$w^2(t, L) \propto \frac{D}{\mu\sigma} \int dq \, q^{d-3} \left(1 - e^{-2\mu\sigma q^2 t}\right)$$

$$\propto \frac{D}{\mu\sigma} \left(\frac{1}{\mu\sigma t}\right)^{\frac{d-2}{2}} \int_{t/t_{\max}}^{t/t_{\min}} dy \, y^{\frac{d-4}{2}} \left(1 - e^{-2y}\right). \tag{9.59}$$

The final integral is convergent for $d < 2$, and dominated by its upper limit for $d \geq 2$. The initial growth of the width is described by another exponent β, defined through $\lim_{t\to 0} w(t, L) \propto t^\beta$, and

$$w^2(t, L) \propto \begin{cases} \dfrac{D}{\mu\sigma} a^{2-d} & \text{for } d > 2, \quad (\beta = 0) \\ \dfrac{D}{\mu\sigma} \ln(t/t_{\min}) & \text{for } d = 2, \quad (\beta = 0^+) \\ \dfrac{D}{(\mu\sigma)^{d/2}} t^{(2-d)/2} & \text{for } d < 2, \quad (\beta = (2-d)/4). \end{cases} \tag{9.60}$$

The exponents χ and β also describe the height–height correlation functions which assumes the *dynamic scaling* form

$$\left\langle [h(\mathbf{x}, t) - h(\mathbf{x}', t')]^2 \right\rangle = |\mathbf{x} - \mathbf{x}'|^{2\chi} g\left(\frac{|t - t'|}{|\mathbf{x} - \mathbf{x}'|^z} \right). \tag{9.61}$$

Since equilibrium equal time correlations only depend on $|\mathbf{x} - \mathbf{x}'|$, $\lim_{y \to 0} g(y)$ should be a constant. On the other hand, correlations at the same point can only depend on time, requiring that $\lim_{y \to \infty} g(y) \propto y^{2\chi/z}$, and leading to the exponent identity $\beta = \chi/z$.

This scale invariance is broken when the gravitational potential energy is added to the Hamiltonian. The correlations now decay as $C(z) \propto \exp(-z/\lambda_c)$ for distances larger than the capillary length. What is the underlying difference between these two cases? The presence of gravity breaks the translational symmetry, $\mathcal{H}[h(\mathbf{x}) + c] = \mathcal{H}[h(\mathbf{x})]$. It is this continuous symmetry that forbids the occurrence of a term proportional to $\int d^d\mathbf{x}\, h(\mathbf{x})^2$ in the Hamiltonian and removes the corresponding length scale. (The coefficient of the quadratic term is usually referred to as a *mass* in field theoretical language.) The presence of a *continuous symmetry* is quite a general condition for obtaining *generic scale invariance* (GSI) [G. Grinstein, S. Sachdev, and D.H. Lee, Phys. Rev. Lett. **64**, 1927 (1990)]. As discussed in previous chapters, there are many low temperature phases of matter in which a continuous symmetry is spontaneously broken. The energy cost of small fluctuations around such a state must obey the *global* symmetry. The resulting excitations are the "massless" *Goldstone modes*. We already discussed such modes in connection with *magnons* in ferromagnets (with broken rotational symmetry), and *phonons* in solids (broken translational symmetry). All these cases exhibit scale invariant fluctuations.

In the realm of dynamics we can ask the more general question of whether temporal correlations, e.g. $C(|\mathbf{x} - \mathbf{x}'|, t - t') = \langle h(\mathbf{x}, t) h(\mathbf{x}', t') \rangle_c$, exhibit a characteristic time scale τ, or are homogeneous in $t - t'$. It is natural to expect that scale invariance in the spacial and temporal domains are closely interlinked. Establishing correlations at large distances requires long times as long as the system follows *local* dynamical rules (typically $(t - t') \propto |\mathbf{x} - \mathbf{x}'|^z$). Spacial scale invariance thus implies the lack of a time scale. The converse is not true as dynamics provides an additional possibility of removing time scales through a *conservation law*. We will, for example, encounter this situation in examining the model B dynamics of the surface Hamiltonian in the presence of gravity. Equation (9.47) indicates that, even though the long wavelength modes are massive, the relaxation time of a mode of wavenumber $\mathbf{q}$ diverges as $1/q^2$ in the $\mathbf{q} \to \mathbf{0}$ limit.

Symmetries and conservation laws are intimately linked in equilibrium systems. Consider a *local* Hamiltonian that is invariant under the symmetry

$\mathcal{H}[h(\mathbf{x}) + c] = \mathcal{H}[h(\mathbf{x})]$. Since $\mathcal{H}$ can only depend on ∇h and higher derivatives,

$$v = \mu F = -\mu \frac{\delta \mathcal{H}}{\delta h(\mathbf{x})} = \mu \nabla \cdot \frac{\partial \mathcal{H}}{\partial \nabla h} + \cdots \equiv \nabla \cdot \vec{j}. \tag{9.62}$$

Even for model A dynamics, the deterministic part of the velocity is the divergence of a current and conserves $\int d^d \mathbf{x} h(\mathbf{x}, t)$. The conservation is only statistical and locally broken by the non-conserved noise in model A. The above result is a consequence of *Noether's theorem*.

9.5 Non-equilibrium dynamics of open systems

We have to be cautious in applying the methods and lessons of near equilibrium dynamics to the various processes in nature which exhibit generic scale invariance. Many such systems, such as a flowing river or a drifting cloud, are very far from equilibrium. Furthermore, they are open and extended systems constantly exchanging particles and constituents with their environment. It is not clear that there is any simple Hamiltonian that governs the dynamics of such processes and hence the traditional approach presented earlier is not necessarily appropriate. However, the robust self-similar correlations observed in these systems [B.B. Mandelbrot, *The Fractal Geometry of Nature* (Freeman, San Francisco, 1982)] suggests that they can be described by stationary scale invariant probability distributions. This section outlines a general approach to the dynamics of open and extended systems that is similar in spirit to the construction of effective coarse-grained field theories described in chapter 2. Let us again consider the dynamics of a static field, $h(\mathbf{x}, t)$:

(1) The starting point in equilibrium statistical mechanics is the Hamiltonian $\mathcal{H}[h]$. Landau's prescription is to include in $\mathcal{H}$ all terms consistent with the symmetries of the problem. The underlying philosophy is that in a generic situation an allowed term is present, and can only vanish by accident. In the case of non-equilibrium dynamics we shall assume that the *equation of motion* is the fundamental object of interest. Over sufficiently long time scales, inertial terms ($\propto \partial_t^2 h$) are irrelevant in the presence of dissipative dynamics, and the evolution of h is governed by

$$\partial_t h(\mathbf{x}, t) = \overbrace{v[h(\mathbf{x}, t)]}^{\text{deterministic}} + \overbrace{\eta(\mathbf{x}, t)}^{\text{stochastic}}. \tag{9.63}$$

(2) If the interactions are short ranged, the velocity at $(\mathbf{x}, t)$ depends only on $h(\mathbf{x}, t)$ and a few derivatives evaluated at $(\mathbf{x}, t)$, i.e.

$$v(\mathbf{x}, t) = v\big(h(\mathbf{x}, t), \nabla h(\mathbf{x}, t), \cdots\big). \tag{9.64}$$

(3) We must next specify the functional form of deterministic velocity, and the correlations in noise. Generalizing Landau's prescription, we assume that all terms consistent with the underlying *symmetries and conservation laws* will generically

appear in v. The noise, $\eta(\mathbf{x}, t)$, may be conservative or non-conservative depending on whether there are only internal rearrangements, or external inputs and outputs.

Corollary: Note that with these set of rules there is no reason for the velocity to be derivable from a potential ($v \neq -\hat{\mu}\delta\mathcal{H}/\delta h$), and there is no fluctuation–dissipation condition ($\hat{D} \neq \hat{\mu}k_{\mathrm{B}}T$). It is even possible for the deterministic velocity to be conservative, while the noise is not. Thus various familiar results of near equilibrium dynamics may no longer hold.

As an example consider the flow of water along a river (or traffic along a highway). The deterministic part of the dynamics is conservative (the amount of water, or the number of cars is unchanged). Hence the velocity is the divergence of a current, $v = -\nabla\vec{j}[h]$. The current, $\vec{j}$, is a vector, and must be constructed out of the other two vectorial quantities in the problem: the gradient operator ∇, and the average transport direction $\hat{t}$. (The unit vector $\hat{t}$ points along the direction of current or traffic flow.) The lowest order terms in the expansion of the current give

$$-\vec{j} = \hat{t}\left(\alpha h - \frac{\lambda}{2}h^2 + \cdots\right) + \nu_1\nabla h + \nu_2\hat{t}(\hat{t}\cdot\nabla)h + \cdots \qquad (9.65)$$

The components of current parallel and perpendicular to the net flow are

$$\begin{cases} -j_{\parallel} = \alpha h - \frac{\lambda}{2}h^2 + (\nu_1 + \nu_2)\partial_{\parallel}h + \cdots \\ -\vec{j}_{\perp} = \nu_1\vec{\partial}_{\perp}h + \cdots \end{cases} \qquad (9.66)$$

The resulting equation of motion is

$$\frac{\partial h(\mathbf{x}, t)}{\partial t} = \partial_{\parallel}\left(\alpha h - \frac{\lambda}{2}h^2\right) + (\nu_1 + \nu_2)\partial_{\parallel}^2 h + \nu_1\partial_{\perp}^2 h + \cdots + \eta(\mathbf{x}, t). \qquad (9.67)$$

In the absence of external inputs and outputs (no rain, drainage, or exits), the noise is also conservative, with correlations,

$$\langle\eta(\mathbf{q}, t)\rangle = 0, \quad \text{and} \quad \langle\eta(\mathbf{q}, t)\eta(\mathbf{q}', t')\rangle = 2(D_{\parallel}q_{\parallel}^2 + D_{\perp}q_{\perp}^2)$$
$$\delta(t - t')(2\pi)^d\delta^d(\mathbf{q} + \mathbf{q}'). \qquad (9.68)$$

Note that the symmetries of the problem allow different noise correlations parallel and perpendicular to the net flow.

Equations (9.67) and (9.68) define a *driven diffusion system* (DDS) [H.K. Janssen and B. Schmitttman, Z. Phys. B **63**, 517 (1986); P.L.Garrido, J. L. Lebowitz, C. Maes, and H. Spohn, Phys. Rev. A **42**, 1954 (1990); Z. Cheng, P.L. Garrido, J.L. Lebowitz, and J.L. Valles, Europhys. Lett. **14**, 507 (1991)]. The first term in Eq. (9.67) can be eliminated by looking at fluctuations in a moving frame,

$$h(x_{\parallel}, \mathbf{x}_{\perp}, t) \to h(x_{\parallel} - \alpha t, \mathbf{x}_{\perp}, t), \qquad (9.69)$$

and shall be ignored henceforth. Neglecting the non-linear terms at first, these fluctuations satisfy the anisotropic noisy diffusion equation

$$\frac{\partial h(\mathbf{x}, t)}{\partial t} = \nu_\| \partial_\|^2 h + \nu_\perp \partial_\perp^2 h + \eta(\mathbf{x}, t), \tag{9.70}$$

where $\nu_\| = \nu_1 + \nu_2$ and $\nu_\perp = \nu_1$. The Fourier modes now relax with characteristic times,

$$\tau(\mathbf{q}) = \frac{1}{\nu_\| q_\|^2 + \nu_\perp q_\perp^2}. \tag{9.71}$$

Following the steps leading to Eq. (9.48),

$$\lim_{t \to \infty} \langle |h(\mathbf{q}, t)|^2 \rangle = \frac{D_\| q_\|^2 + D_\perp q_\perp^2}{\nu_\| q_\|^2 + \nu_\perp q_\perp^2}. \tag{9.72}$$

The stationary correlation functions[2] in real space now behave as [G. Grinstein, S. Sachdev, and D.H. Lee, Phys. Rev. Lett. **64**, 1927 (1990); P.L. Garrido, J. L. Lebowitz, C. Maes, and H. Spohn, Phys. Rev. A **42**, 1954 (1990); Z. Cheng, P.L. Garrido, J.L. Lebowitz, and J.L. Valles, Europhys. Lett. **14**, 507 (1991)].

$$\langle (h(\mathbf{x}) - h(\mathbf{0}))^2 \rangle = \int \frac{\mathrm{d}^{d-1}\mathbf{q}_\perp dq_\|}{(2\pi)^d} \; (2 - 2\cos(q_\| x_\| + \mathbf{q}_\perp \cdot \mathbf{x}_\perp)) \; \frac{D_\| q_\|^2 + D_\perp q_\perp^2}{\nu_\| q_\|^2 + \nu_\perp q_\perp^2}$$

$$\propto \left(\frac{D_\perp}{\nu_\perp} - \frac{D_\|}{\nu_\|} \right) \sqrt{\frac{\nu_\|^{d-1} \nu_\perp}{\left(\nu_\perp x_\|^2 + \nu_\| x_\perp^2 \right)^d}}. \tag{9.73}$$

Note that these correlations are spatially extended and scale invariant. This is again a consequence of the conservation law. Only in the special case where $D_\perp/\nu_\perp = D_\|/\nu_\|$ is the Einstein relation $(D(\mathbf{q}) \propto \nu(\mathbf{q}))$ satisfied, and the fluctuations become uncorrelated $(C(\mathbf{x}) \propto \delta^d(\mathbf{x})D/\nu)$. The results then correspond to model B dynamics with a Hamiltonian $\mathcal{H} \propto \int \mathrm{d}^d\mathbf{x} h^2$. This example illustrates the special nature of near equilibrium dynamics. The fluctuation–dissipation condition is needed to ensure approach to the equilibrium state where there is typically no scale invariance. On the other hand, removing this restriction leads to GSI in a conservative system. $(D_\perp/\nu_\perp \neq D_\|/\nu_\|$ is like having two different temperatures parallel and perpendicular to the flow.)

Let us now break the conservation law stochastically by adding random inputs and outputs to the problem (in the form of rain, drainage, or exits). The properties of the noise are now modified to

$$\langle \eta(\mathbf{x}, t) \rangle = 0, \quad \text{and} \quad \langle \eta(\mathbf{x}, t)\eta(\mathbf{x}', t') \rangle = 2D\delta(x_\| - x_\|')\delta^{d-1}(\mathbf{x}_\perp - \mathbf{x}_\perp')\delta(t - t'). \tag{9.74}$$

---

[2] In non-equilibrium circumstances we shall use the term *stationary* to refer to behavior at long times.

In the stationary state,

$$\lim_{t\to\infty}\langle|h(\mathbf{q}, t)|^2\rangle = \frac{D}{\nu_\parallel q_\parallel^2 + \nu_\perp q_\perp^2} \tag{9.75}$$

in Fourier space, and

$$\langle(h(\mathbf{x}) - h(\mathbf{0}))^2\rangle \propto D\left(\nu_\perp x_\parallel^2 + \nu_\parallel x_\perp^2\right)^{\frac{2-d}{2}} \tag{9.76}$$

in real space. Except for the anisotropy, this is the same result as in Eq. (9.61), with $\chi = (2 - d)/2$.

How are the results modified by the nonlinear term $(-\lambda\partial_\parallel h^2/2)$ in Eq. (9.67)? We first perform a simple *dimensional analysis* by rescaling lengths and time. Allowing for anisotropic scalings, we set $x_\parallel \to bx_\parallel$, accompanied by $t \to b^z t$, $\vec{x}_\perp \to b^\zeta \vec{x}_\perp$, and $h \to b^\chi h$. Eq. (9.67) is now modified to

$$b^{\chi-z}\frac{\partial h}{\partial t} = \nu_\parallel b^{\chi-2}\partial_\parallel^2 h + \nu_\perp b^{\chi-2\zeta}\partial_\perp^2 h - \frac{\lambda}{2}b^{2\chi-1}\partial_\parallel h^2 + b^{-z/2-(d-1)\zeta/2-1/2}\eta, \tag{9.77}$$

where Eq. (9.74) has been used to determine the scaling of η. We thus identify the bare scalings for these parameters as

$$\begin{cases} \nu_\parallel \to b^{z-2}\nu_\parallel \\ \nu_\perp \to b^{z-2\zeta}\nu_\perp \\ \lambda \to b^{\chi+z-1}\lambda \\ D \to b^{z-2\chi-\zeta(d-1)-1}D. \end{cases} \tag{9.78}$$

In the absence of λ, the parameters can be made *scale invariant* (i.e. independent of b) by the choice of $\zeta_0 = 1$, $z_0 = 2$, and $\chi_0 = (2 - d)/2$, as encountered before. However, with this choice, a small λ will grow under rescaling as

$$\lambda \to b^{y_0}\lambda, \quad \text{with} \quad y_0 = \frac{4 - d}{2}. \tag{9.79}$$

Since the nonlinearity grows larger under scaling it cannot be ignored in dimensions $d < 4$.

Equations (9.78) constitute a simple renormalization group (RG) that is valid close to the *fixed point* (scale invariant equation) corresponding to a linearized limit. To calculate the RG equations at finite values of nonlinearity in general requires a perturbative calculation. Sometimes there are exact *non-renormalization conditions* that simplify the calculation and lead to exponent identities. Fortunately there are enough such identities for Eq. (9.77) that the three exponents can be obtained *exactly*.

(1) As the nonlinearity is proportional to $q_\parallel$ in Fourier space, it does not generate under RG any contributions that can modify $\nu_\perp$. The corresponding "bare" scaling of $\nu_\perp$ in Eqs. (9.78) is thus always valid; its fixed point leads to the exponent identity $z - 2\zeta = 0$.

(2) As the nonlinearity is in the conservative part, it does not renormalize the strength of the non-conservative noise. The non-renormalization of D leads to the exponent identity $z - 2\chi - \zeta(d-1) - 1 = 0$. (This condition has a natural counterpart in equilibrium model B dynamics, $z - 2\chi - d - 2 = 0$, leading to the well known relation, $z = 4 - \eta$.)

(3) Equation (9.67) is invariant under an infinitesimal reparameterization $x_{\parallel} \to x_{\parallel} - \delta\lambda t$, $t \to t$, if $h \to h + \delta$. Note that the parameter λ appears both as the coefficient of the nonlinearity in Eq. (9.67) and as an invariant factor relating the $x_{\parallel}$ and h reparameterizations. Hence any renormalization of the driven diffusion equation that preserves this symmetry must leave the coefficient λ unchanged, leading to the exponent identity $z + \chi - 1 = 0$.

The remaining parameter, $\nu_{\parallel}$, does indeed follow a non-trivial evolution under RG. However, the above three exponent identities are sufficient to give the exact exponents in all dimensions $d \leq 4$ as [T. Hwa and M. Kardar, Phys. Rev. A **45**, 7002 (1992)]

$$\chi = \frac{1-d}{7-d}, \quad z = \frac{6}{7-d}, \quad \zeta = \frac{3}{7-d}. \tag{9.80}$$

9.6 Dynamics of a growing surface

The rapid growth of crystals by deposition, or molecular beam epitaxy, is an important technological process. It also provides the simplest example of a non-equilibrium evolution process [*Dynamics of Fractal Surfaces*, edited by F. Family and T. Vicsek, World Scientific, Singapore (1991)]. We would like to understand the dynamic scaling of fluctuations inherent to this type of growth. To construct the dependence of the local, deterministic velocity, v, on the surface height, $h(\mathbf{x}, t)$, note that:

(1) As long as the rearrangements of particles on the surface can result in holes and vacancies, there is no conservation law.

(2) There is a *translation symmetry*, $v[h(\mathbf{x}) + c] = v[h(\mathbf{x})]$, implying that v depends only on gradients of $h(\mathbf{x})$.

(3) For simplicity, we shall focus on *isotropic* surfaces, in which all directions in $\mathbf{x}$ are equivalent [See, however, D. Wolf, Phys. Rev. Lett. **67**, 1783 (1991)].

(4) There is no up–down symmetry, i.e. $v[h(\mathbf{x})] \neq -v[-h(\mathbf{x})]$. The absence of such symmetry allows addition of terms of both parities.

With these conditions, the lowest order terms in the equation of motion give [M. Kardar, G. Parisi, and Y.-C. Zhang, Phys. Rev. Lett. **56**, 889 (1986); E. Medina, T. Hwa, M. Kardar, and Y.-C. Zhang, Phys. Rev. A **39**, 3053 (1989)],

$$\frac{\partial h(\mathbf{x}, t)}{\partial t} = u + \nu\nabla^2 h + \frac{\lambda}{2}(\nabla h)^2 + \cdots + \eta(\mathbf{x}, t), \tag{9.81}$$

with the non-conservative noise satisfying the correlations in Eq. (9.74).

In Eq. (9.81), u is related to the average growth velocity. In fact, the coefficients of all even terms must be proportional to u as they all vanish in the symmetric case with no preferred growth direction. The constant u is easily removed by transforming to a moving frame, $h \rightarrow h - ut$, and will be ignored henceforth. The first non-trivial term is the nonlinear contribution, $\lambda(\nabla h)^2/2$. Geometrically this term can be justified by noting that growth by addition of particles proceeds through a parallel transport of the surface gradient in the normal direction. (See the inset to Fig. 9.1.) This term cannot be generated from the variations of any Hamiltonian, i.e. $v \neq -\mu \delta \mathcal{H}[h]/\delta h$. Thus, contrary to the equilibrium situation (Noether's theorem), the translational symmetry does not imply a conservation law, $v \neq -\nabla j$.

Further evidence of the relevance of Eq. (9.81) to growth phenomena is provided by examining *deterministic growth*. Consider a slow and uniform snowfall, on an initial profile which at $t = 0$ is described by $h_0(\mathbf{x})$. The nonlinear equation can in fact be *linearized* with the aid of a "Cole–Hopf" transformation,

$$W(\mathbf{x}, t) = \exp\left[\frac{\lambda}{2\nu} h(\mathbf{x}, t)\right]. \tag{9.82}$$

The function $W(\mathbf{x}, t)$ evolves according to the diffusion equation with *multiplicative noise*,

$$\frac{\partial W(\mathbf{x}, t)}{\partial t} = \nu \nabla^2 W + \frac{\lambda}{2\nu} W \eta(\mathbf{x}, t). \tag{9.83}$$

In the absence of noise, $\eta(\mathbf{x}, t) = 0$, Eq. (9.83) can be solved subject to the initial condition $W(\mathbf{x}, t = 0) = \exp[\lambda h_0(\mathbf{x})/2\nu]$, and leads to the growth profile,

$$h(\mathbf{x}, t) = \frac{2\nu}{\lambda} \ln\left\{\int d^d\mathbf{x}' \exp\left[-\frac{|\mathbf{x} - \mathbf{x}'|^2}{2\nu t} + \frac{\lambda}{2\nu} h(\mathbf{x}, t)\right]\right\}. \tag{9.84}$$

It is instructive to examine the $\nu \rightarrow 0$ limit, which is indeed appropriate to snow falls since there is not much rearrangement after deposition. In this limit, the integral in Eq. (9.84) can be performed by the saddle point method. For each $\mathbf{x}$ we have to identify a point $\mathbf{x}'$ which maximizes the exponent, leading to a collection of paraboloids described by

$$h(\mathbf{x}, t) = \max_{\mathbf{x}'}\left\{h_0(\mathbf{x}') - \frac{|\mathbf{x} - \mathbf{x}'|^2}{2\lambda t}\right\}. \tag{9.85}$$

Such parabolic sequences are quite common in many layer by layer growth processes in nature, from biological to geological formations. The patterns for $\lambda = 1$ are identical to those obtained by the geometrical method of Huygens, familiar from optics. The growth profile (wave front) is constructed from the outer envelope of circles of radius λt drawn from all points on the initial profile. The nonlinearity in Eq. (9.81) algebraically captures this process of expanding wave fronts.

As growth proceeds, the surface smoothens by the *coarsening* of the parabolas. What is the typical size of these features at time t? In maximizing the exponent in Eq. (9.85), we have to balance a reduction $|\mathbf{x} - \mathbf{x}'|^2/2\lambda t$, by a

Fig. 9.1 Deterministic
growth according to
Eq. (9.81) leads to a
pattern of coarsening
paraboloids. In one
dimension, the slope of
the interface forms
"shock fronts". Inset
depicts projection of
lateral growth on the
vertical direction.

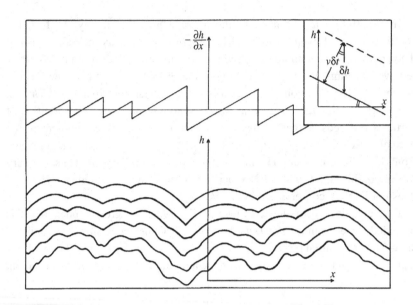

Fig. 9.1 Deterministic growth according to Eq. (9.81) leads to a pattern of coarsening paraboloids. In one dimension, the slope of the interface forms "shock fronts". Inset depicts projection of lateral growth on the vertical direction.

possible gain from $h_0(\mathbf{x}')$ in selecting a point away from $\mathbf{x}$. The final scaling is controlled by the roughness of the initial profile. Let us assume that the original pattern is a *self-affine fractal* of roughness χ, i.e.

$$\overline{|h_0(\mathbf{x}) - h_0(\mathbf{x}')|} \sim |\mathbf{x} - \mathbf{x}'|^{\chi}. \tag{9.86}$$

(According to Mandelbrot, $\chi \approx 0.7$ for mountains [B.B. Mandelbrot, *The Fractal Geometry of Nature* (Freeman, San Francisco, 1982)].) Balancing the two terms in Eq. (9.85) gives

$$(\delta x)^{\chi} \sim \frac{(\delta x)^2}{t} \quad \Longrightarrow \quad \delta x \sim t^{1/z}, \quad \text{with} \quad z + \chi = 2. \tag{9.87}$$

For example, if the initial profile is like a random walk in $d = 1$, $\chi = 1/2$, and $z = 3/2$. This leads to the spreading of information along the profile by a process that is faster than diffusion, $\delta x \sim t^{2/3}$.

Note that the slope, $\vec{v}(\mathbf{x}, t) = -\lambda \vec{\nabla} h(x, t)$, satisfies the equation,

$$\frac{D\vec{v}(\mathbf{x}, t)}{Dt} \equiv \frac{\partial \vec{v}}{\partial t} + \vec{v} \cdot \vec{\nabla} \vec{v} = \nu \nabla^2 \vec{v} - \lambda \nabla \eta. \tag{9.88}$$

The above is the Navier–Stokes equation for the velocity of a fluid of viscosity ν, which is being randomly stirred by a conservative force [D. Forster, D.R. Nelson, and M.J. Stephen, Phys. Rev. A **16**, 732 (1977)], $\vec{f} = -\lambda \nabla \eta$. However, the fluid is vorticity free since

$$\vec{\Omega} = \vec{\nabla} \times \vec{v} = -\lambda \nabla \times \nabla h = 0. \tag{9.89}$$

This is the *Burgers equation* [J.M. Burgers, *The Nonlinear Diffusion Equation* (Riedel, Boston, 1974)], which provides a simple example of the formation of shock waves in a fluid. The gradient of Eq. (9.85) in $d = 1$ gives a saw tooth pattern of shocks which coarsen in time. Further note that in $d = 1$, Eq. (9.88) is also equivalent to the driven diffusion equation of (9.67), with $\vec{v}$ playing the role of h.

To study stochastic roughening in the presence of the nonlinear term, we carry out a scaling analysis as in Eq. (9.77). Under the scaling $\mathbf{x} \to b\mathbf{x}$, $t \to b^z t$, and $h \to b^\chi h$, Eq. (9.81) transforms to

$$b^{\chi-z}\frac{\partial h}{\partial t} = \nu b^{\chi-2}\nabla^2 h + \frac{\lambda}{2}b^{2\chi-2}(\nabla h)^2 + \eta(b\mathbf{x}, b^z t). \tag{9.90}$$

The correlations of the transformed noise, $\eta'(\mathbf{x}, t) = b^{z-\chi}\eta(b\mathbf{x}, b^z t)$, satisfy

$$\begin{aligned}
\langle \eta'(\mathbf{x}, t)\eta'(\mathbf{x}', t')\rangle &= b^{2z-2\chi}2D\,\delta^d(\mathbf{x} - \mathbf{x}')b^{-d}\delta(t - t')b^{-z} \\
&= b^{z-d-2\chi}2D\,\delta^d(\mathbf{x} - \mathbf{x}')\delta(t - t').
\end{aligned} \tag{9.91}$$

Following this scaling the parameters of Eq. (9.81) are transformed to

$$\begin{cases}
\nu \to b^{z-2}\nu \\
\lambda \to b^{\chi+z-2}\lambda \\
D \to b^{z-2\chi-d}D.
\end{cases} \tag{9.92}$$

For $\lambda = 0$, the equation is made scale invariant upon the choice of $z_0 = 2$, and $\chi_0 = (2 - d)/2$. Close to this linear fixed point, λ scales to $b^{z_0+\chi_0-2}\lambda = b^{(2-d)/2}\lambda$, and is a relevant operator for $d < 2$. In fact a perturbative dynamic renormalization group suggests that it is *marginally relevant* at $d = 2$, and that in all dimensions a sufficiently large λ leads to new scaling behavior. (This will be discussed further in the next chapter.)

Are there any non-renormalization conditions that can help in identifying the exponents of the full nonlinear stochastic equation? Note that since Eqs. (9.81) and (9.88) are related by a simple transformation, they must have the same scaling properties. Since the Navier–Stokes equation is derivable from Newton's laws of motion for fluid particles, it has the Galilean invariance of changing to a uniformly moving coordinate frame. This symmetry is preserved under renormalization to larger scales and requires that the ratio of the two terms on the left-hand side of Eq. (9.88) ($\partial_t \vec{v}$ and $\vec{v} \cdot \nabla \vec{v}$) stays at unity. In terms of Eq. (9.81) this implies the non-renormalization of the parameter λ, and leads to the exponent identity

$$\chi + z = 2. \tag{9.93}$$

Unfortunately there is no other non-renormalization condition except in $d = 1$. Following Eq. (9.36), we can write down a Fokker–Planck equation for the evolution of the configurational probability as,

$$\frac{\partial \mathcal{P}([h(\mathbf{x})], t)}{\partial t} = -\int d^d\mathbf{x}\frac{\delta}{\delta h(\mathbf{x})}\left[\left(\nu\nabla^2 h + \frac{\lambda}{2}(\nabla h)^2\right)\mathcal{P} - D\frac{\delta\mathcal{P}}{\delta h(\mathbf{x})}\right]. \tag{9.94}$$

Since Eq. (9.81) was not constructed from a Hamiltonian, in general we do not know the stationary solution at long times. In $d = 1$, we make a guess and try a solution of the form

$$\mathcal{P}_0[h(x)] \propto \exp\left[-\frac{\nu}{2D}\int dx(\partial_x h)^2\right]. \tag{9.95}$$

Since

$$\frac{\delta \mathcal{P}_0}{\delta h(x)} = -\partial_x \frac{\delta \mathcal{P}_0}{\delta(\partial_x h)} = \frac{\nu}{D}(\partial_x^2 h)\,\mathcal{P}_0, \tag{9.96}$$

Eq. (9.94) leads to,

$$\frac{\partial \mathcal{P}_0}{\partial t} = -\int dx \frac{\delta \mathcal{P}_0}{\delta h(x)}\left(\nu \partial_x^2 h + \frac{\lambda}{2}(\partial_x h)^2 - D\frac{\nu}{D}\partial_x^2 h\right)$$

$$= -\frac{\lambda}{2}\mathcal{P}_0\int dx \frac{\nu}{D}(\partial_x^2 h)(\partial_x h)^2 = -\frac{\lambda\nu}{2D}\mathcal{P}_0\int dx\partial_x\left(\frac{(\partial_x h)^3}{3}\right) = 0. \tag{9.97}$$

We have thus identified the stationary state of the one dimensional equation. (This procedure does not work in higher dimensions as it is impossible to write the final result as a total derivative.) Surprisingly, the stationary distribution is the same as the one in equilibrium at a temperature proportional to D/ν. We can thus immediately identify the roughness exponent $\chi = 1/2$, which together with the exponent identity in Eq. (9.93) leads to $z = 3/2$, i.e. superdiffusive behavior.

The values of the exponents in the strongly nonlinear regime are not known exactly in higher dimensions. However, extensive numerical simulations of growth have provided fairly reliable estimates [*Dynamics of Fractal Surfaces*, edited by F. Family and T. Vicsek, World Scientific, Singapore (1991)]. In the physically relevant case ($d = 2$) of a surface grown in three dimensions, $\chi \approx 0.39$ and $z \approx 1.61$ [B.M. Forrest and L.-H. Tang, Phys. Rev. Lett. **64**, 1405 (1990)]. A rather good (but not exact) fit to the exponents in a general dimension d is the following estimate by Kim and Kosterlitz [J.M. Kim and J.M. Kosterlitz, Phys. Rev. Lett. **62**, 2289 (1989)],

$$\chi \approx \frac{2}{d+3} \quad \text{and} \quad z \approx \frac{2(d+2)}{d+3}. \tag{9.98}$$

10
Directed paths in random media

10.1 Introduction

Many physical problems involve calculating sums over paths. Each path could represent one possible physical realization of an object such as a polymer, in which case the weight of the path is the probability of that configuration. The weights themselves could be imaginary as in the case of Feynman paths describing the amplitude for the propagation of a particle. Path integral calculations are now a standard tool of the theoretical physicist, with many excellent books devoted to the subject [R.P. Feynman and A.R. Hibbs, *Quantum Mechanics and Path Integrals* (McGraw-Hill, New York, 1965)]; [F.W. Wiegel, *Introduction to Path-Integral Methods in Physics and Polymer Science* (World Scientific, Singapore, 1986)].

What happens to sums over paths in the presence of quenched disorder in the medium? Individual paths are no longer weighted simply by their length, but are influenced by the impurities along their route. The sum may be dominated by "optimal" paths pinned to the impurities; the optimal paths usually forming complex hierarchical structures. Physical examples are provided by the interface of the random bond Ising model in two dimensions, and by magnetic flux lines in superconductors. The actual value of the sum naturally depends on the particular realization of randomness and varies from sample to sample. I shall initially motivate the problem in the context of the high-temperature expansion for the random bond Ising model. Introducing the sums over paths for such a lattice model avoids the difficulties associated with short distance cutoffs. Furthermore, the Ising model is sufficiently well understood to make the nature of various approximations more evident.

The high-temperature correlation functions of the Ising model are dominated by the shortest paths connecting the spins. Such configurations that exclude loops and overhangs, are referred to as *directed paths*. They dominate the asymptotic behavior of the sum over distances that are much longer than the correlation length. Most of this chapter is devoted to describing the statistical properties of sums over such directed paths. As in all multiplicative noise processes, the probability distribution for the sum is broad. Hence Monte Carlo simulations may not be an appropriate tool for numerical studies, failing to find typical members of the ensemble. Instead, we shall focus on a transfer matrix method that allows a numerical evaluation of the sum in polynomial time in the length of the path. The results indeed show that the sum has a broad probability distribution that resembles (but is not quite) log–normal.

To obtain analytical information about this probability distribution we shall introduce the replica method for examining the moments. It can be shown easily that the one-dimensional sum has a log–normal distribution. The moments of the sum over directed paths in two dimensions can be obtained by using a simple Bethe Ansatz. The implications and limitations of this approach are discussed. There is little analytical information in three and higher dimensions, but a variety of numerical results are available, mostly by taking advantage of a mapping to the growing surfaces introduced in the previous chapter.

The spin glass problem describes a mixture of ferromagnetic and antiferromagnetic bonds. The resulting sums in the high-temperature expansion involve products over a random mixture of positive and negative factors. The calculation of moments is somewhat different from the case of purely positive random bonds. However, we shall demonstrate that the scaling behavior of the distribution is unchanged. A similar sum involving products of random signs is encountered in calculating the probability of an electron tunneling under a random potential. In the strongly localized limit, it is again sufficient to focus on the interference of the forward scattering (directed) paths. A magnetic field introduces *random phases* in the sum; while to describe the tunneling of an electron in the presence of spin–orbit scattering requires examining the evolution of a two component spinor and keeping track of products of *random matrices*. We shall argue that all these cases are in fact described by the *same universal probability distribution* which, however, does retain some remnant of the underlying symmetries of the original electronic Hamiltonian.

Yet another class of directed paths is encountered in the context of light scattering in turbulent media. Assuming that inelastic scattering can be neglected, the intensity of the beam is left unchanged, and the evolution is unitary. Due to the constraint of unitarity the resulting directed paths are described by a probability distribution belonging to a new universality class. We shall introduce a discrete matrix model that explicitly takes care of the unitarity constraint. In this model, several properties of the resulting sum over paths can be calculated exactly.

10.2 High-T expansions for the random-bond Ising model

Consider a d-dimensional hypercubic lattice of N sites. On each site there is an Ising variable $\sigma_i = \pm 1$, and the spins interact through a Hamiltonian

$$\mathcal{H} = -\sum_{\langle ij \rangle} J_{ij} \sigma_i \sigma_j. \tag{10.1}$$

The symbol $\langle ij \rangle$ implies that the sum is restricted to the dN nearest-neighbor bonds on the lattice. The bonds $\{J_{ij}\}$ are *quenched* random variables, independently chosen from a probability distribution $p(J)$. For each realization of random bonds, the partition function is computed as

$$Z[J_{ij}] = \sum_{\{\sigma_i\}} \exp(-\beta \mathcal{H}) = \sum_{\{\sigma_i\}} \prod_{\langle ij \rangle} e^{K_{ij} \sigma_i \sigma_j}, \tag{10.2}$$

where the sums are over the 2^N possible configurations of spins, $\beta = 1/(k_B T)$ and $K_{ij} = \beta J_{ij}$. To obtain a high-temperature expansion, it is more convenient to organize the partition function in powers of $\tanh K_{ij}$. Since $(\sigma_i \sigma_j)^2 = 1$, the Boltzmann factor for each bond can be written as

$$e^{K_{ij} \sigma_i \sigma_j} = \frac{e^{K_{ij}} + e^{-K_{ij}}}{2} + \frac{e^{K_{ij}} - e^{-K_{ij}}}{2} \sigma_i \sigma_j = \cosh K_{ij} (1 + \tau_{ij} \sigma_i \sigma_j), \tag{10.3}$$

where $\tau_{ij} \equiv \tanh K_{ij}$. Applying this transformation to each bond of the lattice results in

$$Z[J_{ij}] = \sum_{\{\sigma_i\}} e^{\sum_{\langle ij \rangle} K_{ij} \sigma_i \sigma_j} = \overline{C} \sum_{\{\sigma_i\}} \prod_{\langle ij \rangle} (1 + \tau_{ij} \sigma_i \sigma_j), \tag{10.4}$$

and

$$\overline{C} \equiv \left(\prod_{\langle ij \rangle} \cosh K_{ij} \right).$$

The term $\overline{C}$ is non-singular, and will be mostly ignored henceforth. The final product in Eq. (10.4) generates 2^{dN} terms which can be represented diagrammatically by drawing lines connecting sites i and j for each factor of $\tau_{ij} \sigma_i \sigma_j$. Each site now obtains a factor of $\sigma_i^{p_i}$, where $0 \leq p_i \leq 2d$ is the number of bonds emanating from i. Summing over the two possible values $\sigma_i = \pm 1$ gives a factor of 2 if p_i is even and 0 if p_i is odd. Thus the only graphs that survive the sum have an even number of lines passing through each site. The contribution of each graph $\mathcal{G}$ is the product of τ_{ij} for the bonds making up the graph, resulting in

$$Z[J_{ij}] = 2^N \times \overline{C} \sum_{\mathcal{G}} \left(\prod_{\langle ij \rangle \in \mathcal{G}} \tau_{ij} \right). \tag{10.5}$$

For a d-dimensional hypercubic lattice the smallest closed graph is a square of four bonds and the next graph has six bonds. Thus,

$$Z[J_{ij}] = 2^N \times \overline{C} \left[1 + \sum_P \tau_{P1} \tau_{P2} \tau_{P3} \tau_{P4} + \mathcal{O}(\tau^6) + \cdots \right], \tag{10.6}$$

where the sum runs over the $Nd(d-1)/2$ plaquettes on the lattice and $\tau_{P\alpha}$ indicates one of the four bonds around plaquette P. A *quench averaged* free energy is now obtained as

$$\frac{\overline{\ln Z}}{N} = \ln 2 + d \overline{\ln \cosh K} + \frac{d(d-1)}{2} \overline{\tau}^4 + \cdots, \tag{10.7}$$

where the over-lines indicate averages over the probability distribution $p(J)$. The expansion up to this order is quite similar to that of the non–random Ising model in Sec. 7.2, with $\bar{\tau}$ in place of the pure τ. However, starting with order of τ^8 we need to evaluate averages of τ_{ij}^2 (and higher moments) quickly removing any semblance to the expansion of the pure Ising model.

The same conclusion applies to expansions for other spin operators. For example the two-spin correlation function is given by

$$\langle \sigma_m \sigma_n \rangle = \sum_{\{\sigma_i\}} \frac{e^{\sum_{\langle ij \rangle} K_{ij} \sigma_i \sigma_j}}{Z} \sigma_m \sigma_n = \frac{\overline{C}}{Z} \sum_{\{\sigma_i\}} \sigma_m \sigma_n \prod_{\langle ij \rangle} (1 + \tau_{ij} \sigma_i \sigma_j) \tag{10.8}$$

The terms in the numerator involve an additional factor of $\sigma_m \sigma_n$. To get a finite value after summing over $\sigma_m = \pm 1$ and $\sigma_n = \pm 1$ we have to examine graphs with an odd number of bonds emanating from these external sites. After canceling the common factors between the numerator and denominator, we obtain

$$\langle \sigma_m \sigma_n \rangle = \frac{\sum \mathcal{G}_{mn} \left(\prod_{\langle ij \rangle \in \mathcal{G}_{mn}} \tau_{ij} \right)}{\sum \mathcal{G} \left(\prod_{\langle ij \rangle \in \mathcal{G}} \tau_{ij} \right)}. \tag{10.9}$$

Whereas the graphs in $\mathcal{G}$ have an even number of bonds going through each site, those of $\mathcal{G}_{mn}$ have an odd number of bonds going through the external points m and n. In calculating the *quench averaged* correlation function, the simplest graphs involving single occurrences of each bond give contributions similar to the corresponding terms for the pure model with τ replaced by $\bar{\tau}$. However, the graphs with multiple factors of a particular τ_{ij} involve higher moments and complicate the computation.

As Eqs. (10.6) and (10.8) indicate, the partition function and correlation functions of the random system are themselves random quantities, dependent on all the bonds J_{ij}. It may not be sufficient to just characterize the mean value of Z (or $\ln Z$), since the full information about these fluctuating quantities is only contained in their respective probability distributions $p(Z)$ and $p(\langle \sigma_m \sigma_n \rangle)$.

A particularly useful way of describing a random variable is through its *characteristic function*, which is simply the Fourier transform of its probability distribution function (PDF), i.e.

$$\tilde{p}(k) = \langle e^{-ikx} \rangle = \int dx p(x) e^{-ikx}. \tag{10.10}$$

Moments of the distribution can be obtained by expanding $\tilde{p}(k)$ in powers of k,

$$\tilde{p}(k) = \left\langle \sum_{n=0}^{\infty} \frac{(-ik)^n}{n!} x^n \right\rangle = \sum_{n=0}^{\infty} \frac{(-ik)^n}{n!} \langle x^n \rangle. \tag{10.11}$$

Similarly, *cumulants* of the random variable are generated by the expansion of the logarithm of the characteristic function as

$$\ln \tilde{p}(k) = \sum_{n=1}^{\infty} \frac{(-ik)^n}{n!} \langle x^n \rangle_c. \tag{10.12}$$

10.3 The one-dimensional chain

The calculations of Section 7.3 are easily generalized to the case of the random bond Ising model in $d = 1$ (and zero field). In particular, for the open chain of N sites, we have

$$Z = 2^N \prod_{\alpha=1}^{N-1} \cosh K_\alpha, \tag{10.13}$$

where $K_\alpha \equiv K_{\alpha\,\alpha+1}$ is the interaction between neighboring sites. There is also only one graph that contributes to the two-point correlation function,

$$\langle \sigma_m \sigma_n \rangle = \sum_{\{\sigma_i\}} \frac{e^{\sum_i K_i \sigma_i \sigma_{i+1}}}{Z} \sigma_m \sigma_n = \prod_{\alpha=m}^{n-1} \tau_\alpha. \tag{10.14}$$

Since the partition function is the sum of $N - 1$ independent variables,

$$\frac{\ln Z}{N} = \ln 2 + \sum_{\alpha=1}^{N-1} \frac{\ln \cosh K_\alpha}{N}, \tag{10.15}$$

we can use the central limit theorem to conclude that as $N \to \infty$ the probability distribution $p(\ln Z/N)$ has a Gaussian distribution with mean

$$\overline{\ln Z} = N \left(\ln 2 + \overline{\ln \cosh K} \right), \tag{10.16}$$

and variance

$$\overline{(\ln Z)_c^2} \equiv \overline{(\ln Z)^2} - \overline{\ln Z}^2 = N \overline{(\ln \cosh K)_c^2}. \tag{10.17}$$

(We have ignored the small difference between N and $N - 1$ in the thermodynamic limit.) Similarly, for the correlation function of two points separated by a distance t, we have

$$\ln \langle \sigma_0 \sigma_t \rangle = \sum_{\alpha=0}^{t-1} \ln \tau_\alpha. \tag{10.18}$$

As long as the random variables on the bonds are independently distributed, the *cumulants* of $\ln \langle \sigma_0 \sigma_t \rangle$ are given by,

$$\begin{cases} \overline{\ln \langle \sigma_0 \sigma_t \rangle} &= t \overline{\ln \tanh K} \\ \overline{(\ln \langle \sigma_0 \sigma_t \rangle)_c^2} &= t \, \overline{(\ln \tanh K)_c^2} \\ \quad \vdots \quad \vdots \\ \overline{(\ln \langle \sigma_0 \sigma_t \rangle)_c^p} &= t \, \overline{(\ln \tanh K)_c^p}. \end{cases} \tag{10.19}$$

In the following sections we shall try to obtain similar information about probability distribution functions for the partition and correlation functions in higher dimensions. To do so, we shall employ the *replica method* for calculating the moments of the distribution. For example, the cumulants of the free energy are given by

$$\overline{Z^n} = \overline{e^{n \ln Z}} = \exp \left[n \overline{\ln Z} + \frac{n^2}{2} \overline{(\ln Z)_c^2} + \cdots + \frac{n^p}{p!} \overline{(\ln Z)_c^p} + \cdots \right], \tag{10.20}$$

where we have taken advantage of Eq. (10.12), replacing $(-ik)$ with n. Usually, the moments on the left-hand side of the above equation are known only for integer n, while the evaluation of the cumulants on the right-hand side relies on an expansion around $n = 0$. This is one of the difficulties associated with the problem of deducing a probability distribution $p(x)$, from the knowledge of its moments $\overline{x^n}$. There is in fact a rigorous theorem that the probability distribution cannot be uniquely inferred if its nth moment increases faster than $n!$ [N.I. Akhiezer, *The Classical Moment Problem* (Oliver and Boyd, London, 1965)]. Most of the distributions of interest to us (such as the above log–normal) do not satisfy this condition! Similar problems are encountered in the replica studies of spin glasses [M. Mézard, G. Parisi, and M.A. Virasoro, *Spin Glass Theory and Beyond* (World Scientific, Singapore, 1987)]. It turns out that many of the difficulties associated with a rigorous inversion are related to the behavior at the tail (extreme values) of the distribution. Most of the information of interest to us is contained in the "bulk" of the distribution which is easier to investigate. Rather than taking a rigorous approach to the problem, we shall illustrate the difficulties and their resolution by examining the above one dimensional example in some detail, since it actually presents a worst case scenario for deducing a distribution from its moments.

We used the central limit theorem to deduce that the probability distribution for $\langle \sigma_0 \sigma_t \rangle$ is log–normal. Its moments are computed from,

$$\overline{\langle \sigma_0 \sigma_t \rangle^n} = \prod_{\alpha=0}^{t-1} \overline{\tau_\alpha^n} = \left(\overline{e^{n \ln \tau}} \right)^t = \exp\left[t \sum_p \frac{n^p}{p!} \overline{(\ln \tau)_c^p} \right]. \tag{10.21}$$

Let us consider a binary distribution in which τ takes two positive values of τ_1 and $\tau_2 > \tau_1$ with equal probability. Then

$$\overline{\tau^n} = \frac{\tau_1^n + \tau_2^n}{2}, \tag{10.22}$$

and the generating function for the cumulants of the correlation function is

$$
\begin{aligned}
\ln \overline{\tau^n} &= n \ln \tau_1 + \ln \left(\frac{1 + (\tau_2/\tau_1)^n}{2} \right) \\
&\underset{n \to 0}{=} n \ln \tau_1 + \ln \left(\frac{1 + 1 + n \ln (\tau_2/\tau_1) + n^2/2 \ln^2 (\tau_2/\tau_1) + \cdots}{2} \right) \\
&= n \ln \tau_1 + \ln \left[1 + \frac{n}{2} \ln \left(\frac{\tau_2}{\tau_1} \right) + \frac{n^2}{4} \ln^2 \left(\frac{\tau_2}{\tau_1} \right) + \cdots \right] \\
&= n \ln \tau_1 + \frac{n}{2} \ln \left(\frac{\tau_2}{\tau_1} \right) + \frac{n^2}{8} \ln^2 \left(\frac{\tau_2}{\tau_1} \right) + \cdots \\
&= n \ln(\sqrt{\tau_1 \tau_2}) + \frac{n^2}{8} \ln^2 \left(\frac{\tau_2}{\tau_1} \right) + \cdots .
\end{aligned}
\tag{10.23}
$$

Combining Eqs. (10.21) and (10.23), the cumulants of the correlation function are given by

$$\begin{cases} \overline{\ln\langle\sigma_0\sigma_t\rangle} & = t\ln(\sqrt{\tau_1\tau_2}) = a_1 t \\ \overline{\ln\langle\sigma_0\sigma_t\rangle_c^2} & = \dfrac{t}{4}\ln^2(\tau_2/\tau_1) = a_2 t \\ \quad\vdots \end{cases} \qquad (10.24)$$

While it is true that $\ln\langle\sigma_0\sigma_t\rangle$ is normally distributed for large t, and that characteristic function of a normal distribution is itself a Gaussian, we should be careful about the order of limits in terminating the power series in the exponent at the second order. If we do so, from

$$\overline{\langle\sigma_0\sigma_t\rangle^n} \approx \exp\left[t\left(na_1 + n^2 a_2/2\right)\right], \qquad (10.25)$$

we should not infer anything about the high moments ($n \to \infty$) and the tail of the distribution. Otherwise (since $a_2 > 0$), we would conclude that sufficiently large moments of $\langle\sigma_0\sigma_t\rangle$ diverge with separation; a clearly false conclusion as $\langle\sigma_0\sigma_t\rangle$ is bounded by unity! The correct result is that

$$\lim_{n\to\infty}\overline{\langle\sigma_0\sigma_t\rangle^n} = \frac{\tau_2^{nt}}{2^t}, \qquad (10.26)$$

i.e. the high moments are almost entirely dominated by the one exceptional sample in which all bonds are equal to τ_2.

We can summarize the situation as follows: The "bulk" of the probability distribution for $\ln\langle\sigma_0\sigma_t\rangle$ is described by the small moments ($n \to 0$), while the tail of the distribution is governed by the large moments ($n \to \infty$). We should have a clear idea of the crossover point n^* in applying the replica method. For the above one dimensional example, an estimate of n^* is given by the ratio of the successive terms in the expansion, i.e.

$$n^* = \frac{1}{\ln(\tau_2/\tau_1)}. \qquad (10.27)$$

Note that as τ_2/τ_1 becomes large, n^* decreases, possibly becoming even smaller than unity. This does not imply that we should conclude that $\ln\langle\sigma_0\sigma_t\rangle$ is not normally distributed, just that the tail of the distribution is more prominent. Failure to appreciate this point is the source of some misunderstandings on the use of the replica method.

Clearly, it is possible to come up with many different microscopic distributions $p(\tau)$, which result in the same first two cumulants in Eqs. (10.24), but different higher cumulants. All these cases lead to the same universal bulk probability distribution for $\ln\langle\sigma_0\sigma_t\rangle$ at large t, but very different tails. Thus the non-uniqueness of the overall probability in this example has to do with the rather uninteresting (and non-universal) behavior of the tail of the distribution. The correct interpretation of Eqs. (10.24) is that the mean value for the logarithm of the correlation function grows linearly with the separation t. In analogy with pure systems, we can regard the coefficient of this decay as the inverse correlation length, i.e. $\xi^{-1} = -\ln\sqrt{\tau_1\tau_2}$. However, due to randomness in the medium, correlations have different decays between different realizations (and between different points in the same realization). The variations in this "inverse correlation length" are scale dependent and fall off as $1/\sqrt{t}$. In the next sections we shall attempt to generalize these results to higher dimensions.

10.4 Directed paths and the transfer matrix

Calculation of the correlation function in higher dimensions is complicated by the presence of an exponentially large number of paths connecting any pair of points. On physical grounds we expect the high temperature phase to be disordered, with correlations that decay exponentially as a function of the separation t. The essence of this exponential decay is captured by the lowest order terms in the high-temperature expansion. The first term in the series comes from the *shortest path* connecting the two points. Actually, along a generic direction on a hypercubic lattice there are many paths that have the same shortest length. (In two dimensions, the length of the shortest path connecting $(0,0)$ to (t,x) is the "Manhattan" distance $|t| + |x|$.) The number of paths grows from a minimum of 1 along a cardinal lattice direction to a maximum of d per step along the diagonal. (The number of paths on the square lattice is $(t+x)!/(t!x!)$.) Thus the decay of correlations depends on orientation, a consequence of the *anisotropy* of the hypercubic lattice. (Note that this anisotropy is absent at distances less than the correlation length. We don't have to worry about anisotropy in discretizing critical (massless) theories on a lattice.)

In a uniform system these shortest paths are sufficient to capture the essence of correlation functions of the high-temperature phase: An exponential decay with separation which is generic to all spin systems with finite interactions. As temperature is reduced, more complicated paths (e.g. with loops and overhangs) start contributing to the sum. Although the contribution of these paths decays exponentially with their length, their number grows exponentially. Ultimately at the critical point this "entropic" increase in the number of paths overcomes the "energetic" decrease due to the factors of $\tau < 1$, and paths of all length become important below T_c. However, throughout the high-temperature phase it is possible to examine the paths at a coarse-grained scale where no loops and overhangs are present. The scale of such structures is roughly the correlation length ξ, and if we use ξ as the unit of a coarse-grained lattice, the graphs contributing to the correlation function are *directed paths*.

Let us define "directed paths" more carefully: Between any pair of points on the lattice we can draw an imaginary line which can be regarded as a "time" axis t. Transverse directions (perpendicular to the t axis) are indicated by $\vec{x}$.

Directed paths are similar to the world-lines of a particle $\vec{x}(t)$ in time; they exclude any path from the initial to the final point that has steps opposite to the main time direction. The validity of this approximation, and the importance of the neglected loops, must be carefully considered. It is certainly not valid in the vicinity of the critical point where loops of all sizes are present and equally important. Away from the critical point, we must distinguish between properties at scales smaller or larger than the correlation length ξ. Limiting the sum for the correlation function to directed paths is only useful for separations $t \gg \xi$. Loops, overhangs, and additional structures occur up to size ξ (the only relevant length scale) and can be removed by coarse graining such that the lattice spacing is larger than or equal to ξ. This is automatically satisfied in a high-temperature expansion since the correlation length is zero at infinite temperature. By the same argument, we may also neglect the closed loops (vacuum bubbles) generated by the denominator of Eq. (10.9).

In the remainder of this section we shall recap how sums over directed paths *in the uniform system* can be calculated exactly by transfer matrix methods. The method is then generalized to random systems, providing an algorithm for summing all paths in polynomial time. For ease of visualization, let us first consider the problem in two dimensions; the results are easily generalizable to higher dimensions. Also to emphasize the general features of the transfer matrix method, we shall compare and contrast the behavior of correlations along the axis and the diagonal of the square lattice.

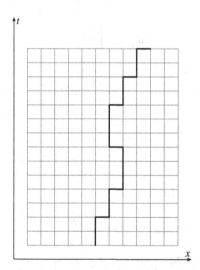

Fig. 10.1 A directed path along the principal axis of the square lattice.

To calculate the correlation function $\langle \sigma_{0,0} \sigma_{0,t} \rangle$, on a non-random square lattice, we shall focus on directed paths oriented along the main axis of the square. These paths are specified by a set of transverse coordinates $(x_0, x_1, x_2, \cdots, x_t)$, with $x_0 = x_t = 0$. Of course, there is only one shortest path with all x_i equal to zero, but we would like to explore the corrections due to longer directed paths. Consider the set of quantities

$$\langle x, t | W | 0, 0 \rangle = \text{sum over paths from } (0, 0) \text{ to } (x, t) \equiv W(x, t). \tag{10.28}$$

The calculation of $W(x, t)$ is easily accomplished by taking advantage of its *Markovian* property: Each step of a path proceeds from its last location and is independent of the previous steps. Hence W can be calculated *recursively* from,

$$W(x, t+1) = \tau \left[W(x, t) + \tau \left(W(x-1, t) + W(x+1, t) \right) + \mathcal{O}(\tau^2) \right]$$
$$\equiv \sum_{x'} \langle x | T | x' \rangle \, W(x', t), \tag{10.29}$$

where we have introduced a *transfer matrix*,

$$\langle x | T | x' \rangle = \tau \delta_{x,x'} + \tau^2 \left(\delta_{x,x'+1} + \delta_{x,x'-1} \right) + \mathcal{O}(\tau^3). \tag{10.30}$$

If we treat the values of W at a particular t as a vector, Eq. (10.29) can be iterated as,

$$\underline{W}(t) = T \underline{W}(t-1) = \cdots = T^t \, \underline{W}(0), \tag{10.31}$$

starting from

$$\underline{W}(0) = \begin{pmatrix} \vdots \\ \tau \\ 1 \\ \tau \\ \vdots \end{pmatrix}. \tag{10.32}$$

The calculations are simplified by diagonalizing the matrix T, using the Fourier basis $\langle x | q \rangle = e^{iq \cdot x} / \sqrt{N}$, as

$$T(q) = \tau (1 + 2\tau \cos q + \cdots) = \tau \exp \left[2\tau \left(1 - \frac{q^2}{2} + \cdots \right) \right]. \tag{10.33}$$

In this basis, W is calculated as

$$W(x, t) = \langle x | T^t | 0 \rangle = \sum_q \langle x | q \rangle T(q)^t \langle q | 0 \rangle$$

$$= \tau^t e^{2\tau t} \int \frac{dq}{2\pi} \exp \left[iqx - q^2 \tau t + \cdots \right] \tag{10.34}$$

$$= \exp \left[t \left(\ln \tau + 2\tau + \mathcal{O}(\tau^2) \right) \right] \times \frac{1}{\sqrt{4\pi\tau t}} \exp \left[-\frac{x^2}{4\tau t} \right].$$

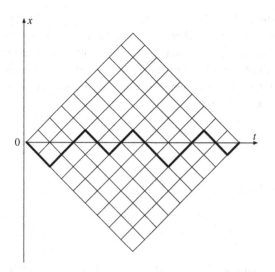

Fig. 10.2 A directed path along the diagonal of the square lattice.

The result is proportional to a Gaussian form in x of width $\sqrt{2\tau t}$. The exponential decay with $\xi^{-1} = \ln(1/\tau) - 2\tau + \mathcal{O}(\tau^2)$ at $x = 0$ is accompanied by a subleading $1/\sqrt{t}$ reflecting the constraint to return to the origin.

The corresponding calculation for paths along the diagonal, contributing to $\langle \sigma_{0,0} \sigma_{0,t} \rangle$ is even simpler. (Note that the t and x axes are rotated by $45°$ compared to the previous example.) At each step the path may proceed up or down, leading to the recursion relation

$$W(x, t+1) = \tau \left(W(x-1, t) + W(x+1, t) \right) \equiv \sum_{x'} \langle x | T | x' \rangle W(x', t), \qquad (10.35)$$

with the transfer matrix

$$\langle x | T | x' \rangle = \tau \left(\delta_{x,x'+1} + \delta_{x,x'-1} \right) \quad \Longrightarrow \quad T(q) = 2\tau \cos q. \qquad (10.36)$$

The calculation of W proceeds as before,

$$W(x, t) = \langle x | T^t | 0 \rangle = \sum_q \langle x | q \rangle T(q)^t \langle q | 0 \rangle$$

$$= \int \frac{dq}{2\pi} (2\tau)^t (\cos q)^t e^{iqx} \qquad (10.37)$$

$$\approx (2\tau)^t \times \frac{1}{\sqrt{2\pi t}} \exp\left[-\frac{x^2}{2t} \right],$$

where the final result is obtained by a saddle point evaluation of the integral, essentially replacing $\cos^t q$ with $\exp\left(-q^2 t/2\right)$.

The similarity between Eqs. (10.34) and (10.37) is apparent. Note that in both cases the leading exponential decay is determined by $T(q = 0)$, i.e.

$$W(0, t) \approx \lambda_{\max}^t = T(q = 0)^t. \qquad (10.38)$$

This is an example of the dominance of the largest eigenvalue in the product of a large number of matrices. There is a corresponding ground state dominance in the evolution of quantum systems. The similarities become further apparent by taking the *continuum limit* of the recursion relations, which are obtained by regarding $W(x, t)$ as a smooth function, and expanding in the derivatives. From Eq. (10.29), we obtain

$$W + \frac{\partial W}{\partial t} + \cdots = \tau W + \tau^2 \left(2W + \frac{\partial^2 W}{\partial x^2} + \cdots \right), \qquad (10.39)$$

while Eq. (10.35) leads to

$$W + \frac{\partial W}{\partial t} + \cdots = 2\tau W + \tau \frac{\partial^2 W}{\partial x^2} + \cdots \qquad (10.40)$$

For large t, the function W decays slowly for adjacent points in the x direction, and it is justified to only consider the lowest order derivatives with respect to x. The decay factor along the t direction is, however, quite big and we shall keep track of all derivatives in this direction, leading to

$$e^{\partial_t} W = \tau \exp\left[2\tau + \tau \partial_x^2 + \cdots \right] W, \qquad (10.41)$$

and

$$e^{\partial_t} W = 2\tau \exp\left[\frac{1}{2} \partial_x^2 + \cdots \right] W, \qquad (10.42)$$

respectively. Both equations can be rearranged (and generalized in higher dimensions) into the differential form,

$$\frac{\partial W}{\partial t} = -\frac{W}{\xi(\theta)} + \nu(\theta) \nabla^2 W, \qquad (10.43)$$

where $\xi(\theta)$ and $\nu(\theta)$ are the orientation dependent correlation length and dispersion coefficient. Equation (10.43) can be regarded as a diffusion equation in the presence of a sink, or an imaginary-time Schrödinger equation.

It is of course quite easy to integrate this linear equation to reproduce the results in Eqs. (10.34) and (10.37). However, it is also possible [R.P. Feynman and A.R. Hibbs, *Quantum Mechanics and Path Integrals* (McGraw-Hill, New York, 1965)] to express the solution in the form of a continuous *path integral*. The solution is trivial in Fourier space,

$$\frac{\partial W(q)}{\partial t} = -(\xi^{-1} + \nu q^2) W \implies W(q, t + \Delta t) = e^{-(\xi^{-1} + \nu q^2)\Delta t} W(q, t), \qquad (10.44)$$

while in real space,

$$
\begin{aligned}
W(x, t + \Delta t) &= \int \frac{dq}{2\pi} e^{iqx} e^{-(\xi^{-1} + \nu q^2)\Delta t} W(q, t) \\
&= \int \frac{dq}{2\pi} e^{iqx} e^{(-\xi^{-1} - \nu q^2)\Delta t} \int dx_t e^{-iqx_t} W(x_t, t) \\
&= \int dx_t \exp\left[-\frac{\Delta t}{\xi} - \frac{(x - x_t)^2}{4\nu \Delta t} \right] W(x_t, t) \\
&= \int dx_t \exp\left[-\frac{\Delta t}{\xi} + \frac{\Delta t}{4\nu} \left(\frac{x_{t+\Delta t} - x_t}{\Delta t} \right)^2 \right] W(x_t, t),
\end{aligned}
\tag{10.45}
$$

which is just a continuum version of Eqs. (10.29) and (10.35). We can subdivide the interval $(0, t)$ into N subintervals of length $\Delta t = t/N$. In the limit of $N \to \infty$, recursion of Eq. (10.45) gives

$$
W(x, t) = \int_{(0.0)}^{(x, t)} \mathcal{D}x(t') \exp\left[\int_0^t dt' \left(-\frac{1}{\xi(\theta)} - \frac{\dot{x}^2}{4\nu(\theta)} \right) \right],
\tag{10.46}
$$

where $\dot{x} = dx/dt'$, and the integration is over all functions $x(t')$.

It is instructive to compare the above path integral with the partition function of a string stretched between $(0, 0)$ and (x, t),

$$
\begin{aligned}
Z(x, t) &= \int_{(0,0)}^{(x, t)} \mathcal{D}x(t') \exp\left[-\beta\sigma \int_0^t dt' \sqrt{1 + \dot{x}^2} \right] \\
&= \int_{(0,0)}^{(x, t)} \mathcal{D}x(t') \exp\left[-\int_0^t dt' \left(\beta\sigma + \frac{\beta\sigma}{2} \dot{x}^2 + \cdots \right) \right],
\end{aligned}
\tag{10.47}
$$

where σ is the *line tension*. Whereas for the string $\xi^{-1} = (2\nu)^{-1} = \beta\sigma$, in general due to the anisotropy of the lattice these quantities need not be equal. By matching solutions at nearby angles of θ and $\theta + d\theta$, it is possible to obtain a relation between $\xi^{-1}(\theta)$ and $\nu^{-1}(\theta)$. (For a similar relation in the context of interfaces of Ising models, see [M.E. Fisher, J. Stat. Phys. **34**, 667 (1984)]. However, $\xi(\theta)^{-1}$ calculated from the shortest paths only is singular along the axis $\theta = 0$. This is why to calculate the parameter ν along this direction it is necessary to include longer directed paths.

10.5 Moments of the correlation function

We now return to the correlation functions in the presence of random bonds. In the high temperature limit, we can still set

$$
W(x, t) \equiv \langle \sigma_{0,0} \sigma_{x,t} \rangle = \sum_P \prod_{i=1}^t \tau_{Pi},
\tag{10.48}
$$

where the sum is over all the diagonally oriented directed paths P from $(0, 0)$ to (x, t), and the τ_{Pi} denote factors of $\tanh K$ encountered for the random bonds along each path. The τ_i are random variables, independently chosen for each bond. We shall assume that the probability distribution $p(\tau)$ is narrowly distributed around a mean value $\bar{\tau}$ with width σ. Clearly, $W(x, t)$ is itself a random variable and we would like to find its probability distribution. Rather than directly calculating $p(W)$, we shall first examine its moments $\overline{W^n}$.

Calculation of the first moment is trivial: Each factor of τ_i occurs at most once in Eq. (10.48), and hence after averaging,

$$\overline{W(x, t)} \equiv \langle \sigma_{0,0}\sigma_{x,t} \rangle = \sum_P \bar{\tau}^t. \tag{10.49}$$

This is precisely the sum encountered in a non-random system, with $\bar{\tau}$ replacing τ. For example, along the square diagonal,

$$\overline{W(x, t)} \approx (2\bar{\tau})^t \times \frac{1}{\sqrt{2\pi t}} \exp\left[-\frac{x^2}{2t}\right], \tag{10.50}$$

and in general, in the continuum limit,

$$\frac{\partial \overline{W}}{\partial t} = -\frac{\overline{W}}{\xi} + \nu \frac{\partial^2 \overline{W}}{\partial x^2}. \tag{10.51}$$

For the calculation of the second moment we need to evaluate

$$\overline{WW} = \sum_{P,P'} \prod_{i=1}^{t} \overline{\tau_{Pi}\tau_{P'i}}. \tag{10.52}$$

For a particular step i, there are two possible averages depending on whether or not the two paths cross the same bond,

$$\overline{\tau_{Pi}\tau_{P'i}} = \begin{cases} \bar{\tau}^2 & \text{if } Pi \neq P'i \\ \overline{\tau^2} & \text{if } Pi = P'i. \end{cases} \tag{10.53}$$

Since $\overline{\tau^2} > \bar{\tau}^2$ there is an enhanced weight favoring paths that intersect to those that don't. This can be regarded as an attraction between the two paths, represented by a Boltzmann weight,

$$U = \frac{\overline{\tau^2}}{\bar{\tau}^2} = \frac{\bar{\tau}^2 + \sigma^2}{\bar{\tau}^2} = 1 + \frac{\sigma^2}{\bar{\tau}^2} \approx e^{\sigma^2/\bar{\tau}^2}. \tag{10.54}$$

Including this attraction, the recursion relation for $\overline{WW}$ is,

$$W_2(x_1, x_2, t) \equiv \overline{W(x_1, t)W(x_2, t)} = \sum_{x_1' x_2'} \langle x_1 x_2 | T_2 | x_1' x_2' \rangle \, W(x_1', x_2', t-1), \tag{10.55}$$

with the two-body transfer matrix

$$\langle x_1 x_2 | T_2 | x_1' x_2' \rangle = \bar{\tau}^2 \left(\delta_{x_1, x_1'+1} + \delta_{x_1, x_1'-1} \right)$$
$$\times \left(\delta_{x_2, x_2'+1}x + \delta_{x_2, x_2'-1} \right) \left(1 + (U-1)\delta_{x_1, x_2}\delta_{x_1', x_2'} \right). \tag{10.56}$$

The significance of the attraction in Eq. (10.54) is as follows. In the random system the paths prefer to pass through regions with particularly favorable values of τ. After performing the average there are no preferred locations. The tendency for replicas of the original paths to bunch up at favorable spots is instead mimicked by the uniform attraction which tends to bundle together multiple paths representing the higher moments.

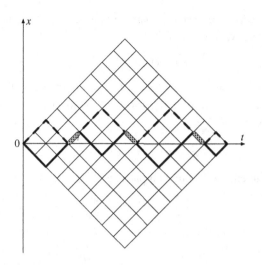

Fig. 10.3 A pair of paths (dashed and solid) contributing to calculation of the second moment of the sum over diagonally directed paths. The averaging over random bonds enhances the weight when the two paths cross the same bond.

In the continuum limit, Eq. (10.55) goes over to a differential equation of the form,

$$\frac{\partial W_2(x_1, x_2, t)}{\partial t} = -\frac{2W_2}{\xi} + \nu \frac{\partial^2 W_2}{\partial x_1^2} + \nu \frac{\partial^2 W_2}{\partial x_2^2} + u\, \delta(x_1 - x_2) W_2 \equiv -H_2 W_2, \quad (10.57)$$

with $u \approx \sigma^2/\bar{\tau}^2$. Alternatively, we could have obtained Eq. (10.57) from the continuum version of the path integral,

$$W_2(x_1, x_2, t) = \int_{(0,0)}^{(x_1,t)} \mathcal{D}x_1(t') \int_{(0,0)}^{(x_2,t)} \mathcal{D}x_2(t') \exp\left[\int_0^t \mathrm{d}t'\, u\, \delta\left(x_1(t') - x_2(t')\right)\right]$$

$$\exp\left[\int_0^t \mathrm{d}t'\left(-\frac{1}{\xi} - \frac{\dot{x}_1^2}{4\nu}\right)\right] \exp\left[\int_0^t \mathrm{d}t'\left(-\frac{1}{\xi} - \frac{\dot{x}_2^2}{4\nu}\right)\right]. \quad (10.58)$$

Formally integrating Eq. (10.57) yields $W_2 \propto \exp(-tH_2)$, which can be evaluated in the basis of eigenvalues of H_2 as

$$W_2(x_1, x_2, t) = \langle x_1 x_2 | T_2^t | 00 \rangle = \sum_m \langle x_1 x_2 | m \rangle e^{-\varepsilon_m t} \langle m | 00 \rangle \approx_{t \to \infty} e^{-\varepsilon_0 t}, \quad (10.59)$$

where $\{\varepsilon_m\}$ are the eigenenergies of H_2, regarded as a quantum Hamiltonian. The exponential growth of W_2 for $t \to \infty$ is dominated by the ground state "energy" of ε_0.

The two-body Hamiltonian depends only on the relative separation of the two particles. After transforming to the center of mass coordinates,

$$\begin{cases} r = x_1 - x_2 \\ R = (x_1 + x_2)/2 \end{cases} \implies \partial_1^2 + \partial_2^2 = \frac{1}{2}\partial_R^2 + 2\partial_r^2, \quad (10.60)$$

the Hamiltonian reads,

$$H_2 = \frac{2}{\xi} - \frac{\nu}{2}\partial_R^2 - 2\nu\partial_r^2 - u\delta(r). \quad (10.61)$$

The relative coordinate describes a particle in a delta-function potential, which has a ground state wavefunction

$$\psi_0(r, R) \propto e^{-\kappa|r|}.$$ (10.62)

The value of κ is obtained by integrating $H_2\psi_0$ from 0^- to 0^+, and requiring the discontinuity in the logarithmic derivative of ψ_0 to match the strength of the potential; hence

$$-2\nu(-\kappa - \kappa) = u \implies \kappa = \frac{u}{4\nu} \approx \frac{\sigma^2}{2\bar{\tau}^2}.$$ (10.63)

The ground state energy of this two particle system is

$$\varepsilon_0 = +\frac{2}{\xi} - 2\nu\kappa^2 \approx \frac{2}{\xi} - \frac{u^2}{8\nu}.$$ (10.64)

The inequality

$$\overline{W^2(t)} = \exp\left[-\frac{2t}{\xi} + \frac{u^2 t}{8\nu}\right] = \overline{W(t)}^2 \exp\left(\frac{u^2 t}{8\nu}\right) \gg \overline{W(t)}^2,$$ (10.65)

implies that the probability distribution for $W(t)$ is quite broad, and becomes progressively wider distributed as $t \to \infty$.

Higher moments of the sum are obtained from

$$\overline{W^n} = \sum_{P_1, \cdots, P_n} \prod_{i=1}^{t} \overline{\tau_{P_1 i} \cdots \tau_{P_n i}}.$$ (10.66)

At a particular "time" slice there may or may not be intersections amongst the paths. Let us assume that τ is Gaussian distributed with a mean $\bar{\tau}$, and a *narrow width* σ; then,

$$\overline{\tau^m} \approx \int \frac{dx\, x^m}{\sqrt{2\pi\sigma^2}} \exp\left[-\frac{(x-\bar{\tau})^2}{2\sigma^2}\right] \qquad (\text{set } x = \bar{\tau} + \epsilon \text{ and expand in } \epsilon)$$

$$\approx \int \frac{d\epsilon}{\sqrt{2\pi\sigma^2}} \left(\bar{\tau}^m + m\bar{\tau}^{m-1}\epsilon + \frac{m(m-1)}{2}\bar{\tau}^{m-2}\epsilon^2 + \cdots\right) \exp\left[-\frac{\epsilon^2}{2\sigma^2}\right]$$

$$\approx \bar{\tau}^m + \frac{m(m-1)}{2}\bar{\tau}^{m-2}\sigma^2 + \cdots \approx \bar{\tau}^m\left(1 + \frac{m(m-1)}{2}\frac{\sigma^2}{\bar{\tau}^2} + \cdots\right)$$

$$\approx \bar{\tau}^m \exp\left[\frac{m(m-1)}{2}u\right].$$ (10.67)

Since there are $m(m-1)/2$ possibilities for pairing m particles, the above result represents the Boltzmann factor with a pairwise attraction of u for particles in contact. Since τ is bounded by unity, the approximations leading to Eq. (10.67) must break down for sufficiently large m. This implies the presence of three and higher body interactions. Such interactions are usually of less importance at low densities and can be safely ignored. For a discussion of these higher order interactions in a similar context see [M. Lässig, Phys. Rev. Lett. **73**, 561 (1994)].

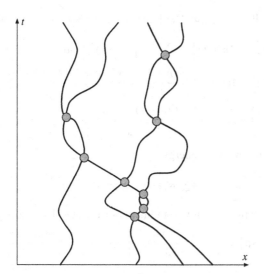

Fig. 10.4 A configuration of four paths with attraction upon contact.

The continuum version of the resulting path integral is

$$
W_n(x_1, \cdots, x_n, t) \equiv \overline{W(x_1, t) \cdots W(x_n, t)} = \int_{(0,0,\cdots,0)}^{(x_1,\cdots,x_n,t)} \mathcal{D}x_1(t') \cdots \mathcal{D}x_n(t')
$$
$$
\exp\left[-\frac{nt}{\xi} - \int_0^t dt' \left(\sum_\alpha \frac{\dot{x}_\alpha^2}{4\nu} - \frac{u}{2} \sum_{\alpha \neq \beta} \delta\left(x_\alpha(t') - x_\beta(t')\right) \right) \right],
\tag{10.68}
$$

and evolves according to

$$
\frac{\partial W_n}{\partial t} = -\frac{n W_n}{\xi} + \nu \sum_{\alpha=1}^n \frac{\partial^2 W_n}{\partial x_\alpha^2} + \frac{u}{2} \sum_{\alpha \neq \beta} \delta\left(x_\alpha(\tau) - x_\beta(\tau)\right) W_n \equiv -H_n W_n.
\tag{10.69}
$$

The asymptotic behavior of W_n at large t is controlled by the ground state of H_n. The corresponding wavefunction is obtained by a simple *Bethe Ansatz* [H.B. Thacker, Rev. Mod. Phys. **53**, 253 (1981)] which generalizes Eq. (10.62) to

$$
\psi_0(x_1, \cdots, x_n) \propto \exp\left[-\frac{\kappa}{2} \sum_{\alpha \neq \beta} |x_\alpha - x_\beta| \right] \quad \text{with} \quad \kappa = \frac{u}{4\nu}.
\tag{10.70}
$$

For each ordering of particles on the line, the wave function can be written as a product of exponentials $\psi_0 \propto \exp[\kappa_\alpha x_\alpha]$, with the "momenta" κ_α getting permuted for different orderings. For example, if $x_1 < x_2 < \cdots < x_n$, the momenta are

$$
\kappa_\alpha = \kappa\left[2\alpha - (n+1)\right],
\tag{10.71}
$$

forming a so-called *n-string*. The kinetic energy is proportional to

$$
\begin{aligned}
S &= \sum_{\alpha=1}^{n} \left[2\alpha - (n+1)\right]^2 = \sum_{\alpha=1}^{n} \left[(n+1)^2 - 4\alpha(n+1) + 4\alpha^2\right] \\
&= n(n+1)^2 - 4(n+1) \cdot \frac{n(n+1)}{2} + \frac{4\,n(n+1)(2n+1)}{6} \\
&= n(n+1)\left[-(n+1) + \frac{2(2n+1)}{3}\right] = \frac{n(n+1)(n-1)}{3},
\end{aligned}
\tag{10.72}
$$

leading to the ground state energy

$$
\varepsilon_0 = \frac{n}{\xi} - \nu \sum_{\alpha=1}^{n} \kappa_\alpha^2 = \frac{n}{\xi} - \frac{\nu\kappa^2}{3} n(n^2 - 1).
\tag{10.73}
$$

Thus the asymptotic behavior of moments of the sum has the form

$$
\lim_{t\to\infty} \overline{W^n(t)} = \exp\left[-\frac{nt}{\xi} + \frac{n(n^2-1)\nu\kappa^2 t}{3}\right] = \overline{W(t)}^{\,n} \exp\left(\frac{n(n^2-1)u^2 t}{48\nu}\right).
\tag{10.74}
$$

10.6 The probability distribution in two dimensions

It is tempting to use Eq. (10.74) in conjunction with

$$
\lim_{n\to 0} \ln\left(\overline{W^n(t)}\right) = n \overline{\ln W} + \frac{n^2}{2} \overline{(\ln W)_c^2} + \cdots + \frac{n^p}{p!} \overline{(\ln W)_c^p} + \cdots,
\tag{10.75}
$$

to read off the cumulants for the probability distribution for $\ln W$. The key point is the absence of the n^2 term and the presence of the $n^3 t$ factor in the exponent of Eq. (10.74), suggesting a third cumulant, and hence fluctuations in $\ln W$ that grow as $t^{1/3}$ [M. Kardar, Nucl. Phys. B **290**, 582 (1987)]. However, as discussed before, there are subtleties in trying to deduce a probability distribution from the knowledge of its moments which we need to consider first. Since $W(t)$ is bounded by unity, Eq. (10.74) cannot be valid for arbitrarily large n. Our first task is to identify the crossover point n^* beyond which this result is no longer correct.

Equation (10.73) is obtained for the ground state of n particles subject to a two body interaction in the continuum limit. A simple argument can be used to understand the origin of the n^3 term in the energy, as well as the limitations of the continuum approach. Let us assume that the n particles form a bound state of size R. For large n, the energy of such a state can be estimated as

$$
\varepsilon \approx \frac{n}{\xi} + \frac{\nu n}{R^2} - \frac{u n^2}{R}.
\tag{10.76}
$$

A variational estimate is obtained by minimizing the above expression with respect to R, resulting in $R \propto \nu/(un)$ and $\varepsilon \propto u^2 n^3/\nu$. The size of the bound state decreases with increasing n, and the continuum approximation breaks down when it becomes of the order of the lattice spacing for $n^* \propto \nu/u \approx \overline{\tau}^2/\sigma^2$. For $n \gg n^*$ all the paths collapse together and

$$
\lim_{n\to\infty} \overline{W^n(t)} \simeq \left(2\overline{\tau^n}\right)^t.
\tag{10.77}
$$

This asymptotic behavior is *non-universal* and depends on the extreme values of the local probability distribution for τ. Depending on the choice of parameters, n^* can be large or small. However, as discussed in the context of the one dimensional problem, its value controls only the relative importance of the tail and the bulk of the probability distribution for $\ln W$. The behavior of the bulk of the distribution is expected to be *universal*. The crossover at n^* is explicitly demonstrated in a related model in [E. Medina and M. Kardar, J. Stat. Phys. **71**, 967 (1992)].

Another important consideration is the order of limits. Equation (10.74) is obtained by taking the $t \to \infty$ limit at fixed n, while the cumulant series in Eq. (10.75) relies on an expansion around $n \to 0$ for fixed t. The two limits do not commute. In fact, we would naively deduce from Eq. (10.74) that only the third cumulant of $\ln W$ is non-zero. This is incorrect as it is impossible to have a probability distribution with only a third cumulant [W. Feller, *An Introduction to Probability Theory and its Applications* (Wiley, 1971)]. The correct procedure [E. Medina and M. Kardar, J. Stat. Phys. **71**, 967 (1992)] is to *assume* that the *singular* behavior associated with $n \to 0$ and $t \to \infty$ is described by a scaling function of the form $g_s(nt^\omega)$. (This is similar to a singular form at a critical point with n behaving as a relevant operator with scaling exponent of ω.) If $t \to \infty$ at fixed n, extensivity of the free energy of the n particle system forces $\ln \overline{W^n(t)}$ to be proportional to t. At the other limit of $n \to 0$ at fixed t, the result is a power series in n, i.e.

$$\ln \overline{W^n(t)} = ant + g_s(nt^\omega) = \begin{cases} ant + \rho n^{1/\omega} t & \text{for } t \to \infty \text{ at fixed } n \\ ant + g_1 nt^\omega + g_2 (nt^\omega)^2 + \cdots \text{for fixed } t \text{ as } n \to 0. \end{cases} \quad (10.78)$$

(Note the inclusion of a *non-singular* term, ant.) Similar considerations have been put forward in [R. Friedberg and Y.-K. Yu, Phys. Rev. E **49**, 4157 (1994)]. Comparison with Eq. (10.74) gives $\omega = 1/3$, and we can read off the cumulants of $\ln W$ as

$$\begin{cases} \overline{\ln W(t)} & = at + g_1 t^{1/3} \\ \overline{\ln W^2(t)}_c & = 2g_2 t^{2/3} \\ \quad \vdots \\ \overline{\ln W^p(t)}_c & = p! g_p t^{p/3}. \end{cases} \quad (10.79)$$

The existence of $t^{1/3}$ corrections to the quench averaged value of $\ln W(t)$ was first proposed by Bouchaud and Orland [J.-P. Bouchaud and H. Orland, J. Stat. Phys. **61**, 877 (1990)] and has been numerically verified [E. Medina and M. Kardar, Phys. Rev. B **46**, 9984 (1992)]. The $t^{2/3}$ growth of the variance of the probability distribution was obtained by Huse and Henley [D.A. Huse and C.L. Henley, Phys. Rev. Lett. **54**, 2708 (1985)] in the context of interfaces of Ising models at zero temperature where an optimal path dominates the sum. The results remain valid at finite temperatures [M. Kardar, Phys. Rev. Lett. **55**,

2923 (1985)]. Simulations are performed by implementing the transfer matrix method numerically. For example, along the diagonal of the square lattice, the recursion relation

$$W(x, t+1) = \tau_{x,t,-} W(x-1, t) + \tau_{x,t,+} W(x+1, t), \qquad (10.80)$$

is iterated starting from $W(x, 0) = \delta_{x,0}$. The random numbers $\tau_{x,t,\sigma}$ are generated as the iteration proceeds. The memory requirement (the arrays $W(x)$) depend on the final length t; each update requires t operations, and the total execution time grows as t^2. Thus for a given realization of randomness, *exact* results are obtained in *polynomial* time. Of course the results have to be averaged over many realizations of randomness. The typical values of t used in the transfer matrix simulations range from 10^3 to 10^4, with 10^2 to 10^3 realizations. Calculating higher cumulants becomes progressively more difficult. The existence of the third cumulant was verified by Halpin-Healy [J. Krug, P. Meakin and T. Halpin-Healy, Phys. Rev. A, **45** 638 (1992)]. A fourth cumulant, growing as $t^{4/3}$ was observed by Kim *et al.* [J.M. Kim, M.A. Moore, and A.J. Bray, Phys. Rev. A **44**, 2345 (1991)]. Starting from the replica result, Zhang [Y.-C. Zhang, Phys. Rev. Lett. **62**, 979 (1989); Europhys. Lett. **9**, 113 (1989); J. Stat. Phys. **57**, 1123 (1989)] proposed an analytical form, $p(\ln W, t) \sim \exp(-a|\ln W - \overline{\ln W}|^{3/2}/t^{1/2})$. While this form captures the correct scaling of free energy fluctuations, it is symmetric about the average value precluding the observed finite third cumulant. This deficiency was remedied by Crisanti *et al.* [A. Crisanti, G. Paladin, H.-J. Sommers, and A. Vulpiani, J. Phys. I (France) **2**, 1325 (1992)] who generalized the above probability to one with different coefficients $a_\pm$ on the two sides of the mean value.

So far, we have focused on the asymptotic behavior of $W(x, t)$ at large t, ignoring the dependence on the transverse coordinate. For the pure problem, the dependence of W on the transverse coordinate is a Gaussian, centered at the origin, with a width that grows as $t^{1/2}$. The full dependence is obtained in the non-random case by including the band of eigenvalues with energies close to the ground state. Unfortunately, determining the appropriate eigenvalues for the interacting problem is rather difficult. In addition to the eigenvalues obtained by simply multiplying Eq. (10.70) by $\exp[iq(x_1 + \cdots + x_n)]$, there are other states with broken replica symmetry [G. Parisi, J. Physique (France) **51**, 1595 (1990)]. A treatment by Bouchaud and Orland [J.-P. Bouchaud and H. Orland, J. Stat. Phys. **61**, 877 (1990)] includes some of the effects of such excitations but is not fully rigorous. It does predict that the extent of transverse fluctuations grows as t^ζ with $\zeta = 2/3$ as observed numerically [D.A. Huse and C.L. Henley, Phys. Rev. Lett. **54**, 2708 (1985)] [M. Kardar, Phys. Rev. Lett. **55**, 2923 (1985)]. There is in fact a relation between the exponents ζ and ω which follows from simple physical considerations: By analogy with a string, the energy to stretch a path by a distance x grows as x^2/t. The path wanders away from the origin, only if the cost of this stretching can be made up by favorable configurations of bonds. Since the typical fluctuations in (free) energy at scale t grow as t^ω, we have

$$\frac{x^2}{t} \propto t^\omega \implies \omega = 2\zeta - 1. \tag{10.81}$$

This relation remains valid in higher dimensions and has been verified in many numerical simulations. The first (indirect) proof of $\omega = 1/3$ was based on a replica analysis of a problem with many interacting paths [M. Kardar and D.R. Nelson, Phys. Rev. Lett. **55**, 1157 (1985)]. It was soon followed by a more direct proof [D.A. Huse, C.L. Henley, and D.S. Fisher, Phys. Rev. Lett. **55**, 2924 (1985)] based on a completely different approach: the Cole–Hopf transformation and the mapping to the interface problem (see below) described in the previous chapter.

10.7 Higher dimensions

The approach described in the previous sections is easily generalized to higher dimensions. The directed path in $d = D+1$ is described by $\vec{x}(t)$, where $\vec{x}$ is a D-dimensional vector. Along a diagonal, Eq. (10.80) is generalized to

$$W(\vec{x}, t+1) = \sum_{i=1}^{d} \tau_{\vec{x}-\vec{e}_i, t} W\left(\vec{x} - \vec{e}_i, t\right), \tag{10.82}$$

where $\vec{e}_i$ are unit vectors. The recursion relation is easily iterated on a computer, but the memory requirement and execution time now grow as t^D and t^{D+1}, respectively. The continuum limit of this recursion relation is

$$\frac{\partial W(\vec{x}, t)}{\partial t} = -\frac{W}{\xi} + \nu \nabla^2 W + \mu(\vec{x}, t) W, \tag{10.83}$$

where $\mu(\vec{x}, t)$ represents the fluctuations of $\tau(\vec{x}, t)$ around its average. Thus it has zero mean, and a variance

$$\overline{\mu(\vec{x}, t)\mu(\vec{x}', t')} = \sigma^2 \delta^D\left(\vec{x} - \vec{x}'\right) \delta(t - t'). \tag{10.84}$$

(In a more general anisotropic situation, Eq. (10.83) has to be generalized to include different diffusivities ν_α along different directions. Such anisotropy is easily removed by rescaling the coordinates x_α.)

Equation (10.83) can be regarded as the imaginary time Schrödinger equation for a particle in a random *time dependent* potential. It can be integrated to yield the continuous path integral

$$W(\vec{x}, t) = \int_{(0,0)}^{(\vec{x}, t)} \mathcal{D}\vec{x}(t') \exp\left[-\int_0^t dt' \left(\frac{1}{\xi} + \frac{\dot{\vec{x}}^2}{4\nu} - \mu\left(\vec{x}(t'), t'\right)\right)\right], \tag{10.85}$$

describing the fluctuations of a directed polymer in a random medium (DPRM) [M. Kardar and Y.-C. Zhang, Phys. Rev. Lett. **58**, 2087 (1987)]. The nth moment of W is computed by replicating the above path integral and averaging over $\mu(\vec{x}, t)$. This generalizes Eq. (10.58) to

$$W_n(\{\vec{x}_\alpha\}, t) = \int_{(\{\vec{0}\},0)}^{(\{\vec{x}_\alpha\},t)} \mathcal{D}\vec{x}_1(t') \cdots \mathcal{D}\vec{x}_n(t')$$

$$\exp\left[-\int_0^t dt' \left(\sum_\alpha \frac{1}{\xi} + \frac{\dot{\vec{x}}_\alpha^2}{4\nu} - \frac{u}{2} \sum_{\alpha \neq \beta} \delta^D\left(\vec{x}_\alpha(t') - \vec{x}_\beta(t')\right)\right)\right],$$

$$(10.86)$$

with $u \propto \sigma^2$. The differential equation governing the evolution of $W_n(t)$ is,

$$\frac{\partial W_n}{\partial t} = -\frac{n}{\xi} W_n + \nu \sum_\alpha \nabla_\alpha^2 W_n + \frac{u}{2} \sum_{\alpha \neq \beta} \delta^D(\vec{x}_\alpha - \vec{x}_\beta) W_n \equiv -H_n W_n. \qquad (10.87)$$

Evaluating the asymptotic behavior of $W_n(t)$ requires knowledge of the ground state energy of the Hamiltonian H_n. Unfortunately, the exact dependence of the bound state energy on n is known only for $D = 0$ ($\varepsilon \propto n(n-1)$) and $D = 1$ ($\varepsilon \propto n(n^2-1)$). As discussed earlier, these two results can then be used to deduce the behavior of the bulk of the probability distribution for $\ln W(t)$. Elementary results from quantum mechanics tell us that an arbitrarily small attraction leads to the formation of a bound state in $D \leq 2$, but that a finite strength of the potential is needed to form a bound state in $D > 2$. Thus, in the most interesting case of $2+1$ dimensions we expect a non-trivial probability distribution, while the replica method provides no information on its behavior. In higher dimensions, there is a phase transition between weak and strong randomness regimes. For weak randomness there is no bound state and asymptotically $\overline{W^n(t)} = \overline{W(t)}^{\,n}$, indicating a sharp probability distribution. This statement has also been established by more rigorous methods [J. Imbrie and T. Spencer, J. Stat. Phys. **52**, 609 (1988)]. There is another phase for strong randomness where the probability distribution for $W(t)$ becomes broad. The resulting bound state has been analytically studied in a $1/D$ expansion valid for large D [Y.Y. Goldschmidt, Nucl. Phys. B **393**, 507 (1993)]. The ground state wavefunction is rather complex, involving replica symmetry breaking. Note that the phase transition in the probability distribution of the correlation function occurs in the high-temperature phase of the random bond Ising model. The implications of this phase transition for bulk properties are not known. As the stiffness associated with line tension decreases on approaching the order/disorder phase transition of the Ising model, close to this transition the probability distribution for $W(t)$ is likely to be broad.

As one of the simplest models of statistical mechanics in random systems (a "toy" spin glass), the problem of DPRM has been rather extensively studied [D.S. Fisher and D.A. Huse, Phys. Rev. B **43**, 10728 (1991)]. The model has been generalized to manifolds of arbitrary internal dimensions in random media [T. Halpin-Healy, Phys. Rev. Lett. **62**, 442 (1989)], and treated by functional

RG methods [L. Balents and D.S. Fisher, Phys. Rev. B **48**, 5949 (1993)]. The same model has also been studied by a variational approach that involves replica symmetry breaking [M. Mèzard and G. Parisi, J. Physique I **1**, 809 (1991); J. Phys. A **23**, L1229 (1990); J. Physique I **2**, 2231 (1992)]. The latter is also exact in the $D \to \infty$ limit. Directed paths have been examined on non-Euclidean lattices: In particular, the problem can be solved exactly on the Cayley tree [B. Derrida and H. Spohn, J. Stat. Phys. **51**, 817 (1988)] where it has a transition between a "free" and a glassy state. There are also quite a few treatments based on a position space renormalization group scheme [B. Derrida and R.B. Griffiths, Europhys. Lett. **8**, 111 (1989); J. Cook and B. Derrida, J. Stat. Phys. **57**, 89 (1989)] which becomes exact on a *hierarchical* lattice [see A.N. Berker and S. Ostlund, J. Phys. C **12**, 4961 (1979); and references therein]. This lattice has no loops, and at the $(m+1)$th level is constructed by putting together 2^D branches, each containing two lattices of the mth level. Starting from a set of random bonds at the first level, the values of the sum $W(m = \log_2 t)$ are constructed recursively from

$$W(m+1, \beta) = \sum_{\alpha=1}^{2^D} W(m, \alpha_1) W(m, \alpha_2), \qquad (10.88)$$

where the Greek indices are used to indicate specific bonds for a particular realization. Alternatively, these recursion relations can be used to study the evolution of the probability distribution for W [S. Roux, A. Hansen, L.R. di Silva, L.S. Lucena, and R.B. Pandey, J. Stat. Phys. **65**, 183 (1991)]. The exponent $\omega \approx 0.30$ for $D = 1$ is not too far off from the exact value of $1/3$.

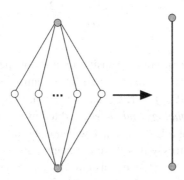

Fig. 10.5 The hierarchical lattice corresponding to Eq. (10.88).

Additional information about the higher dimensional DPRM is obtained by taking advantage of a mapping to the non-equilibrium problem of kinetic roughening of growing interfaces. Using the Cole–Hopf transformation [E. Hopf, Comm. Pure Appl. Math. **3**, 201 (1950); J.D. Cole, Quart. Appl. Math. **9**, 225 (1951)],

$$W(\vec{x}, t) = \exp\left[-\frac{\lambda h(\vec{x}, t)}{2\nu}\right], \qquad (10.89)$$

Eq. (10.83) is transformed to the Kardar, Parisi, Zhang (KPZ) [M. Kardar, G. Parisi, and Y.-C. Zhang, Phys. Rev. Lett. **56**, 889 (1986)] equation,

$$\frac{\partial h}{\partial t} = \frac{2\nu}{\lambda \xi} + \nu \nabla^2 h - \frac{\lambda}{2} (\nabla h)^2 - \frac{2\nu}{\lambda} \mu(\vec{x}, t), \tag{10.90}$$

describing the fluctuations in height $h(\vec{x}, t)$ of a growing interface (see the discussion in Sec 9.6). A dynamical renormalization group (RG) analysis at the one-loop level [D. Forster, D.R. Nelson, and M.J. Stephen, Phys. Rev. A **16**, 732 (1977)] [E. Medina, T. Hwa, M. Kardar, and Y.-C. Zhang, Phys. Rev. A **39**, 3053 (1989)] of this equation indicates that the effective coupling constant $g = 4\sigma^2/\nu$, satisfies the rescaling relation

$$\frac{dg}{d\ell} = (2 - D)g + C(D)g^2, \tag{10.91}$$

where $C(D) = K_D(2D - 3)/D$ and K_D is the D-dimensional solid angle divided by $(2\pi)^D$. The RG equation merely confirms the expectations based on the replica analysis: there is flow to strong coupling for $D \leq 2$ (formation of a bound state), while there is a transition between weak (unbound) and strong (bound) coupling behavior in higher dimensions. Can a perturbation analysis at higher order provide information about the scaling behavior in the strong coupling regime [T. Sun and M. Plischke, Phys. Rev. E **49**, 5046 (1994)]; [E. Frey and U.C. Täuber, Phys. Rev. E **50**, 1024 (1994)]? The above analogy to attracting particles in quantum mechanics negates this possibility. In fact Wiese [K.J. Wiese, J. Stat. Phys. **93**, 143 (1998)] has shown that *to all orders in perturbation theory*, the above form remains unchanged, with an exact value of

$$C(D) = \frac{2\Gamma(2 - d/2)}{(8\pi)^{d/2}}.$$

Since there are several comprehensive reviews of the KPZ equation [see, e.g. *Dynamics of Fractal Surfaces*, edited by F. Family and T. Vicsek (World Scientific, Singapore, 1991); J. Krug and H. Spohn, in *Solids Far From Equilibrium: Growth, Morphology and Defects*, edited by C. Godreche (Cambridge University Press, Cambridge, 1991)]. I will not discuss its properties in any detail here. It suffices to say that there are many numerical models of growth that fall in the universality class of this equation. They are in complete agreement with the exactly known results for $D = 1$. The estimates for the exponent ζ in higher dimensions are $\zeta = 0.624 \pm 0.001$ for $D = 2$ [B.M. Forrest and L.-H. Tang, Phys. Rev. Lett. **64**, 1405 (1990)] and $\zeta \approx 0.59$ for $D = 3$ [J.M. Kim, A.J. Bray, and M.A. Moore, Phys. Rev. A **44**, 2345 (1991)]. The numerical results in higher dimensions are consistent with an exponent ζ that gets closer to 1/2 as $D \to \infty$. It is not presently known whether there is a finite upper critical dimension [T. Halpin-Healy, Phys. Rev. Lett. **62**, 442 (1989); M.A. Moore, T. Blum, J.P. Doherty, M. Marsili, J.-P. Bouchaud, and P. Claudin, Phys. Rev. Lett. **74**, 4257 (1995)] beyond which $\zeta = 1/2$ exactly.

10.8 Random signs

So far we have focused on nearest-neighbor bonds $\{K_{ij}\}$, which though random, are all positive. For such couplings the ground state is uniform and ferromagnetic. The study of low temperature states is considerably more complicated for the random *spin glass* which describes a mixture of ferromagnetic and anti-ferromagnetic bonds. The competition between the bonds leads to *frustration*, resulting in quite complicated landscapes for the low energy states [M. Mézard, G. Parisi, and M.A. Virasoro, *Spin Glass Theory and Beyond* (World Scientific, Singapore, 1987)]. Here we shall explore the high temperature properties of spin glass models. To focus on the effects of the randomness in sign, we study a simple binary probability distribution in which negative and positive bonds *of equal* magnitude occur with probabilities p and $1 - p$, respectively.

The computation of the high-temperature series for the correlation function (along the diagonal) proceeds as before, and

$$W(\vec{x}, t) \equiv \langle \sigma_{0,0} \sigma_{\vec{x},t} \rangle = \tau^t \sum_P \prod_{i=1}^{t} \eta_{Pi}, \tag{10.92}$$

where τ indicates the fixed magnitude of $\tanh K$, while $\eta_{Pi} = \pm 1$ are random signs. Since the elements of the sum can be both positive and negative, the first question is whether the system maintains a coherence in sign (at least for small p), i.e. what is the likelihood that the two spins separated by a distance t have a preference to have the same sign? This question can be answered definitively only in one and high enough dimensions.

For the one-dimensional chain the moments of $W(t)$ are easily calculated as

$$\overline{W^n(t)} = \tau^{nt} \times \begin{cases} (1 - 2p)^t & \text{for } n \text{ odd,} \\ 1 & \text{for } n \text{ even.} \end{cases} \tag{10.93}$$

As all odd moments asymptotically decay to zero, at large distances $W(t)$ is equally likely to be positive or negative. This is expected since the sign of the effective bond depends only on the product of the intermediate bonds and the appearance of a few negative bonds is sufficient to remove any information about the overall sign. From Eq. (10.93), we can define a characteristic sign correlation length $\xi_s = -1/\ln(1 - 2p)$.

There is also a "mean-field" type of approach to the sign coherence problem [see the contribution by S. Obukhov in *Hopping Conduction in Semiconductors*, by M. Pollak and B.I. Shklovskii (North Holland, 1990)] which is likely to be exact in high dimensions. For paths along the diagonal of the hypercubic lattice, the mean value of $W(t)$ is

$$\overline{W(t)} \approx [d\tau(1 - 2p)]^t. \tag{10.94}$$

Calculating the variance of W is complicated due to the previously encountered problem of intersecting paths. We can *approximately* evaluate it by considering a subset of paths contributing to the second moment as,

$$\overline{W^2} \approx [d\tau(1-2p)]^{2t} + (d\tau^2)[d\tau(1-2p)]^{2(t-1)} + (d\tau^2)^2$$

$$[d\tau(1-2p)]^{2(t-2)} + \cdots + (d\tau^2)^t = \tau^{2t}\frac{[d(1-2p)]^{2(t+1)} - d^{t+1}}{[d(1-2p)]^2 - d}. \quad (10.95)$$

The first term in the above sum comes from two distinct paths between the end points; the second term from two paths that have their first step in common and then proceed independently. The mth term in the series describes two paths that take m steps together before becoming separated. The underlying assumption is that once the two paths have separated they will not come back together again. This independent path approximation (IPA) is better justified in higher dimensions and leads to

$$\frac{\overline{W^2(t)}}{\overline{W(t)}^2} = \frac{d(1-2p)^2 - [d(1-2p)^2]^{-t}}{d(1-2p)^2 - 1}. \quad (10.96)$$

For small p, such that $d(1-2p)^2 > 1$, the above ratio converges to a constant as $t \to \infty$; the distribution is asymptotically sharp and the correlations preserve sign information. However, if the concentration of negative bonds exceeds $p_c = \left(1 - 1/\sqrt{d}\right)/2$, the ratio diverges exponentially in t, indicating a broad distribution. This has been interpreted [V.L. Nguyen, B.Z. Spivak, and B.I. Shklovskii, Pis'ma Zh. Eksp. Teor. Fiz. **41**, 35 (1985)]; [JETP Lett. **41**, 42 (1985)]; Zh. Eksp. Teor. Fiz. **89**, 11 (1985)]; [JETP Sov. Phys. **62**, 1021 (1985)] as signaling a *sign transition*. The IPA suggests that there is a finite p_c for all $d > 1$. However, it is important to note that the approximation ignores important correlations between the paths. Shapir and Wang [Y. Shapir and X.-R. Wang, Europhys. Lett. **4**, 1165 (1987)] criticize the assumption of independent paths and suggest that as intersections are important for $d \leq 3$, there should be no phase transition in these dimensions. However, the identification of the *lower critical dimension* for the sign transition is not completely settled. Numerical simulations based on the transfer matrix method for t of up to 600 [E. Medina and M. Kardar, Phys. Rev. B **46**, 9984 (1992)] as well as exact enumeration studies [X.-R. Wang, Y. Shapir, E. Medina, and M. Kardar, Phys. Rev. B **42**, 4559 (1990)] for $t \leq 10$, fail to find a phase transition in $d = 2$. The results suggest that if there is a phase transition in $d = 2$ it occurs for $p_c < 0.05$. The phase diagram of a generalized model with complex phases has also been studied in higher dimensions [J. Cook and B. Derrida, J. Stat. Phys. **61**, 961 (1990)]; [Y.Y. Goldschmidt and T. Blum, J. Physique I **2**, 1607 (1992)] [T. Blum and Y.Y. Goldschmidt, J. Phys. A: Math. Gen. **25**, 6517 (1992)].

For $p > p_c$, the information on sign is lost beyond a coherence length ξ_s. If the system is coarse grained beyond this scale, the effective bonds are equally likely to be positive or negative. Thus we shall concentrate on the symmetric case of $p = 1/2$ in the rest of this section. This corresponds to the much studied $\pm J$ Ising spin glass [G. Toulouse, Commun. Phys. **2**, 115 (1977)]. We performed [E. Medina and M. Kardar, Phys. Rev. B **46**, 9984 (1992)] transfer matrix computations on systems of up to size $t = 2000$, and averaged over 2000 realizations of randomness. The random numbers ($+1$ or -1) were generated by a well tested random number generator [W.H. Press, B.P. Flannery, S.A. Teukolsky, and W.T. Vetterling, *Numerical Recipes*, (Cambridge University Press, 1986)]. Since W grows exponentially in t, $\ln|W|$ has a well defined probability distribution; we examined its mean $\overline{\ln|W(t)|}$, and variance $\overline{\ln|W(t)|^2} - \overline{\ln|W(t)|}^2$, for $p = 1/2$ (both signs equally probable). We also computed the typical excursions of the paths in the lateral direction as defined by

$$\overline{[x(t)^2]_{\mathrm{av}}} \equiv \overline{\frac{\sum_x x^2 |W(x,t)|^2}{\sum_x |W(x,t)|^2}}, \tag{10.97}$$

and

$$\overline{[x(t)]_{\mathrm{av}}^2} \equiv \overline{\left(\frac{\sum_x x|W(x,t)|^2}{\sum_x |W(x,t)|^2}\right)^2}, \tag{10.98}$$

where $[\cdot]_{\mathrm{av}}$ denotes an average over the lateral coordinate at a fixed t, using a weight $|W(x,t)|^2$.

The simulations confirm that the average of $\ln|W(t)|$ is extensive $(\overline{\ln|W(t)|} = (0.322 \pm 0.001)t)$, while its fluctuations satisfy a power law growth t^ω, with $\omega = 0.33 \pm 0.05$. For several choices of t we also checked in detail that $W(t)$ is positive or negative with equal probability. For lateral excursions, we examined simulations with $t = 4000$, and with 200 realizations of randomness (reasonable data for fluctuations of $\ln|W(t)|$ are only obtained from higher averaging). The results for $\overline{[x^2]_{\mathrm{av}}}$ and $\overline{[x]_{\mathrm{av}}^2}$ appear to converge to a common asymptotic limit; fitted to a power law $t^{2\zeta}$ with $\zeta = 0.68 \pm 0.05$. The scaling properties of $|W(x,t)|$ thus appear identical to those of directed polymers with positive random weights! It should be noted, however, that using a similar procedure, Zhang [Y.-C. Zhang, Phys. Rev. Lett. **62**, 979 (1989); Europhys. Lett. **9**, 113 (1989); J. Stat. Phys. **57**, 1123 (1989)] concluded from fits to his numerical results a value of $\zeta = 0.74 \pm 0.01$. Using a variety of theoretical arguments [Y.-C. Zhang, Phys. Rev. Lett. **62**, 979 (1989); Europhys. Lett. **9**, 113 (1989); J. Stat. Phys. **57**, 1123 (1989)], he suggests $\omega = 1/2$ and $\zeta = 3/4$. The exponent $\omega = 1/2$ is clearly inconsistent with our data, while $\zeta = 3/4$ can be obtained if one fits only to $\overline{[x]_{\mathrm{av}}^2}$. Two subsequent, rather extensive, numerical studies [M.P. Gelfand, *Physica A* **177**, 67 (1991)]; [Y.Y. Goldschmidt and T. Blum, Nucl. Phys. B **380**, 588 (1992)] shed more light on this problem. Both simulations seem to equivocally point to the importance of including corrections to scaling in the fits. In 1+1 dimensions they indeed find $\omega = 1/3$ for the variance, and $\zeta = 2/3$ (with a large correction to scaling term) for transverse fluctuations.

Fig. 10.6 Upon contact the paired paths can exchange partners leading to an effective attraction, as the weight is increased by a factor of 3.

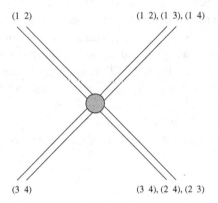

(1 2)   (1 2), (1 3), (1 4)

(3 4)   (3 4), (2 4), (2 3)

The similarity in the probability distributions of random weight and random sign problems can be understood by examination of the moments. The terms in W^n correspond to the product of contributions from n independent paths. Upon averaging, if m paths cross a particular bond $(0 \leq m \leq n)$, we obtain a factor of $[1 + (-1)^m]/2$, which is 0 or 1 depending on the parity of m. For odd n there must be bonds with m odd, and hence $\overline{W^{2n+1}} = 0$; which of course implies and follows from the symmetry $p(W) = p(-W)$. For even moments $\overline{W^{2n}}$, the only configurations that survive averaging are those in which the $2n$ replicated paths are arranged such that each bond is crossed an even number of times. The simplest configurations satisfying this constraint correspond to drawing n independent paths between the end points and assigning two replica indices to each. The above constraint is also satisfied by forming groups of four or higher even numbers, but such configurations are statistically unlikely and we shall henceforth consider only doublet paths. There is an important subtlety in calculating $\overline{W^{2n}}$ from the n doublet paths: After two such paths cross, the outgoing doublets can either carry the same replica labels as the ingoing ones, or they can exchange one label (e.g. $(12)(34) \rightarrow (12)(34)$ $(13)(24)$ or (14) (23)). Therefore, after summing over all possible ways of labeling the doublet paths, there is a multiplicity of three for each intersection. The n paired paths attract each other through the exchange of replica partners!

Although the origin of the attraction between paths is very different from the case of random weights, the final outcome is the same. The even moments in $1 + 1$ dimension are related by an expression similar to Eq. (10.74),

$$\lim_{t \to \infty} \overline{W^{2n}(t)} = \overline{W(t)^2}^{\,n} \exp\left[\rho n(n^2 - 1)t\right], \qquad (10.99)$$

and the conclusions regarding $\ln W(t)$ are the same as before. If, rather than having only one possible value for the magnitude of the random bond, we start with a symmetric distribution $p(\tau)$, there will be an additional attraction between the paired paths coming from the variance of τ^2. This increases the bound state energy (and the factor ρ) in Eq. (10.99) but does not affect the universal properties.

10.9 Other realizations of DPRM

So far we have focused on sums over DPRM as encountered in high-temperature series of Ising models. In fact several other realizations of such paths have been discussed in the literature, and others are likely to emerge in the future.

- *Random-bond interface:* One of the original motivations was to understand the domain wall of an Ising model in the presence of random bond impurities [D.A. Huse and C.L. Henley, Phys. Rev. Lett. **54**, 2708 (1985)]. As mentioned in the previous section, if all the random bonds are ferromagnetic, in the ground state all spins are up or down. Now consider a mixed state in which a domain wall is forced into the system by fixing spins at opposite edges of the lattice to $+$ and $-$. Bonds are broken at the interface of the two domains, and the total energy of the defect is twice the sum of all the K_{ij} crossed by the interface. In the solid-on-solid approximation, configurations of the domain wall are restricted to directed paths. The resulting partition function $Z(t)$, can be computed by exactly the same transfer matrix method used to calculate $W(t)$. Rather than looking at the finite temperature partition function, Huse and Henley [D.A. Huse and C.L. Henley, Phys. Rev. Lett. **54**, 2708 (1985)] worked directly with the zero temperature configuration of the interface.

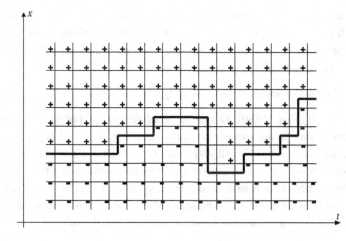

Fig. **10.7** Configuration of an interface separating + and − domains of an Ising model.

Denoting by $E(x, t)$ the minimum in the energy of all domain boundaries passing through the vertical bonds at $(0,0)$ (x, t), oriented along the horizontal axis of the square lattice, it is possible to construct the recursion relation,

$$E(x, t+1) = J_{x,t}^{v} + \min \left\{ E(x-1, t) - 2J_{x,t}^{h}, \ E(x+1, t) - 2J_{x+1,t}^{h}, \cdots \right\}, \quad (10.100)$$

where J^{v} and J^{h} denote bonds oriented in the vertical and horizonatal directions, respectively. The ellipsis refers to the interface jumping by two or more steps in the vertical direction, the corresponding bond energies have to be added in this case. In practice, such jumps are unlikely, and the same universal characteristics are obtained by allowing only one-step jumps.

The statistics of the $E(x, t)$ at $T = 0$ are identical to those of $\ln W(x, t)$: the optimal path wanders as $t^{2/3}$, while the fluctuations in $E(t)$ scale as $t^{1/3}$. The scale of energy fluctuations also sets the scale of energy barriers that the interface must cross in going from one optimal state to another [L.V. Mikheev, B. Drossel, and M. Kardar, Phys. Rev. Lett. **75**, 1170 (1995)]. Since such barriers grow with t, any activated process is slowed down to a logarithmic crawl.

Fig. 10.8 A path directed along the diagonal of the square lattice. A random varaible $\mu_{x,t}$ is assigned to each bond of this lattice.

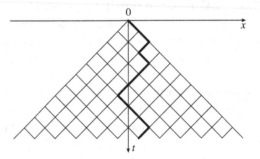

- *Optimal paths:* The above interface is one example of a path optimizing a local energy function. As another example, consider paths directed along the diagonal of the square lattice as in Fig. 10.8. Let us assume that a random variable $\mu_{x,t}$ is assigned to each bond of the lattice, and we would like to find the configuration of the path that maximizes the accumulated sum of random variables in the path. We can again denote by $M(x, t)$ the maximum value obtained for all paths connecting $(0,0)$ to (x, t), and calculate it recursively as

$$M(x, t+1) = \max \left\{ M(x-1, t) + \mu_{x-1,t}, \ M(x+1, t) + \mu_{x+1,t} \right\}, \quad (10.101)$$

closely related to Eq. (10.80). To find the actual configuration of the path, it is also necessary to store in memory one bit of information at each point (x, t), indicating whether the maximum in Eq. (10.101) comes from the first or second term. This bit of information indicates the direction of arrival for the optimal path at (x, t). After the recursion relations have been iterated forward to "time" step t, the optimal location is obtained from the maximum of the array $\{M(x, t)\}$. From this location the optimal path is reconstructed by stepping backward along the stored directions.

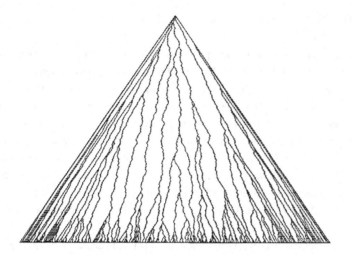

Fig. 10.9 A collection of optimal paths connecting the apex to the points on the base after $t = 500$ steps.

This is how the pictures of optimal paths in Fig. 10.9 (from [M. Kardar and Y.-C. Zhang, Phys., Rev. Lett. **58**, 2087 (1987)]; [E. Medina, T. Hwa, M. Kardar, and Y.-C. Zhang, Phys. Rev. A **39**, 3053 (1989)]) were constructed. These optimal paths have a beautiful hierarchical structure that resembles the deltas of river basins, and many other natural branching patterns. Finding the optimal path is reminiscent of the traveling salesman problem of finding the minimal route through a given set of points. However, in the former case, although the number of possible paths grow as 2^t, their directed nature allows us to find the best solution in polynomial time.

- It has been suggested that optimal paths are relevant to fracture and failure phenomena [*Disorder and Fracture*, edited by J.C. Charmet, S. Roux, and E. Guyon, (Plenum Press, New York, 1990)]. Imagine a two-dimensional elastic medium with impurities, e.g. a network of springs of different strengths and extensions [see, e.g. the article by H.J. Herrmann and L. de Arcangelis, in the above book; and references therein]. If the network is subjected to external shear, a complicated stress field is set up in the material. It is possible that nonlinear effects in combination with randomness enhance the stress field along particular paths in the medium. Such bands of enhanced stress are visible by polarized light in a sheet of plexiglas. The localization of deformation is nicely demonstrated in a two dimensional packing of straws [C. Poirier, M. Ammi, D. Bideau, and J.-P. Troadec, Phys. Rev. Lett. **68**, 216 (1992)]. The roughness of the localization band is characterized by the exponent $\zeta = 0.73 \pm 0.07$, not inconsistent with the value of 2/3 for DPRM. The experiment was inspired by random fuse models [B. Kahng, G.G. Batrouni, S. Redner, L. de Arcangelis, and H.J. Herrmann, Phys. Rev. B **37**, 7625 (1988)] which apply a similar procedure to describe the failure of an electrical network. Hansen *et al.* [A. Hansen, E.L. Hinrichsen, and S. Roux, Phys. Rev. Lett. **66**, 2476 (1991)] suggest that at the threshold in all such models, failure occurs along an optimal path with statistics similar to a DPRM. Their numerical results obtain a roughness exponent of $\zeta = 0.7$ for the crack interface with a precision of about 10%.

In fact, the minimal directed path was proposed in 1964 [P.A. Tydeman and A.M. Hiron, B.P. & B.I.R.A. Bulletin **35**, 9 (1964)] as a model for tensile rupture of paper. The variations in brightness of a piece of paper held in front of a light source are indicative of nonuniformities in local thickness and density $\rho(\mathbf{x})$. Tydeman and Hiron suggested that rupture occurs along the weakest line for which the sum of $\rho(\mathbf{x})$ is minimum. This is clearly just a continuum version of the optimal energy path in a random medium. (Since the average of $\rho(\mathbf{x})$ is positive, the optimal path will be directed.) This model was tested by Kertész et al. [J. Kertész, V.K. Horváth, and F. Weber, Fractals **1**, 67 (1993)] who used a tensile testing machine to gradually tear apart many sheets of paper. They found that the resulting rupture lines are self-affine, characterized by $0.63 < \zeta < 0.72$.

- The three dimensional DPRM was introduced [M. Kardar and Y.-C. Zhang, Phys., Rev. Lett. **58**, 2087 (1987)] as a model for a polyelectrolyte in a gel matrix. Probably a better realization is provided by defect lines, such as dislocations or vortices, in a medium with impurities. In fact, flux lines in high temperature ceramic super-conductors are highly flexible, and easily pinned by the oxygen impurities that are usually present in these materials [D.R. Nelson, Phys. Rev. Lett. **60**, 1973 (1988); D.R. Nelson and H.S. Seung, Phys. Rev. B **39**, 9153 (1989)]. Pinning by impurities is crucial for any application, as otherwise the flux lines drift due to the Lorentz force giving rise to flux flow resistivity [see, for example, G. Blatter et al., Rev. Mod. Phys. **66**, 1125 (1994), and references therein].

- *Sequence alignment:* The ability to rapidly sequence the DNA from different organ-isms has made a large body of data available, and created a host of challenges in the emerging field of *bioinformatics*. Let us suppose that the sequence of bases for a gene, or (equivalently) the sequence of amino acids for a protein, has been newly discovered. Can one obtain any information about the potential functions of this pro-tein given the existing data on the many sequenced proteins whose functions are (at least partially) known? A commonly used method is to try to match the new sequence to the existing ones, finding the best possible alignment(s) based on some method of scoring similarities. Biostatisticians have constructed efficient (so-called dynamic programming) algorithms whose output is the optimal alignment, and a corresponding score. How can one be sure that the resulting alignment is significant, and not due to pure chance? The common way of assessing this significance for a given scoring scheme is to numerically construct a probability distribution for the score by simu-lations of matchings between random sequences. This is time consuming (especially in the relevant tails of the distribution), and any analytical information is a valuable guide.

It was noted by Hwa and Lässig [T. Hwa and M. Lässig, Phys. Rev. Lett. **76**, 2591 (1996); T. Hwa, Nature **399**, 43 (1999)] that finding the optimal alignment of two sequences $\{s_i\}$ $(i = 1, 2, \cdots, I)$ and $\{s_j\}$ $(j = 1, 2, \cdots, J)$ is similar to finding the lowest energy directed path on an $I \times J$ lattice. Each diagonal bond [from (i, j)

to $(i+1, j+1)$] is assigned an "energy" equal to the score of the local pairing [s_i and s_j], and there are additional costs associated with segments along the axes (corresponding to insertions and/or deletions). The "dynamic programming" algorithm is an appropriate variant of Eq. (10.101) used to recursively obtain the directed path (alignment) of optimal energy (score). Some aspects of the probability distribution for the score can then be gleaned from the knowledge of the distribution for the energy of the directed paths in random media. Indeed, some of these results are to be implemented in the widely used alignment algorithms (PSI-BLAST) disseminated by the National Center for Biotechnology Information.

10.10 Quantum interference of strongly localized electrons

The wavefunctions for non-interacting electrons in a regular solid are *extended* Bloch states. In the presence of disorder and impurities, gradually more and more of these states become *localized*. This was first pointed out by Anderson [P.W. Anderson, Phys. Rev. **109**, 1492 (1958)] who studied a random tight-binding Hamiltonian

$$\mathcal{H} = \sum_i \varepsilon_i a_i^\dagger a_i + \sum_{\langle ij \rangle} V_{ij} a_i^\dagger a_j. \tag{10.102}$$

Here ε_i are the site energies and V_{ij} represent the nearest neighbor couplings or transfer terms. For simplicity we shall focus on

$$V_{ij} = \begin{cases} V & \text{if } i, j \text{ are nearest neighbors} \\ 0 & \text{otherwise,} \end{cases}$$

so that all the randomness is in the site energies. This is just a discretized version of the continuum Hamiltonian $H = \nu \nabla^2 + \varepsilon(\vec{x})$, for a quantum particle in a random potential $\varepsilon(\vec{x})$. For a uniform ε, the Hamiltonian is diagonalized by extended Fourier modes $a_{\vec{q}}^\dagger = \sum_{\vec{x}} \exp\left(i\vec{q} \cdot \vec{x}\right) a_{\vec{x}}^\dagger / \sqrt{N}$, resulting in a band of energies $\varepsilon(\vec{q}) = \varepsilon + 2V(\cos q_1 + \cos q_2 + \cdots + \cos q_d)$. (The lattice spacing has been set to unity.) As long as the fermi energy falls within this band of excited states the system is metallic.

In the random system the wave functions become distorted, and local-ized to the vicinity of low energy impurities. This localization starts with the states at the edge of the band and proceeds to include all states as random-ness is increased. In fact in $d \le 2$, as the diffusing path of a non-localized electron will always encounter an impurity, *all* states are localized by even weak randomness. The original ideas of Anderson localization, and a heuris-tic scaling approach by Thouless [D.J. Thouless, Phys. Rep. **13**, 93 (1974); Phys. Rev. Lett. **39**, 1167 (1977)], have been placed on more rigorous footing by perturbative RG studies [E. Abrahams, P.W. Anderson, D.C. Licciardello, and T.V. Ramakrishnan, Phys. Rev. Lett. **42**, 673 (1979)]; [L.P. Gor'kov,

A.I. Larkin, and D.E. Khmel'nitskii, Zh. Eksp. Teor. Fiz. Pis'ma Red. **30**, 248 (1979)] [JETP Lett. **30**, 248 (1979)]; [F.J. Wegner, Z. Phys. B **35**, 207 (1979)]. The perturbative approach emphasizes the importance of quantum interference effects in the weakly disordered metal. *Weak localization* phenomena include the effects of magnetic fields, spin–orbit (SO) scattering (corresponding, respectively, to interactions breaking time reversal and spin space symmetries) on the conductivity [P.A. Lee and T.V. Ramakrishnan, Rev. Mod. Phys. **57**, 287 (1985)], as well as predicting a universal value of the order of $e^2/\hbar$ for conductance fluctuations [B.L. Altshuler, JEPT Lett. **41**, 648 (1985)]; [P.A. Lee and A.D. Stone, Phys. Rev. Lett. **55**, 1622 (1985)]. These phenomena can be traced to the quantum interference of time reversed paths in *backscattering* loops and their suppression by magnetic fields and SO [G. Bergmann, Phys. Rep. **107**, 1 (1984)]: In the of absence SO, a magnetic field causes an increase in the localization length, and a factor of 2 decrease in the conductance fluctuations; with SO, it has the opposite effect of decreasing the localization length, while still reducing the conductance fluctuations [B.L. Altshuler and B.I. Shklovskii, Zh. Eksp. Teor. Fiz. **91**, 220 (1986)]; [Sov. Phys. JETP **64**, 127 (1986)]; [P.A. Lee, A.D. Stone, and H. Fukuyama, Phys. Rev. B **35**, 1039 (1987)]. An alternative description of these phenomena is based on the theory of random matrices [N. Zannon and J.-L. Pichard, J. Phys. (Paris) **49**, 907 (1988)], where the only input is the symmetries of the underlying Hamiltonian and their modification by a magnetic field. Mesoscopic devices at low temperature have provided many experimental verifications of *weak localization* theory [G. Bergmann, Phys. Rep. **107**, 1 (1984)]; [R.A. Webb and S. Washburn, Physics Today **41**, No. 12, 46 (1988)] and there are many excellent reviews on the subject [P.A. Lee and T.V. Ramakrishnan, Rev. Mod. Phys. **57**, 287 (1985)] [G. Bergmann, Phys. Rep. **107**, 1 (1984)]; [P.A. Lee and B.L. Altshuler, Physics Today **41**, No. 12, 36 (1988)].

When the electronic states at the fermi energy are localized, the material is an insulator and there is no conductivity at zero temperature. However, at finite temperatures there is a small conductivity that originates from the quantum tunneling of electrons between localized states, described by Mott's variable range hopping (VRH) process [N.F. Mott, J. Non-Cryst. Solids **1**, 1 (1968). The probability for tunneling a distance t is the product of two factors

$$p(t) \propto \exp\left(-\frac{2t}{\xi}\right) \times \exp\left(-\frac{\delta\varepsilon}{k_B T}\right). \tag{10.103}$$

The first factor is the quantum tunneling probability and assumes that the overlap of the two localized states decays with a characteristic *localization length* ξ. The second factor recognizes that the different localized states must have different energies $\delta\varepsilon$ (otherwise a new state is obtained by their mixture using degenerate perturbation theory). The difference in energy must be provided by inelastic processes such as phonon scattering, and is governed by the

Boltzmann weight at temperature T. The most likely tunneling sites must be close in energy. If there is a uniform density of states $N(\varepsilon_f)$ in the vicinity of the Fermi energy, there are roughly $N(\varepsilon_f)t^d$ candidate states in a volume of linear size t in d dimensions, with the smallest energy difference of the order of $\delta\varepsilon \propto \left(N(\varepsilon_f)t^d\right)^{-1}$. Thus the two exponential factors in Eq. (10.103) oppose each other, encouraging the electron to travel shorter and longer distances, respectively. The optimal distance scales as

$$t \approx \xi(T_0/T)^{\frac{1}{d+1}},\tag{10.104}$$

with $T_0 \propto \left(k_B N(\varepsilon_f)\xi^d\right)^{-1}$, diverging at zero temperature.

In the strongly localized regime, the optimal hopping length is many times greater than the localization length ξ. The localized sites are then assumed to be connected by a classical random resistor network [A. Miller and E. Abrahams, Phys. Rev. **120**, 745 (1960)]. Since the individual resistors are taken from a very wide distribution, it is then argued [V. Ambegaokar, B.I. Halperin, and J.S. Langer, Phys. Rev. B **4**, 2612 (1971)] that the resistance of the whole sample is governed by the critical resistor that makes the network percolate. This leads to a dependence

$$\sigma(T) = \sigma_0 \exp[-(T_0/T)^{\frac{1}{d+1}}],\tag{10.105}$$

for the conductivity. This behavior has been verified experimentally both in two and three dimensions [M. Pollak and B.I. Shklovskii, *Hopping Conduction in Semiconductors* (North Holland, 1990)]. Due to the difficulty of measuring variations in the much smaller conductivities of insulators, there have been relatively few studies of the conductivity and its fluctuations for *strongly localized* electrons. Nonetheless, experiments [O. Faran and Z. Ovadyahu, Phys. Rev. **B38**, 5457 (1988)] find a positive magneto-conductance (MC) in Si-inversion layers, GaAs and In_2O_{3-x} films. Furthermore, the observed reproducible conductance fluctuations are quite suggestive of quantum interference (QI) effects. However, the magnitudes of these fluctuations grow with lowering temperature, and are about 100 times larger than $e^2/\hbar$ at the lowest temperatures.

Clearly a different theory is needed to account for QI effects in the *strong localization* regime. The most natural candidate is the quantum overlap factor in Eq. (10.103). Nguyen, Spivak, and Shklovskii (NSS) have proposed a model that accounts for QI of multiply scattered tunneling paths in the hopping probability: In between the phonon assisted tunneling events the electron preserves its phase memory. However, at low temperatures it tunnels over very large distances according to Eq. (10.104), and encounters many impurities. The overall tunneling amplitude is then obtained from the sum over all trajectories between the initial and final sites. NSS emphasized that since the contribution of each trajectory is exponentially small in its length, the dominant contributions to the sum come from the shortest or *forward scattering* paths. The traditional explanations of weak localization phenomena which rely on the QI of *back scattering* paths are therefore inappropriate to this regime. This picture is clearly reminiscent of the directed paths and will be developed more formally in the next section.

10.11 The locator expansion and forward scattering paths

The overlaps in the insulating regime can be studied by performing a "locator" expansion [P.W. Anderson, Phys. Rev. **109**, 1492 (1958)], valid in the limit $|V_{ij}| = V \ll (E - \varepsilon_i)$, where E is the electron energy. Indeed, for $V = 0$, the eigenfunctions are just the single site states, and the localization length is zero (no transfer term). For $V/(E - \varepsilon_i) \ll 1$, various quantities can be obtained perturbatively around this solution, as expressed by the Lippman–Schwinger equation [K. Gottfried, *Quantum Mechanics* (Addison-Wesley, 1990)]

$$|\Psi^+\rangle = |\Phi\rangle + \frac{1}{E - H_0 + i\delta} \mathcal{V} |\Psi^+\rangle. \tag{10.106}$$

The bare Hamiltonian,

$$H_0 = \sum_i \varepsilon_i a_i^+ a_i,$$

has no nearest-neighbor coupling, while the perturbation

$$\mathcal{V} = \sum_{\langle ij \rangle} V_{ij} a_i^+ a_j$$

describes the small transfer terms. $|\Phi\rangle$ represents the state with a localized electron at the initial site (or incident wave), $|\Psi^+\rangle$ the state where a localized electron is at the final site. In the coordinate representation, the wavefunctions are exponentially localized around the impurity sites and there are no propagating waves since electrons can only tunnel under a potential barrier. (This situation was first addressed in detail by Lifshits and Kirpichenko [I.M. Lifshits and V.Ya. Kirpichenko, Sov. Phys. JETP **50**, 499 (1979)].) We can now iterate this implicit equation to obtain an expansion in powers of the ratio $\mathcal{V}/(E - \varepsilon_i)$ as

$$|\Psi^+\rangle = |\Phi\rangle + \frac{1}{E - H_0 + i\delta} \mathcal{V} |\Phi\rangle + \frac{1}{E - H_0 + i\delta} \mathcal{V} \frac{1}{E - H_0 + i\delta} \mathcal{V} |\Phi\rangle + \cdots \tag{10.107}$$

Acting with $\langle \Psi^+|$ on the left and taking δ to zero, we obtain the overlap between the two states

$$\langle \Psi^+ | \Psi^+ \rangle = \langle \Psi^+ | \Phi \rangle + \left\langle \Psi^+ \left| \frac{1}{E - H_0} \mathcal{V} \right| \Phi \right\rangle$$
$$+ \left\langle \Psi^+ \left| \frac{1}{E - H_0} \mathcal{V} \frac{1}{E - H_0} \mathcal{V} \right| \Phi \right\rangle + \cdots \tag{10.108}$$

For a more general transfer term $\mathcal{V}$ connecting all sites, the first term represents an electron starting from the initial site and ending at the final site without scattering (the overlap $\langle \Psi^+ | \Phi \rangle$); the second term represents electrons scattering *once* off intermediate sites, the third, scattering twice, etc. The operator $\mathcal{V}$ acting on $|\Phi\rangle$ produces a factor V for each segment crossed, and H_0 acting on a particular site i results in ε_i, the bare site energy. Thus we finally arrive at a simple expression for the amplitude or the Green's function between the initial and final states as

$$\langle \Psi^+ | \Psi^+ \rangle = \langle \Phi | G(E) | \Psi^+ \rangle = V \sum_\Gamma \prod_{i_\Gamma} \frac{V}{E - \varepsilon_{i_\Gamma}}. \tag{10.109}$$

The terms in the above perturbation series correspond to all paths Γ connecting the end points; i_Γ label the sites along each path. Except that the random variables appear on the sites rather than the bonds of the lattice, this sum over paths is quite reminiscent of the corresponding one for the correlation functions of the random bond Ising model. There is, however, one complication that distinguishes the localization problem: The energy denominators in Eq. (10.109) may accidentally be zero, invalidating the perturbation series. Physically, this corresponds to intermediate sites that are at the same energy as the external points. Presumably in this case a degenerate perturbation theory has to be used to construct the wavefunction. NSS [V.L. Nguyen, B.Z. Spivak, and B.I. Shklovskii, Pis'ma Zh. Eksp. Teor. Fiz. **41**, 35 (1985)]; [JETP Lett. **41**, 42 (1985); Zh. Eksp. Teor. Fiz. **89**, 11 (1985)] [JETP Sov. Phys. **62**, 1021 (1985)] circumvent this issue by considering initial and final sites of approximately the same energy $\varepsilon_F = E = 0$, while the intermediate sites have energies $\varepsilon_i = \pm U$ with equal probability. All the energy denominators in Eq. (10.109) now contribute the same finite magnitude U, but random signs $\eta_{i_\Gamma} = \varepsilon_{i_\Gamma}/U$. The justification is that the Mott argument implicitly assumes that the lowest energy difference $\delta\varepsilon$ occurs at a distance t, and that there are no intermediate sites that are more favorable. However, it is not clear that due to the very same considerations, we should not include some dependence of the effective energy gap U on t. We shall set aside such considerations and focus on the properties of the NSS model in the remainder.

A path of length ℓ now contributes an amplitude $U(V/U)^\ell$ to the sum, as well as an overall sign. In the localized regime the sum is rapidly convergent, dominated by its lowest order terms. In general, the sum is bounded by one in which all terms make a *positive* contribution, i.e. by a lattice random walk which is convergent for $z(V/U) < 1$, where z is the lattice coordination number. This provides a lower bound for the delocalization transition, and the series is certainly convergent for smaller values of V/U. As in the Ising model we expect loops to become important only after the transition, while in the localized phase typical paths are directed beyond the localization length ξ. For $(V/U) \ll 1$, the localization length is less than a single lattice spacing, and only directed (*forward scattering*) paths need to be considered. Loops (*back scattering* paths) are irrelevant in the renormalization group sense. For sites separated by a distance $t+1$ along a diagonal of the square lattice, Eq. (10.109) is now simplified to

$$\langle i|G(E)|f\rangle = V\left(\frac{V}{U}\right)^t \sum_P \prod_{i=1}^t \eta_{P_i}, \qquad (10.110)$$

which is identical to Eq. (10.92) with (V/U) replacing τ. The diagonal geometry maximizes possible interference by having a large number of shortest paths. For tunneling along the axes rather than the diagonal of a square lattice there is only one shortest path. Then, including longer paths with kinks is essential to the interference phenomena. However, the analogy to previous results suggests that the universal behavior is the same in the two cases while the approach to asymptotic behavior is much slower in the latter.

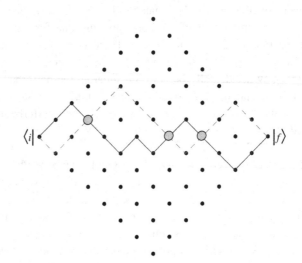

Fig. 10.10 Two paths (solid and dashed) contributing to the locator expansion for tunneling between sites i and f.

Using the equivalence to Eq. (10.99), in conjunction with Eq. (10.78), results in

$$\lim_{t\to\infty} \overline{\ln|\langle i|G|f\rangle|^2} = \ln\left[2\left(\frac{V}{U}\right)^2\right] t - \rho t \equiv -2t\left(\xi_0^{-1} + \xi_g^{-1}\right), \qquad (10.111)$$

where we have defined *local* and *global* contributions to the effective localization length, respectively given by

$$\xi_0 = \left[\ln\left(\frac{U}{\sqrt{2}V}\right)\right]^{-1} \quad \text{and} \quad \xi_g^{-1} = \frac{\rho}{2}. \qquad (10.112)$$

The QI information is encoded in $2\xi_g^{-1} = \rho$. Numerical estimates indicate that for the NSS model $\xi_g \approx 40$, and confirm that the width of the distribution scales as

$$\delta \ln|\langle i|G|f\rangle| \sim \left|\frac{t}{\xi_g}\right|^{1/3}. \qquad (10.113)$$

Since $t \propto T^{-1/3}$ in Mott VRH, we expect fluctuations in log–conductivity to grow as $T^{-1/9}$ for $T \to 0$, in qualitative agreement with the experimental results [O. Faran and Z. Ovadyahu, Phys. Rev. **B38**, 5457 (1988)]. A quantitative test of this dependence has not been performed.

10.12 Magnetic field response

All that is needed to include a magnetic field B in the tight binding Hamiltonian of Eq. (10.102) is to multiply the transfer elements V_{ij} by $\exp(A_{ij})$, where A_{ij} is the line integral of the gauge field along the bond from i to j. Due to these factors, the Hamiltonian becomes complex and is no longer time reversal symmetric ($H^* \neq H$). In the parlance of random matrix theory [N. Zannon and J.-L. Pichard, J. Phys. (Paris) **49**, 907 (1988)], the Hamiltonian with $B = 0$ belongs to the *orthogonal matrix ensemble*, while a finite field places it in the *unitary matrix ensemble*. Actually, random matrix theory recognizes a third (*symplectic matrix ensemble*) of Hamiltonians which are time reversal symmetric, but not invariant under rotations in spin space. Up to this point we had not mentioned the spin of the electron: The states of Eq. (10.102) are thus doubly degenerate and can be occupied by (non-interacting) up or down spin states. We can remove this degeneracy by including *spin–orbit* (SO) scattering, which rotates the spin of the electron as it moves through the lattice.

The generalized tight binding Hamiltonian that includes both the effects of SO scattering and magnetic field is

$$H = \sum_{i,\sigma} \varepsilon_i a_{i,\sigma}^\dagger a_{i,\sigma} + \sum_{\langle ij \rangle, \sigma\sigma'} V_{ij,\sigma\sigma'} e^{iA_{ij}} a_{i,\sigma}^\dagger a_{j,\sigma'}. \tag{10.114}$$

The constant, nearest-neighbor only hopping, elements V in Eq. (10.102) are no longer diagonal in spin space. Instead, each is multiplied by $\mathcal{U}_{ij}$, a randomly chosen $SU(2)$ matrix which describes the spin rotation due to strong SO scatterers on each bond. Equation (10.109) for the overlap of wavefunctions at the two end-points must now include the initial and final spins, and Eq. (10.110) for the sum of directed paths generalizes to

$$\mathcal{A} = \langle i\sigma | G(0) | f\sigma' \rangle = V(V/U)^t J(t); \quad J(t) = \sum_P \prod_{j=1}^t \eta_{Pj} e^{iA_{Pj,P(j+1)}} \mathcal{U}_{Pj,P(j+1)}. \tag{10.115}$$

After averaging over the initial spin, and summing over the final spin, the tunneling probability is

$$T = \frac{1}{2}\mathrm{tr}(\mathcal{A}^\dagger \mathcal{A}) = V^2(V/U)^{2t} I(t); \quad I(t) = \frac{1}{2}\mathrm{tr}(J^\dagger J). \tag{10.116}$$

We numerically studied the statistical properties of $I(t)$, using a transfer matrix method to exactly calculate I up to $t = 1000$, for over 2000 realizations of the random Hamiltonian. We found that the distribution is broad (almost log–normal), and that the appropriate variable to consider is $\ln I(t)$. In all cases the mean of $\ln I(t)$ scaled linearly with t, while its fluctuations scaled as t^{ω} with $\omega \approx 1/3$ [E. Medina, M. Kardar, Y. Shapir, and X.-R. Wang, Phys. Rev. Lett. **62**, 941 (1989)]; [E. Medina, M. Kardar, Y. Shapir, and X.-R. Wang, Phys. Rev. Lett. **64**, 1816 (1990)]; [E. Medina and M. Kardar, Phys. Rev. Lett. **66**, 3187 (1991)]. For the sake of comparison with experiments we define a log–magneto-conductance (MC) by

$$MC(t, B) \equiv \overline{\ln I(t, B)} - \overline{\ln I(t, 0)}. \tag{10.117}$$

We find numerically that the magnetic field always causes an enhancement in tunneling (a positive MC), but that the asymptotic behavior is quite distinct in the presence or absence of SO scattering.

(1) In the absence of SO, the MC is unbounded and grows linearly with t. This can be interpreted as an increase in the global contribution to the localization length. The numerical results indicate that for small B, the change in slope is proportional to $B^{1/2}$. Indeed the data for different t and B can be collapsed together, using the fit

$$MC(t, B) = (0.15 \pm 0.03) \left(\frac{\phi}{\phi_0} \right)^{1/2} t, \tag{10.118}$$

where $\phi = Ba^2$ is the flux per plaquette, and ϕ_0 is the elementary flux quantum.

(2) In the presence of SO, the MC quickly saturates with t and there is no change in the localization length. The data can still be collapsed, but by using $Bt^{3/2}$ as the scaling argument, and we find

$$MC_{SO}(t, B) = \begin{cases} cB^2 t^3 & \text{if } B^2 t^3 < 1 \\ C \approx 0.25 & \text{if } B^2 t^3 > 1. \end{cases} \tag{10.119}$$

Fig. 10.11 The behavior of the magneto-conductance in the presence of SO scattering.

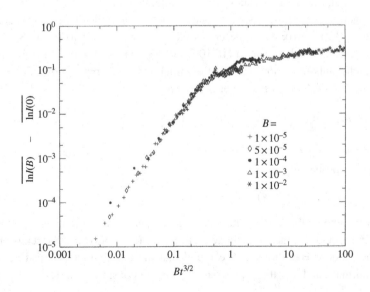

We can gain some analytic understanding of the distribution function for $I(t, B)$ by examining the moments $\overline{I(t)^n}$. From Eqs. (10.115) and (10.116) we see that each $I(t)$ represents a forward path from i to f, and a time reversed path from f to i. For $\overline{I(t)^n}$, we have to average over the contributions of n such pairs of paths. Averaging over the random signs of the site energies forces a *pairing* of the $2n$ paths (since any site crossed by an odd number of paths leads to a zero contribution). To understand the MC, it is useful to distinguish two classes of pairings:

(1) *Neutral paths* in which one member is selected from J and the other from $J^\dagger$. Such pairs do not feel the field since the phase factors of e^{iA} picked up by one member on each bond are canceled by the conjugate factors e^{-iA} collected by its partner.

(2) *Charged paths* in which both elements are taken from J or from $J^\dagger$. Such pairs couple to the magnetic field like particles of charge $\pm 2e$.

In the presence of SO, we must also average over the random $SU(2)$ matrices. From the orthogonality relation for group representations [H.F. Jones, *Groups, Representations and Physics* (Adam Hilger, Bristol and New York, 1990)], we have

$$\int \Gamma^k(g)_{ij}^* \Gamma^{k'}(g)_{i'j'} W(\alpha_1, \cdots, \alpha_n) d\alpha_1 \cdots d\alpha_n$$
$$= \frac{\delta_{ii'} \delta_{jj'} \delta_{kk'}}{\lambda_k} \int W(\alpha_1, \cdots, \alpha_n) d\alpha_1 \cdots d\alpha_n, \tag{10.120}$$

where $\Gamma^k(g)_{ij}$ is the ij matrix element of a representation of the group element g, $W(\alpha_1, \cdots, \alpha_n)$ is an appropriate weight function so that the matrix space is sampled uniformly as the continuous parameters $\alpha_1, \cdots, \alpha_n$ vary (e.g. Euler angles for a representation of $SU(2)$). Finally λ_k is the order of the representation k. Choosing the Euler angle parameterization of $SU(2)$ it can be shown that the only non-zero paired averages are

$$\overline{\mathcal{U}_{\alpha\beta} \mathcal{U}_{\alpha\beta}^*} = \frac{1}{2}, \quad \overline{\mathcal{U}_{\uparrow\uparrow} \mathcal{U}_{\downarrow\downarrow}} = \frac{1}{2}, \quad \overline{\mathcal{U}_{\uparrow\downarrow} \mathcal{U}_{\downarrow\uparrow}} = -\frac{1}{2}, \tag{10.121}$$

and their complex conjugates. Thus SO averaging forces neutral paths to carry parallel spins, while the spins on the two partners of charged paths must be anti-parallel.

We next consider the statistical weights associated with the intersections of paths. These weights depend crucially on the symmetries of the Hamiltonian in Eq. (10.114): For $B = 0$ and without SO, the Hamiltonian has *orthogonal* symmetry. All pairings are allowed and the attraction factor is 3, since an incoming (12) (34) can go to (12) (34), (13) (24), or (14) (23). Note that even if both incoming paths are neutral, one of the exchanged configurations is charged. A magnetic field breaks time reversal symmetry, discourages charged configurations, and reduces the exchange attraction. The limiting case of a

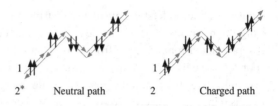

"large" magnetic field is mimicked by replacing the gauge factors with random phases. In this extreme, the Hamiltonian has *unitary* symmetry and only neutral paths are allowed. The exchange factor is now reduced to 2; from $(11^*)(22^*) \rightarrow (11^*)(22^*)$, or $(12^*)(21^*)$. With SO averaging, we must also take into account the allowed spin exchanges: Two neutral paths entering the intersection can have indices $(\alpha\alpha), (\alpha\alpha)$ or $(\alpha\alpha), (\overline{\alpha}\overline{\alpha})$; there are 2 possibilities for the first ($\alpha = \uparrow$ or $\downarrow$) and two for the second ($\overline{\alpha}$ is antiparallel to α). In the former case, however, there are two exchanges preserving neutrality, while in the latter only one exchange is possible satisfying this constraint. Hence an overall multiplicity of $[2 \times 2 + 2 \times 1] \times (1/2)^2 = 3/2$ is obtained, where the $(1/2)^2$ comes from the averages in Eq. (10.121). Thus the intersection of two paired paths results in an exchange attraction of 3/2; a signature of the *symplectic symmetry*.

Based on the above symmetry dependent statistical attraction factors, we can provide an understanding of the numerical results for MC. The sum over n attracting paths again leads to

$$\langle I(t)^n \rangle = A(n) 2^{nt} \exp[\rho n (n^2 - 1)t], \tag{10.122}$$

where we have also allowed for an overall n-dependent amplitude. Without SO, the magnetic field *gradually* reduces the attraction factor from 3 to 2 leading to the increase in slope. Addition of SO to the Hamiltonian has the effect of *suddenly* decreasing the attraction factor to 3/2. Why does the addition of the magnetic field lead to no further change in ρ in the presence of SO? Without SO, the origin of the continuous change in the attraction factor is a charged bubble that may appear in between successive intersections of two neutral paths. In the presence of SO, from the averages in Eq. (10.121) we find the contribution of such configurations to be zero. To produce intermediate charged paths (with their antiparallel spins), the entering pair must have indices of the type (ii), $(\overline{i}\,\overline{i})$ (where $\overline{\downarrow} = \uparrow$, and $\overline{\uparrow} = \downarrow$). Within the bubble we can have intermediate sites labeled $(j\overline{j})$ and $(k\overline{k})$ which must be summed over due to matrix contractions. It is easy to check that, independent of the choice of j, if the incoming and outgoing spins (i and m) are the same on a branch it contributes a positive sign, while if they are opposite the overall sign is negative. However, for any choice of i and m, one may choose similar (e.g. $i \rightarrow m$ on both branches), or opposite (e.g. $i \rightarrow m$ on top and $i \rightarrow \bar{m}$ on lower branch) connections. The difference in sign between the two choices thus cancels their overall contributions. Hence the neutral paths traverse the system without being affected by charged segments. In a magnetic field, their attraction factor stays at 3/2 and $\rho = \xi_g^{-1}$ is unchanged.

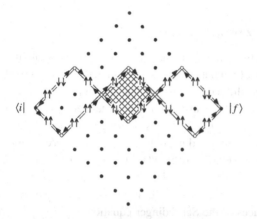

$\langle i|$ $|f\rangle$

Fig. 10.13
Magneto-conductance can be traced to the influence of loops of charged paths.

The smaller positive MC observed in the simulations is due to changes in the amplitude $A(n)$ in Eq. (10.122). This originates from the charged paths that contribute to tunneling at small B but are quenched at higher B. However, due to their lack of interactions, we may treat the charged and neutral paths as independent. At zero field any of the pairings into charged and neutral paths is acceptable, while at finite fields only neutral pairs survive. This leads to a reduction in the amplitude $A(n)$ for $n \geq 2$, but an increase in $\ln I$ (a positive MC). The typical value of $\ln I$ thus increases by a t independent amount. This behavior is similar to the predictions of IPA, and is indeed due to the independence of charged and neutral paths. Since the typical scale of decay for charged paths depends on the combination $Bt^{3/2}$ (typical flux through a random walk of length t), we can explain the scaling obtained numerically in Eq. (10.119).

The exchange attraction between neutral paths can also be computed for (unphysical) $SU(n)$ impurities and equals $1 + 1/n$, which reproduces 2 for $U(1)$ or random phases, and 3/2 for $SU(2)$ or SO scattering. The attraction vanishes in the $n \rightarrow \infty$ limit, where the paths become independent. The statistical exchange factors are thus universal numbers, simply related to the symmetries of the underlying Hamiltonian. The attractions in turn are responsible for the formation of bound states in replica space, and the universal scaling of the moments in Eq. (10.122). In fact, since the single parameter ρ completely characterizes the distribution, the variations in the mean and variance of $\ln I(t)$ should be perfectly correlated. This can be tested numerically by examining respectively coefficients of the mean and the variance for different cases. All results do indeed fall on a single line, parameterized by ρ. The largest value corresponds to the NSS model for $B = 0$ and no SO (orthogonal symmetry, exchange attraction 3). Introduction of a field gradually reduces ρ until saturated at the limit of random phases (unitary symmetry, exchange attraction 2). SO scattering reduces ρ further (symplectic symmetry, exchange attraction 3/2).

10.13 Unitary propagation

We can put together the results discussed so far by generalizing Eq. (10.83) to allow for complex (and matrix valued) parameters. In the originally encountered directed polymer, the parameters $\nu > 0$ and μ appearing in this equation were both real. To discuss the wavefunction in a magnetic field, we have to allow μ to take complex values. Finally, SO scattering is included by generalizing W to a two component spinor, and using matrix valued μ. We found that in all these cases the statistical behavior of $\ln W(\mathbf{x}, t)$ is the same. Is this a general property of Eq. (10.83), independent of the choice of parameters? A special limit of this equation is when both $\mu \to -\mathrm{i}\mu$ and $\nu \to -\mathrm{i}\nu$ are purely imaginary. Then Eq. (10.83) reduces to the Schrödinger equation

$$\mathrm{i}\frac{\partial W}{\partial t} = \left[\nu\nabla^2 + \mu(\vec{x}, t)\right] W, \tag{10.123}$$

for a particle in a random *time-dependent* potential. This equation has been considered in the context of particle diffusion in crystals at finite temperature [J.-P. Bouchaud, Europhys. Lett. **11**, 505 (1990)]; [A.A. Ovchinnikov and N.S. Erikhman, Sov. Phys. JETP **40**, 733 (1975)]; [A.M. Jayannavar and N. Kumar, Phys. Rev. Lett. **48**, 553 (1982)], and to model the environment of a light test particle in a gas of much heavier particles [L. Golubovic, S. Feng, and F. Zeng, Phys. Rev. Lett. **67**, 2115 (1991)]. Several authors [R. Dashen, J. Math Phys. **20**, 894 (1979)]; [H. de Raedt, A. Lagendijk, and P. de Vries, Phys. Rev. Lett. **62**, 47 (1988)]; [S. Feng, L. Golubovic, and Y.-C. Zhang, Phys. Rev. Lett. **65**, 1028 (1990)] have also suggested that the diffusion of *directed* wave fronts in disordered media are described by Eq. (10.123).

The path-integral solution to Eq. (10.123) is

$$W(x, t) = \int_{(0,0)}^{(\vec{x}, t)} \mathcal{D}\vec{x}(t') \exp\left\{-\mathrm{i}\int_0^t \mathrm{d}t'\left[\frac{1}{4\nu}\left(\frac{\mathrm{d}\vec{x}}{\mathrm{d}t'}\right)^2 + \mu(\vec{x}(t'), t')\right]\right\}, \tag{10.124}$$

where $x(t')$ now describes a path in $d - 1$ dimensions. In writing Eq. (10.124), we have chosen the standard initial condition that at time $t = 0$ the "wavefunction" is localized at the origin. The beam positions $\overline{\langle x^2 \rangle}$ and $\overline{\langle x \rangle^2}$ characterize the transverse fluctuations of a directed beam $W(\vec{x}, t)$ about the forward path of least scattering. Here we use $\langle \cdots \rangle$ to indicate an average with the weight

$|W(x, t)|^2$ for a given realization, and $\overline{\cdots}$ to indicate quenched averaging over all realizations of randomness. Roughly speaking, $\overline{\langle x \rangle}^2$ describes the wandering of the beam center, while $\overline{\langle x^2 \rangle} - \overline{\langle x \rangle}^2$ provides a measure of the beam width.

A special property of Eq. (10.123) which is valid only for real ν and μ is *unitarity*, i.e. the norm $\int d\vec{x} |W(\vec{x}, t)|^2$ is preserved at all times. (In the directed polymer and tunneling problems, the norm clearly decays with the length t.) This additional conservation law sets apart the random directed wave problem from directed polymers, and in a sense makes its solution more tractable. Unitarity is of course a natural consequence of particle conservation for the Schrödinger equation, but has no counterpart for directed wave propagation. It is likely that a beam of light propagating in a random medium will suffer a loss of intensity, due to either back-reflection, inelastic scattering, or localization phenomena [for a review of localization of light, see S. John, Physics Today, 32 (May 1991)].

A number of efforts at understanding unitary propagation in random media have focused on the scaling behavior of the beam positions $\overline{\langle x \rangle}^2$ and $\overline{\langle x \rangle^2}$ at large t. *Lattice* models have been used here with some success. It has been shown using density-matrix techniques, for instance, that $\overline{\langle x \rangle^2}$ scales linearly in time as a consequence of unitarity [A.A. Ovchinnikov and N.S. Erikhman, Sov. Phys. JETP **40**, 733 (1975)]; numerical simulations [E. Medina, M. Kardar, and H. Spohn, Phys. Rev. Lett. **66**, 2176 (1991)]; [J.-P. Bouchaud, D. Touati, and D. Sornette, Phys. Rev. Lett. **68**, 1787 (1992)] also support this view. The scaling behavior of $\overline{\langle x \rangle}^2$ at large t, however, is somewhat more complicated. An early numerical simulation in [S. Feng, L. Golubovic, and Y.-C. Zhang, Phys. Rev. Lett. **65**, 1028 (1990)] employed a discretization procedure in which the norm of the wave function was *not* strictly preserved. In $2d$, this leads to $\overline{\langle x \rangle}^2$ growing super-diffusively as t^ζ with $\zeta \approx 3/4$, and in $3d$ a phase transition separating regimes of weak and strong disorder. However, subsequent numerical studies [E. Medina, M. Kardar, and H. Spohn, Phys. Rev. Lett. **66**, 2176 (1991)] on directed waves, when the time evolution is strictly unitary, indicate that $\overline{\langle x \rangle}^2$ scales subdiffusively in $2d$ with $\zeta \approx 0.3$.

Somewhat surprising is the fact that a *continuum* formulation of the wave problem leads to different results. An exact treatment of the continuum Schrödinger equation (10.102) has been given by Jayannavar and Kumar [A.M. Jayannavar and N. Kumar, Phys. Rev. Lett. **48**, 553 (1982)]. They show that for a random potential δ-correlated in time, $\overline{\langle x \rangle}^2 \sim t^3$ as $t \to \infty$. This behavior is modified when there are short-range correlations in time [L. Golubovic, S. Feng, and F. Zeng, Phys. Rev. Lett. **67**, 2115 (1991)] but the motion remains non-diffusive in that the particle is accelerated indefinitely as $t \to \infty$. Lattice models introduce a momentum cutoff $p_{max} \sim a^{-1}$, where a is the lattice spacing, and therefore do not exhibit this effect. The momentum cutoff generated by the lattice discretization is in some sense artificial. Nevertheless, in a real fluctuating medium, we do expect on large time scales to recover the lattice result, i.e. normal diffusion. The reason is that dissipative effects do generate an effective momentum cutoff in most physical systems. (Strictly speaking, even in the absence of dissipation, relativistic constraints lead to a velocity cutoff $v = c$.) The presence of such a cutoff for the wave propagation problem, and hence the physical relevance of lattice versus continuum models, is still a matter of debate. While there is no underlying lattice, one suspects on physical grounds that there does exist an effective momentum cutoff for propagating waves, related to the speed of light in the background medium.

Standard numerical investigations of this problem start with a discretization of the parabolic wave equation in Eq. (10.123). Alternatively, one can treat the path integral representation as more fundamental and provide a direct discretization of Eq. (10.124) that preserves unitarity [L. Saul, M. Kardar, and N. Read, Phys. Rev. A **45**, 8859 (1992)]. For concreteness, we introduce the model in $2d$. A discussion of its generalization to higher dimensions is taken up later. As usual, we identify the time axis with the primary direction of propagation and orient it along the diagonal of the square lattice. The wave function is defined on the *bonds* of this lattice. We use $W_{\pm}(x, t)$ to refer to the amplitude for arriving at the site (x, t) from the $\pm x$ direction. At $t = 0$, the wave function is localized at the origin, with $W_{\pm}(0, 0) = 1/\sqrt{2}$. Transfer matrix techniques are then used to simulate diffusion in the presence of disorder. At time t, we imagine that a random scattering event occurs at each site on the lattice at which either $W_{+}(x, t)$ or $W_{-}(x, t)$ is non-zero. We implement these events by assigning to each scattering site a 2×2 unitary matrix $S(x, t)$. The values of the wave function at time $t + 1$ are then computed from the recursion relation:

$$\begin{pmatrix} W_{-}(x+1, t+1) \\ W_{+}(x-1, t+1) \end{pmatrix} = \begin{pmatrix} S_{11}(x, t) & S_{12}(x, t) \\ S_{21}(x, t) & S_{22}(x, t) \end{pmatrix} \begin{pmatrix} W_{-}(x, t) \\ W_{+}(x, t) \end{pmatrix}. \quad (10.125)$$

The S-matrices are required to be unitary in order to locally preserve the norm of the wave function. As a particular instance, we may consider the rotation matrix

$$S(\theta, \phi) = \begin{pmatrix} \cos(\theta/2)\,e^{i\phi} & \sin(\theta/2)\,e^{-i\phi} \\ -\sin(\theta/2)\,e^{i\phi} & \cos(\theta/2)\,e^{-i\phi} \end{pmatrix}. \quad (10.126)$$

A physical realization of this model is obtained by placing semi-polished mirrors of variable thickness, parallel to the t axis, on the sites of a square lattice. Within this framework, it should be clear that the value of $W_{\pm}(x, t)$ is obtained by summing the individual amplitudes of all directed paths which start at the origin and arrive at the point (x, t) from the $\pm x$ direction. We thus have a unitary discretization of the path integral in Eq. (10.124) in which the phase change from the potential $\mu(x, t)$ is replaced by an element of the matrix $S(x, t)$. A lattice S-matrix approach for the study of electron localization and the quantum Hall effect has been used by Chalker and Coddington [J.T. Chalker and P.D. Coddington, J. Phys. **C21**, 2665 (1988)]. A related model has also been used [C. Vanneste, P. Sebbah, and D. Sornette, Europhys. Lett. **17**, 715 (1992)] to investigate the localization of wave packets in random media. These models also include back scattering and hence involve a larger matrix at each site.

10.14 Unitary averages

A particularly nice feature of unitary propagation is that the weights $W(x, t)$ are automatically normalized. In particular, we are interested in the beam positions

$$\overline{\langle x^2(t) \rangle} = \sum_x \overline{P(x, t)} \, x^2, \tag{10.127}$$

and

$$\overline{\langle x(t) \rangle^2} = \sum_{x_1, x_2} \overline{P(x_1, t) \, P(x_2, t)} \, x_1 x_2, \tag{10.128}$$

where $P(x, t)$ is the probability distribution function on the lattice at time t, defined by

$$P(x, t) = | W_+ Plus(x, t) |^2 + | W_-(x, t) |^2. \tag{10.129}$$

(Defining the weights directly on the bonds does not substantially change the results.) Note that unlike the directed polymer problem, $P(x, t)$ is properly normalized, i.e.

$$\sum_x P(x, t) = 1,$$

and Eqs. (10.127) and (10.128) are not divided by normalizations such as $\sum_x P(x, t)$. This simplification is a consequence of unitarity and makes the directed wave problem tractable.

The average $\overline{\cdots}$, in Eqs. (10.127) and (10.128) is to be performed over a distribution of S-matrices that closely resembles the corresponding distribution for μ in the continuum problem. However, by analogy to the directed polymer problem we expect any disorder to be relevant. Hence, to obtain the asymptotic scaling behavior, we consider the extreme limit of strong scattering in which each matrix $S(x, t)$ is an *independently* chosen random element of the group $U(2)$. With such a distribution we lose any pre-asymptotic behavior associated with weak scattering [L. Golubovic, S. Feng, and F. Zeng, Phys. Rev. Lett. **67**, 2115 (1991)]. The results are expected to be valid over a range of length scales $a \ll x \ll \xi$, where a is a distance over which the change of phase due to randomness is around 2π, and ξ is the characteristic length for the decay of intensity and breakdown of unitarity. In the language of path integrals, the quantity $\overline{P(x, t)}$ represents the average over a conjugate pair of paths (from $W_\pm$ and $W_\pm^*$ respectively.) As in the random sign problem, the paths must be exactly paired to make a non-zero contribution (since $\overline{S_{\alpha\beta}} = 0$). In the strong disorder limit, each step along the paired paths contributes a factor of 1/2. (It can be easily checked from Eq. (10.126) that $\overline{|S_{\alpha\beta}|^2} = \overline{\cos^2(\theta/2)} = \overline{\sin^2(\theta/2)} = 1/2$.) Thus, in this limit, the effect of an impurity at (x, t) is to redistribute the incident probability flux $P(x, t)$ at random in the $+x$ and $-x$ directions. On average, the flux is scattered symmetrically so that the disorder-averaged probability describes the event space of a classical random walk, i.e.

$$\overline{P(x, t)} = \frac{t!}{(\frac{t-x}{2})!(\frac{t+x}{2})!}. \tag{10.130}$$

Substituting this into Eq. (10.127), we find $\overline{\langle x^2(t) \rangle} = t$, in agreement with previous studies [A.A. Ovchinnikov and N.S. Erikhman, Sov. Phys. JETP **40**, 733 (1975)].

Consider now the position of the beam center $\overline{\langle x(t) \rangle^2}$, given by Eq. (10.128). Unlike $\overline{P(x, t)}$, the correlation function $\overline{P(x_1, t)P(x_2, t)}$ does not have a simple form. It involves a sum over four paths, collapsed into two pairs by randomness averaging. The center of mass coordinate $R = (x_1 + x_2)/2$ performs a random walk with $\overline{R^2} = t/2$. Let us define a new correlation function for the relative coordinate $r = x_2 - x_1$ as

$$W_2(r, t) = \sum_R \overline{P(R - r/2, t)P(R + r/2, t)}, \tag{10.131}$$

with the initial condition

$$W_2(r, t = 0) = \delta_{r,0}. \tag{10.132}$$

The value of $W_2(r, t)$ is the disorder-averaged probability that two paired paths, evolved *in the same realization of randomness*, are separated by a distance r at time t, and can be computed as a sum over all configurations that meet this criterion. Consider now the evolution of two such pairs from time t to $t + 1$. Clearly, at times when $r \neq 0$, the two pairs behave as independent random walks. On the other hand, when $r = 0$, there is an increased probability that the paths move together as a result of participating in the same scattering event. An event in which the pairs stay together is enhanced (since $\overline{|S_{\alpha\beta}|^4} = \overline{\cos^4(\theta/2)} = \overline{\sin^4(\theta/2)} = 3/8$), while one in which the pairs separate is diminished (since $\overline{\sin^2(\theta/2)\cos^2(\theta/2)} = 1/8$). These observations lead to the following recursion relation for the evolution of $W_2(r, t)$,

$$W_2(r, t+1) = \left(\frac{1 + \epsilon\delta_{r,0}}{2}\right) W_2(r, t) + \left(\frac{1 - \epsilon\delta_{r,2}}{4}\right) W_2(r - 2, t)$$
$$+ \left(\frac{1 - \epsilon\delta_{r,-2}}{4}\right) W_2(r + 2, t). \tag{10.133}$$

The parameter $\epsilon \geq 0$ measures the tendency of the paths to stick together on contact. (If the S-matrix is uniformly distributed over the group $U(2)$, then $\epsilon = 1/4$.) Note that $\sum_r W_2(r)$ is preserved, as required by unitarity.

Using Eq. (10.133), we evolved $W_2(r, t)$ numerically for various values of $0 < \epsilon < 1$ up to $t \leq 15\,000$. The position of the beam center was then calculated from

$$\overline{\langle x(t) \rangle^2} = \sum_{R,r} \left(R^2 - \frac{r^2}{4} \right) \overline{P(R - r/2, t) P(R + r/2, t)} = \frac{t}{2} - \frac{1}{4} \sum_r W_2(r, t) \, r^2. \quad (10.134)$$

The results suggest quite unambiguously that $\overline{\langle x(t) \rangle}^2$ scales as $t^{2\zeta}$, with $\zeta = 1/4$. We emphasize here the utility of the S-matrix model for directed waves in random media. Not only does our final algorithm compute averages over disorder in an exact way, but it requires substantially less time to do so than simulations which perform averages by statistical sampling as in DPRM. We have in fact confirmed our $2d$ results with these slower methods on smaller lattices ($t < 2000$).

The model is easily extended to higher dimensions. The wave function takes its values on the bonds of a lattice in d dimensions. Random $d \times d$ dimensional S-matrices are then used to simulate scattering events at the sites of the lattice. When the matrices $S(\vec{x}, t)$ are distributed uniformly over the group $U(d)$, the same considerations as before permit one to perform averages over disorder in an exact way. In addition, one obtains the general result for $d \geq 2$ that $\overline{\langle x \rangle^2}$ scales linearly in time. The computation of $\overline{\langle x \rangle^2}$ in $d > 2$, of course, requires significantly more computer resources. We have computed $\overline{\langle x \rangle^2}$ on a $d = 3$ body-centered cubic lattice, starting from the appropriate generalization of Eq. (10.133). The results for $t < 3000$, indicate that $\overline{\langle x \rangle^2}$ scales logarithmically in time.

The above numerical results can be understood by appealing to some well-known properties of random walks. Consider a random walker on a $D = d - 1$ dimensional hypercubic lattice. We suppose, as usual, that the walker starts out at the origin, and that at times $t = 0, 1, 2, \cdots$, the walker has probability $0 < p \leq 1/2D$ to move one step in any lattice direction and probability $1 - 2Dp$ to pause for a rest. The mean time t_0 spent by the walker at the origin grows as [J.-P. Bouchaud and A. Georges, Phys. Rep. **195**, 128 (1990)]

$$t_0 \sim \begin{cases} t^{\frac{1}{2}} & (D = 1) \\ \ln t & (D = 2) \\ \text{constant} & (D = 3). \end{cases} \quad (10.135)$$

The numerical results indicate a similar scaling for the wandering of the beam center $\overline{\langle x \rangle^2}$ in $d = D + 1$ dimensions, for $d = 2$ and $d = 3$. We now show that this equivalence is not coincidental; moreover, it strongly suggests that $d_u = 3$ is a critical dimension for directed waves in random media. To this end, let us consider a continuum version of Eq. (10.133), which in general dimensions takes the form

$$W_2(\mathbf{r}, t + 1) = W_2(\mathbf{r}, t) + \nabla^2 \left[W_2 \left(1 - \epsilon \delta^D(\mathbf{r}) + \cdots \right) \right]. \quad (10.136)$$

The asymptotic solution for $\epsilon = 0$ is just a Gaussian packet of width $\overline{r^2} = 2t$. We can next perform a perturbative calculation in ϵ. However, simple dimensional analysis shows the corrections scale as powers of $\epsilon/r^D \sim \epsilon t^{-D/2}$, and thus

$$\lim_{t \to \infty} W_2(\mathbf{r}, t) = \frac{1}{(4\pi t)^{D/2}} \exp\left(-\frac{r^2}{4t}\right) \left[1 + \mathcal{O}\left(\epsilon t^{-(d-1)/2}\right)\right]. \tag{10.137}$$

Applying the above results to the continuum version of Eq. (10.133) gives

$$\overline{\langle x \rangle_{t+1}^2} - \overline{\langle x \rangle_t^2} = \frac{1}{2} - \frac{1}{4} \sum_r [W_2(r, t+1) - W_2(r, t)] \, r^2$$

$$\simeq \frac{1}{2} - \frac{1}{4} \int d^D \mathbf{r} \, r^2 \nabla^2 \left[W_2 \left(1 - \epsilon \delta^D(\mathbf{r}) + \cdots\right)\right] \tag{10.138}$$

$$\simeq \frac{1}{2} - \frac{1}{2} \int d^D \mathbf{r} W_2 \left(1 - \epsilon \delta^D(\mathbf{r})\right) = \epsilon W_2(0, t).$$

Summing both sides of this equation over t, one finds

$$\overline{\langle x(t+1) \rangle^2} = \epsilon \sum_{t'=0}^{t} W_2(0, t') \approx \int_0^t dt' (4\pi t')^{-D/2}. \tag{10.139}$$

The final integral is proportional to the time a random walker spends at the origin, and reproduces the results in Eq. (10.135).

We can also regard $W_2(r, t)$ as a probability distribution function for the relative coordinate between two interacting random walkers. In this interpretation, the value of ϵ in Eq. (10.133) parameterizes the strength of a contact interaction between the walkers. If $\epsilon = 0$, the walkers do not interact at all; if $\epsilon = 1$, the walkers bind on contact. According to Eq. (10.139), the wandering of the beam center $\overline{\langle x(t) \rangle^2}$ is proportional to the mean number of times that the paths of these walkers intersect during time t. If $\epsilon = 0$, the number of intersections during time t obeys the scaling law in Eq. (10.135), since in this case, the relative coordinate between the walkers performs a simple random walk. Numerical results indicate that the same scaling law applies when $0 < \epsilon < 1$: the contact attraction does *not* affect the asymptotic properties of the random walk. In summary, three classes of behavior are possible in this model. For $\epsilon = 0$, i.e. no randomness, the incoming beam stays centered at the origin, while its width grows diffusively. For $0 < \epsilon < 1$, the beam center, $\overline{\langle x \rangle^2}$, also fluctuates, but with a dimension dependent behavior as in Eq. (10.135). In the limit of $\epsilon = 1$, interference phenomena disappear completely. In this case, the beam width is zero and the beam center performs a simple random walk.

We conclude by comparing the situation here to that of the DPRM. In the replica approach to DPRM, the n-th moment of the weight $W(x, t)$ is obtained from the statistics of n directed paths. Disorder-averaging produces an attractive interaction between these paths with the result that they may form a bound state. In $d \leq 2$, any amount of randomness (and hence attraction) leads to the formation of a bound state. The behavior of the bound state energy can then be used to extract an exponent of $\zeta = 2/3$ for superdiffusive wandering. By contrast, the replicated paths encountered in the directed wave problem (such as the two paths considered for Eq. 10.131), although interacting, cannot form a bound state, as such is inconsistent with unitarity. This result also emerges in a natural way from our model of directed waves. In $d = 2$, for instance, it is easy to check that $W_2(\mathbf{r}) \sim (1 - \epsilon \delta_{r,0})^{-1}$ is the eigenstate of largest eigenvalue for the evolution of the relative coordinate. Hence, as $t \to \infty$, for randomness δ-correlated in space and time, there is no bound state. This conclusion holds in $d \geq 2$ and is not modified by short-range correlations in the randomness. The probability-conserving nature of Eq. (10.133) is crucial in this regard [F. Igloi, Europhys. Lett. **16**, 171 (1991)] as it precludes a $u\delta^D(\mathbf{r})$ attraction in Eq. (10.136). Small perturbations that violate the conservation of probability lead to the formation of a bound state. In the language of the renormalization group, the scaling of directed waves in random media is governed by a fixed point that is unstable with respect to changes that do not preserve a strictly unitary evolution.

In subsequent studies a number of authors have obtained additional results from the random S-matrix model. Friedberg and Yu [R. Friedberg and Y.-K. Yu, Phys. Rev. E **49**, 5755 (1994)] calculated the leading terms in the scaling laws for the beam center in $d \geq 2$, and also the next-order corrections. The analytical results are in agreement with those presented above. Cule and Shapir [D. Cule and Y. Shapir, Phys. Rev. B **50**, 5119 (1994)] extended the methods of this section to compute the higher moments of the probability distribution for directed waves in random media. If this probability distribution is multifractal, as claimed in [J.-P. Bouchaud, D. Touati, and D. Sornette, Phys. Rev. Lett. **68**, 1787 (1992)] the higher moments should obey new scaling laws whose exponents are not simply related to those of the lower moments. Within the framework of the S-matrix model, Cule and Shapir did not find evidence for multifractal scaling, while suggesting that certain aspects of the scaling behavior may be sensitive to details of the unitary time evolution. The above model has also found applications in diverse contexts such as force chains in granular media, and surface of quantum Hall multilayers [M. Lewandowska, H. Mathur, and Y.-K. Yu, Phys. Rev. E **64**, 026107 (2001)].

Solutions to selected problems from chapter 1

1. *The binary alloy*: A binary alloy (as in β brass) consists of N_A atoms of type A, and N_B atoms of type B. The atoms form a simple cubic lattice, each interacting only with its six nearest neighbors. Assume an attractive energy of $-J$ $(J > 0)$ between like neighbors $A - A$ and $B - B$, but a repulsive energy of $+J$ for an $A - B$ pair.

   (a) What is the minimum energy configuration, or the state of the system at zero temperature?

   - *The minimum energy configuration has as few A–B bonds as possible. Thus, at zero temperature atoms A and B phase separate, e.g. as indicated below.*

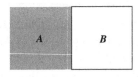

   (b) Estimate the total interaction energy assuming that the atoms are randomly distributed among the N sites; i.e. each site is occupied independently with probabilities $p_A = N_A/N$ and $p_B = N_B/N$.

   - *In a mixed state, the average energy is obtained from*

$$E = \text{(number of bonds)} \times \text{(average bond energy)}$$

$$= 3N \cdot \left(-Jp_A^2 - Jp_B^2 + Jp_Ap_B\right)$$

$$= -3JN \left(\frac{N_A - N_B}{N}\right)^2.$$

   (c) Estimate the mixing entropy of the alloy with the same approximation. Assume $N_A, N_B \gg 1$.

   - *From the number of ways of randomly mixing N_A and N_B particles, we obtain the mixing entropy of*

$$S = k_\text{B} \ln \left(\frac{N!}{N_A!N_B!}\right).$$

Using Stirling's approximation for large N ($\ln N! \approx N \ln N - N$), the above expression can be written as

$$S \approx k_{\mathrm{B}} \left(N \ln N - N_A \ln N_A - N_B \ln N_B \right) = -N k_{\mathrm{B}} \left(p_A \ln p_A + p_B \ln p_B \right).$$

(d) Using the above, obtain a free energy function $F(x)$, where $x = (N_A - N_B)/N$. Expand $F(x)$ to the fourth order in x, and show that the requirement of convexity of F breaks down below a critical temperature T_c. For the remainder of this problem use the expansion obtained in (d) in place of the full function $F(x)$.

- *In terms of $x = p_A - p_B$, the free energy can be written as*

$$F = E - TS$$

$$= -3JNx^2 + N k_{\mathrm{B}} T \left\{ \left(\frac{1+x}{2} \right) \ln \left(\frac{1+x}{2} \right) + \left(\frac{1-x}{2} \right) \ln \left(\frac{1-x}{2} \right) \right\}.$$

Expanding about $x = 0$ to fourth order gives

$$F \simeq -N k_{\mathrm{B}} T \ln 2 + N \left(\frac{k_{\mathrm{B}} T}{2} - 3J \right) x^2 + \frac{N k_{\mathrm{B}} T}{12} x^4.$$

Clearly, the second derivative of F,

$$\frac{\partial^2 F}{\partial x^2} = N \left(k_{\mathrm{B}} T - 6J \right) + N k_{\mathrm{B}} T x^2,$$

becomes negative for T small enough. Upon decreasing the temperature, F becomes concave first at $x = 0$, at a critical temperature $T_c = 6J/k_{\mathrm{B}}$.

(e) Sketch $F(x)$ for $T > T_c$, $T = T_c$, and $T < T_c$. For $T < T_c$ there is a range of compositions $x < |x_{\mathrm{sp}}(T)|$ where $F(x)$ is not convex and hence the composition is locally unstable. Find $x_{\mathrm{sp}}(T)$.

- *The function $F(x)$ is concave if $\partial^2 F/\partial x^2 < 0$, i.e. if*

$$x^2 < \left(\frac{6J}{k_{\mathrm{B}} T} - 1 \right).$$

This occurs for $T < T_c$, at the spinodal line given by

$$x_{\mathrm{sp}}(T) = \sqrt{\frac{6J}{k_{\mathrm{B}} T} - 1},$$

as indicated by the dashed line in the figure.

(f) The alloy globally minimizes its free energy by separating into A rich and B rich phases of compositions $\pm x_{\mathrm{eq}}(T)$, where $x_{\mathrm{eq}}(T)$ minimizes the function $F(x)$. Find $x_{\mathrm{eq}}(T)$.

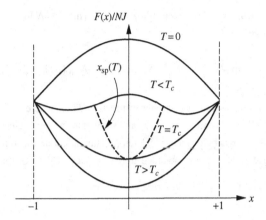

- *Setting the first derivative of* $dF(x)/dx = Nx\left\{(k_B T - 6J) + k_B T x^2/3\right\}$ *to zero yields the equilibrium value of*

$$
x_{eq}(T) = \begin{cases} \pm\sqrt{3}\sqrt{\dfrac{6J}{k_B T} - 1} & \text{for } T < T_c \\[2mm] 0 & \text{for } T > T_c. \end{cases}
$$

(g) In the (T, x) plane sketch the phase separation boundary $\pm x_{eq}(T)$; and the so-called spinodal line $\pm x_{sp}(T)$. (The spinodal line indicates onset of metastability and hysteresis effects.)

- *The spinodal and equilibrium curves are indicated in the figure below. In the interval between the two curves, the system is locally stable, but globally unstable. The formation of ordered regions in this regime requires nucleation, and is very slow. The dashed area is locally unstable, and the system easily phase separates to regions rich in A and B.*

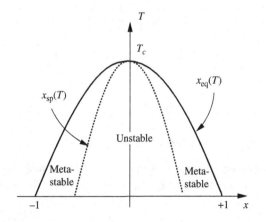

2. *The Ising model of magnetism:* The local environment of an electron in a crystal sometimes forces its spin to stay parallel or anti-parallel to a given lattice direction. As a model of magnetism in such materials we denote the direction of the spin by a single variable $\sigma_i = \pm 1$ (an Ising spin). The energy of a configuration $\{\sigma_i\}$ of spins is then given by

$$\mathcal{H} = \frac{1}{2} \sum_{i,j=1}^{N} J_{ij} \sigma_i \sigma_j - h \sum_i \sigma_i;$$

where h is an external magnetic field, and J_{ij} is the interaction energy between spins at sites i and j.

(a) For N spins we make the drastic *approximation* that the interaction between all spins is the same, and $J_{ij} = -J/N$ (the equivalent neighbor model). Show that the energy can now be written as $E(M, h) = -N[Jm^2/2 + hm]$, with a magnetization $m = \sum_{i=1}^{N} \sigma_i/N = M/N$.

- *For $J_{ij} = -J/N$, the energy of each configuration is only a function of $m = \sum_i \sigma_i/N$, given by*

$$E(M, h) = -\frac{J}{2N} \sum_{i,j=1}^{N} \sigma_i \sigma_j - h \sum_{i=1}^{N} \sigma_i$$

$$= -N\frac{J}{2} \left(\sum_{i=1}^{N} \sigma_i/N \right) \left(\sum_{j=1}^{N} \sigma_j/N \right) - Nh \left(\sum_{i=1}^{N} \sigma_i/N \right)$$

$$= -N \left(\frac{J}{2} m^2 + hm \right).$$

(b) Show that the partition function $Z(h, T) = \sum_{\{\sigma_i\}} \exp(-\beta\mathcal{H})$ can be rewritten as $Z = \sum_M \exp[-\beta F(m, h)]$; with $F(m, h)$ easily calculated by analogy to problem (1). For the remainder of the problem work only with $F(m, h)$ expanded to fourth order in m.

- *Since the energy depends only on the number of up spins N_+, and not on their configuration, we have*

$$Z(h, T) = \sum_{\{\sigma_i\}} \exp(-\beta\mathcal{H})$$

$$= \sum_{N_+=0}^{N} (\text{number of configurations with } N_+ \text{ fixed}) \cdot \exp[-\beta E(M, h)]$$

$$= \sum_{N_+=0}^{N} \left[\frac{N!}{N_+!(N-N_+)!} \right] \exp[-\beta E(M, h)]$$

$$= \sum_{N_+=0}^{N} \exp\left\{ -\beta \left[E(M, h) - k_B T \ln \left(\frac{N!}{N_+!(N-N_+)!} \right) \right] \right\}$$

$$= \sum_M \exp[-\beta F(m, h)].$$

By analogy to the previous problem ($N_+ \leftrightarrow N_A$, $m \leftrightarrow x$, $J/2 \leftrightarrow 3J$),

$$\frac{F(m, h)}{N} = -k_{\mathrm{B}}T \ln 2 - hm + \frac{1}{2}(k_{\mathrm{B}}T - J)m^2 + \frac{k_{\mathrm{B}}T}{12}m^4 + \mathcal{O}(m^6).$$

(c) By saddle point integration show that the actual free energy $F(h, T) = -k_{\mathrm{B}}T \ln Z(h, T)$ is given by $F(h, T) = \min[F(m, h)]_m$. When is the saddle point method valid? Note that $F(m, h)$ is an analytic function but not convex for $T < T_c$, while the true free energy $F(h, T)$ is convex but becomes non-analytic due to the minimization.

- *Let $m^*(h, T)$ minimize $F(m, h)$, i.e. $\min[F(m, h)]_m = F(m^*, h)$. Since there are N terms in the sum for Z, we have the bounds*

$$\exp(-\beta F(m^*, h)) \leq Z \leq N \exp(-\beta F(m^*, h)),$$

or, taking the logarithm and dividing by $-\beta N$,

$$\frac{F(m^*, h)}{N} \geq \frac{F(h, T)}{N} \geq \frac{F(m^*, h)}{N} + \frac{\ln N}{N}.$$

Since F is extensive, we have therefore

$$\frac{F(m^*, h)}{N} = \frac{F(h, T)}{N}$$

in the $N \to \infty$ limit.

(d) For $h = 0$ find the critical temperature T_c below which spontaneous magnetization appears; and calculate the magnetization $\overline{m}(T)$ in the low temperature phase.

- *From the definition of the actual free energy, the magnetization is given by*

$$\overline{m} = -\frac{1}{N}\frac{\partial F(h, T)}{\partial h},$$

i.e.

$$\overline{m} = -\frac{1}{N}\frac{dF(m, h)}{dh} = -\frac{1}{N}\left\{ \frac{\partial F(m, h)}{\partial h} + \frac{\partial F(m, h)}{\partial m}\frac{\partial m}{\partial h} \right\}.$$

Thus, if m^* minimizes $F(m, h)$, i.e. if $\partial F(m, h)/\partial m|_{m^*} = 0$, then

$$\overline{m} = -\frac{1}{N}\frac{\partial F(m, h)}{\partial h}\bigg|_{m^*} = m^*.$$

For $h = 0$,

$$m^{*2} = \frac{3(J - k_{\mathrm{B}}T)}{k_{\mathrm{B}}T},$$

yielding

$$T_c = \frac{J}{k_{\mathrm{B}}},$$

and

$$\overline{m} = \begin{cases} \pm\sqrt{\dfrac{3\,(J - k_{\mathrm{B}}T)}{k_{\mathrm{B}}T}} & \text{if } T < T_c \\[4mm] 0 & \text{if } T > T_c. \end{cases}$$

(e) Calculate the singular (non-analytic) behavior of the response functions

$$C = \left.\frac{\partial E}{\partial T}\right|_{h=0} \quad \text{and} \quad \chi = \left.\frac{\partial \overline{m}}{\partial h}\right|_{h=0}.$$

- *The heat capacity is given by*

$$C = \left.\frac{\partial E}{\partial T}\right|_{h=0,\,m=m^*} = -\frac{NJ}{2}\frac{\partial m^{*2}}{\partial T} = \begin{cases} \dfrac{3NJT_c}{2T^2} & \text{if } T < T_c \\[4mm] 0 & \text{if } T > T_c, \end{cases}$$

i.e. $\alpha = 0$, indicating a discontinuity. To calculate the susceptibility, we use

$$h = (k_{\mathrm{B}}T - J)\,\overline{m} + \frac{k_{\mathrm{B}}T}{3}\overline{m}^3.$$

Taking a derivative with respect to h,

$$1 = \left(k_{\mathrm{B}}T - J + k_{\mathrm{B}}T\overline{m}^2\right)\frac{\partial \overline{m}}{\partial h},$$

which gives

$$\chi = \left.\frac{\partial \overline{m}}{\partial h}\right|_{h=0} = \begin{cases} \dfrac{1}{2k_B\,(T_c - T)} & \text{if } T < T_c \\[4mm] \dfrac{1}{k_B\,(T - T_c)} & \text{if } T > T_c. \end{cases}$$

From the above expression we obtain $\gamma_\pm = 1$, and $A_+/A_- = 2$.

3. *The lattice-gas model:* Consider a gas of particles subject to a Hamiltonian

$$\mathcal{H} = \sum_{i=1}^{N}\frac{\vec{p}_i{}^2}{2m} + \frac{1}{2}\sum_{i,j}\mathcal{V}(\vec{r}_i - \vec{r}_j), \quad \text{in a volume } V.$$

(a) Show that the grand partition function Ξ can be written as

$$\Xi = \sum_{N=0}^{\infty}\frac{1}{N!}\left(\frac{e^{\beta\mu}}{\lambda^3}\right)^N \int \prod_{i=1}^{N}\mathrm{d}^3\vec{r}_i \exp\left[-\frac{\beta}{2}\sum_{i,j}\mathcal{V}(\vec{r}_i - \vec{r}_j)\right].$$

- *The grand partition function is calculated as*

$$
\Xi = \sum_{N=0}^{\infty} \frac{e^{N\beta\mu}}{N!} Z_N
$$

$$
= \sum_{N=0}^{\infty} \frac{e^{N\beta\mu}}{N!} \int \prod_{i=1}^{N} \frac{d^3 p_i d^3 r_i}{h^3} e^{-\beta \mathcal{H}}
$$

$$
= \sum_{N=0}^{\infty} \frac{e^{N\beta\mu}}{N!} \left(\prod_{i=1}^{N} \int \frac{d^3 p_i}{h^3} e^{-\beta p_i^2/2m} \right) \int \prod_{i=1}^{N} d^3 r_i \exp\left(-\frac{\beta}{2} \sum_{i,j} V_{ij} \right)
$$

$$
= \sum_{N=0}^{\infty} \frac{1}{N!} \left(\frac{e^{N\beta}}{\lambda^3} \right)^N \int \prod_{i=1}^{N} d^3 r_i \exp\left(-\frac{\beta}{2} \sum_{i,j} V_{ij} \right),
$$

where $\lambda^{-1} = \sqrt{2\pi m k_B T}/h$.

(b) The volume V is now subdivided into $\mathcal{N} = V/a^3$ cells of volume a^3, with the spacing a chosen small enough so that each cell α is either empty or occupied by one particle; i.e. the cell occupation number n_α is restricted to 0 or 1 ($\alpha = 1, 2, \cdots, \mathcal{N}$). After approximating the integrals $\int d^3 \vec{r}$ by sums $a^3 \sum_{\alpha=1}^{\mathcal{N}}$, show that

$$
\Xi \approx \sum_{\{n_\alpha=0,1\}} \left(\frac{e^{\beta\mu} a^3}{\lambda^3} \right)^{\sum_\alpha n_\alpha} \exp\left[-\frac{\beta}{2} \sum_{\alpha,\beta=1}^{\mathcal{N}} n_\alpha n_\beta V(\vec{r}_\alpha - \vec{r}_\beta) \right].
$$

- *Since*

$$
\int \prod_{i=1}^{N} d^3 r_i \exp\left(-\frac{\beta}{2} \sum_{i,j} V_{ij} \right) \approx a^{3N} \sum' \exp\left\{ -\frac{\beta}{2} \sum_{\alpha,\beta=1}^{\mathcal{N}} n_\alpha n_\beta V(\vec{r}_\alpha - \vec{r}_\beta) \right\} \cdot N!,
$$

where the primed sum is over the configurations $\{n_\alpha = 0, 1\}$ *with fixed N, and*

$$
N = \sum_{\alpha=1}^{\mathcal{N}} n_\alpha,
$$

we have

$$
\Xi \approx \sum_{\{n_\alpha=0,1\}} \left(\frac{e^{\beta\mu} a^3}{\lambda^3} \right)^{\sum_\alpha n_\alpha} \exp\left\{ -\frac{\beta}{2} \sum_{\alpha,\beta=1}^{\mathcal{N}} n_\alpha n_\beta V(\vec{r}_\alpha - \vec{r}_\beta) \right\}.
$$

(c) By setting $n_\alpha = (1 + \sigma_\alpha)/2$ and approximating the potential by $V(\vec{r}_\alpha - \vec{r}_\beta) = -J/\mathcal{N}$, show that this model is identical to the one studied in problem (2). What does this imply about the behavior of this imperfect gas?

- *With* $n_\alpha = (1 + \sigma_\alpha)/2$, *and* $V(\vec{r}_\alpha - \vec{r}_\beta) = -J/\mathcal{N}$,

$$
\Xi = \sum_{\{n_\alpha=0,1\}} \exp\left\{ \left(\beta\mu + 3\ln\frac{a}{\lambda} \right) \sum_{\alpha=1}^{\mathcal{N}} \left(\frac{1+\sigma_\alpha}{2} \right) + \frac{\beta J}{2\mathcal{N}} \sum_{\alpha,\beta=1}^{\mathcal{N}} \left(\frac{1+\sigma_\alpha}{2} \right) \left(\frac{1+\sigma_\beta}{2} \right) \right\}.
$$

Setting $m \equiv \sum_\alpha \sigma_\alpha / \mathcal{N}$, $h' = \frac{1}{2} \left(\mu + \frac{3}{\beta} \ln \frac{a}{\lambda} + \frac{J}{2} \right)$, and $J' = J/4$, the grand partition function is written

$$\Xi = \text{const.} \sum_{\{n_\alpha = 0, 1\}} \exp \left\{ \mathcal{N} \beta \left(J' m^2 / 2 + h' m \right) \right\}.$$

The phase diagram of the lattice gas can thus be mapped onto the phase diagram of the Ising model of problem 2. In particular, at a chemical potential μ such that $h' = 0$, there is a continuous "condensation" transition at a critical temperature $T_c = J/4k_B$. (Note that

$$m = \sum_\alpha \sigma_\alpha / \mathcal{N} = \sum_\alpha (2 n_\alpha - 1) / \mathcal{N} = 2 a^3 \rho - 1,$$

where $\rho = N/V$ is the density of the gas.)

• The manifest equivalence between these three systems is a straightforward consequence of their mapping onto the same (Ising) Hamiltonian. However, there is a more subtle equivalence relating the critical behavior of systems that cannot be so easily mapped onto each other due to the universality principle.

Solutions to selected problems from chapter 2

1. *Cubic invariants:* When the order parameter m, goes to zero discontinuously, the phase transition is said to be first order (discontinuous). A common example occurs in systems where symmetry considerations do not exclude a cubic term in the Landau free energy, as in

$$\beta \mathcal{H} = \int d^d \mathbf{x} \left[\frac{K}{2} (\nabla m)^2 + \frac{t}{2} m^2 + cm^3 + um^4 \right] \qquad (K, c, u > 0).$$

(a) By plotting the energy density $\Psi(m)$, for uniform m at various values of t, show that as t is reduced there is a discontinuous jump to $\overline{m} \neq 0$ for a positive $\overline{t}$ in the saddle-point approximation.

- *To simplify the algebra, let us rewrite the energy density $\Psi(m)$, for uniform m, in terms of the rescaled quantity*

$$m_r = \frac{u}{c} m.$$

In this way, we can eliminate the constant parameters c, and u, to get the expression of the energy density as

$$\Psi_r(m_r) = \frac{1}{2} t_r m_r^2 + m_r^3 + m_r^4,$$

where we have defined

$$\Psi_r = \left(\frac{c^4}{u^3} \right) \Psi, \quad \text{and} \quad t_r = \left(\frac{u}{c^2} \right) t.$$

To obtain the extrema of Ψ_r, we set the first derivative with respect to m_r to zero, i.e.

$$\frac{d\Psi_r(m_r)}{dm_r} = m_r \left(t_r + 3m_r + 4m_r^2 \right) = 0.$$

The trivial solution of this equation is $m_r^* = 0$. But if $t_r \leq 9/16$, the derivative vanishes also at $m_r^* = (-3 \pm \sqrt{9 - 16t_r})/8$. Provided that $t_r > 0$, $m_r^* = 0$ is

268

a minimum of the function $\Psi_r(m_r)$*. In addition, if* $t_r < 9/16$, $\Psi_r(m_r)$ *has another minimum at*

$$m_r^* = -\frac{3+\sqrt{9-16t_r}}{8},$$

and a maximum, located in between the two minima, at

$$m_r^* = \frac{-3+\sqrt{9-16t_r}}{8}.$$

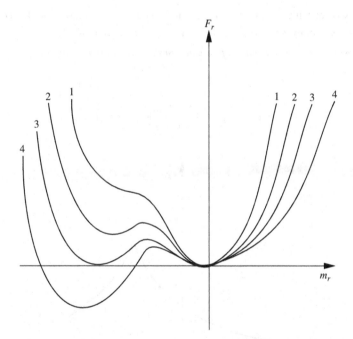

The figure above depicts the behavior of $\Psi_r(m_r)$ *for different values of* t_r.

1. *For* $t_r > 9/16$, *there is only one minimum* $m_r^* = 0$.
2. *For* $0 < \bar{t}_r < t_r < 9/16$, *there are two minima, but* $\Psi_r(m_r^*) > \Psi_r(0) = 0$.
3. *For* $0 < t_r = \bar{t}_r$, $\Psi_r(m_r^*) = \Psi_r(0) = 0$.
4. *For* $0 < t_r < \bar{t}_r$, $\Psi_r(m_r^*) < \Psi_r(0) = 0$.

The discontinuous transition occurs when the local minimum at $m_r^* < 0$ *becomes the absolute minimum. There is a corresponding jump of* m_r, *from* $m_r^* = 0$ *to* $m_r^* = \bar{m}_r$, *where* $\bar{m}_r = m_r^*(t_r = \bar{t}_r)$.

(b) By writing down the two conditions that $\bar{m}$ and $\bar{t}$ must satisfy at the transition, solve for $\bar{m}$ and $\bar{t}$.

- *To determine $\overline{m}_r$ and $\overline{t}_r$, we have to simultaneously solve the equations*

$$\frac{d\Psi_r(m_r)}{dm_r} = 0, \quad \text{and} \quad \Psi_r(m_r) = \Psi_r(0) = 0.$$

Excluding the trivial solution $m_r^* = 0$, from

$$\begin{cases} t_r + 3m_r + 4m_r^2 = 0 \\ \dfrac{t_r}{2} + m_r + m_r^2 = 0, \end{cases}$$

we obtain $\overline{t}_r = -m_r = 1/2$, or in the original units,

$$\overline{t} = \frac{c^2}{2u}, \quad \text{and} \quad \overline{m} = -\frac{c}{2u}.$$

(c) Note that the correlation length ξ is related to the curvature of $\Psi(m)$ at its minimum by $K\xi^{-2} = \partial^2\Psi/\partial m^2|_{eq}$. Plot ξ as a function of t.

- *The equilibrium value of $m = m_{eq}$ in the original units equals to*

$$m_{eq} = \begin{cases} 0 & \text{for} \quad t > \overline{t} = \dfrac{c^2}{2u}, \\ -\left(\dfrac{c}{u}\right)\dfrac{3+\sqrt{9-16ut/c^2}}{8} & \text{for} \quad t < \overline{t}. \end{cases}$$

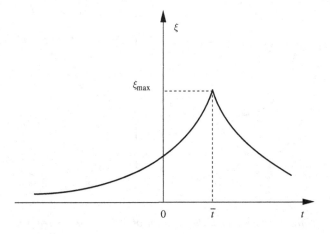

The correlation length ξ, is related to the curvature of $\Psi(m)$ at its equilibrium value by

$$K\xi^{-2} = \left.\frac{\partial^2\Psi}{\partial m^2}\right|_{m_{eq}} = t + 6cm_{eq} + 12um_{eq}^2,$$

which is equal to

$$\xi = \begin{cases} \left(\dfrac{K}{t}\right)^{1/2} & \text{if} \quad t > \overline{t}, \\ \left(-\dfrac{K}{2t+3cm_{eq}}\right)^{1/2} & \text{if} \quad t < \overline{t}. \end{cases}$$

(To arrive to the last expression, we have used $\mathrm{d}\Psi(m)/\mathrm{d}m|_{m=m_{eq}} = 0$.) *Note that, as indicated in the figure, the correlation length* ξ *is finite at the discontinuous phase transition, attaining a maximum value of*

$$\xi_{\max} = \xi(\bar{t}) = \frac{\sqrt{2Ku}}{c}.$$

2. *Tricritical point:* By tuning an additional parameter, a second order transition can be made first order. The special point separating the two types of transitions is known as a tricritical point, and can be studied by examining the Landau–Ginzburg Hamiltonian

$$\beta\mathcal{H} = \int \mathrm{d}^d\mathbf{x}\left[\frac{K}{2}(\nabla m)^2 + \frac{t}{2}m^2 + um^4 + vm^6 - hm\right],$$

where u can be positive or negative. For $u < 0$, a positive v is necessary to insure stability.

(a) By sketching the energy density $\Psi(m)$, for various t, show that in the saddle point approximation there is a first order transition for $u < 0$ and $h = 0$.

* *If we consider $h = 0$, the energy density $\Psi(m)$, for uniform m, is*

$$\Psi(m) = \frac{t}{2}m^2 + um^4 + vm^6.$$

As in the previous problem, to obtain the extrema of Ψ, let us set the first derivative with respect to m to zero. Again, provided that $t > 0$, $\Psi(m)$ has a minimum at $m^* = 0$. But the derivative also vanishes for other non-zero values of m as long as certain conditions are satisfied. In order to find them, we have to solve the following equation

$$t + 4um^2 + 6vm^4 = 0,$$

from which,

$$m^{*2} = -\frac{u}{3v} \pm \frac{\sqrt{4u^2 - 6tv}}{6v}.$$

Thus, we have real and positive solutions provided that

$$u < 0, \quad \text{and} \quad t < \frac{2u^2}{3v}.$$

Under these conditions $\Psi(m)$ has another two minima at

$$m^{*2} = \frac{|u|}{3v} + \frac{\sqrt{4u^2 - 6tv}}{6v},$$

and two maxima at

$$m^{*2} = \frac{|u|}{3v} - \frac{\sqrt{4u^2 - 6tv}}{6v},$$

as depicted in the figure below.

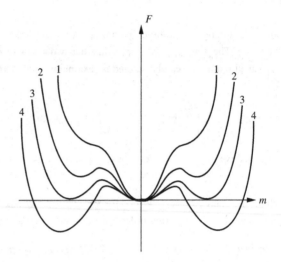

The different behaviors of the function $\Psi(m)$ *are as follows:*

1. *For* $t > 2u^2/3v$, *there is only one minimum* $m^* = 0$.
2. *For* $0 < \bar{t} < t < 2u^2/3v$, *there are three minima, but* $\Psi(\pm m^*) > \Psi(0) = 0$.
3. *For* $0 < t = \bar{t}$, $\Psi(\pm m^*) = \Psi(0) = 0$.
4. *For* $0 < t < \bar{t}$, $\Psi(\pm m^*) < \Psi(0) = 0$.

There is thus a discontinuous phase transition for $u < 0$, *and* $t = \bar{t}(u)$.

(b) Calculate $\bar{t}$ and the discontinuity $\overline{m}$ at this transition.

- *To determine* $\bar{t}$, *and* $\overline{m} = m^*(t = \bar{t})$, *we again have to simultaneously solve the equations*

$$\frac{d\Psi(m)}{dm^2} = 0, \quad \text{and} \quad \Psi(m^2) = \Psi(0) = 0,$$

or equivalently,

$$\begin{cases} \dfrac{t}{2} + 2um^2 + 3vm^4 = 0 \\ \dfrac{t}{2} + um^2 + vm^4 = 0, \end{cases}$$

from which we obtain

$$\bar{t} = \frac{u^2}{2v}, \quad \text{and} \quad \overline{m}^2 = -\frac{u}{2v} = \frac{|u|}{2v}.$$

(c) For $h = 0$ and $v > 0$, plot the phase boundary in the (u, t) plane, identifying the phases, and order of the phase transitions.

- *In the (u, t) plane, the line $t = u^2/2v$ for $u < 0$ is a first order phase transition boundary. In addition, the line $t = 0$ for $u > 0$ defines a second order phase transition boundary, as indicated in the figure.*

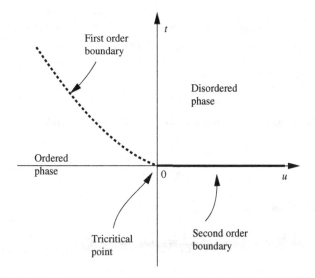

(d) The special point $u = t = 0$, separating first and second order phase boundaries, is a *tricritical* point. For $u = 0$, calculate the tricritical exponents β, δ, γ, and α, governing the singularities in magnetization, susceptibility, and heat capacity. (Recall: $C \propto t^{-\alpha}$; $\overline{m}(h = 0) \propto t^{\beta}$; $\chi \propto t^{-\gamma}$; and $\overline{m}(t = 0) \propto h^{1/\delta}$.)

- *For $u = 0$, let us calculate the tricritical exponents α, β, γ, and δ. In order to calculate α and β, we set $h = 0$, so that*

$$\Psi(m) = \frac{t}{2}m^2 + vm^6.$$

Thus from

$$\frac{\partial \Psi}{\partial m}\overline{m} = \overline{m}\left(t + 6v\overline{m}^4\right) = 0,$$

we obtain,

$$\overline{m} = \begin{cases} 0 & \text{for } t > \overline{t} = 0, \\ \left(-\dfrac{t}{6v}\right)^{1/4} & \text{for } t < 0, \end{cases}$$

resulting in,

$$\overline{m}(h=0) \propto t^{\beta}, \quad \text{with} \quad \beta = \frac{1}{4}.$$

The corresponding free energy density scales as

$$\Psi(\overline{m}) \sim \overline{m}^6 \propto (-t)^{3/2}.$$

The tricritical exponent α characterizes the non-analytic behavior of the heat capacity $C \sim (\partial^2 \Psi / \partial T^2)|_{h=0,\overline{m}}$, and since $t \propto (T-T_c)$,

$$C \sim \left. \frac{\partial^2 \Psi}{\partial t^2} \right|_{h=0,\overline{m}} \propto t^{-\alpha}, \quad \text{with} \quad \alpha = \frac{1}{2}.$$

To calculate the tricritical exponent δ, we set $t=0$ while keeping $h \neq 0$, so that

$$\Psi(m) = vm^6 - hm.$$

Thus from

$$\frac{\partial \Psi}{\partial m} \overline{m} = 6v\overline{m}^5 - h = 0,$$

we obtain,

$$\overline{m} \propto h^{1/\delta}, \quad \text{with} \quad \delta = 5.$$

Finally, for $h \neq 0$ and $t \neq 0$,

$$\frac{\partial \Psi}{\partial m} \overline{m} = t\overline{m} + 6v\overline{m}^5 - h = 0,$$

so that the susceptibility scales as

$$\chi = \left. \frac{\partial \overline{m}}{\partial h} \right|_{h=0} \propto |t|^{-1}, \quad \text{for both} \quad t<0 \quad \text{and} \quad t>0,$$

i.e. with the exponents $\gamma_{\perp} = 1$.

3. *Transverse susceptibility:* An n-component magnetization field $\vec{m}(\mathbf{x})$ is coupled to an external field $\vec{h}$ through a term $-\int d^d x \, \vec{h} \cdot \vec{m}(\mathbf{x})$ in the Hamiltonian $\beta\mathcal{H}$. If $\beta\mathcal{H}$ for $\vec{h}=0$ is invariant under rotations of $\vec{m}(\mathbf{x})$, then the free energy density $(f = -\ln Z/V)$ only depends on the absolute value of $\vec{h}$; i.e. $f(\vec{h}) = f(h)$, where $h = |\vec{h}|$.

(a) Show that $m_\alpha = \langle \int d^d x m_\alpha(\mathbf{x}) \rangle / V = -h_\alpha f'(h)/h$.

- The magnetic work is the product of the magnetic field and the magnetization density, and appears as the argument of the exponential weight in the (Gibbs) canonical ensemble. We can thus "lower" the magnetization $M = \int d^d x \, m_\alpha(\mathbf{x})$ "inside the average" by taking derivatives of the (Gibbs) partition function with respect to h_α, as

$$m_\alpha = \frac{1}{V}\left\langle \int d^d x\, m_\alpha(\mathbf{x})\right\rangle = \frac{1}{V}\frac{\int \mathcal{D}\mathbf{m}(\mathbf{x})\left(\int d^d x' m_\alpha(\mathbf{x}')\right)e^{-\beta\mathcal{H}}}{\int \mathcal{D}\mathbf{m}(\mathbf{x})\,e^{-\beta\mathcal{H}}}$$

$$= \frac{1}{V}\frac{1}{Z}\frac{1}{\beta}\frac{\partial}{\partial h_\alpha}\int \mathcal{D}\mathbf{m}(\mathbf{x})\,e^{-\beta\mathcal{H}} = \frac{1}{\beta V}\frac{\partial}{\partial h_\alpha}\ln Z = -\frac{\partial f}{\partial h_\alpha}.$$

For an otherwise rotationally symmetric system, the (Gibbs) free energy depends only on the magnitude of $\vec{h}$, and using

$$\frac{\partial h}{\partial h_\alpha} = \frac{\partial \sqrt{h_\beta h_\beta}}{\partial h_\alpha} = \frac{1}{2}\frac{2\delta_{\alpha\beta}h_\beta}{\sqrt{h_\beta h_\beta}} = \frac{h_\alpha}{h},$$

we obtain

$$m_\alpha = -\frac{\partial f}{\partial h_\alpha} = -\frac{df}{dh}\frac{\partial h}{\partial h_\alpha} = -f'\frac{h_\alpha}{h}.$$

(b) Relate the susceptibility tensor $\chi_{\alpha\beta} = \partial m_\alpha/\partial h_\beta$ to $f''(h)$, $\vec{m}$, and $\vec{h}$.

- The susceptibility tensor is now obtained as

$$\chi_{\alpha\beta} = \frac{\partial m_\alpha}{\partial h_\beta} = \frac{\partial}{\partial h_\beta}\left(-\frac{h_\alpha}{h}f'(h)\right) = -\frac{\partial h_\alpha}{\partial h_\beta}\frac{1}{h}f' - \frac{\partial h^{-1}}{\partial h_\beta}h_\alpha f' - \frac{h_\alpha}{h}\frac{\partial f'}{\partial h_\beta}$$

$$= -\left(\delta_{\alpha\beta} - \frac{h_\alpha h_\beta}{h^2}\right)\frac{f'}{h} - \frac{h_\alpha h_\beta}{h^2}f''.$$

In order to express f' in terms of the magnetization, we take the magnitude of the result of part (a),

$$m = |f'(h)| = -f'(h)$$

from which we obtain

$$\chi_{\alpha\beta} = \left(\delta_{\alpha\beta} - \frac{h_\alpha h_\beta}{h^2}\right)\frac{m}{h} + \frac{h_\alpha h_\beta}{h^2}\frac{dm}{dh}.$$

(c) Show that the transverse and longitudinal susceptibilities are given by $\chi_t = m/h$ and $\chi_\ell = -f''(h)$, where m is the magnitude of $\vec{m}$.

- Since the matrix $\left(\delta_{\alpha\beta} - h_\alpha h_\beta/h^2\right)$ removes the projection of any vector along the magnetic field, we conclude

$$\begin{cases} \chi_\ell = -f''(h) = \dfrac{dm}{dh} \\ \chi_t = \dfrac{m}{h}. \end{cases}$$

Alternatively, we can choose the coordinate system such that $h_i = h\delta_{i1}$ ($i = 1, \ldots, d$), to get

$$\begin{cases} \chi_\ell = \chi_{11} = \left(\delta_{11} - \dfrac{h_1 h_1}{h^2} \right) \dfrac{m}{h} - \dfrac{h_1 h_1}{h^2} f''(h) = \dfrac{dm}{dh} \\ \chi_t = \chi_{22} = \left(\delta_{11} - \dfrac{h_2 h_2}{h^2} \right) \dfrac{m}{h} - \dfrac{h_2 h_2}{h^2} f''(h) = \dfrac{m}{h}. \end{cases}$$

(d) Conclude that χ_t diverges as $\vec{h} \to 0$, whenever there is a spontaneous magnetization. Is there any similar a priori reason for χ_ℓ to diverge?

- *Provided that $\lim_{h\to 0} m \neq 0$, the transverse susceptibility clearly diverges for $h \to 0$. There is no similar reason, on the other hand, for the longitudinal susceptibility to diverge. In the saddle point approximation of the Landau–Ginzburg model, for example, we have*

$$tm + 4um^3 + h = 0,$$

implying (since $4um^2 = -t$ at $h = 0$, for $t < 0$) that

$$\chi_\ell|_{h=0} = \left(\dfrac{dh}{dm} \right)^{-1} \Bigg|_{h=0} = (t - 3t)^{-1}, \quad i.e. \quad \chi_\ell = \dfrac{1}{2|t|},$$

at zero magnetic field, in the ordered phase ($t < 0$).

Note: *Another, more pictorial approach to this problem is as follows. Since the Hamiltonian is invariant under rotations about $\vec{h}$, $\vec{m}$ must be parallel to $\vec{h}$, i.e.*

$$m_\alpha = \dfrac{h_\alpha}{h} \varphi(h),$$

where φ is some function of the magnitude of the magnetic field. For simplicity, let $\vec{h} = h\mathbf{e}_1$, with $\mathbf{e}_1$ a unit vector, implying that

$$\vec{m} = m\mathbf{e}_1 = \varphi(h)\mathbf{e}_1.$$

The longitudinal susceptibility is then calculated as

$$\chi_\ell = \dfrac{\partial m_1}{\partial h_1} \Bigg|_{\mathbf{h}=h\mathbf{e}_1} = \dfrac{dm}{dh} = \varphi'(h).$$

To find the transverse susceptibility, we first note that if the system is perturbed by a small external magnetic field $\delta h\mathbf{e}_2$, the change in m_1 is, by symmetry, the same for $\delta h > 0$ and $\delta h < 0$, implying

$$m_1(h\mathbf{e}_1 + \delta h\mathbf{e}_2) = m_1(h\mathbf{e}_1) + \mathcal{O}\left(\delta h^2\right).$$

Hence

$$\frac{\partial m_1}{\partial h_2}\bigg|_{\vec{h}=h\mathbf{e}_1} = 0.$$

Furthermore, since $\vec{m}$ and $\vec{h}$ are parallel,

$$\frac{m_1\left(h\mathbf{e}_1 + \delta h\mathbf{e}_2\right)}{h} = \frac{m_2\left(h\mathbf{e}_1 + \delta h\mathbf{e}_2\right)}{\delta h},$$

from which

$$m_2\left(h\mathbf{e}_1 + \delta h\mathbf{e}_2\right) = \frac{m_1\left(h\mathbf{e}_1\right)}{h}\delta h + \mathcal{O}\left(\delta h^3\right),$$

yielding

$$\chi_t = \frac{\partial m_2}{\partial h_2}\bigg|_{\vec{h}=h\mathbf{e}_1} = \frac{m}{h}.$$

Solutions to selected problems from chapter 3

1. *Spin waves:* In the XY model of $n = 2$ magnetism, a unit vector $\vec{s} = (s_x, s_y)$ (with $s_x^2 + s_y^2 = 1$) is placed on each site of a d-dimensional lattice. There is an interaction that tends to keep nearest neighbors parallel, i.e. a Hamiltonian

$$-\beta\mathcal{H} = K \sum_{\langle ij \rangle} \vec{s}_i \cdot \vec{s}_j.$$

The notation $\langle ij \rangle$ is conventionally used to indicate summing over all *nearest-neighbor* pairs (i, j).

(a) Rewrite the partition function $Z = \int \prod_i d\vec{s}_i \exp(-\beta\mathcal{H})$ as an integral over the set of angles $\{\theta_i\}$ between the spins $\{\vec{s}_i\}$ and some arbitrary axis.

- *The partition function is*

$$Z = \int \prod_i d^2\vec{s}_i \exp\left(K \sum_{\langle ij \rangle} \vec{s}_i \cdot \vec{s}_j\right) \delta\left(\vec{s}_i^{\,2} - 1\right).$$

Since $\vec{s}_i \cdot \vec{s}_j = \cos\left(\theta_i - \theta_j\right)$, and $d^2\vec{s}_i \delta\left(\vec{s}_i^{\,2} - 1\right) = d\theta_i$, we obtain

$$Z = \int \prod_i d\theta_i \exp\left(K \sum_{\langle ij \rangle} \cos\left(\theta_i - \theta_j\right)\right).$$

(b) At low temperatures ($K \gg 1$), the angles $\{\theta_i\}$ vary slowly from site to site. In this case expand $-\beta\mathcal{H}$ to get a quadratic form in $\{\theta_i\}$.

- *Expanding the cosines to quadratic order gives*

$$Z = e^{N_b K} \int \prod_i d\theta_i \exp\left(-\frac{K}{2} \sum_{\langle ij \rangle} (\theta_i - \theta_j)^2\right),$$

where N_b is the total number of bonds. Higher order terms in the expansion may be neglected for large K, since the integral is dominated by $|\theta_i - \theta_j| \approx \sqrt{2/K}$.

(c) For $d = 1$, consider L sites with periodic boundary conditions (i.e. forming a closed chain). Find the normal modes θ_q that diagonalize the quadratic form (by

Fourier transformation), and the corresponding eigenvalues $K(q)$. Pay careful attention to whether the modes are real or complex, and to the allowed values of q.

- *For a chain of L sites, we can change to Fourier modes by setting*

$$\theta_j = \sum_q \theta(q) \frac{e^{iqj}}{\sqrt{L}}.$$

Since θ_j are real numbers, we must have

$$\theta(-q) = \theta(q)^*,$$

and the allowed q values are restricted, for periodic boundary conditions, by the requirement of

$$\theta_{j+L} = \theta_j, \quad \Rightarrow \quad qL = 2\pi n, \quad \text{with} \quad n = 0, \pm 1, \pm 2, \dots, \pm \frac{L}{2}.$$

Using

$$\theta_j - \theta_{j-1} = \sum_q \theta(q) \frac{e^{iqj}}{\sqrt{L}} \left(1 - e^{-iq}\right),$$

the one dimensional Hamiltonian, $\beta\mathcal{H} = \frac{K}{2} \sum_j \left(\theta_j - \theta_{j-1}\right)^2$, can be rewritten in terms of Fourier components as

$$\beta\mathcal{H} = \frac{K}{2} \sum_{q,q'} \theta(q)\theta(q') \sum_j \frac{e^{i(q+q')j}}{L} \left(1 - e^{-iq}\right) \left(1 - e^{-iq'}\right).$$

Using the identity $\sum_j e^{i(q+q')j} = L\delta_{q,-q'}$, we obtain

$$\beta\mathcal{H} = K \sum_q |\theta(q)|^2 \left[1 - \cos(q)\right].$$

(d) Generalize the results from the previous part to a d-dimensional simple cubic lattice with periodic boundary conditions.

- *In the case of a d-dimensional system, the index j is replaced by a vector*

$$j \mapsto \mathbf{j} = (j_1, \dots, j_d),$$

which describes the lattice. We can then write

$$\beta\mathcal{H} = \frac{K}{2} \sum_{\mathbf{j}} \sum_\alpha \left(\theta_{\mathbf{j}} - \theta_{\mathbf{j}+\mathbf{e}_\alpha}\right)^2,$$

where the e_α's are unit vectors $\{e_1 = (1, 0, \cdots, 0), \cdots, e_d = (0, \cdots, 0, 1)\}$, generalizing the one-dimensional result to

$$\beta \mathcal{H} = \frac{K}{2} \sum_{q,q'} \theta(q) \theta(q') \sum_\alpha \sum_j \frac{e^{i(q+q')\cdot j}}{L^d} \left(1 - e^{-iq\cdot e_\alpha}\right) \left(1 - e^{-iq'\cdot e_\alpha}\right).$$

Again, summation over j constrains q and $-q'$ to be equal, and

$$\beta \mathcal{H} = K \sum_q |\theta(q)|^2 \sum_\alpha [1 - \cos(q_\alpha)].$$

(e) Calculate the contribution of these modes to the free energy and heat capacity. (Evaluate the *classical* partition function, i.e. do not quantize the modes.)

- *With $K(q) \equiv 2K \sum_\alpha [1 - \cos(q_\alpha)]$,*

$$Z = \int \prod_q d\theta(q) \exp\left[-\frac{1}{2} K(q) |\theta(q)|^2\right] = \prod_q \sqrt{\frac{2\pi}{K(q)}},$$

and the corresponding free energy is

$$F = -k_B T \ln Z = -k_B T \left[\text{constant} - \frac{1}{2} \sum_q \ln K(q)\right],$$

or, in the continuum limit (using the fact that the density of states in q space is $(L/2\pi)^d$),

$$F = -k_B T \left[\text{constant} - \frac{1}{2} L^d \int \frac{d^d q}{(2\pi)^d} \ln K(q)\right].$$

As $K \sim 1/T$ at $T \to \infty$, we can write

$$F = -k_B T \left[\text{constant}' - \frac{1}{2} L^d \ln T\right],$$

and the heat capacity per site is given by

$$C = -T \frac{\partial^2 F}{\partial T^2} \cdot \frac{1}{L^d} = \frac{k_B}{2}.$$

This is because there is one degree of freedom (the angle) per site that can store potential energy.

(f) Find an expression for $\langle \vec{s}_0 \cdot \vec{s}_x \rangle = \Re\langle \exp[i\theta_x - i\theta_0] \rangle$ by adding contributions from different Fourier modes. Convince yourself that for $|x| \to \infty$, only $q \to 0$ modes contribute appreciably to this expression, and hence calculate the asymptotic limit.

- *We have*

$$\theta_{\mathbf{x}} - \theta_0 = \sum_{\mathbf{q}} \theta(\mathbf{q}) \frac{e^{i\mathbf{q}\cdot\mathbf{x}} - 1}{L^{d/2}},$$

and by completing the square for the argument of the exponential in $\langle e^{i(\theta_{\mathbf{x}}-\theta_0)}\rangle$, i.e. for

$$-\frac{1}{2} K(\mathbf{q}) |\theta(\mathbf{q})|^2 + i\theta(\mathbf{q}) \frac{e^{i\mathbf{q}\cdot\mathbf{x}} - 1}{L^{d/2}},$$

it follows immediately that

$$\langle e^{i(\theta_{\mathbf{x}}-\theta_0)}\rangle = \exp\left\{ -\frac{1}{L^d} \sum_{\mathbf{q}} \frac{|e^{i\mathbf{q}\cdot\mathbf{x}} - 1|^2}{2K(\mathbf{q})} \right\} = \exp\left\{ -\int \frac{d^d q}{(2\pi)^d} \frac{1 - \cos(\mathbf{q}\cdot\mathbf{x})}{K(\mathbf{q})} \right\}.$$

For x larger than 1, the integrand has a peak of height $\sim x^2/2K$ at $q = 0$ (as seen by expanding the cosine for small argument). Furthermore, the integrand has a first node, as q increases, at $q \sim 1/x$. From these considerations, we can obtain the leading behavior for large x:

- *In $d = 1$, we have to integrate $\sim x^2/2K$ over a length $\sim 1/x$, and thus*

$$\langle e^{i(\theta_{\mathbf{x}}-\theta_0)}\rangle \sim \exp\left(-\frac{|x|}{2K}\right).$$

- *In $d = 2$, we have to integrate $\sim x^2/2K$ over an area $\sim (1/x)^2$. A better approximation, at large x, than merely taking the height of the peak, is given by*

$$\int \frac{d^2 q}{(2\pi)^2} \frac{1 - \cos(\mathbf{q}\cdot\mathbf{x})}{K(\mathbf{q})} \approx \int \frac{dq d\varphi q}{(2\pi)^2} \frac{1 - \cos(qx\cos\varphi)}{Kq^2}$$

$$= \int \frac{dq d\varphi}{(2\pi)^2} \frac{1}{Kq} - \int \frac{dq d\varphi}{(2\pi)^2} \frac{\cos(qx\cos\varphi)}{Kq},$$

or, doing the angular integration in the first term,

$$\int \frac{d^2 q}{(2\pi)^2} \frac{1 - \cos(\mathbf{q}\cdot\mathbf{x})}{K(\mathbf{q})} \approx \int^{1/|x|} \frac{dq}{2\pi} \frac{1}{Kq} + \text{subleading in } x,$$

resulting in

$$\langle e^{i(\theta_{\mathbf{x}}-\theta_0)}\rangle \sim \exp\left(-\frac{\ln|x|}{2\pi K}\right) = |x|^{-\frac{1}{2\pi K}}, \quad \text{as } x \to \infty.$$

- *In $d \geq 3$, we have to integrate $\sim x^2/2K$ over a volume $\sim (1/x)^3$. Thus, as $x \to \infty$, the x dependence of the integral is removed, and*

$$\langle e^{i(\theta_{\mathbf{x}}-\theta_0)}\rangle \to \text{constant},$$

implying that correlations don't disappear at large x.

The results can also be obtained by noting that the fluctuations are important only for small q. Using the expansion of $K(\mathbf{q}) \approx Kq^2/2$ then reduces the problem to calculation of the Coulomb kernel $\int d^d\mathbf{q}e^{i\mathbf{q}\cdot\mathbf{x}}/q^2$, as described in the text.

(g) Calculate the transverse susceptibility from $\chi_t \propto \int d^d\mathbf{x}\langle\vec{s}_0 \cdot \vec{s}_\mathbf{x}\rangle_c$. How does it depend on the system size L?

- *We have*

$$\langle e^{i(\theta_\mathbf{x}-\theta_0)}\rangle = \exp\left\{-\int \frac{d^d q}{(2\pi)^d} \frac{1-\cos(\mathbf{q}\cdot\mathbf{x})}{K(\mathbf{q})}\right\}$$

and, similarly,

$$\langle e^{i\theta_\mathbf{x}}\rangle = \exp\left\{-\int \frac{d^d q}{(2\pi)^d} \frac{1}{2K(\mathbf{q})}\right\}.$$

Hence the connected correlation function

$$\langle\vec{s}_\mathbf{x} \cdot \vec{s}_0\rangle_c = \langle e^{i(\theta_\mathbf{x}-\theta_0)}\rangle_c = \langle e^{i(\theta_\mathbf{x}-\theta_0)}\rangle - \langle e^{i\theta_\mathbf{x}}\rangle\langle e^{i\theta_0}\rangle$$

is given by

$$\langle\vec{s}_\mathbf{x} \cdot \vec{s}_0\rangle_c = e^{-\int \frac{d^d q}{(2\pi)^d} \frac{1}{K(\mathbf{q})}}\left\{\exp\left[\int \frac{d^d q}{(2\pi)^d} \frac{\cos(\mathbf{q}\cdot\mathbf{x})}{K(\mathbf{q})}\right] - 1\right\}.$$

In $d > 2$, the x dependent integral vanishes at $x \to \infty$. We can thus expand its exponential, for large x, obtaining

$$\langle\vec{s}_\mathbf{x} \cdot \vec{s}_0\rangle_c \sim \int \frac{d^d q}{(2\pi)^d} \frac{\cos(\mathbf{q}\cdot\mathbf{x})}{K(\mathbf{q})} \approx \int \frac{d^d q}{(2\pi)^d} \frac{\cos(\mathbf{q}\cdot\mathbf{x})}{Kq^2} = \frac{1}{K}C_d(x) \sim \frac{1}{K|x|^{d-2}}.$$

Thus, the transverse susceptibility diverges as

$$\chi_t \propto \int d^d x \langle\vec{s}_\mathbf{x} \cdot \vec{s}_0\rangle_c \sim \frac{L^2}{K}.$$

(h) In $d = 2$, show that χ_t only diverges for K larger than a critical value $K_c = 1/(4\pi)$.

- *In $d = 2$, there is no long range order, $\langle\vec{s}_\mathbf{x}\rangle = 0$, and*

$$\langle\vec{s}_\mathbf{x} \cdot \vec{s}_0\rangle_c = \langle\vec{s}_\mathbf{x} \cdot \vec{s}_0\rangle \sim |x|^{-1/(2\pi K)}.$$

The susceptibility

$$\chi_t \sim \int^L d^2 x \, |x|^{-1/(2\pi K)}$$

thus converges for $1/(2\pi K) > 2$, for K below $K_c = 1/(4\pi)$. For $K > K_c$, the susceptibility diverges as

$$\chi_t \sim L^{2-2K_c/K}.$$

2. *Capillary waves:* A reasonably flat surface in d dimensions can be described by its height h, as a function of the remaining $(d-1)$ coordinates $\mathbf{x} = (x_1, \ldots x_{d-1})$. Convince yourself that the generalized "area" is given by $\mathcal{A} = \int d^{d-1}\mathbf{x}\sqrt{1+(\nabla h)^2}$. With a surface tension σ, the Hamiltonian is simply $\mathcal{H} = \sigma\mathcal{A}$.

(a) At sufficiently low temperatures, there are only slow variations in h. Expand the energy to quadratic order, and write down the partition function as a functional integral.

- *For a surface parametrized by the height function*

$$x_d = h(x_1, \ldots, x_{d-1}),$$

an area element can be calculated as

$$dA = \frac{1}{\cos\alpha}dx_1\cdots dx_{d-1},$$

where α is the angle between the dth direction and the normal

$$\vec{n} = \frac{1}{\sqrt{1+(\nabla h)^2}}\left(-\frac{\partial h}{\partial x_1}, \ldots, -\frac{\partial h}{\partial x_{d-1}}, 1\right)$$

to the surface ($n^2 = 1$). Since $\cos\alpha = n_d = \left[1+(\nabla h)^2\right]^{-1/2} \approx 1 - \frac{1}{2}(\nabla h)^2$, we obtain

$$\mathcal{H} = \sigma\mathcal{A} \approx \sigma\int d^{d-1}x\left\{1 + \frac{1}{2}(\nabla h)^2\right\},$$

and, dropping a multiplicative constant,

$$Z = \int \mathcal{D}h(\mathbf{x})\exp\left\{-\beta\frac{\sigma}{2}\int d^{d-1}x\,(\nabla h)^2\right\}.$$

(b) Use Fourier transformation to diagonalize the quadratic Hamiltonian into its normal modes $\{h_\mathbf{q}\}$ (capillary waves).

- *After changing variables to the Fourier modes,*

$$h(\mathbf{x}) = \int \frac{d^{d-1}q}{(2\pi)^{d-1}}h(\mathbf{q})\,e^{i\mathbf{q}\cdot\mathbf{x}},$$

the partition function is given by

$$Z = \int \mathcal{D}h(\mathbf{q})\exp\left\{-\beta\frac{\sigma}{2}\int \frac{d^{d-1}q}{(2\pi)^{d-1}}q^2\,|h(\mathbf{q})|^2\right\}.$$

(c) What symmetry breaking is responsible for these Goldstone modes?

- *By selecting a particular height, the ground state breaks the translation symmetry in the dth direction. The transformation $h(\mathbf{x}) \rightarrow h(\mathbf{x}) + f(\mathbf{x})$ leaves the energy unchanged if $f(\mathbf{x})$ is constant. By continuity, we can have an arbitrarily small change in the energy by varying $f(\mathbf{x})$ sufficiently slowly.*

(d) Calculate the height–height correlations $\langle (h(\mathbf{x}) - h(\mathbf{x}'))^2 \rangle$.

- *From*

$$h(\mathbf{x}) - h(\mathbf{x}') = \int \frac{d^{d-1}q}{(2\pi)^{d-1}} h(\mathbf{q}) \left(e^{i\mathbf{q}\cdot\mathbf{x}} - e^{i\mathbf{q}\cdot\mathbf{x}'} \right),$$

we obtain

$$\left\langle (h(\mathbf{x}) - h(\mathbf{x}'))^2 \right\rangle = \int \frac{d^{d-1}q}{(2\pi)^{d-1}} \frac{d^{d-1}q'}{(2\pi)^{d-1}} \langle h(\mathbf{q}) h(\mathbf{q}') \rangle \left(e^{i\mathbf{q}\cdot\mathbf{x}} - e^{i\mathbf{q}\cdot\mathbf{x}'} \right)$$
$$\left(e^{i\mathbf{q}'\cdot\mathbf{x}} - e^{i\mathbf{q}'\cdot\mathbf{x}'} \right).$$

The height–height correlations thus behave as

$$G(\mathbf{x} - \mathbf{x}') \equiv \left\langle (h(\mathbf{x}) - h(\mathbf{x}'))^2 \right\rangle$$

$$= \frac{2}{\beta\sigma} \int \frac{d^{d-1}q}{(2\pi)^{d-1}} \frac{1 - \cos[\mathbf{q} \cdot (\mathbf{x} - \mathbf{x}')]}{q^2} = \frac{2}{\beta\sigma} C_{d-1}(\mathbf{x} - \mathbf{x}').$$

(e) Comment on the form of the result (d) in dimensions $d = 4, 3, 2$, and 1.

- *We can now discuss the asymptotic behavior of the Coulomb kernel for large $|\mathbf{x} - \mathbf{x}'|$, either using the results from problem 1(f), or the exact form given in the text.*

  - *In $d \geq 4$, $G(\mathbf{x} - \mathbf{x}') \rightarrow$ constant, and the surface is flat.*
  - *In $d = 3$, $G(\mathbf{x} - \mathbf{x}') \sim \ln|\mathbf{x} - \mathbf{x}'|$, and we come to the surprising conclusion that there are no asymptotically flat surfaces in three dimensions. While this is technically correct, since the logarithm grows slowly, very large surfaces are needed to detect appreciable fluctuations.*
  - *In $d = 2$, $G(\mathbf{x} - \mathbf{x}') \sim |\mathbf{x} - \mathbf{x}'|$. This is easy to comprehend, once we realize that the interface $h(x)$ is similar to the path $x(t)$ of a random walker, and has similar ($x \sim \sqrt{t}$) fluctuations.*
  - *In $d = 1$, $G(\mathbf{x} - \mathbf{x}') \sim |\mathbf{x} - \mathbf{x}'|^2$. The transverse fluctuation of the "point" interface are very big, and the approximations break down as discussed next.*

(f) By estimating typical values of ∇h, comment on when it is justified to ignore higher order terms in the expansion for $\mathcal{A}$.

- *We can estimate $(\nabla h)^2$ as*

$$\frac{\left\langle (h(\mathbf{x}) - h(\mathbf{x}'))^2 \right\rangle}{(\mathbf{x} - \mathbf{x}')^2} \propto |\mathbf{x} - \mathbf{x}'|^{1-d}.$$

For dimensions $d \geq d_\ell = 1$, the typical size of the gradient decreases upon coarse graining. The gradient expansion of the area used before is then justified. For dimensions $d \leq d_\ell$, the whole idea of the gradient expansion fails to be sensible.

3. *Gauge fluctuations in superconductors:* The Landau–Ginzburg model of superconductivity describes a complex superconducting order parameter $\Psi(\mathbf{x}) = \Psi_1(\mathbf{x}) + i\Psi_2(\mathbf{x})$, and the electromagnetic vector potential $\vec{A}(\mathbf{x})$, which are subject to a Hamiltonian

$$\beta\mathcal{H} = \int d^3\mathbf{x}\left[\frac{t}{2}|\Psi|^2 + u|\Psi|^4 + \frac{K}{2}D_\mu\Psi D_\mu^*\Psi^* + \frac{L}{2}(\nabla \times A)^2\right].$$

The gauge-invariant derivative $D_\mu \equiv \partial_\mu - ieA_\mu(\mathbf{x})$ introduces the coupling between the two fields. (In terms of Cooper pair parameters, $e = e^*c/\hbar$, $K = \hbar^2/2m^*$.)

(a) Show that the above Hamiltonian is invariant under the *local gauge symmetry*:

$$\Psi(\mathbf{x}) \mapsto \Psi(x)\exp(i\theta(\mathbf{x})), \quad \text{and} \quad A_\mu(\mathbf{x}) \mapsto A_\mu(\mathbf{x}) + \frac{1}{e}\partial_\mu\theta.$$

- *Under a local gauge transformation, $\beta\mathcal{H}$ changes to*

$$\int d^3x\left\{\frac{t}{2}|\Psi|^2 + u|\Psi|^4 + \frac{K}{2}\left[(\partial_\mu - ieA_\mu - i\partial_\mu\theta)\Psi e^{i\theta}\right]\left[(\partial_\mu + ieA_\mu\right.\right.$$

$$\left.\left. + i\partial_\mu\theta)\Psi^*e^{-i\theta}\right] + \frac{L}{2}\left(\nabla \times \vec{A} + \nabla \times \frac{1}{e}\nabla\theta\right)^2\right\}.$$

But this is none other than $\beta\mathcal{H}$ again, since

$$(\partial_\mu - ieA_\mu - i\partial_\mu\theta)\Psi e^{i\theta} = e^{i\theta}(\partial_\mu - ieA_\mu)\Psi = e^{i\theta}D_\mu\Psi,$$

and

$$\nabla \times \frac{1}{e}\nabla\theta = 0.$$

(b) Show that there is a saddle point solution of the form $\Psi(\mathbf{x}) = \overline{\Psi}$, and $\vec{A}(\mathbf{x}) = 0$, and find $\overline{\Psi}$ for $t > 0$ and $t < 0$.

- *The saddle point solutions are obtained from*

$$\frac{\delta\mathcal{H}}{\delta\Psi^*} = 0, \quad \Longrightarrow \quad \frac{t}{2}\Psi + 2u\Psi|\Psi|^2 - \frac{K}{2}D_\mu D_\mu\Psi = 0,$$

and

$$\frac{\delta\mathcal{H}}{\delta A_\mu} = 0, \quad \Longrightarrow \quad \frac{K}{2}\left(-ie\Psi D_\mu^*\Psi^* + ie\Psi^* D_\mu\Psi\right) - L\epsilon_{\alpha\beta\mu}\epsilon_{\alpha\gamma\delta}\partial_\beta\partial_\gamma A_\delta = 0.$$

The Ansatz $\Psi(\mathbf{x}) = \overline{\Psi}$, $\vec{A} = 0$, *clearly solves these equations. The first equation then becomes*

$$t\overline{\Psi} + 4u\overline{\Psi}\left|\overline{\Psi}\right|^2 = 0,$$

yielding (for $u > 0$) $\overline{\Psi} = 0$ for $t > 0$, whereas $\left|\overline{\Psi}\right|^2 = -t/4u$ for $t < 0$.

(c) For $t < 0$, calculate the cost of fluctuations by setting

$$\begin{cases} \Psi(\mathbf{x}) = \left(\overline{\Psi} + \phi(\mathbf{x})\right)\exp\left(i\theta(\mathbf{x})\right), \\ A_\mu(\mathbf{x}) = a_\mu(\mathbf{x}) \quad \text{(with } \partial_\mu a_\mu = 0 \text{ in the Coulomb gauge)} \end{cases}$$

and expanding $\beta\mathcal{H}$ to quadratic order in ϕ, θ, and $\vec{a}$.

- *For simplicity, let us choose $\overline{\Psi}$ to be real. From the Hamiltonian term*

$$D_\mu\Psi D_\mu^*\Psi^* = \left[\left(\partial_\mu - iea_\mu\right)\left(\overline{\Psi} + \phi\right)e^{i\theta}\right]\left[\left(\partial_\mu + iea_\mu\right)\left(\overline{\Psi} + \phi\right)e^{-i\theta}\right],$$

we get the following quadratic contribution

$$\overline{\Psi}^2(\nabla\theta)^2 + (\nabla\phi)^2 - 2e\overline{\Psi}^2 a_\mu\partial_\mu\theta + e^2\overline{\Psi}^2\left|\vec{a}\right|^2.$$

The third term in the above expression integrates to zero (as can be seen through integrating by parts and invoking the Coulomb gauge condition $\partial_\mu a_\mu = 0$). Thus, the quadratic terms read

$$\beta\mathcal{H}^{(2)} = \int d^3x\left\{\left(\frac{t}{2} + 6u\overline{\Psi}^2\right)\phi^2 + \frac{K}{2}(\nabla\phi)^2 + \frac{K}{2}\overline{\Psi}^2(\nabla\theta)^2\right.$$

$$\left. + \frac{K}{2}e^2\overline{\Psi}^2\left|\vec{a}\right|^2 + \frac{L}{2}(\nabla\times\vec{a})^2\right\}.$$

(d) Perform a Fourier transformation, and calculate the expectation values of $\langle|\phi(\mathbf{q})|^2\rangle$, $\langle|\theta(\mathbf{q})|^2\rangle$, and $\langle|\vec{a}(\mathbf{q})|^2\rangle$.

- *In terms of Fourier transforms, we obtain*

$$\beta\mathcal{H}^{(2)} = \sum_{\mathbf{q}}\left\{\left(\frac{t}{2} + 6u\overline{\Psi}^2 + \frac{K}{2}q^2\right)|\phi(\mathbf{q})|^2 + \frac{K}{2}\overline{\Psi}^2 q^2|\theta(\mathbf{q})|^2\right.$$

$$\left. + \frac{K}{2}e^2\overline{\Psi}^2\left|\vec{a}(\mathbf{q})\right|^2 + \frac{L}{2}(\mathbf{q}\times\vec{a})^2\right\}.$$

In the Coulomb gauge, $\mathbf{q} \perp \vec{a}(\mathbf{q})$, and so $\left[\mathbf{q}\times\vec{a}(\mathbf{q})\right]^2 = q^2|\vec{a}(\mathbf{q})|^2$. This diagonal form then yields immediately (for $t < 0$)

$$\left\langle|\phi(\mathbf{q})|^2\right\rangle = \left(t + 12u\overline{\Psi}^2 + Kq^2\right)^{-1} = \frac{1}{Kq^2 - 2t},$$

$$\left\langle|\theta(\mathbf{q})|^2\right\rangle = \left(K\overline{\Psi}^2 q^2\right)^{-1} = -\frac{4u}{Ktq^2},$$

$$\left\langle|\vec{a}(\mathbf{q})|^2\right\rangle = 2\left(Ke^2\overline{\Psi}^2 + Lq^2\right)^{-1} = \frac{2}{Lq^2 - Ke^2t/4u} \quad (\vec{a}\text{ has two components}).$$

Note that the gauge field, "massless" in the original theory, acquires a "mass" $|Ke^2t/4u|$ through its coupling to the order parameter. This is known as the Higgs mechanism.

4. *Fluctuations around a tricritical point*: As shown in a previous problem, the Hamiltonian

$$\beta\mathcal{H} = \int d^d x \left[\frac{K}{2}(\nabla m)^2 + \frac{t}{2}m^2 + um^4 + vm^6\right],$$

with $u = 0$ and $v > 0$ describes a tricritical point.

(a) Calculate the heat capacity singularity as $t \to 0$ by the saddle point approximation.

- *As already calculated in a previous problem, the saddle point minimum of the free energy $\vec{m} = \overline{m}\hat{e}_\ell$, can be obtained from*

$$\frac{\partial\Psi}{\partial m} = \overline{m}\left(t + 6v\overline{m}^4\right) = 0,$$

yielding,

$$\overline{m} = \begin{cases} 0 & \text{for } t > \overline{t} = 0 \\ \left(-\dfrac{t}{6v}\right)^{1/4} & \text{for } t < 0. \end{cases}$$

The corresponding free energy density equals to

$$\Psi(\overline{m}) = \frac{t}{2}\overline{m}^2 + v\overline{m}^6 = \begin{cases} 0 & \text{for } t > 0 \\ -\dfrac{1}{3}\dfrac{(-t)^{3/2}}{(6v)^{1/2}} & \text{for } t < 0. \end{cases}$$

Therefore, the singular behavior of the heat capacity is given by

$$C = C_{\text{s.p.}} \sim -T_c \left.\frac{\partial^2\Psi}{\partial t^2}\right|_{\overline{m}} = \begin{cases} 0 & \text{for } t > 0 \\ \dfrac{T_c}{4}(-6vt)^{-1/2} & \text{for } t < 0, \end{cases}$$

as sketched in the figure below.

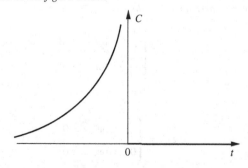

(b) Include both longitudinal and transverse fluctuations by setting

$$\vec{m}(\mathbf{x}) = (\overline{m} + \phi_\ell(\mathbf{x}))\hat{e}_\ell + \sum_{\alpha=2}^{n} \phi_t^\alpha(\mathbf{x})\hat{e}_\alpha,$$

and expanding $\beta\mathcal{H}$ to quadratic order in ϕ.

- *Let us now include both longitudinal and transversal fluctuations by setting*

$$\vec{m}(\mathbf{x}) = (\overline{m} + \phi_\ell(\mathbf{x}))\hat{e}_\ell + \sum_{\alpha=2}^{n} \phi_t^\alpha(\mathbf{x})\hat{e}_\alpha,$$

where $\hat{e}_\ell$ and $\hat{e}_\alpha$ form an orthonormal set of n vectors. Consequently, the free energy $\beta\mathcal{H}$ is a function of ϕ_ℓ and ϕ_t. Since $\overline{m}\hat{e}_\ell$ is a minimum, there are no linear terms in the expansion of $\beta\mathcal{H}$ in ϕ. The contributions of each factor in the free energy to the quadratic term in the expansion are

$$(\nabla\vec{m})^2 \Longrightarrow (\nabla\phi_\ell)^2 + \sum_{\alpha=2}^{n}(\nabla\phi_t^\alpha)^2,$$

$$(\vec{m})^2 \Longrightarrow (\phi_\ell)^2 + \sum_{\alpha=2}^{n}(\phi_t^\alpha)^2,$$

$$(\vec{m})^6 = ((\vec{m})^2)^3 = (\overline{m}^2 + 2\overline{m}\phi_\ell + \phi_\ell^2 + \sum_{\alpha=2}^{n}(\phi_t^\alpha)^2)^3 \Longrightarrow 15\overline{m}^4(\phi_\ell)^2 + 3\overline{m}^4\sum_{\alpha=2}^{n}(\phi_t^\alpha)^2.$$

The expansion of $\beta\mathcal{H}$ to second order now gives

$$\beta\mathcal{H}(\phi_\ell, \phi_t^\alpha) = \beta\mathcal{H}(0,0) + \int d^d\mathbf{x}\left\{\left[\frac{K}{2}(\nabla\phi_\ell)^2 + \frac{\phi_\ell^2}{2}\left(t + 30v\overline{m}^4\right)\right]\right.$$
$$\left. + \sum_{\alpha=2}^{n}\left[\frac{K}{2}(\nabla\phi_t^\alpha)^2 + \frac{(\phi_t^\alpha)^2}{2}\left(t + 6v\overline{m}^4\right)\right]\right\}.$$

We can formally rewrite it as

$$\beta\mathcal{H}(\phi_\ell, \phi_t^\alpha) = \beta\mathcal{H}(0,0) + \beta\mathcal{H}_\ell(\phi_\ell) + \sum_{\alpha=2}^{n}\beta\mathcal{H}_{t_\alpha}(\phi_t^\alpha),$$

where $\beta\mathcal{H}_i(\phi_i)$, with $i = \ell, t_\alpha$, is in general given by

$$\beta\mathcal{H}_i(\phi_i) = \frac{K}{2}\int d^d\mathbf{x}\left[(\nabla\phi_i)^2 + \frac{\phi_i^2}{\xi_i^2}\right],$$

with the inverse correlation lengths

$$\xi_\ell^{-2} = \begin{cases} \dfrac{t}{K} & \text{for} \quad t > 0 \\ \dfrac{-4t}{K} & \text{for} \quad t < 0, \end{cases}$$

and

$$\xi_{t_\alpha}^{-2} = \begin{cases} \dfrac{t}{K} & \text{for} \quad t > 0 \\ 0 & \text{for} \quad t < 0. \end{cases}$$

As discussed in the text, near the critical point of a magnet, for $t > 0$ there is no difference between longitudinal and transverse components, whereas for $t < 0$, there is no restoring force for the Goldstone modes ϕ_t^α due to the rotational symmetry of the ordered state.

(c) Calculate the longitudinal and transverse correlation functions.

- *Since in the harmonic approximation $\beta\mathcal{H}$ turns out to be a sum of the Hamiltonians of the different fluctuating components ϕ_ℓ, ϕ_t^α, these quantities are independent of each other, i.e.*

$$\langle\phi_\ell\phi_t^\alpha\rangle = 0, \quad \text{and} \quad \langle\phi_t^\gamma\phi_t^\alpha\rangle = 0 \quad \text{for } \alpha \neq \gamma.$$

To determine the longitudinal and transverse correlation functions, we first express the free energy in terms of Fourier modes, so that the probability of a particular fluctuation configuration is given by

$$\mathcal{P}(\{\phi_\ell, \phi_t^\alpha\}) \propto \prod_{\mathbf{q},\alpha} \exp\left\{-\frac{K}{2}\left(q^2 + \xi_\ell^{-2}\right)|\phi_{\ell,\mathbf{q}}|^2\right\} \cdot \exp\left\{-\frac{K}{2}\left(q^2 + \xi_{t_\alpha}^{-2}\right)|\phi_{t,\mathbf{q}}^\alpha|^2\right\}.$$

Thus, as also shown in the text, the correlation function is

$$\langle\phi_\alpha(\mathbf{x})\phi_\beta(0)\rangle = \frac{\delta_{\alpha,\beta}}{VK}\sum_\mathbf{q}\frac{e^{i\mathbf{q}\cdot\mathbf{x}}}{(q^2 + \xi_\alpha^{-2})} = -\frac{\delta_{\alpha,\beta}}{K}I_d(\mathbf{x}, \xi_\alpha),$$

i.e.

$$\langle\phi_\ell(\mathbf{x})\phi_\ell(0)\rangle = -\frac{1}{K}I_d(\mathbf{x}, \xi_\ell),$$

and

$$\langle\phi_t^\alpha(\mathbf{x})\phi_t^\beta(0)\rangle = -\frac{\delta_{\alpha,\beta}}{K}I_d(\mathbf{x}, \xi_{t_\alpha}).$$

(d) Compute the first correction to the saddle point free energy from fluctuations.

- *The partition function has contributions of the form*

$$Z = e^{-\beta\mathcal{H}(0,0)}\int \mathcal{D}\phi(\mathbf{x})\exp\left\{-\frac{K}{2}\int d^d\mathbf{x}\left[(\nabla\phi)^2 + \xi^{-2}\phi^2\right]\right\}$$

$$- e^{-\beta\mathcal{H}(0,0)}\int\prod_\mathbf{q}d\psi_\mathbf{q}\exp\left\{-\frac{K}{2}\sum_\mathbf{q}\left(q^2 + \xi^{-2}\right)\phi_\mathbf{q}\phi_\mathbf{q}^*\right\}$$

$$= \prod_\mathbf{q}\left[K\left(q^2 + \xi^{-2}\right)\right]^{-1/2} = \exp\left\{-\frac{1}{2}\sum_\mathbf{q}\left(Kq^2 + K\xi^{-2}\right)\right\},$$

and the free energy density equals to

$$\beta f = \frac{\beta\mathcal{H}(0,0)}{V} + \begin{cases} \dfrac{n}{2}\int\dfrac{d^d\mathbf{q}}{(2\pi)^d}\ln\left(Kq^2 + t\right) & \text{for } t > 0 \\[3mm] \dfrac{1}{2}\int\dfrac{d^d\mathbf{q}}{(2\pi)^d}\ln\left(Kq^2 - 4t\right) + \dfrac{n-1}{2}\int\dfrac{d^d\mathbf{q}}{(2\pi)^d}\ln\left(Kq^2\right) & \text{for } t < 0. \end{cases}$$

Note that the first term is the saddle point free energy, and that there are n contributions to the free energy from fluctuations.

(e) Find the fluctuation correction to the heat capacity.

- *As $C = -T(d^2 f/dT^2)$, the fluctuation corrections to the heat capacity are given by*

$$C - C_{\text{s.p.}} \propto \begin{cases} \dfrac{n}{2} \int \dfrac{d^d \mathbf{q}}{(2\pi)^d} \left(Kq^2 + t\right)^{-2} & \text{for} \quad t > 0 \\[2mm] \dfrac{16}{2} \int \dfrac{d^d \mathbf{q}}{(2\pi)^d} \left(Kq^2 - 4t\right)^{-2} & \text{for} \quad t < 0. \end{cases}$$

These integrals change behavior at $d = 4$. For $d > 4$, the integrals diverge at large $\mathbf{q}$, and are dominated by the upper cutoff $\Delta \simeq 1/a$. That is why fluctuation corrections to the heat capacity add just a constant term on each side of the transition, and the saddle point solution keeps its qualitative form. On the other hand, for $d < 4$, the integrals are proportional to the corresponding correlation length ξ^{4-d}. Due to the divergence of ξ, the fluctuation corrections diverge as

$$C_{\text{fl.}} = C - C_{\text{s.p.}} \propto K^{-d/2} |t|^{d/2-2}.$$

(f) By comparing the results from parts (a) and (e) for $t < 0$ obtain a Ginzburg criterion, and the upper critical dimension for validity of mean-field theory at a tricritical point.

- *To obtain a Ginzburg criterion, let us consider $t < 0$. In this region, the saddle point contribution already diverges as $C_{\text{s.p.}} \propto (-vt)^{-1/2}$, so that*

$$\frac{C_{\text{fl.}}}{C_{\text{s.p.}}} \propto (-t)^{\frac{d-3}{2}} \left(\frac{v}{K^d}\right)^{1/2}.$$

Therefore at $t < 0$, the saddle point contribution dominates the behavior of this ratio provided that $d > 3$. For $d < 3$, the mean field result will continue being dominant far enough from the critical point, i.e. if

$$(-t)^{d-3} \gg \left(\frac{K^d}{v}\right), \quad \text{or} \quad |t| \gg \left(\frac{K^d}{v}\right)^{1/(d-3)}.$$

Otherwise, i.e. if

$$|t| < \left(\frac{K^d}{v}\right)^{1/(d-3)},$$

the fluctuation contribution to the heat capacity becomes dominant. The upper critical dimension for the tricritical point is then $d = 3$.

(g) A generalized multicritical point is described by replacing the term vm^6 with $u_{2n}m^{2n}$. Use simple power counting to find the upper critical dimension of this multicritical point.

- *If instead of the term vm^6 we have a general factor of the form $u_{2n}m^{2n}$, we can easily generalize our results to*

$$\overline{m} \propto (-t)^{1/(2n-2)}, \qquad \Psi(\overline{m}) \propto (-t)^{n/(n-1)}, \qquad C_{\text{s.p.}} \propto (-t)^{n/(n-1)-2}.$$

Moreover, the fluctuation correction to the heat capacity for any value of n is the same as before

$$C_{\text{fl.}} \propto (-t)^{d/2-2}.$$

Hence the upper critical dimension is, in general, determined by the equation

$$\frac{d}{2} - 2 = \frac{n}{n-1} - 2, \quad \text{or} \quad d_u = \frac{2n}{n-1}.$$

Solutions to selected problems from chapter 4

1. *Scaling in fluids*: Near the liquid–gas critical point, the free energy is assumed to take the scaling form $F/N = t^{2-\alpha} g(\delta\rho/t^{\beta})$, where $t = |T - T_c|/T_c$ is the reduced temperature, and $\delta\rho = \rho - \rho_c$ measures deviations from the critical point density. The leading singular behavior of any thermodynamic parameter $Q(t, \delta\rho)$ is of the form t^x on approaching the critical point along the isochore $\rho = \rho_c$; or $\delta\rho^y$ for a path along the isotherm $T = T_c$. Find the exponents x and y for the following quantities:

* *Any homogeneous function (thermodynamic quantity)* $Q(t, \delta\rho)$ *can be written in the scaling form*

$$Q(t, \delta\rho) = t^{x_Q} g_Q \left(\frac{\delta\rho}{t^{\beta}} \right).$$

Thus, the leading singular behavior of Q *is of the form* t^{x_Q} *if* $\delta\rho = 0$, *i.e. along the critical isochore. In order for any* Q *to be independent of* t *along the critical isotherm as* $t \to 0$, *the scaling function for a large enough argument should be of the form*

$$\lim_{x \to \infty} g_Q(x) = x^{x_Q/\beta},$$

so that

$$Q(0, \delta\rho) \propto (\delta\rho)^{y_Q}, \quad \text{with} \quad y_Q = \frac{x_Q}{\beta}.$$

(a) The internal energy per particle $\langle H \rangle / N$, and the entropy per particle $s = S/N$.

* *Let us assume that the free energy per particle is*

$$f = \frac{F}{N} = t^{2-\alpha} g \left(\frac{\delta\rho}{t^{\beta}} \right),$$

and that $T < T_c$, *so that* $\dfrac{\partial}{\partial T} = -\dfrac{1}{T_c} \dfrac{\partial}{\partial t}$. *The entropy is then given by*

$$s = -\left. \frac{\partial f}{\partial T} \right|_V = \frac{1}{T_c} \left. \frac{\partial f}{\partial t} \right|_\rho = \frac{t^{1-\alpha}}{T_c} g_s \left(\frac{\delta\rho}{t^{\beta}} \right),$$

so that $x_S = 1 - \alpha$, and $y_S = (1 - \alpha)/\beta$. *For the internal energy, we have*

$$f = \frac{\langle \mathcal{H} \rangle}{N} - Ts, \quad \text{or} \quad \frac{\langle \mathcal{H} \rangle}{N} \sim T_c \, s(1 + t) \sim t^{1-\alpha} g_{\mathcal{H}} \left(\frac{\delta\rho}{t^\beta} \right),$$

therefore, $x_{\mathcal{H}} = 1 - \alpha$ *and* $y_{\mathcal{H}} = (1 - \alpha)/\beta$.

(b) The heat capacities $C_V = T \partial s / \partial T \, |_V$, and $C_P = T \partial s / \partial T \, |_P$.

- *The heat capacity at constant volume*

$$C_V = T \frac{\partial S}{\partial T} \bigg|_V = - \frac{\partial s}{\partial t} \bigg|_\rho = \frac{t^{-\alpha}}{T_c} g_{C_V} \left(\frac{\delta\rho}{t^\beta} \right),$$

so that $x_{C_V} = -\alpha$ *and* $y_{C_V} = -\alpha/\beta$.

To calculate the heat capacity at constant pressure, we need to determine first the form of $\delta\rho(t)$ at constant P. For that purpose we will use the chain rule identity

$$\frac{\partial \delta\rho}{\partial t} \bigg|_P = - \frac{\frac{\partial P}{\partial t} |_\rho}{\frac{\partial P}{\partial \delta\rho} |_t}.$$

The pressure P is determined as

$$P = - \frac{\partial F}{\partial V} = \rho^2 \frac{\partial f}{\partial \delta\rho} \sim \rho_c^2 \, t^{2 - \alpha - \beta} g_P \left(\frac{\delta\rho}{t^\beta} \right),$$

which for $\delta\rho \ll t^\beta$ goes like

$$P \propto t^{2-\alpha-\beta} \left(1 + A \frac{\delta\rho}{t^\beta} \right), \quad \text{and consequently} \quad \begin{cases} \frac{\partial P}{\partial t} |_\rho \propto t^{1-\alpha-\beta} \\ \frac{\partial P}{\partial \delta\rho} |_t \propto t^{2-\alpha-2\beta}. \end{cases}$$

In the other extreme of $\delta\rho \gg t^\beta$,

$$P \propto \delta\rho^{(2-\alpha-\beta)/\beta} \left(1 + B \frac{t}{\delta\rho^{1/\beta}} \right), \quad \text{and} \quad \begin{cases} \frac{\partial P}{\partial t} |_\rho \propto \delta\rho^{(1-\alpha-\beta)/\beta} \\ \frac{\partial P}{\partial \delta\rho} |_t \propto \delta\rho^{(2-\alpha-2\beta)/\beta}, \end{cases}$$

where we have again required that P does not depend on $\delta\rho$ when $\delta\rho \to 0$, and on t if $t \to 0$.

From the previous results, we can now determine

$$\frac{\partial \delta\rho}{\partial t} \bigg|_P \propto \begin{cases} t^{\beta-1} & \implies \delta\rho \propto t^\beta \\ \delta\rho^{(\beta-1)/\beta} & \implies t \propto \delta\rho^{1/\beta}. \end{cases}$$

From any of these relationships it follows that $\delta\rho \propto t^\beta$, and consequently the entropy is $s \propto t^{1-\alpha}$. The heat capacity at constant pressure is then given by

$$C_P \propto t^{-\alpha}, \quad \text{with} \quad x_{C_P} = -\alpha \quad \text{and} \quad y_{C_P} = -\frac{\alpha}{\beta}.$$

(c) The isothermal compressibility $\kappa_T = \partial\rho/\partial P \mid_T /\rho$, and the thermal expansion coefficient $\alpha = \partial V/\partial T \mid_P /V$.

Check that your results for parts (b) and (c) are consistent with the thermodynamic identity $C_P - C_V = TV\alpha^2/\kappa_T$.

- *The isothermal compressibility and the thermal expansion coefficient can be computed using some of the relations obtained previously*

$$\kappa_T = \frac{1}{\rho}\frac{\partial\rho}{\partial P}\bigg|_T = \frac{1}{\rho_c}\frac{\partial P}{\partial\rho}\bigg|_T^{-1} = \frac{1}{\rho_c^3}t^{\alpha+2\beta-2}g_\kappa\left(\frac{\delta\rho}{t^\beta}\right),$$

with $x_\kappa = \alpha + 2\beta - 2$, and $y_\kappa = (\alpha + 2\beta - 2)/\beta$. And

$$\alpha = \frac{1}{V}\frac{\partial V}{\partial T}\bigg|_P = \frac{1}{\rho T_c}\frac{\partial\rho}{\partial t}\bigg|_P \propto t^{\beta-1},$$

with $x_\alpha = \beta - 1$, and $y_\alpha = (\beta - 1)/\beta$. So clearly, these results are consistent with the thermodynamic identity,

$$(C_P - C_V)(t, 0) \propto t^{-\alpha}, \quad \text{or} \quad (C_P - C_V)(0, \delta\rho) \propto \delta\rho^{-\alpha/\beta},$$

and

$$\frac{\alpha^2}{\kappa_T}(t, 0) \propto t^{-\alpha}, \quad \text{or} \quad \frac{\alpha^2}{\kappa_T}(0, \delta\rho) \propto \delta\rho^{-\alpha/\beta}.$$

(d) Sketch the behavior of the latent heat per particle L, on the coexistence curve for $T < T_c$, and find its singularity as a function of t.

- *The latent heat*

$$L = T\left(s_+ - s_-\right)$$

is defined at the coexistence line, and as we have seen before

$$Ts_\pm = t^{1-\alpha}g_s\left(\frac{\delta\rho_\pm}{t^\beta}\right).$$

The density difference between the two coexisting phases is the order parameter, and vanishes as t^β, as do each of the two deviations $\delta\rho_+ = \rho_c - 1/v_+$ and $\delta\rho_- = \rho_c - 1/v_-$ of the gas and liquid densities from the critical value. (More precisely, as seen in (b), $\delta\rho\mid_{P=\text{constant}} \propto t^\beta$.) The argument of g in the above expression is thus evaluated at a finite value, and since the latent heat goes to zero on approaching the critical point, we get

$$L \propto t^{1-\alpha}, \quad \text{with} \quad x_L = 1 - \alpha.$$

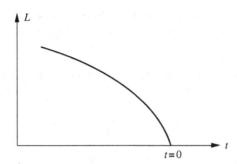

2. *The Ising model*: The differential recursion relations for temperature T, and magnetic field h, of the Ising model in $d = 1 + \epsilon$ dimensions are (for $b = e^{\ell}$)

$$\begin{cases} \dfrac{\mathrm{d}T}{\mathrm{d}\ell} = -\epsilon T + \dfrac{T^2}{2} \\[2mm] \dfrac{\mathrm{d}h}{\mathrm{d}\ell} = dh. \end{cases}$$

(a) Sketch the renormalization group flows in the (T, h) plane (for $\epsilon > 0$), marking the fixed points along the $h = 0$ axis.

- *The fixed points of the flow occur along the $h = 0$ axis, which is mapped to itself under RG. On this axis, there are three fixed points: (i) $T^* = 0$ is the stable sink for the low temperature phase. (ii) $T^* \to \infty$ is the stable sink for the high-temperature phase. (iii) There is a critical fixed point at $(T^* = 2\epsilon, h^* = 0)$, which is unstable. All fixed points are unstable in the field direction.*

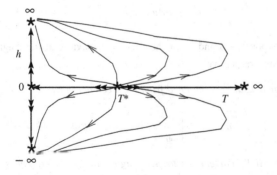

(b) Calculate the eigenvalues y_t and y_h, at the critical fixed point, to order of ϵ.

- *Linearizing $T = T^* + \delta T$ around the critical fixed point yields*

$$\begin{cases} \dfrac{\mathrm{d}\delta T}{\mathrm{d}\ell} = -\epsilon\,\delta T + T^*\delta T = \epsilon\,\delta T \\[2mm] \dfrac{\mathrm{d}h}{\mathrm{d}\ell} = (1 + \epsilon)h \end{cases} \qquad \Longrightarrow \qquad \begin{cases} y_t = +\epsilon \\[1mm] y_h = 1 + \epsilon. \end{cases}$$

(c) Starting from the relation governing the change of the correlation length ξ under renormalization, show that $\xi(t, h) = t^{-\nu} g_\xi \left(h/|t|^\Delta \right)$ (where $t = T/T_c - 1$), and find the exponents ν and Δ.

- *Under rescaling by a factor of b, the correlation length is reduced by b, resulting in the homogeneity relation*

$$\xi(t, h) = b\xi(b^{y_t} t, b^{y_h} h).$$

Upon selecting a rescaling factor such that $b^{y_t} t \sim 1$, we obtain

$$\xi(t, h) = t^{-\nu} g_\xi \left(h/|t|^\Delta \right),$$

with

$$\nu = \frac{1}{y_t} = \frac{1}{\epsilon}, \quad \text{and} \quad \Delta = \frac{y_h}{y_t} = \frac{1}{\epsilon} + 1.$$

(d) Use a hyperscaling relation to find the singular part of the free energy $f_{\text{sing}}(t, h)$, and hence the heat capacity exponent α.

- *According to hyperscaling*

$$f_{\text{sing}}(t, h) \propto \xi(t, h)^{-d} = t^{d/y_t} g_f \left(h/|t|^\Delta \right).$$

Taking two derivatives with respect to t leads to the heat capacity, whose singularity for $h = 0$ is described by the exponent

$$\alpha = 2 - d\nu = 2 - \frac{1 + \epsilon}{\epsilon} = -\frac{1}{\epsilon} + 1.$$

(e) Find the exponents β and γ for the singular behaviors of the magnetization and susceptibility, respectively.

- *The magnetization is obtained from the free energy by*

$$m = -\left. \frac{\partial f}{\partial h} \right|_{h=0} \sim |t|^\beta, \quad \text{with} \quad \beta = \frac{d - y_h}{y_t} = 0.$$

(There will be corrections to β at higher orders in ϵ.) The susceptibility is obtained from a derivative of the magnetization, or

$$\chi = -\left. \frac{\partial^2 f}{\partial h^2} \right|_{h=0} \sim |t|^{-\gamma}, \quad \text{with} \quad \gamma = \frac{2y_h - d}{y_t} = \frac{1 + \epsilon}{\epsilon} = \frac{1}{\epsilon} + 1.$$

(f) Starting with the relation between susceptibility and correlations of local magnetizations, calculate the exponent η for the critical correlations ($\langle m(\mathbf{0})m(\mathbf{x}) \rangle \sim |\mathbf{x}|^{-(d-2+\eta)}$).

- *The magnetic susceptibility is related to the connected correlation function via*

$$\chi = \int d^d \mathbf{x} \, \langle m(\mathbf{0})m(\mathbf{x}) \rangle_c \, .$$

Close to criticality, the correlations decay as a power law $\langle m(\mathbf{0})m(\mathbf{x}) \rangle \sim |\mathbf{x}|^{-(d-2+\eta)}$, *which is cut off at the correlation length* ξ, *resulting in*

$$\chi \sim \xi^{(2-\eta)} \sim |t|^{-(2-\eta)\nu}.$$

From the corresponding exponent identity, we find

$$\gamma = (2-\eta)\nu, \quad \Longrightarrow \quad \eta = 2 - y_t \gamma = 2 - 2y_h + d = 2 - d = 1 - \epsilon.$$

(g) How does the correlation length diverge as $T \to 0$ (along $h = 0$) for $d = 1$?

- *For $d = 1$, the recursion relation for temperature can be rearranged and integrated, i.e.*

$$\frac{1}{T^2} \frac{dT}{d\ell} = \frac{1}{2}, \quad \Longrightarrow \quad d\left(-\frac{2}{T}\right) = d\ell.$$

We can integrate the above expression from a low temperature with correlation length $\xi(T)$ to a high temperature where $1/T \approx 0$, and at which the correlation length is of the order of the lattice spacing, to get

$$-\frac{2}{T} = \ln\left(\frac{\xi}{a}\right) \quad \Longrightarrow \quad \xi(T) = a \exp\left(\frac{2}{T}\right).$$

Solutions to selected problems from chapter 5

1. *Longitudinal susceptibility:* While there is no reason for the longitudinal susceptibility to diverge at the mean-field level, it in fact does so due to fluctuations in dimensions $d < 4$. This problem is intended to show you the origin of this divergence in perturbation theory. There are actually a number of subtleties in this calculation which you are instructed to ignore at various steps. You may want to think about why they are justified.

Consider the Landau–Ginzburg Hamiltonian:

$$\beta \mathcal{H} = \int d^d \mathbf{x} \left[\frac{K}{2} (\nabla \vec{m})^2 + \frac{t}{2} \vec{m}^2 + u(\vec{m}^2)^2 \right],$$

describing an n-component magnetization vector $\vec{m}(\mathbf{x})$, in the ordered phase for $t < 0$.

(a) Let $\vec{m}(\mathbf{x}) = (\overline{m} + \phi_\ell(\mathbf{x}))\hat{e}_\ell + \vec{\phi}_t(\mathbf{x})\hat{e}_t$, and expand $\beta \mathcal{H}$ *keeping all terms in the expansion.*

- *With* $\vec{m}(\mathbf{x}) = (\overline{m} + \phi_\ell(\mathbf{x}))\hat{e}_\ell + \vec{\phi}_t(\mathbf{x})\hat{e}_t$, *and* $\overline{m}$ *the minimum of* $\beta \mathcal{H}$,

$$\begin{aligned}
\beta \mathcal{H} = \; & V \left(\frac{t}{2} \overline{m}^2 + u\overline{m}^4 \right) + \int d^d x \left\{ \frac{K}{2} \left[(\nabla \phi_\ell)^2 + \left(\nabla \vec{\phi}_t \right)^2 \right] + \left(\frac{t}{2} + 6u\overline{m}^2 \right) \phi_\ell^2 \right. \\
& + \left(\frac{t}{2} + 2u\overline{m}^2 \right) \vec{\phi}_t^{\,2} + 4u\overline{m} \left(\phi_\ell^3 + \phi_\ell \vec{\phi}_t^{\,2} \right) \\
& \left. + u \left[\phi_\ell^4 + 2\phi_\ell^2 \vec{\phi}_t^{\,2} + \left(\vec{\phi}_t^{\,2} \right)^2 \right] \right\}.
\end{aligned}$$

Since $\overline{m}^2 = -t/4u$ *in the ordered phase* ($t < 0$), *this expression can be simplified, upon dropping the constant term, as*

$$\begin{aligned}
\beta \mathcal{H} = \int d^d x \left\{ \frac{K}{2} \left[(\nabla \phi_\ell)^2 + \left(\nabla \vec{\phi}_t \right)^2 \right] - t\phi_\ell^2 + 4u\overline{m} \left(\phi_\ell^3 + \phi_\ell \vec{\phi}_t^{\,2} \right) \right. \\
\left. + u \left[\phi_\ell^4 + 2\phi_\ell^2 \vec{\phi}_t^{\,2} + \left(\vec{\phi}_t^{\,2} \right)^2 \right] \right\}.
\end{aligned}$$

(b) Regard the quadratic terms in ϕ_ℓ and $\vec{\phi}_t$ as an unperturbed Hamiltonian $\beta \mathcal{H}_0$, and the lowest order term coupling ϕ_ℓ and $\vec{\phi}_t$ as a perturbation U; i.e.

$$U = 4u\overline{m} \int d^d \mathbf{x} \, \phi_\ell(\mathbf{x}) \vec{\phi}_t(\mathbf{x})^2.$$

Write U in Fourier space in terms of $\phi_\ell(\mathbf{q})$ and $\vec{\phi}_t(\mathbf{q})$.

- *We shall focus on the cubic term as a perturbation*

$$U = 4u\overline{m} \int d^d x \phi_\ell(\mathbf{x})\, \vec{\phi}_t(\mathbf{x})^2,$$

which can be written in Fourier space as

$$U = 4u\overline{m} \int \frac{d^d q}{(2\pi)^d} \frac{d^d q'}{(2\pi)^d} \phi_\ell(-\mathbf{q}-\mathbf{q}')\, \vec{\phi}_t(\mathbf{q}) \cdot \vec{\phi}_t(\mathbf{q}').$$

(c) Calculate the Gaussian (bare) expectation values $\langle \phi_\ell(\mathbf{q})\phi_\ell(\mathbf{q}')\rangle_0$ and $\langle \phi_{t,\alpha}(\mathbf{q})\phi_{t,\beta}(\mathbf{q}')\rangle_0$, and the corresponding momentum dependent susceptibilities $\chi_\ell(\mathbf{q})_0$ and $\chi_t(\mathbf{q})_0$.

- *From the quadratic part of the Hamiltonian,*

$$\beta \mathcal{H}_0 = \int d^d x \frac{1}{2} \left\{ K \left[(\nabla \phi_\ell)^2 + \left(\nabla \vec{\phi}_t\right)^2 \right] - 2t\phi_\ell^2 \right\},$$

we read off the expectation values

$$\begin{cases} \langle \phi_\ell(\mathbf{q})\, \phi_\ell(\mathbf{q}')\rangle_0 = \dfrac{(2\pi)^d\, \delta^d(\mathbf{q}+\mathbf{q}')}{Kq^2 - 2t} \\[2mm] \langle \phi_{t,\alpha}(\mathbf{q})\, \phi_{t,\beta}(\mathbf{q}')\rangle_0 = \dfrac{(2\pi)^d\, \delta^d(\mathbf{q}+\mathbf{q}')\, \delta_{\alpha\beta}}{Kq^2}, \end{cases}$$

and the corresponding susceptibilities

$$\begin{cases} \chi_\ell(\mathbf{q})_0 = \dfrac{1}{Kq^2 - 2t} \\[2mm] \chi_t(\mathbf{q})_0 = \dfrac{1}{Kq^2}. \end{cases}$$

(d) Calculate $\langle \vec{\phi}_t(\mathbf{q}_1) \cdot \vec{\phi}_t(\mathbf{q}_2)\, \vec{\phi}_t(\mathbf{q}_1') \cdot \vec{\phi}_t(\mathbf{q}_2')\rangle_0$ using Wick's theorem. (Don't forget that $\vec{\phi}_t$ is an $(n-1)$ component vector.)

- *Using Wick's theorem,*

$$\left\langle \vec{\phi}_t(\mathbf{q}_1) \cdot \vec{\phi}_t(\mathbf{q}_2) \vec{\phi}_t(\mathbf{q}_1') \cdot \vec{\phi}_t(\mathbf{q}_2') \right\rangle_0 \equiv \left\langle \phi_{t,\alpha}(\mathbf{q}_1) \phi_{t,\alpha}(\mathbf{q}_2) \phi_{t,\beta}(\mathbf{q}_1') \phi_{t,\beta}(\mathbf{q}_2') \right\rangle_0$$

$$= \left\langle \phi_{t,\alpha}(\mathbf{q}_1) \phi_{t,\alpha}(\mathbf{q}_2)\right\rangle_0 \left\langle \phi_{t,\beta}(\mathbf{q}_1') \phi_{t,\beta}(\mathbf{q}_2')\right\rangle_0$$

$$+ \left\langle \phi_{t,\alpha}(\mathbf{q}_1) \phi_{t,\beta}(\mathbf{q}_1')\right\rangle_0 \left\langle \phi_{t,\alpha}(\mathbf{q}_2) \phi_{t,\beta}(\mathbf{q}_2')\right\rangle_0$$

$$+ \left\langle \phi_{t,\alpha}(\mathbf{q}_1) \phi_{t,\beta}(\mathbf{q}_2')\right\rangle_0 \left\langle \phi_{t,\alpha}(\mathbf{q}_2) \phi_{t,\beta}(\mathbf{q}_1')\right\rangle_0.$$

Then, from part (c),

$$\left\langle \vec{\phi}_t(\mathbf{q}_1) \cdot \vec{\phi}_t(\mathbf{q}_2) \vec{\phi}_t(\mathbf{q}_1') \cdot \vec{\phi}_t(\mathbf{q}_2') \right\rangle_0 = \frac{(2\pi)^{2d}}{K^2} \left\{ (n-1)^2 \frac{\delta^d(\mathbf{q}_1+\mathbf{q}_2)\, \delta^d(\mathbf{q}_1'+\mathbf{q}_2')}{q_1^2 q_1'^2} \right.$$

$$+ (n-1) \frac{\delta^d(\mathbf{q}_1+\mathbf{q}_1')\, \delta^d(\mathbf{q}_2+\mathbf{q}_2')}{q_1^2 q_2^2}$$

$$\left. + (n-1) \frac{\delta^d(\mathbf{q}_1+\mathbf{q}_2')\, \delta^d(\mathbf{q}_1'+\mathbf{q}_2)}{q_1^2 q_2^2} \right\},$$

since $\delta_{\alpha\alpha}\delta_{\beta\beta} = (n-1)^2$, and $\delta_{\alpha\beta}\delta_{\alpha\beta} = (n-1)$.

(e) Write down the expression for $\langle \phi_\ell(\mathbf{q}) \phi_\ell(\mathbf{q}') \rangle$ to second order in the perturbation U. Note that since U is odd in ϕ_ℓ, only two terms at the second order are non-zero.

- *Including the perturbation U in the calculation of the correlation function, we have*

$$\langle \phi_\ell(\mathbf{q}) \phi_\ell(\mathbf{q}') \rangle = \frac{\langle \phi_\ell(\mathbf{q}) \phi_\ell(\mathbf{q}') e^{-U} \rangle_0}{\langle e^{-U} \rangle_0} = \frac{\langle \phi_\ell(\mathbf{q}) \phi_\ell(\mathbf{q}') (1 - U + U^2/2 + \cdots) \rangle_0}{\langle (1 - U + U^2/2 + \cdots) \rangle_0}.$$

Since U is odd in ϕ_ℓ, $\langle U \rangle_0 = \langle \phi_\ell(\mathbf{q}) \phi_\ell(\mathbf{q}') U \rangle_0 = 0$. Thus, after expanding the denominator to second order,

$$\frac{1}{1 + \langle U^2/2 \rangle_0 + \cdots} = 1 - \left\langle \frac{U^2}{2} \right\rangle_0 + \mathcal{O}(U^3),$$

we obtain

$$\langle \phi_\ell(\mathbf{q}) \phi_\ell(\mathbf{q}') \rangle = \langle \phi_\ell(\mathbf{q}) \phi_\ell(\mathbf{q}') \rangle_0 + \frac{1}{2} \left(\langle \phi_\ell(\mathbf{q}) \phi_\ell(\mathbf{q}') U^2 \rangle_0 \right.$$

$$\left. - \langle \phi_\ell(\mathbf{q}) \phi_\ell(\mathbf{q}') \rangle_0 \langle U^2 \rangle_0 \right).$$

(f) Using the form of U in Fourier space, write the correction term as a product of two four-point expectation values similar to those of part (d). Note that only connected terms for the longitudinal four-point function should be included.

- *Substituting for U its expression in terms of Fourier transforms from part (b), the fluctuation correction to the correlation function reads*

$$G_F(\mathbf{q}, \mathbf{q}') \equiv \langle \phi_\ell(\mathbf{q}) \phi_\ell(\mathbf{q}') \rangle - \langle \phi_\ell(\mathbf{q}) \phi_\ell(\mathbf{q}') \rangle_0$$

$$= \frac{1}{2} (4u\bar{m})^2 \int \frac{d^d q_1}{(2\pi)^d} \frac{d^d q_2}{(2\pi)^d} \frac{d^d q_1'}{(2\pi)^d} \frac{d^d q_2'}{(2\pi)^d}$$

$$\times \left\langle \phi_\ell(\mathbf{q}) \phi_\ell(\mathbf{q}') \phi_\ell(-\mathbf{q}_1 - \mathbf{q}_2) \vec{\phi}_t(\mathbf{q}_1) \right.$$

$$\times \vec{\phi}_t(\mathbf{q}_2) \times \phi_\ell(-\mathbf{q}_1' - \mathbf{q}_2') \vec{\phi}_t(\mathbf{q}_1') \cdot \vec{\phi}_t(\mathbf{q}_2') \Big\rangle_0$$

$$- \frac{1}{2} \langle \phi_\ell(\mathbf{q}) \phi_\ell(\mathbf{q}') \rangle_0 \langle U^2 \rangle_0,$$

i.e. $G_F(\mathbf{q}, \mathbf{q}')$ is calculated as the connected part of

$$\frac{1}{2} (4u\bar{m})^2 \int \frac{d^d q_1}{(2\pi)^d} \frac{d^d q_2}{(2\pi)^d} \frac{d^d q_1'}{(2\pi)^d} \frac{d^d q_2'}{(2\pi)^d}$$

$$\times \langle \phi_\ell(\mathbf{q}) \phi_\ell(\mathbf{q}') \phi_\ell(-\mathbf{q}_1 - \mathbf{q}_2) \phi_\ell(-\mathbf{q}_1' - \mathbf{q}_2') \rangle_0$$

$$\times \left\langle \vec{\phi}_t(\mathbf{q}_1) \cdot \vec{\phi}_t(\mathbf{q}_2) \vec{\phi}_t(\mathbf{q}_1') \cdot \vec{\phi}_t(\mathbf{q}_2') \right\rangle_0,$$

where we have used the fact that the unperturbed averages of products of longitudinal and transverse fields factorize. Hence

$$
\begin{aligned}
G_F(\mathbf{q}, \mathbf{q}') = {} & \frac{1}{2} (4u\overline{m})^2 \int \frac{\mathrm{d}^d q_1}{(2\pi)^d} \frac{\mathrm{d}^d q_2}{(2\pi)^d} \frac{\mathrm{d}^d q_1'}{(2\pi)^d} \frac{\mathrm{d}^d q_2'}{(2\pi)^d} \\
& \times \left\langle \vec{\phi}_t(\mathbf{q}_1) \cdot \vec{\phi}_t(\mathbf{q}_2)\, \vec{\phi}_t(\mathbf{q}_1') \cdot \vec{\phi}_t(\mathbf{q}_2') \right\rangle_0 \\
& \times \Big\{ \langle \phi_\ell(\mathbf{q})\, \phi_\ell(-\mathbf{q}_1 - \mathbf{q}_2) \rangle_0 \, \langle \phi_\ell(\mathbf{q}')\, \phi_\ell(-\mathbf{q}_1' - \mathbf{q}_2') \rangle_0 \\
& \quad + \langle \phi_\ell(\mathbf{q})\, \phi_\ell(-\mathbf{q}_1' - \mathbf{q}_2') \rangle_0 \, \langle \phi_\ell(\mathbf{q}')\, \phi_\ell(-\mathbf{q}_1 - \mathbf{q}_2) \rangle_0 \Big\} \\
= {} & 2 \times \frac{1}{2} (4u\overline{m})^2 \int \frac{\mathrm{d}^d q_1}{(2\pi)^d} \frac{\mathrm{d}^d q_2}{(2\pi)^d} \frac{\mathrm{d}^d q_1'}{(2\pi)^d} \frac{\mathrm{d}^d q_2'}{(2\pi)^d} \\
& \times \left\langle \vec{\phi}_t(\mathbf{q}_1) \cdot \vec{\phi}_t(\mathbf{q}_2)\, \vec{\phi}_t(\mathbf{q}_1') \cdot \vec{\phi}_t(\mathbf{q}_2') \right\rangle_0 \\
& \times \langle \phi_\ell(\mathbf{q})\, \phi_\ell(-\mathbf{q}_1 - \mathbf{q}_2) \rangle_0 \, \langle \phi_\ell(\mathbf{q}')\, \phi_\ell(-\mathbf{q}_1' - \mathbf{q}_2') \rangle_0 .
\end{aligned}
$$

Using the results of parts (c) and (d) for the two and four point correlation functions, and since $u^2 \overline{m}^2 = -ut/4$, we obtain

$$
\begin{aligned}
G_F(\mathbf{q}, \mathbf{q}') = {} & 4u(-t) \int \frac{\mathrm{d}^d q_1}{(2\pi)^d} \frac{\mathrm{d}^d q_2}{(2\pi)^d} \frac{\mathrm{d}^d q_1'}{(2\pi)^d} \frac{\mathrm{d}^d q_2'}{(2\pi)^d} \frac{(2\pi)^{2d}}{K^2} \left\{ (n-1)^2 \frac{\delta^d(\mathbf{q}_1 + \mathbf{q}_2)\, \delta^d(\mathbf{q}_1' + \mathbf{q}_2')}{q_1^2 q_1'^2} \right. \\
& \left. + (n-1) \frac{\delta^d(\mathbf{q}_1 + \mathbf{q}_1')\, \delta^d(\mathbf{q}_2 + \mathbf{q}_2') + \delta^d(\mathbf{q}_1 + \mathbf{q}_2')\, \delta^d(\mathbf{q}_1' + \mathbf{q}_2)}{q_1^2 q_2^2} \right\} \\
& \times \frac{(2\pi)^d \delta^d(\mathbf{q} - \mathbf{q}_1 - \mathbf{q}_2)}{Kq^2 - 2t} \frac{(2\pi)^d \delta^d(\mathbf{q}' - \mathbf{q}_1' - \mathbf{q}_2')}{Kq^2 - 2t} ,
\end{aligned}
$$

which, after doing some of the integrals, reduces to

$$
\begin{aligned}
G_F(\mathbf{q}, \mathbf{q}') = {} & \frac{4u(-t)}{K^2} \left\{ (n-1)^2 \frac{\delta^d(\mathbf{q})\, \delta^d(\mathbf{q}')}{4t^2} \left(\int \frac{\mathrm{d}^d q_1}{q_1^2} \right)^2 \right. \\
& \left. + 2(n-1) \frac{\delta^d(\mathbf{q} + \mathbf{q}')}{(Kq^2 - 2t)^2} \int \frac{\mathrm{d}^d q_1}{q_1^2 (\mathbf{q} + \mathbf{q}_1)^2} \right\} .
\end{aligned}
$$

(g) Ignore the disconnected term obtained in (d) (i.e. the part proportional to $(n-1)^2$), and write down the expression for $\chi_\ell(\mathbf{q})$ in second order perturbation theory.

- *From the dependence of the first term (proportional to $\delta^d(\mathbf{q})\, \delta^d(\mathbf{q}')$), we deduce that this term is actually a correction to the unperturbed value of the magnetization, i.e.*

$$
\overline{m} \to \overline{m} \left[1 - \frac{2(n-1)u}{Kt} \left(\int \frac{\mathrm{d}^d q_1}{q_1^2} \right) \right],
$$

and does not contribute to the correlation function at non-zero separation. The spatially varying part of the connected correlation function is thus

$$
\langle \phi_\ell(\mathbf{q})\, \phi_\ell(\mathbf{q}') \rangle = \frac{(2\pi)^d \delta^d(\mathbf{q} + \mathbf{q}')}{Kq^2 - 2t} + \frac{8u(-t)}{K^2} (n-1) \frac{\delta^d(\mathbf{q} + \mathbf{q}')}{(Kq^2 - 2t)^2} \int \frac{\mathrm{d}^d q_1}{q_1^2 (\mathbf{q} + \mathbf{q}_1)^2} ,
$$

leading to

$$\chi_{\ell}(\mathbf{q}) = \frac{1}{Kq^2 - 2t} + \frac{8u(-t)}{K^2} \frac{(n-1)}{(Kq^2 - 2t)^2} \int \frac{\mathrm{d}^d q_1}{(2\pi)^d} \frac{1}{q_1^2 (\mathbf{q} + \mathbf{q}_1)^2}.$$

(h) Show that for $d < 4$, the correction term diverges as q^{d-4} for $q \to 0$, implying an infinite longitudinal susceptibility.

• In $d > 4$, the above integral converges and is dominated by the large q cutoff Λ. In $d < 4$, on the other hand, the integral clearly diverges as $q \to 0$, and is thus dominated by small q_1 values. Changing the variable of integration to $\mathbf{q}_1' = \mathbf{q}_1/q$, the fluctuation correction to the susceptibility reads

$$\chi_{\ell}(\mathbf{q})_F \sim q^{d-4} \int_0^{\Lambda/q} \frac{\mathrm{d}^d q_1'}{(2\pi)^d} \frac{1}{q_1'^2 (\hat{\mathbf{q}} + \mathbf{q}_1')^2} = q^{d-4} \int_0^{\infty} \frac{\mathrm{d}^d q_1'}{(2\pi)^d} \frac{1}{q_1'^2 (\hat{\mathbf{q}} + \mathbf{q}_1')^2} + \mathcal{O}(q^0),$$

which diverges as q^{d-4} for $q \to 0$.

Note: *For a translationally invariant system,*

$$\langle \phi(\mathbf{x}) \phi(\mathbf{x}') \rangle = \varphi(\mathbf{x} - \mathbf{x}'),$$

which implies

$$\langle \phi(\mathbf{q}) \phi(\mathbf{q}') \rangle = \int \mathrm{d}^d x \mathrm{d}^d x' e^{i\mathbf{q}\cdot\mathbf{x} + i\mathbf{q}'\cdot\mathbf{x}'} \langle \phi(\mathbf{x}) \phi(\mathbf{x}') \rangle$$

$$= \int \mathrm{d}^d(x - x') \mathrm{d}^d x' e^{i\mathbf{q}\cdot(\mathbf{x} - \mathbf{x}') + i(\mathbf{q} + \mathbf{q}')\cdot\mathbf{x}'} \varphi(\mathbf{x} - \mathbf{x}')$$

$$= (2\pi)^d \delta^d(\mathbf{q} + \mathbf{q}') \psi(\mathbf{q}).$$

Consider the Hamiltonian

$$-\beta\mathcal{H}' = -\beta\mathcal{H} + \int \mathrm{d}^d x h(\mathbf{x}) \phi(\mathbf{x}) = -\beta\mathcal{H} + \int \frac{\mathrm{d}^d q}{(2\pi)^d} h(\mathbf{q}) \phi(-\mathbf{q}),$$

where $-\beta\mathcal{H}$ is a translationally invariant functional of ϕ (a one-component field for simplicity), independent of $h(\mathbf{x})$. We have

$$m(\mathbf{x} = 0) = \langle \phi(0) \rangle = \int \frac{\mathrm{d}^d q}{(2\pi)^d} \langle \phi(\mathbf{q}) \rangle,$$

and, taking a derivative,

$$\frac{\partial m}{\partial h(\mathbf{q})} = \int \frac{\mathrm{d}^d q'}{(2\pi)^d} \langle \phi(\mathbf{q}') \phi(\mathbf{q}) \rangle.$$

At $h = 0$, the system is translationally invariant, and

$$\left. \frac{\partial m}{\partial h(\mathbf{q})} \right|_{h=0} = \psi(\mathbf{q}).$$

Also, for a uniform external magnetic field, the system is translationally invariant, and

$$-\beta \mathcal{H}' = -\beta \mathcal{H} + h \int \mathrm{d}^d x \phi(\mathbf{x}) = -\beta \mathcal{H} + h \phi(\mathbf{q} = 0),$$

yielding

$$\chi = \frac{\partial m}{\partial h} = \int \frac{\mathrm{d}^d q'}{(2\pi)^d} \langle \phi(\mathbf{q}') \phi(\mathbf{q} = 0) \rangle = \psi(\mathbf{0}).$$

2. *Crystal anisotropy:* Consider a ferromagnet with a tetragonal crystal structure. Coupling of the spins to the underlying lattice may destroy their full rotational symmetry. The resulting anisotropies can be described by modifying the Landau–Ginzburg Hamiltonian to

$$\beta \mathcal{H} = \int \mathrm{d}^d x \left[\frac{K}{2} (\nabla \vec{m})^2 + \frac{t}{2} \vec{m}^2 + u (\vec{m}^2)^2 + \frac{r}{2} m_1^2 + v\, m_1^2\, \vec{m}^2 \right],$$

where $\vec{m} \equiv (m_1, \cdots, m_n)$, and $\vec{m}^2 = \sum_{i=1}^n m_i^2$ ($d = n = 3$ for magnets in three dimensions). Here $u > 0$, and *to simplify calculations we shall set $v = 0$ throughout.*

(a) For a fixed magnitude $|\vec{m}|$, what directions in the n component magnetization space are selected for $r > 0$, and for $r < 0$?

- $r > 0$ *discourages ordering along direction 1, and leads to order along the remaining $(n-1)$ directions.*
  $r < 0$ *encourages ordering along direction 1.*

(b) Using the saddle point approximation, calculate the free energies ($\ln Z$) for phases uniformly magnetized *parallel* and *perpendicular* to direction 1.

- *In the saddle point approximation for $\vec{m}(\mathbf{x}) = m \hat{e}_1$, we have*

$$\ln Z_{\mathrm{sp}} = -V \min \left[\frac{t+r}{2} m^2 + u m^4 \right]_m,$$

where $V = \int \mathrm{d}^d \mathbf{x}$, is the system volume. The minimum is obtained for

$$(t+r)\overline{m} + 4u\overline{m}^3 = 0, \implies \overline{m} = \begin{cases} 0 & \text{for } t+r > 0 \\ \sqrt{-(t+r)/4u} & \text{for } t+r < 0. \end{cases}$$

For $t + r < 0$, the free energy is given by

$$f_{\mathrm{sp}} = -\frac{\ln Z_{\mathrm{sp}}}{V} = -\frac{(t+r)^2}{16u}.$$

When the magnetization is perpendicular to direction 1, i.e. for $\vec{m}(\mathbf{x}) = m \hat{e}_i$ for $i \neq 1$, the corresponding expressions are

$$\ln Z_{\mathrm{sp}} = -V \min \left[\frac{t}{2} m^2 + u m^4 \right]_m, \quad t\overline{m} + 4u\overline{m}^3 = 0, \quad \overline{m} = \begin{cases} 0 & \text{for } t > 0 \\ \sqrt{-t/4u} & \text{for } t < 0, \end{cases}$$

and the free energy for $t < 0$ is

$$f_{\text{sp}} = -\frac{t^2}{16u}.$$

(c) Sketch the phase diagram in the (t, r) plane, and indicate the phases (type of order), and the nature of the phase transitions (continuous or discontinuous).

- *The saddle point phase diagram is sketched in the figure.*

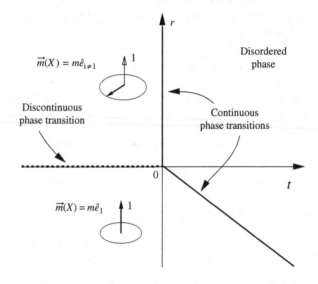

(d) Are there Goldstone modes in the ordered phases?

- *There are no Goldstone modes in the phase with magnetization aligned along direction 1, as the broken symmetry in this case is discrete. However, there are $(n-2)$ Goldstone modes in the phase where magnetization is perpendicular to direction 1.*

(e) For $u = 0$, and positive t and r, calculate the unperturbed averages $\langle m_1(\mathbf{q})m_1(\mathbf{q}')\rangle_0$ and $\langle m_2(\mathbf{q})m_2(\mathbf{q}')\rangle_0$, where $m_i(\mathbf{q})$ indicates the Fourier transform of $m_i(\mathbf{x})$.

- *The Gaussian part of the Hamiltonian can be decomposed into Fourier modes as*

$$\beta \mathcal{H}_0 = \int \frac{d^d\mathbf{q}}{(2\pi)^d} \left[\frac{K}{2} q^2 |\vec{m}(\mathbf{q})|^2 + \frac{t+r}{2} |m_1(\mathbf{q})|^2 + \sum_{i=2}^{n} \frac{t}{2} |m_i(\mathbf{q})|^2 \right].$$

From this form we can easily read off the covariances

$$\begin{cases} \langle m_1(\mathbf{q})m_1(\mathbf{q}')\rangle_0 = \dfrac{(2\pi)^d\delta^d(\mathbf{q}+\mathbf{q}')}{t+r+Kq^2} \\[4mm] \langle m_2(\mathbf{q})m_2(\mathbf{q}')\rangle_0 = \dfrac{(2\pi)^d\delta^d(\mathbf{q}+\mathbf{q}')}{t+Kq^2}. \end{cases}$$

(f) Write the fourth order term $\mathcal{U} \equiv u \int d^d\mathbf{x}(\vec{m}^2)^2$, in terms of the Fourier modes $m_i(\mathbf{q})$.

- *Substituting $m_i(\mathbf{x}) = \int \frac{d^d\mathbf{q}}{(2\pi)^d}\exp(i\mathbf{q}\cdot\mathbf{x})m_i(\mathbf{q})$ in the quartic term, and integrating over $\mathbf{x}$ yields*

$$\mathcal{U} = u\int d^d\mathbf{x}(\vec{m}^2)^2 = u\int \frac{d^dq_1 d^dq_2 d^dq_3}{(2\pi)^{3d}}\sum_{i,j=1}^n m_i(\mathbf{q}_1)m_i(\mathbf{q}_2)m_j(\mathbf{q}_3)m_j(-\mathbf{q}_1-\mathbf{q}_2-\mathbf{q}_3).$$

(g) Treating $\mathcal{U}$ as a perturbation, calculate the *first order* correction to $\langle m_1(\mathbf{q})m_1(\mathbf{q}')\rangle$. (You can leave your answers in the form of some integrals.)

- *In first order perturbation theory $\langle \mathcal{O}\rangle = \langle \mathcal{O}\rangle_0 - (\langle \mathcal{O}\mathcal{U}\rangle_0 - \langle \mathcal{O}\rangle_0 \langle \mathcal{U}\rangle_0)$, and hence*

$$\langle m_1(\mathbf{q})m_1(\mathbf{q}')\rangle = \langle m_1(\mathbf{q})m_1(\mathbf{q}')\rangle_0 - u\int\frac{d^dq_1 d^dq_2 d^dq_3}{(2\pi)^{3d}}$$
$$\times \sum_{i,j=1}^n \left(\langle m_1(\mathbf{q})m_1(\mathbf{q}')m_i(\mathbf{q}_1)m_i(\mathbf{q}_2)m_j(\mathbf{q}_3)m_j(-\mathbf{q}_1-\mathbf{q}_2-\mathbf{q}_3)\rangle_0^c\right)$$
$$= \frac{(2\pi)^d\delta^d(\mathbf{q}+\mathbf{q}')}{t+r+Kq^2}\left\{1 - \frac{u}{t+r+Kq^2}\right.$$
$$\left.\times \int\frac{d^d\mathbf{k}}{(2\pi)^d}\left[\frac{4(n-1)}{t+Kk^2}+\frac{4}{t+r+Kk^2}+\frac{8}{t+r+Kk^2}\right]\right\}.$$

The last result is obtained by listing all possible contractions, and keeping track of how many involve m_1 versus $m_{i\neq1}$. The final result can be simplified to

$$\langle m_1(\mathbf{q})m_1(\mathbf{q}')\rangle = \frac{(2\pi)^d\delta^d(\mathbf{q}+\mathbf{q}')}{t+r+Kq^2}$$
$$\times \left\{1 - \frac{u}{t+r+Kq^2}\int\frac{d^d\mathbf{k}}{(2\pi)^d}\left[\frac{n-1}{t+Kk^2}+\frac{3}{t+r+Kk^2}\right]\right\}.$$

(h) Treating $\mathcal{U}$ as a perturbation, calculate the *first order* correction to $\langle m_2(\mathbf{q})m_2(\mathbf{q}')\rangle$.

- *Similar analysis yields*

$$\langle m_2(\mathbf{q})m_2(\mathbf{q}')\rangle = \langle m_2(\mathbf{q})m_2(\mathbf{q}')\rangle_0 - u \int \frac{d^d q_1 d^d q_2 d^d q_3}{(2\pi)^{3d}}$$

$$\times \sum_{i,j=1}^{n} (\langle m_2(\mathbf{q})m_2(\mathbf{q}')m_i(\mathbf{q}_1)m_i(\mathbf{q}_2)m_j(\mathbf{q}_3)m_j(-\mathbf{q}_1-\mathbf{q}_2-\mathbf{q}_3)\rangle_0^c)$$

$$= \frac{(2\pi)^d \delta^d(\mathbf{q}+\mathbf{q}')}{t+Kq^2} \left\{ 1 - \frac{u}{t+Kq^2} \int \frac{d^d k}{(2\pi)^d} \right.$$

$$\times \left[\frac{4(n-1)}{t+Kk^2} + \frac{4}{t+r+Kk^2} + \frac{8}{t+Kk^2} \right] \right\}$$

$$= \frac{(2\pi)^d \delta^d(\mathbf{q}+\mathbf{q}')}{t+Kq^2} \left\{ 1 - \frac{u}{t+Kq^2} \int \frac{d^d k}{(2\pi)^d} \left[\frac{n+1}{t+Kk^2} + \frac{1}{t+r+Kk^2} \right] \right\}.$$

(i) Using the above answer, identify the inverse susceptibility χ_{22}^{-1}, and then find the transition point, t_c, from its vanishing to first order in u.

- *Using the fluctuation–response relation, the susceptibility is given by*

$$\chi_{22} = \int d^d x \langle m_2(\mathbf{x})m_2(\mathbf{0})\rangle = \int \frac{d^d q}{(2\pi)^d} \langle m_2(\mathbf{q})m_2(\mathbf{q}=0)\rangle$$

$$= \frac{1}{t} \left\{ 1 - \frac{u}{t} \int \frac{d^d k}{(2\pi)^d} \left[\frac{n+1}{t+Kk^2} + \frac{1}{t+r+Kk^2} \right] \right\}.$$

Inverting the correction term gives

$$\chi_{22}^{-1} = t + 4u \int \frac{d^d k}{(2\pi)^d} \left[\frac{n+1}{t+Kk^2} + \frac{1}{t+r+Kk^2} \right] + O(u^2).$$

The susceptibility diverges at

$$t_c = -4u \int \frac{d^d k}{(2\pi)^d} \left[\frac{n+1}{Kk^2} + \frac{1}{r+Kk^2} \right] + O(u^2).$$

(j) Is the critical behavior different from the isotropic $O(n)$ model in $d < 4$? In RG language, is the parameter r *relevant* at the $\mathcal{O}(n)$ fixed point? In either case indicate the universality classes expected for the transitions.

- *The parameter r changes the symmetry of the ordered state, and hence the universality class of the disordering transition. The transition belongs to the $O(n-1)$ universality class for $r > 0$, and to the Ising class for $r < 0$. Any RG transformation must thus find r to be a relevant perturbation to the $O(n)$ fixed point.*

3. *Cubic anisotropy – Mean-field treatment:* Consider the modified Landau–Ginzburg Hamiltonian

$$\beta \mathcal{H} = \int d^d x \left[\frac{K}{2} (\nabla \vec{m})^2 + \frac{t}{2} \vec{m}^2 + u(\vec{m}^2)^2 + v \sum_{i=1}^{n} m_i^4 \right],$$

for an n-component vector $\vec{m}(\mathbf{x}) = (m_1, m_2, \cdots, m_n)$. The "cubic anisotropy" term $\sum_{i=1}^{n} m_i^4$ breaks the full rotational symmetry and selects specific directions.

(a) For a fixed magnitude $|\vec{m}|$, what directions in the n component magnetization space are selected for $v > 0$ and for $v < 0$? What is the degeneracy of easy magnetization axes in each case?

- *In the figures below, we indicate the possible directions of the magnetization which are selected depending upon the sign of the coefficient v, for the simple case of $n = 2$:*

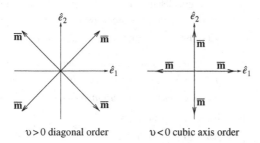

$\upsilon > 0$ diagonal order $\upsilon < 0$ cubic axis order

This qualitative behavior can be generalized for an n-component vector: For $v > 0$, $\overline{\mathbf{m}}$ lies along the diagonals of an n-dimensional hypercube, which can be labeled as

$$\overline{\mathbf{m}} = \frac{\overline{m}}{\sqrt{n}}(\pm 1, \pm 1, \ldots, \pm 1),$$

and are consequently 2^n-fold degenerate. Conversely, for $v < 0$, $\overline{\mathbf{m}}$ can point along any of the cubic axes $\hat{e}_i$, yielding

$$\overline{\mathbf{m}} = \pm \overline{m}\hat{e}_i,$$

which is 2n-fold degenerate.

(b) What are the restrictions on u and v for $\beta\mathcal{H}$ to have finite minima? Sketch these regions of stability in the (u, v) plane.

- *The Landau–Ginzburg Hamiltonian for each of these cases evaluates to*

$$\begin{cases} \beta\mathcal{H} = \dfrac{t}{2}\overline{m}^2 + u\overline{m}^4 + \dfrac{v}{n}\overline{m}^4, & \text{if } v > 0 \\[2mm] \beta\mathcal{H} = \dfrac{t}{2}\overline{m}^2 + u\overline{m}^4 + v\overline{m}^4, & \text{if } v < 0. \end{cases}$$

implying that there are finite minima provided that

$$\begin{cases} u + \dfrac{v}{n} > 0, & \text{for } v > 0, \\[2mm] u + v > 0, & \text{for } v < 0. \end{cases}$$

The figure below depicts these regions in the (u, v) plane.

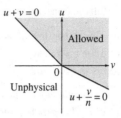

(c) In general, higher order terms (e.g. $u_6(\vec{m}^2)^3$ with $u_6 > 0$) are present and insure stability in the regions not allowed in part (b) (as in the case of the tricritical point discussed in earlier problems). With such terms in mind, sketch the saddle point phase diagram in the (t, v) plane for $u > 0$, clearly identifying the phases, and order of the transition lines.

- *We need to take into account higher order terms to ensure stability in the regions not allowed in part (b). There is a tricritical point which can be obtained after simultaneously solving the equations*

$$\begin{cases} t + 4(u+v)m^2 + 6u_6 m^4 = 0 \\ t + 2(u+v)m^2 + 2u_6 m^4 = 0 \end{cases} \implies t^* = \frac{(u+v)^2}{2u_6}, \quad \overline{m}^2 = -\frac{(u+v)^2}{2u_6}.$$

The saddle point phase diagram in the (t, v) plane is then as follows:

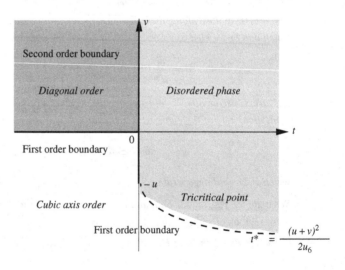

(d) Are there any Goldstone modes in the ordered phases?

- *There are no Goldstone modes in the ordered phases because the broken symmetry is discrete rather than continuous. We can easily calculate the estimated value of the transverse fluctuations in Fourier space as*

$$\langle \phi_t(\mathbf{q})\phi_t(-\mathbf{q}) \rangle = \frac{(2\pi)^d}{Kq^2 + \frac{vt}{u+v}},$$

from which we can see that indeed these modes become massless only when $v = 0$, i.e. when we retrieve the $O(n)$ symmetry.

4. Cubic anisotropy – ε-expansion:

(a) By looking at diagrams in a second order perturbation expansion in both u and v show that the recursion relations for these couplings are

$$\begin{cases} \dfrac{du}{d\ell} = \varepsilon u - 4C\left[(n+8)u^2 + 6uv \right] \\ \dfrac{dv}{d\ell} = \varepsilon v - 4C\left[12uv + 9v^2 \right], \end{cases}$$

where $C = K_d \Lambda^d / (t + K\Lambda^2)^2 \approx K_4/K^2$ is approximately a constant.

- *Let us write the Hamiltonian in terms of Fourier modes*

$$\beta \mathcal{H} = \int \frac{d^d\mathbf{q}}{(2\pi)^d} \frac{t + Kq^2}{2} \, \vec{m}(\mathbf{q}) \cdot \vec{m}(-\mathbf{q})$$

$$+ u \int \frac{d^d\mathbf{q}_1 d^d\mathbf{q}_2 d^d\mathbf{q}_3}{(2\pi)^{3d}} \, m_i(\mathbf{q}_1) m_i(\mathbf{q}_2) \, m_j(\mathbf{q}_3) m_j(-\mathbf{q}_1 - \mathbf{q}_2 - \mathbf{q}_3)$$

$$+ v \int \frac{d^d\mathbf{q}_1 d^d\mathbf{q}_2 d^d\mathbf{q}_3}{(2\pi)^{3d}} \, m_i(\mathbf{q}_1) m_i(\mathbf{q}_2) \, m_i(\mathbf{q}_3) m_i(-\mathbf{q}_1 - \mathbf{q}_2 - \mathbf{q}_3),$$

where, as usual, we assume summation over repeated indices. After the three steps of the RG transformation, we obtain the renormalized parameters:

$$\begin{cases} t' = b^{-d} z^2 \tilde{t} \\ K' = b^{-d-2} z^2 \tilde{K} \\ u' = b^{-3d} z^4 \tilde{u} \\ v' = b^{-3d} z^4 \tilde{v}, \end{cases}$$

where $\tilde{t}$, $\tilde{K}$, $\tilde{u}$, and $\tilde{v}$, are the parameters in the coarse grained Hamiltonian. The dependence of $\tilde{u}$ and $\tilde{v}$ on the original parameters can be obtained by looking at diagrams in a second order perturbation expansion in both u and v. Let us introduce diagrammatic representations of u and v, as

Contributions to u Contributions to v

where, again we have set $b = e^{\delta\ell}$. The new coarse grained parameters are

$$\begin{cases} \tilde{u} = u - 4C[(n+8)u^2 + 6uv]\delta\ell \\ \tilde{v} = v - 4C[9v^2 + 12uv]\delta\ell, \end{cases}$$

which after introducing the parameter $\epsilon = 4 - d$, rescaling, and renormalizing, yield the recursion relations

$$\begin{cases} \dfrac{du}{d\ell} = \epsilon u - 4C[(n+8)u^2 + 6uv] \\ \dfrac{dv}{d\ell} = \epsilon v - 4C[9v^2 + 12uv]. \end{cases}$$

(b) Find all fixed points in the (u, v) plane, and draw the flow patterns for $n < 4$ and $n > 4$. Discuss the relevance of the cubic anisotropy term near the stable fixed point in each case.

- *From the recursion relations, we can obtain the fixed points (u^*, v^*). For the sake of simplicity, from now on we will refer to the rescaled quantities $u = 4Cu$, and $v = 4Cv$, in terms of which there are four fixed points located at*

$$\begin{cases} u^* = v^* = 0 & \text{Gaussian fixed point} \\ u^* = 0, \quad v^* = \frac{\epsilon}{9} & \text{Ising fixed point} \\ u^* = \frac{\epsilon}{(n+8)}, \quad v^* = 0 & \mathcal{O}(n) \text{ fixed point} \\ u^* = \frac{\epsilon}{3n}, \quad v^* = \frac{\epsilon(n-4)}{9n} & \text{Cubic fixed point.} \end{cases}$$

Linearizing the recursion relations in the vicinity of the fixed point gives

$$A = \frac{d}{d\ell}\begin{pmatrix} \delta u \\ \delta v \end{pmatrix}_{u^*, v^*} = \begin{pmatrix} \epsilon - 2(n+8)u^* - 6v^* & -6u^* \\ -12v^* & \epsilon - 12u^* - 18v^* \end{pmatrix}\begin{pmatrix} \delta u \\ \delta v \end{pmatrix}.$$

As usual, a positive eigenvalue corresponds to an unstable direction, whereas negative ones correspond to stable directions. For each of the four fixed points, we obtain:

1. *Gaussian fixed point:* $\lambda_1 = \lambda_2 = \epsilon$, *i.e. this fixed point is doubly unstable for $\epsilon > 0$, as*

$$A = \begin{pmatrix} \epsilon & 0 \\ 0 & \epsilon \end{pmatrix}.$$

2. *Ising fixed point: This fixed point has one stable and one unstable direction, as*

$$A = \begin{pmatrix} \dfrac{\epsilon}{3} & 0 \\ -4\dfrac{\epsilon}{3} & -\epsilon \end{pmatrix}$$

corresponding to $\lambda_1 = \epsilon/3$ and $\lambda_2 = -\epsilon$. Note that for $u = 0$, the system decouples into n non-interacting one-component Ising spins.

3. $\mathcal{O}(n)$ *fixed point: The matrix*

$$A = \begin{pmatrix} -\epsilon & -6\dfrac{\epsilon}{(n+8)} \\ 0 & \dfrac{(n-4)}{(n+8)}\epsilon \end{pmatrix}$$

has eigenvalues $\lambda_1 = -\epsilon$, and $\lambda_2 = \epsilon(n-4)/(n+8)$. Hence for $n > 4$ this fixed point has one stable and one unstable direction, while for $n < 4$ both eigendirections are stable. This fixed point thus controls the critical behavior of the system for $n < 4$.

4. *Cubic fixed point: The eigenvalues of*

$$A = \begin{pmatrix} -\dfrac{(n+8)}{3} & -2 \\ -4\dfrac{(n-4)}{3} & 4-n \end{pmatrix}\dfrac{\epsilon}{n}$$

are $\lambda_1 = \epsilon(4-n)/3n$, $\lambda_2 = -\epsilon$. Thus for $n < 4$, this fixed point has one stable and one unstable direction, and for $n > 4$ both eigendirections are stable. This fixed point controls critical behavior of the system for $n > 4$.

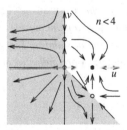 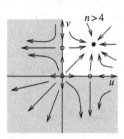

In the (u,v) plane, for $n < 4$ the flows terminate at a fixed point with $v^* = 0$, at which full rotation symmetry is restored (due to fluctuations). For $n > 4$, the RG flows culminate on a fixed point at finite v as indicated in the figure.

(c) Find the recursion relation for the reduced temperature, t, and calculate the exponent ν at the stable fixed points for $n < 4$ and $n > 4$.

• *The diagrams contributing to the determination of $\tilde{t}$ at order of ϵ and the corresponding recursion relation are presented in the figure.*

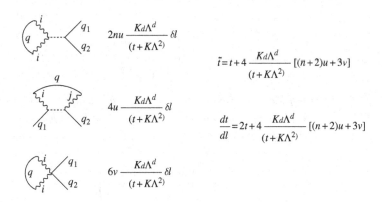

$$\tilde{t} = t + 4\,\frac{K_d \Lambda^d}{(t + K\Lambda^2)}\,[(n+2)u + 3v]$$

$$\frac{dt}{dl} = 2t + 4\,\frac{K_d \Lambda^d}{(t + K\Lambda^2)}\,[(n+2)u + 3v]$$

After linearizing in the vicinity of the stable fixed points, the exponent y_t is given by

$$y_t = 2 - 4C[(n+2)u^* + 3v^*], \Longrightarrow \nu = \frac{1}{y_t} = \begin{cases} \dfrac{1}{2} + \dfrac{(n+2)}{4(n+8)}\epsilon + \mathcal{O}(\epsilon^2) & \text{for } n < 4 \\[2mm] \dfrac{1}{2} + \dfrac{(n-1)}{6n}\epsilon + \mathcal{O}(\epsilon^2) & \text{for } n > 4. \end{cases}$$

(d) Is the region of stability in the (u, v) plane calculated in part (b) of the previous problem enhanced or diminished by inclusion of fluctuations? Since in reality higher order terms will be present, what does this imply about the nature of the phase transition for a small negative v and $n > 4$?

• *All fixed points are located within the allowed region calculated in 1(b). However, not all flows starting in classically stable regions are attracted to stable fixed point. If the RG flows take a point outside the region of stability, then fluctuations decrease the region of stability. The domains of attraction of the fixed points for $n < 4$ and $n > 4$ are indicated in the figure.*

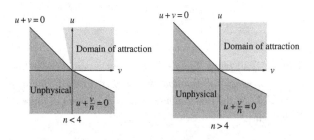

Flows which are not originally located within these domains of attraction flow into the unphysical regions. The coupling constants u and v become more negative. This is the signal of a fluctuation induced discontinuous transition, by what is known as the Coleman–Weinberg mechanism. Fluctuations are responsible for the change of order of the transition in the regions of the (u,v) plane outside the domain of attraction of the stable fixed points.

(e) Draw schematic phase diagrams in the (t, v) plane $(u > 0)$ for $n > 4$ and $n < 4$, identifying the ordered phases. Are there Goldstone modes in any of these phases close to the phase transition?

* *From the recursion relation obtained in part (c) for the parameter t, we obtain the following non-trivial t^*49*

$$t^* = -\frac{1}{2}[(n+2)u^* + 3v^*] \propto -\epsilon.$$

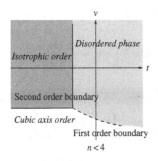 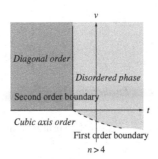

The corresponding phase diagrams in the (t,v) plane are schematically represented in the figure.

For $n < 4$ fluctuations restore the full rotational symmetry at the transition point where v is renormalized to zero. However, within the ordered phases, the renormalized value of v is finite, albeit small, indicating that there are no Goldstone modes.

5. *Exponents:* Two critical exponents at second order are,

$$\begin{cases} \nu = \tfrac{1}{2} + \dfrac{(n+2)}{4(n+8)}\epsilon + \dfrac{(n+2)(n^2 + 23n + 60)}{8(n+8)^3}\epsilon^2 \ , \\ \eta = \dfrac{(n+2)}{2(n+8)^2}\epsilon^2. \end{cases}$$

Use scaling relations to obtain ϵ–expansions for two or more of the remaining exponents α, β, γ, δ and Δ. Make a table of the results obtained by setting $\epsilon = 1$, 2 for $n = 1$, 2 and 3; and compare to the best estimates of these exponents that you can find by other sources (series, experiments, etc.).

• *The divergence of the correlation length ξ is controlled by the exponent $\nu = 1/y_t$. Since $\chi \sim \xi^{2-\eta} \sim t^{-\nu(2-\eta)}$ and $\chi \sim t^{-\gamma}$, one obtains from the Fisher exponent identity*

$$\gamma = (2-\eta)\nu, \quad \text{that} \quad \gamma = 1 + \frac{(n+2)}{2(n+8)}\epsilon + \frac{(n+2)(n^2+22n+52)}{4(n+8)^3}\epsilon^2.$$

From the hyperscaling relation we obtain the heat capacity exponent α as

$$\alpha = 2 - d\nu, \quad \text{so that} \quad \alpha = \frac{(4-n)}{2(n+8)}\epsilon - \frac{(n+2)^2(n+28)}{4(n+8)^3}\epsilon^2.$$

Using the Rushbrooke and Widom Identities (see the text), we can obtain the exponents β and δ, respectively, as

$$\alpha + 2\beta + \gamma = 2, \quad \text{yielding} \quad \beta = \frac{1}{2} - \frac{3}{2(n+8)}\epsilon + \frac{(n+2)(2n+1)}{2(n+8)^3}\epsilon^2,$$

and

$$\delta - 1 = \frac{\gamma}{\beta}, \quad \text{yielding} \quad \delta = 3 + \epsilon + \frac{(n^2+14n+60)}{2(n+8)^2}\epsilon^2.$$

Furthermore,

$$\Delta = \delta\beta, \quad \text{resulting in} \quad \Delta = \frac{3}{2} + \frac{(n-1)}{2(n+8)}\epsilon + \frac{(n+2)(n^2+26n+54)}{4(n+8)^3}\epsilon^2.$$

Substituting $\epsilon = 1, 2$ in the above expressions, we obtain for $n = 1, 2, 3$,

n	α	β	γ	δ	Δ	ν	η	$d = 4 - \epsilon$
1	0.0771	0.3395	1.2438	4.4630	1.5833	0.6265	0.0185	$3d$ Ising
2	−0.020	0.3600	1.3000	4.4600	1.6600	0.6550	0.0200	$3d$ XY
3	−0.1001	0.3768	1.3465	4.4587	1.7233	0.6784	0.0207	$3d$ Heisenberg
1	−0.0247	0.1914	1.6420	6.8518	1.8333	0.8395	0.0740	$2d$ Ising
2	−0.2800	0.2400	1.8000	6.8400	2.0400	0.9200	0.0800	$2d$ XY
2	−0.4914	0.2799	1.9316	6.8347	2.2115	0.9865	0.0826	$2d$ Heisenberg

Results from the ϵ-expansion up to five loops are:

n	β	γ	ν	η	
1	0.3260	1.2360	0.6293	0.0360	$3d$ Ising
2	0.3472	1.3120	0.6685	0.0385	$3d$ XY
3	0.3660	1.3830	0.7050	0.0380	$3d$ Heisenberg

*[From R. Guida and J. Zinn-Justin, J. Phys. **A 31**, 8130 (1998).]*

Note that these ϵ-expansion results pertain to continuum analogues of isotropic models, which correspond to an FCC lattice for $d = 3$, and to a triangular lattice for $d = 2$. No $d = 3$ model has been solved exactly. Table 8.3 of Stanley (Introduction to Phase Transitions and Critical Phenomena, Oxford, 1971) has high temperature series results for an FCC lattice with n-component spins:

n	α	β	γ	δ	ν	η
1	0.125	0.3125	1.25	5	0.638	0.041
2	0.02	0.33	1.32	5	0.675	0.04
3	−0.07	0.345	1.38	5	0.70	0.03

Some experimentally measured critical exponents are:

n	α	β	γ	ν	η	d
1	0.11	0.32	1.24	0.63	0.03	3
3	0.1	0.34	1.4	0.7		3
1	0.0	0.3	1.82	1.02		2

Experiments on $d = 3$ and $n = 1$ are compiled from liquid-gas, binary fluid, ferromagnetic, and antiferromagnetic transitions; on $d = 3$ and $n = 3$ compiled from some ferromagnetic and antiferromagnetic transitions; and on $d = 2$ and $n = 1$ compiled from some antiferromagnetic transitions.

The exponents with values close to zero (α, η) are the most problematic. The other exponents agree to within 10%.

The exact exponents for the two dimensional Ising model are

α	β	γ	δ	ν	η
0	1/8	7/4	15	1	1/4

The 2d XY model was solved exactly by Kosterlitz and Thouless (see chapter 8). It has a correlation length that diverges exponentially, $\xi \sim e^{bt^{-1/2}}$ (ν is not defined). Thus all the usual scaling laws hold provided they are expressed in terms of the correlation length:

$$\chi \sim \xi^{2-\eta}, \quad \text{with} \quad \eta = \frac{1}{4},$$

$$\chi \sim \xi^{\tilde{\gamma}}, \quad \text{with} \quad \tilde{\gamma} = \frac{\gamma}{\nu}, \quad \text{and} \quad \tilde{\gamma} = \frac{7}{4},$$

$$m \sim h^{1/\delta}, \quad \text{with} \quad \delta = 15,$$

$$C \sim \xi^{\tilde{\alpha}}, \quad \text{with} \quad \tilde{\alpha} = \frac{\alpha}{\nu} \quad \text{and} \quad \tilde{\alpha} = -2.$$

For the 2d XY model there is no spontaneous magnetization, so the exponent β is not defined. The 2d Heisenberg model does not have a finite temperature phase transition, and the extrapolated exponents are spurious.

Solutions to selected problems from chapter 6

1. *Cumulant method:* Apply the Niemeijer–van Leeuwen first order cumulant expansion to the Ising model on a *square* lattice with $-\beta\mathcal{H} = K\sum_{\langle ij\rangle}\sigma_i\sigma_j$, by following these steps:

   (a) For an RG with $b = 2$, divide the bonds into *intracell* components $\beta\mathcal{H}_0$ and *intercell* components $\mathcal{U}$.

   - *The N sites of the square lattice are partitioned into N/4 cells as indicated in the figure below (the intracell and intercell bonds are represented by solid and dashed lines respectively).*

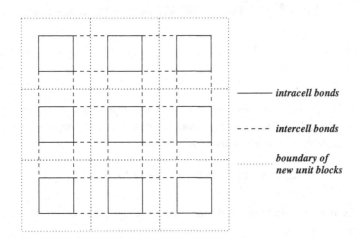

   The renormalized Hamiltonian $\beta\mathcal{H}'[\sigma'_\alpha]$ is calculated from

   $$\beta\mathcal{H}'[\sigma'_\alpha] = -\ln Z_0[\sigma'_\alpha] + \langle \mathcal{U}\rangle_0 - \frac{1}{2}\left(\langle \mathcal{U}^2\rangle_0 - \langle \mathcal{U}\rangle_0^2\right) + \mathcal{O}\left(\mathcal{U}^3\right),$$

   where $\langle\rangle_0$ indicates expectation values calculated with the weight $\exp(-\beta\mathcal{H}_0)$ at fixed $[\sigma'_\alpha]$.

(b) For each cell α, define a renormalized spin $\sigma'_\alpha = \text{sign}(\sigma^1_\alpha + \sigma^2_\alpha + \sigma^3_\alpha + \sigma^4_\alpha)$. This choice becomes ambiguous for configurations such that $\sum^4_{i=1}\sigma^i_\alpha = 0$. Distribute the weight of these configurations equally between $\sigma'_\alpha = +1$ and -1 (i.e. put a factor of 1/2 in addition to the Boltzmann weight). Make a table for all possible configurations of a cell, the internal probability $\exp(-\beta\mathcal{H}_0)$, and the weights contributing to $\sigma'_\alpha = \pm 1$.

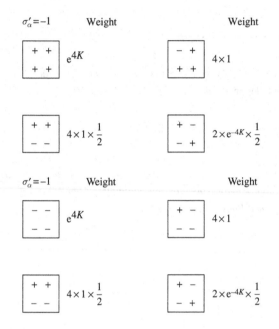

- *The possible intracell configurations compatible with a renormalized spin $\sigma'_\alpha = \pm 1$, and their corresponding contributions to the intracell probability $\exp(-\beta\mathcal{H}_0)$, are given above, resulting in*

$$Z_0\left[\sigma'_\alpha\right] = \prod_\alpha \left(e^{4K} + 6 + e^{-4K}\right) = \left(e^{4K} + 6 + e^{-4K}\right)^{N/4}.$$

(c) Express $\langle \mathcal{U}\rangle_0$ in terms of the cell spins σ'_α, and hence obtain the recursion relation $K'(K)$.

- *The first cumulant of the interaction term is*

$$-\langle \mathcal{U}\rangle_0 = K\sum_{\langle\alpha,\beta\rangle}\left\langle\sigma_{\alpha 2}\sigma_{\beta 1} + \sigma_{\alpha 3}\sigma_{\beta 4}\right\rangle_0 = 2K\sum_{\langle\alpha,\beta\rangle}\left\langle\sigma_{\alpha 2}\right\rangle_0\left\langle\sigma_{\beta 1}\right\rangle_0,$$

where, for $\sigma'_\alpha = 1$,

$$\langle \sigma_{\alpha i} \rangle_0 = \frac{e^{4K} + (3-1) + 0 + 0}{(e^{4K} + 6 + e^{-4K})} = \frac{e^{4K} + 2}{(e^{4K} + 6 + e^{-4K})}.$$

Clearly, for $\sigma'_\alpha = -1$ we obtain the same result with a global negative sign, and thus

$$\langle \sigma_{\alpha i} \rangle_0 = \sigma'_\alpha \frac{e^{4K} + 2}{(e^{4K} + 6 + e^{-4K})}.$$

As a result,

$$-\beta \mathcal{H}' [\sigma'_\alpha] = \frac{N}{4} \ln \left(e^{4K} + 6 + e^{-4K} \right) + 2K \left(\frac{e^{4K} + 2}{e^{4K} + 6 + e^{-4K}} \right)^2 \sum_{\langle \alpha, \beta \rangle} \sigma'_\alpha \sigma'_\beta,$$

corresponding to the recursion relation $K'(K)$,

$$K' = 2K \left(\frac{e^{4K} + 2}{e^{4K} + 6 + e^{-4K}} \right)^2.$$

(d) Find the fixed point K^*, and the thermal eigenvalue y_t.

- *To find the fixed point with $K' = K = K^*$, we introduce the variable $x = e^{4K^*}$. Hence, we have to solve the equation*

$$\frac{x+2}{x+6+x^{-1}} = \frac{1}{\sqrt{2}}, \quad \text{or} \quad \left(\sqrt{2} - 1 \right) x^2 - \left(6 - 2\sqrt{2} \right) x - 1 = 0,$$

  whose only meaningful solution is $x \simeq 7.96$, resulting in $K^* \simeq 0.52$. To obtain the thermal eigenvalue, let us linearize the recursion relation around this non-trivial fixed point,

$$\left. \frac{\partial K'}{\partial K} \right|_{K^*} = b^{y_t} \quad \Rightarrow \quad 2^{y_t} = 1 + 8K^* \left[\frac{e^{4K^*}}{e^{4K^*} + 2} - \frac{e^{4K^*} - e^{-4K^*}}{e^{4K^*} + 6 + e^{-4K^*}} \right] \quad \Rightarrow \quad y_t \simeq 1.006.$$

(e) In the presence of a small magnetic field $h \sum_i \sigma_i$, find the recursion relation for h, and calculate the magnetic eigenvalue y_h at the fixed point.

- *In the presence of a small magnetic field, we will have an extra contribution to the Hamiltonian*

$$h \sum_{\alpha,i} \langle \sigma_{\alpha,i} \rangle_0 = 4h \frac{e^{4K} + 2}{(e^{4K} + 6 + e^{-4K})} \sum_\alpha \sigma'_\alpha.$$

Therefore,

$$h' = 4h \frac{e^{4K} + 2}{(e^{4K} + 6 + e^{-4K})}.$$

(f) Compare K^*, y_t, and y_h to their exact values.

- *The cumulant method gives a value of $K^* = 0.52$, while the critical point of the Ising model on a square lattice is located at $K_c \approx 0.44$. The exact values of y_t and y_h for the two dimensional Ising model are respectively 1 and 1.875, while the cumulant method yields $y_t \approx 1.006$ and $y_h \approx 1.5$. As in the case of a triangular lattice, y_h is lower than the exact result. Nevertheless, the thermal exponent y_t is fortuitously close to its exact value.*

2. *Migdal–Kadanoff method:* Consider Potts spins $s_i = (1, 2, \cdots, q)$, on sites i of a hypercubic lattice, interacting with their nearest neighbors via a Hamiltonian

$$-\beta \mathcal{H} = K \sum_{\langle ij \rangle} \delta_{s_i, s_j} .$$

(a) In $d = 1$ find the exact recursion relations by a $b = 2$ renormalization/decimation process. Indentify all fixed points and note their stability.

- *In $d = 1$, if we average over the q possible values of s_1, we obtain*

$$\sum_{s_1=1}^{q} e^{K(\delta_{\sigma_1 s_1} + \delta_{s_1 \sigma_2})} = \begin{cases} q - 1 + e^{2K} & \text{if } \sigma_1 = \sigma_2 \\ q - 2 + 2e^{K} & \text{if } \sigma_1 \neq \sigma_2 \end{cases} = e^{g' + K' \delta_{\sigma_1 \sigma_2}},$$

from which we arrive at the exact recursion relations:

$$e^{K'} = \frac{q - 1 + e^{2K}}{q - 2 + 2e^{K}}, \qquad e^{g'} = q - 2 + 2e^{K}.$$

To find the fixed points we set $K' = K = K^*$. As in the previous problem, let us introduce the variable $x = e^{K^*}$. Hence, we have to solve the equation

$$x = \frac{q - 1 + x^2}{q - 2 + 2x}, \quad \text{or} \quad x^2 + (q-2)x - (q-1) = 0,$$

whose only meaningful solution is $x = 1$, resulting in $K^* = 0$. To check its stability, we consider $K \ll 1$, so that

$$K' \simeq \ln\left(\frac{q + 2K + 2K^2}{q + 2K + K^2}\right) \simeq \frac{K^2}{q} \ll K,$$

which indicates that this fixed point is stable.

In addition, $K^* \to \infty$ is also a fixed point. If we consider $K \gg 1$,

$$e^{K'} \simeq \frac{1}{2}e^{K} \implies K' = K - \ln 2 < K,$$

which implies that this fixed point is unstable.

(b) Write down the recursion relation $K'(K)$ in d dimensions for $b = 2$, using the Migdal–Kadanoff bond moving scheme.

- *In the Migdal–Kadanoff approximation, moving bonds strengthens the remaining bonds by a factor 2^{d-1}. Therefore, in the decimated lattice we have*

$$e^{K'} = \frac{q - 1 + e^{2 \times 2^{d-1}K}}{q - 2 + 2e^{2^{d-1}K}}.$$

(c) By considering the stability of the fixed points at zero and infinite coupling, prove the existence of a non-trivial fixed point at finite K^* for $d > 1$.

- *In the vicinity of the fixed point $K^* = 0$, i.e. for $K \ll 1$,*

$$K' \simeq \frac{2^{2d-2}K^2}{q} \ll K,$$

and consequently, this point is again stable. However, for $K^* \to \infty$, we have

$$e^{K'} \simeq \frac{1}{2}\exp\left[\left(2^d - 2^{d-1}\right)K\right] \implies K' = 2^{d-1}K - \ln 2 \gg K,$$

which implies that this fixed point is now stable provided that $d > 1$.
As a result, there must be a finite K^* fixed point, which separates the flows to the other fixed points.

(d) For $d = 2$, obtain K^* and y_t for $q = 3$, 1, and 0.

- *Let us now discuss a few particular cases in $d = 2$. For instance, if we consider $q = 3$, the non-trivial fixed point is a solution of the equation*

$$x = \frac{2 + x^4}{1 + 2x^2}, \quad \text{or} \quad x^4 - 2x^3 - x + 2 = (x - 2)(x^3 - 1) = 0,$$

which clearly yields a non-trivial fixed point at $K^* = \ln 2 \simeq 0.69$. The thermal exponent for this point

$$\left.\frac{\partial K'}{\partial K}\right|_{K^*} = 2^{y_t} = 4\left[\frac{e^{4K^*}}{e^{4K^*} + 2} - \frac{e^{2K^*}}{1 + 2e^{2K^*}}\right] = \frac{16}{9}, \implies y_t \simeq 0.83,$$

which can be compared to the exact values, $K^* = 1.005$, and $y_t = 1.2$.
By analytic continuation for $q \to 1$, we obtain

$$e^{K'} = \frac{e^{4K}}{-1 + 2e^{2K}}.$$

The non-trivial fixed point is a solution of the equation

$$x = \frac{x^4}{-1 + 2x^2}, \quad \text{or} \quad (x^3 - 2x^2 + 1) = (x - 1)(x^2 - x - 1) = 0,$$

whose only non-trivial solution is $x = (1 + \sqrt{5})/2 = 1.62$, resulting in $K^* = 0.48$. The thermal exponent for this point

$$\left.\frac{\partial K'}{\partial K}\right|_{K^*} = 2^{y_t} = 4\left[1 - e^{-K^*}\right] \quad \Longrightarrow \quad y_t \simeq 0.61.$$

The Potts model for $q \to 1$ can be mapped onto the problem of bond percolation, which despite being a purely geometrical phenomenon, shows many features completely analogous to those of a continuous thermal phase transition.

And finally for $q \to 0$, relevant to lattice animals, we obtain

$$e^{K'} = \frac{-1 + e^{4K}}{-2 + 2e^{2K}},$$

for which we have to solve the equation

$$x = \frac{-1 + x^4}{-2 + 2x^2}, \quad \text{or} \quad x^4 - 2x^3 + 2x - 1 = (x - 1)^3(x + 1) = 0,$$

whose only finite solution is the trivial one, $x = 1$. For $q \to 0$, if $K \ll 1$, we obtain

$$K' \simeq K + \frac{K^2}{2} > K,$$

indicating that this fixed point is now unstable. Note that the first correction only indicates marginal stability ($y_t = 0$). Nevertheless, for $K^* \to \infty$, we have

$$e^{K'} \simeq \frac{1}{2}\exp[2K], \quad \Longrightarrow \quad K' = 2K - \ln 2 \gg K,$$

which implies that this fixed point is stable.

$$\overset{*}{\underset{0}{\rule{0pt}{0pt}}} \longrightarrow \qquad \longrightarrow \overset{*}{\underset{\infty}{\rule{0pt}{0pt}}}$$

3. *The Potts model:* The *transfer matrix* procedure can be extended to Potts models, where the spin s_i on each site takes q values $s_i = (1, 2, \cdots, q)$; and the Hamiltonian is $-\beta\mathcal{H} = K\sum_{i=1}^{N} \delta_{s_i, s_{i+1}} + K\delta_{s_N, s_1}$.

(a) Write down the transfer matrix and diagonalize it. Note that you do not have to solve a qth order secular equation as it is easy to guess the eigenvectors from the symmetry of the matrix.

 • *The partition function is*

$$Z = \sum_{\{s_i\}} \langle s_1|T|s_2\rangle\langle s_2|T|s_3\rangle \cdots \langle s_{N-1}|T|s_N\rangle\langle s_N|T|s_1\rangle = \mathrm{tr}(T^N),$$

where $\langle s_i|T|s_j\rangle = \exp\left(K\delta_{s_i,s_j}\right)$ *is a* $q \times q$ *transfer matrix. The diagonal elements of the matrix are* e^K, *while the off-diagonal elements are unity. The eigenvectors of the matrix are easily found by inspection. There is one eigenvector with all elements equal; the corresponding eigenvalue is* $\lambda_1 = e^K + q - 1$. *There are also* $(q-1)$ *eigenvectors orthogonal to the first, i.e. the sum of whose elements is zero. The corresponding eigenvalues are degenerate and equal to* $e^K - 1$. *Thus*

$$Z = \sum_\alpha \lambda_\alpha^N = \left(e^K + q - 1\right)^N + (q-1)\left(e^K - 1\right)^N.$$

(b) Calculate the free energy per site.

- *Since the largest eigenvalue dominates for* $N \gg 1$,

$$\frac{\ln Z}{N} = \ln\left(e^K + q - 1\right).$$

(c) Give the expression for the correlation length ξ (you don't need to provide a detailed derivation), and discuss its behavior as $T = 1/K \to 0$.

- *Correlations decay as the ratio of the eigenvalues to the power of the separation. Hence the correlation length is*

$$\xi = \left[\ln\left(\frac{\lambda_1}{\lambda_2}\right)\right]^{-1} = \left[\ln\left(\frac{e^K + q - 1}{e^K - 1}\right)\right]^{-1}.$$

In the limit of $K \to \infty$, *expanding the above result gives*

$$\xi \simeq \frac{e^K}{q} = \frac{1}{q}\exp\left(\frac{1}{T}\right).$$

Solutions to selected problems from chapter 7

1. *Continuous spins:* In the standard $\mathcal{O}(n)$ model, n component unit vectors are placed on the sites of a lattice. The nearest-neighbor spins are then connected by a bond $J\vec{s}_i \cdot \vec{s}_j$. In fact, if we are only interested in universal properties, any generalized interaction $f(\vec{s}_i \cdot \vec{s}_j)$ leads to the same critical behavior. By analogy with the Ising model, a suitable choice is

$$\exp\left[f(\vec{s}_i \cdot \vec{s}_j)\right] = 1 + (nt)\vec{s}_i \cdot \vec{s}_j,$$

resulting in the so called *loop model*.

(a) Construct a high temperature expansion of the loop model (for the partition function Z) in the parameter t, on a two-dimensional *hexagonal* (honeycomb) lattice.

- *The partition function for the loop model has the form*

$$Z = \int \{\mathcal{D}\mathbf{s}_i\} \prod_{\langle ij \rangle} \left[1 + (nt)\mathbf{s}_i \cdot \mathbf{s}_j\right],$$

which we can expand in powers of the parameter t. If the total number of nearest-neighbor bonds on the lattice is N_B, the above product generates 2^{N_B} possible terms. Each term may be represented by a graph on the lattice, in which a bond joining spins i and j is included if the factor $\mathbf{s}_i \cdot \mathbf{s}_j$ appears in the term considered. Moreover, each included bond carries a factor of nt. As in the Ising model, the integral over the variables $\{\mathbf{s}_i\}$ leaves only graphs with an even number of bonds emanating from each site, because

$$\int \mathrm{d}\mathbf{s}\, s_\alpha = \int \mathrm{d}\mathbf{s}\, s_\alpha s_\beta s_\gamma = \cdots = 0.$$

In a honeycomb lattice, as depicted below, there are only one, two, or three bonds emerging from each site. Thus the only contributing graphs are those with two bonds at each site, which, as any bond can only appear once, are closed self-avoiding loops.

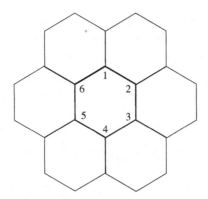

While the honeycomb lattice has the advantage of not allowing intersections of loops at a site, the universal results are equally applicable to other lattices.

We shall rescale all integrals over spin by the n-dimensional solid angle, such that $\int d\mathbf{s} = 1$. Since $s_\alpha s_\alpha = 1$, it immediately follows that

$$\int d\mathbf{s}\, s_\alpha s_\beta = \frac{\delta_{\alpha\beta}}{n},$$

resulting in

$$\int d\mathbf{s}'\, (s_\alpha s_\alpha')(s_\beta' s_\beta'') = \frac{1}{n} s_\alpha s_\alpha''.$$

A sequence of such integrals forces the components of the spins around any loop to be the same, and there is a factor n when integrating over the last spin in the loop, for instance

$$\int \{\mathcal{D}\mathbf{s}_i\}\, (s_{1\alpha}s_{2\alpha})(s_{2\beta}s_{3\beta})(s_{3\gamma}s_{4\gamma})(s_{4\delta}s_{5\delta})(s_{5\eta}s_{6\eta})(s_{6\nu}s_{1\nu})$$
$$= \frac{\delta_{\alpha\beta}\delta_{\beta\gamma}\delta_{\gamma\delta}\delta_{\delta\eta}\delta_{\eta\nu}\delta_{\alpha\nu}}{n^6} = \frac{n}{n^6}.$$

Since each bond carries a factor of nt, each loop finally contributes a factor $n \times t^\ell$, where ℓ is the number of bonds in the loop. The partition function may then be written as

$$Z = \sum_{\text{self-avoiding loops}} n^{N_\ell} t^{N_b},$$

where the sum runs over distinct disconnected or self-avoiding loop collections with a bond fugacity t, and N_ℓ, N_b are the number of loops, and the number of bonds in the graph, respectively. Note that, as we are only interested in the critical behavior of the model, any global analytic prefactor is unimportant.

(b) Show that the limit $n \to 0$ describes the configurations of a single self-avoiding polymer on the lattice.

- *While $Z = 1$, at exactly $n = 0$, one may obtain non-trivial information by considering the limit $n \to 0$. The leading term ($\mathcal{O}(n^1)$) when $n \to 0$ picks out just those configurations with a single self-avoiding loop, i.e. $N_\ell = 1$.*

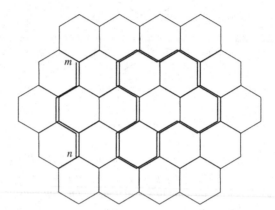

The correlation function can also be calculated graphically from

$$G_{\alpha\beta}(n - m) = \langle s_{n\alpha} s_{m\beta} \rangle = \frac{1}{Z} \int \{\mathcal{D}\mathbf{s}_i\} s_{n\alpha} s_{m\beta} \prod_{\langle ij \rangle} \left[1 + (nt)\mathbf{s}_i \cdot \mathbf{s}_j \right].$$

After disregarding any global prefactor, and taking the limit $n \to 0$, the only surviving graph consists of a single line going from n to m, and the index of all the spins along the line is fixed to be the same. All other possible graphs disappear in the limit $n \to 0$. Therefore, we are left with a sum over self-avoiding walks that go from n to m, each carrying a factor t^ℓ, where ℓ indicates the length of the walk. If we denote by $W_\ell(R)$ the number of self-avoiding walks of length ℓ whose end-to-end distance is R, we can write that

$$\sum_\ell W_\ell(R) t^\ell = \lim_{n \to 0} G(R).$$

As in the case of phantom random walks, we expect that for small t, small paths dominate the behavior of the correlation function. As t increases, larger paths dominate the sum, and, ultimately, we will find a singularity at a particular t_c, at which arbitrarily long paths become possible.

Although we presented the mapping of self-avoiding walks to the $n \to 0$ limit of the $\mathcal{O}(n)$ model for a honeycomb lattice, the critical behavior should be universal, and therefore independent of this lattice choice. What is more, various scaling properties of self-avoiding walks can be deduced from the

$O(n)$ model with $n \to 0$. Let us, for instance, characterize the mean square end-to-end distance of a self-avoiding walk, defined as

$$\langle R^2 \rangle = \frac{1}{W_\ell} \sum_R R^2 W_\ell(R),$$

where $W_\ell = \sum_R W_\ell(R)$ is the total number of self-avoiding walks of length ℓ.

The singular part of the correlation function decays with separation R as $G \propto |R|^{-(d-2+\eta)}$, up to the correlation length ξ, which diverges as $\xi \propto (t_c - t)^{-\nu}$. Hence,

$$\sum_R R^2 G(R) \propto \xi^{d+2-(d-2+\eta)} = (t_c - t)^{-\nu(4-\eta)} = (t_c - t)^{-\gamma-2\nu}.$$

We noted above that $G(t, R)$ is the generating function of $W_\ell(R)$, in the sense that $\sum_\ell W_\ell(R)t^\ell = G(t, R)$. Similarly $\sum_\ell W_\ell t^\ell$ is the generating function of W_ℓ, and is related to the susceptibility χ by

$$\sum_\ell W_\ell t^\ell = \sum_R G(R) = \chi \propto (t_c - t)^{-\gamma}.$$

To obtain the singular behavior of W_ℓ from its generating function, we perform a Taylor expansion of $(t_c - t)^{-\gamma}$, as

$$\sum_\ell W_\ell t^\ell = t_c^{-\gamma} \left(1 - \frac{t}{t_c}\right)^{-\gamma} = t_c^{-\gamma} \sum_\ell \frac{\Gamma(1-\gamma)}{\Gamma(1+\ell)\Gamma(1-\gamma-\ell)} \left(\frac{t}{t_c}\right)^\ell,$$

which results in

$$W_\ell = \frac{\Gamma(1-\gamma)}{\Gamma(1+\ell)\Gamma(1-\gamma-\ell)} t_c^{-\ell-\gamma}.$$

After using that $\Gamma(p)\Gamma(1-p) = \pi/\sin p\pi$, considering $\ell \to \infty$, and the asymptotic expression of the gamma function, we obtain

$$W_\ell \propto \frac{\Gamma(\gamma+\ell)}{\Gamma(1+\ell)} t_c^{-\ell} \propto \ell^{\gamma-1} t_c^{-\ell},$$

and similarly one can estimate $\sum_R R^2 W_\ell(R)$ from $\sum_R R^2 G(R)$, yielding

$$\langle R^2 \rangle \propto \frac{\ell^{2\nu+\gamma-1} t_c^{-\ell}}{\ell^{\gamma-1} t_c^{-\ell}} = \ell^{2\nu}.$$

Setting $n = 0$ in the results of the ϵ-expansion for the $O(n)$ model, for instance, gives the exponent $\nu = 1/2 + \epsilon/16 + O(\epsilon^2)$, characterizing the mean square end-to-end distance of a self-avoiding polymer as a function of its length ℓ, rather than $\nu_0 = 1/2$ which describes the scaling of phantom random walks. Because of self-avoidance, the (polymeric) walk is swollen, giving a larger exponent ν. The results of the first order expansion for $\epsilon = 1, 2$, and 3, in $d = 3, 2$, and 1 are 0.56, 0.625, and 0.69, to be compared to 0.59, 3/4 (exact), and 1 (exact).

2. *Potts model I:* Consider Potts spins $s_i = (1, 2, \cdots, q)$, interacting via the Hamiltonian $-\beta\mathcal{H} = K \sum_{\langle ij \rangle} \delta_{s_i, s_j}$.

(a) To treat this problem graphically at high temperatures, the Boltzmann weight for each bond is written as

$$\exp\left(K\delta_{s_i, s_j}\right) = C(K)\left[1 + T(K)g(s_i, s_j)\right],$$

with $g(s, s') = q\delta_{s, s'} - 1$. Find $C(K)$ and $T(K)$.

- *To determine the two unknowns $C(K)$ and $T(K)$, we can use the expressions*

$$\begin{cases} e^K = C[1 + T(q-1)] & \text{if} \quad s_i = s_j \\ 1 = C[1 - T] & \text{if} \quad s_i \neq s_j, \end{cases}$$

from which we obtain

$$T(K) = \frac{e^K - 1}{e^K + q - 1}, \quad \text{and} \quad C(K) = \frac{e^K + q - 1}{q}.$$

(b) Show that

$$\sum_{s=1}^{q} g(s, s') = 0, \quad \sum_{s=1}^{q} g(s_1, s)g(s, s_2) = qg(s_1, s_2), \quad \text{and}$$

$$\sum_{s, s'}^{q} g(s, s')g(s', s) = q^2(q-1).$$

- *It is easy to check that*

$$\sum_{s=1}^{q} g(s, s') = q - 1 - (q-1) = 0,$$

$$\sum_{s=1}^{q} g(s_1, s)g(s, s_2) = \sum_{s=1}^{q} \left[q^2 \delta_{s_1 s} \delta_{s_2 s} - q(\delta_{s_1 s} + \delta_{s_2 s}) + 1\right]$$

$$= q\left(q\delta_{s_1 s_2} - 1\right) = qg(s_1, s_2),$$

$$\sum_{s, s'=1}^{q} g(s, s')g(s, s') = \sum_{s, s'=1}^{q} \left[q^2 \delta_{ss'} \delta_{ss'} - 2q\delta_{ss'} + 1\right]$$

$$= q^3 - 2q^2 + q^2 = q^2(q-1).$$

(c) Use the above results to calculate the free energy, and the correlation function $\langle g(s_m, s_n) \rangle$ for a one-dimensional chain.

- *The factor $T(K)$ will be our high temperature expansion parameter. Each bond contributes a factor $Tg(s_i, s_j)$ and, since $\sum_s g(s, s') = 0$, there can not be only one bond per any site. As in the Ising case considered in the text, each bond can only be considered once, and the only graphs that survive have no dangling bonds. As a result, for a one-dimensional chain, with for instance open boundary conditions, it is impossible to draw any acceptable graph, and we obtain*

$$Z = \sum_{\{s_i\}} \prod_{\langle ij \rangle} C(K) \left[1 + T(K)g(s_i, s_j) \right] = C(K)^{N-1} q^N = q \left(e^K + q - 1 \right)^{N-1}.$$

Ignoring the boundary effects, i.e. that there are $N-1$ bonds in the chain, the free energy per site is obtained as

$$-\frac{\beta F}{N} = \ln \left(e^K + q - 1 \right).$$

With the same method, we can also calculate the correlation function $\langle g(s_n s_m) \rangle$. To get a non-zero contribution, we have to consider a graph that directly connects these two sites. Assuming that $n > m$, this gives

$$\begin{aligned}
\langle g(s_n s_m) \rangle &= \frac{C(K)^N}{Z} \sum_{\{s_i\}} g(s_n s_m) \prod_{\langle ij \rangle} \left[1 + T(K)g(s_i, s_j) \right] \\
&= \frac{C(K)^N}{Z} T(K)^{n-m} \sum_{\{s_i\}} g(s_n s_m) g(s_m, s_{m+1}) \cdots g(s_{n-1}, s_n) \\
&= \frac{C(K)^N}{Z} T(K)^{n-m} q^{n-m+1} (q-1) q^{N-(n-m)-1} \\
&= T^{n-m}(q-1)
\end{aligned}$$

where we have used the relationships obtained in (b).

(d) Calculate the partition function on the square lattice to order of T^4. Also calculate the first term in the low-temperature expansion of this problem.

- *The first term in the high temperature series for a square lattice comes from a square of four bonds. There are a total of N such squares. Therefore,*

$$Z = \sum_{\{s_i\}} \prod_{\langle ij \rangle} C(K) \left[1 + T(K)g(s_i, s_j) \right] = C(K)^{2N} q^N \left[1 + NT(K)^4(q-1) + \cdots \right].$$

Note that any closed loop involving ℓ bonds without intersections contributes $T^\ell q^\ell (q-1)$.

On the other hand, at low temperatures, the energy is minimized by the spins all being in one of the q possible states. The lowest energy excitation is a single spin in a different state, resulting in an energy cost of $K \times 4$ with a degeneracy factor $N \times (q-1)$, resulting in

$$Z = q e^{2NK} \left[1 + N(q-1)e^{-4K} + \cdots \right].$$

(e) By comparing the first terms in low- and high-temperature series, find a duality rule for Potts models. Don't worry about higher order graphs, they will work out! Assuming a single transition temperature, find the value of $K_c(q)$.

- Comparing these expansions, we find the following duality condition for the Potts model

$$e^{-\tilde{K}} = T(K) = \frac{e^K - 1}{e^K + q - 1}.$$

This duality rule maps the low temperature expansion to a high temperature series, or vice versa. It also maps pairs of points, $\tilde{K} \Leftrightarrow K$, since we can rewrite the above relationship in a symmetric way

$$\left(e^{\tilde{K}} - 1\right)\left(e^K - 1\right) = q,$$

and consequently, if there is a single singular point K_c, it must be at the self-dual point,

$$K_c = \tilde{K}_c, \quad \Longrightarrow \quad K_c = \ln\left(\sqrt{q} + 1\right).$$

(f) How do the higher order terms in the high-temperature series for the Potts model differ from those of the Ising model? What is the fundamental difference that sets apart the graphs for $q = 2$? (This is ultimately the reason why only the Ising model is solvable.)

- As indicated in the text, the Potts model with $q = 2$ can be mapped to the Ising model by noticing that $\delta_{ss'} = (1 + ss')/2$. However, higher order terms in the high-temperature series of the Potts model involve, in general, graphs with three or more bonds emanating from each site. These configurations do not correspond to a random walk, not even a constrained one as introduced in the text for the 2d Ising model on a square lattice. The quantity

$$\sum_{s_1=1}^{q} g(s_1, s_2)g(s_1, s_3)g(s_1, s_4) = q^3 \delta_{s_2 s_3} \delta_{s_2 s_4} - q^2(\delta_{s_2 s_3} + \delta_{s_2 s_4} + \delta_{s_3 s_4}) + 2q$$

is always zero when $q = 2$ (as can be easily checked for any possible state of the spins s_2, s_3 and s_4), but is in general different from zero for $q > 2$. This is the fundamental difference that ultimately sets apart the case $q = 2$. Note that the corresponding diagrams in the low temperature expansion involve adjacent regions in 3 (or more) distinct states.

3. *Potts model II:* An alternative expansion is obtained by starting with

$$\exp\left[K\delta(s_i, s_j)\right] = 1 + v(K)\delta(s_i, s_j),$$

where $v(K) = e^K - 1$. In this case, the sum over spins *does not* remove any graphs, and all choices of distributing bonds at random on the lattice are acceptable.

(a) Including a magnetic field $h\sum_i \delta_{s_i,1}$, show that the partition function takes the form

$$Z(q,K,h) = \sum_{\text{all graphs}} \prod_{\text{clusters } c \text{ in graph}} \left[v^{n_b^c} \times \left(q-1+e^{hn_s^c} \right) \right],$$

where n_b^c and n_s^c are the numbers of bonds and sites in cluster c. This is known as the *random cluster expansion*.

- *Including a symmetry breaking field along direction 1, the partition function*

$$Z = \sum_{\{s_i\}} \prod_{\langle ij \rangle} \left[1 + v(K)\delta(s_i, s_j) \right] \prod_i e^{h\delta_{s_i,1}}$$

can be expanded in powers of $v(K)$ as follows. As usual, if there is a total number N_B of nearest-neighbor bonds on the lattice, the product over bonds generates 2^{N_B} possible terms. Each term may be represented by a graph on the lattice, in which a bond joining sites i and j is included if the factor $v\delta(s_i, s_j)$ appears in the term considered. Each included bond carries a factor $v(K)$, as well as a delta function enforcing the equality of the spins on the sites which it connects. In general, these bonds form clusters of different sizes and shapes, and within each cluster, the delta functions force the spins at each vertex to be the same. The sum $\sum_{\{s_i\}}$ therefore gives a factor of $(q-1)+e^{hn_s^c}$ for each cluster c, where n_s^c is the number of sites in the cluster. The partition function may then be written as

$$Z(q,v,h) = \sum_{\text{all graphs}} \prod_{\text{clusters in graph}} \left[v(K)^{n_b^c} \left(q-1+e^{hn_s^c} \right) \right],$$

where n_b^c is the number of bonds in cluster c, and the sum runs over all distinct cluster collections. Note that an isolated site is also included in this definition of a cluster. While the Potts model was originally defined for integer q, using this expansion, we can evaluate Z for all values of q.

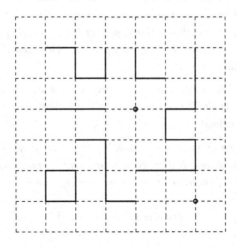

(b) Show that the limit $q \to 1$ describes a *percolation* problem, in which bonds are randomly distributed on the lattice with probability $p = v/(v+1)$. What is the percolation threshold on the square lattice?

- *In the problem of bond percolation, bonds are independently distributed on the lattice, with a probability p of being present. The weight for a given configuration of occupied and absent bonds bonds is therefore*

$$W(\text{graph}) = (1-p)^{zN} \prod_{\text{clusters in graph}} \left(\frac{p}{1-p}\right)^{n_b^c}.$$

The prefactor of $(1-p)^{zN}$ is merely the weight of the configuration with no bonds. The above weights clearly become identical to those appearing in the random cluster expansion of the Potts model for $q = 1$ (and $h = 0$). Clearly, we have to set $p = v/(v+1)$, and neglect an overall factor of $(1+v)^N$, which is analytic in v, and does not affect any singular behavior. The partition function itself is trivial in this limit as $Z(1, v, h) = (1+v)^{zN}e^{hN}$. On the other hand, we can obtain information on the number of clusters by considering the limit of $q \to 1$ from

$$\left.\frac{\partial \ln Z(q, v)}{\partial q}\right|_{q=1} = \sum_{\text{all graphs}} [\text{probablility of graph}] \sum_{\text{clusters in graph}} e^{-hn_s^c}.$$

Various properties of interest to percolation can then be calculated from the above generating function. This mapping enables us to extract the scaling laws at the percolation point, which is a continuous geometrical phase transition. The analog of the critical temperature is played by the percolation threshold p_c, which we can calculate using the expression obtained in problem 2 as $p_c = 1/2$ (after noting that $v^* = 1$).

An alternative way of obtaining this threshold is to find a duality rule for the percolation problem itself: One can similarly think of the problem in terms of empty bonds with a corresponding probability q. As p plays the role of temperature, there is a mapping of low p to high q or vice versa, and such that $q = 1 - p$. The self-dual point is then obtained by setting $p^* = 1 - p^*$, resulting in $p^* = 1/2$.

(c) Show that in the limit $q \to 0$, only a single connected cluster contributes to leading order. The enumeration of all such clusters is known as listing *branched lattice animals*.

- *The partition function $Z(q, v, h)$ goes to zero at $q = 0$, but again information about geometrical lattice structure can be obtained by taking the limit $q \to 0$ in an appropriate fashion. In particular, if we set $v = q^a x$, then*

$$Z(q, v = xq^a, h = 0) = \sum_{\text{all graphs}} x^{N_b} q^{N_c + aN_b},$$

where N_b and N_c are the total number of bonds and clusters. The leading dependence on q as $q \to 0$ comes from graphs with the lowest number of $N_c + aN_b$, and depends on the value of a. For $0 < a < 1$, these are the spanning trees, which connect all sites of the lattice (hence $N_c = 1$) and that enclose no loops (hence $N_b = N - 1$). Such spanning trees have a power of $x^{a(N-1)}q^{aN-a+1}$, and all other graphs have higher powers of q. For $a = 0$ one can add bonds to the spanning cluster (creating loops) without changing the power, as long as all sites remain connected in a single cluster. These have a relation to a problem referred to as branched lattice animals.

4. *Ising model in a field:* Consider the partition function for the Ising model ($\sigma_i = \pm 1$) on a square lattice, *in a magnetic field h*; i.e.

$$Z = \sum_{\{\sigma_i\}} \exp\left[K \sum_{\langle ij \rangle} \sigma_i \sigma_j + h \sum_i \sigma_i \right].$$

(a) Find the general behavior of the terms in a low-temperature expansion for Z.

- *At low temperatures, almost all the spins are oriented in the same direction; low-energy excitations correspond to islands of flipped spins. The general behavior of the terms in a low-temperature expansion of the partition function is then of the form*

$$e^{-2KL}e^{-2hA},$$

where L is the perimeter of the island, or the number of unsatisfied bonds, and A is the area of the island, or the number of flipped spins.

$$L = 4, A = 1 \qquad L = 6, A = 2$$

(b) Think of a model whose high-temperature series reproduces the generic behavior found in (a); and hence obtain the Hamiltonian, and interactions of the dual model.

- *To reproduce terms that are proportional to the area, we may consider plaquette-like interactions of the form*

$$\tilde{h} \sum_{\text{plaquettes}} \sigma_1 \sigma_2 \sigma_3 \sigma_4,$$

where σ is a spin variable defined on each bond of the lattice. Thus, we can define a (lattice gauge) model whose Hamiltonian is given by

$$-\beta \mathcal{H} = \tilde{K} \sum_{\text{bonds}} \sigma + \tilde{h} \sum_{\text{plaquettes}} \sigma_1 \sigma_2 \sigma_3 \sigma_4,$$

and, as usual, rewrite

$$e^{\tilde{K}\sigma} = \cosh \tilde{K} \left(1 + \tanh \tilde{K} \sigma\right),$$

$$e^{\tilde{h}\sigma_1 \sigma_2 \sigma_3 \sigma_4} = \cosh \tilde{h} \left(1 + \tanh \tilde{h} \, \sigma_1 \sigma_2 \sigma_3 \sigma_4\right).$$

The high-temperature series of the partition function will be then of the form,

$$Z \propto \sum_{\text{bonds}} \sum_{\text{plaquettes}} \left(\tanh \tilde{K}\right)^{N_b} \left(\tanh \tilde{h}\right)^{N_p},$$

where N_b and N_p are the number of bonds and plaquettes involved in each term of the expansion. Nevertheless, we will only have non-zero contributions when the bond variables σ appear an even number of times, either in a plaquette or as a bond. For example, the lowest order contributions contain one plaquette (so that $A = 1$), and four extra bonds ($L = 4$), or two plaquettes ($A = 2$), and six bonds ($L = 6$), etc.

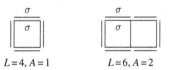

$$L = 4, A = 1 \qquad\qquad L = 6, A = 2$$

In general, we obtain terms of the form

$$\left(\tanh \tilde{K}\right)^L \left(\tanh \tilde{h}\right)^A,$$

with L and A, the perimeter and the area of the graph considered. Hence, as in the Ising model, we can write the duality relations

$$e^{-2K} = \tanh \tilde{K}, \qquad \text{and} \qquad e^{-2h} = \tanh \tilde{h}.$$

5. *Potts duality:* Consider Potts spins, $s_i = (1, 2, \cdots, q)$, placed on the sites of a *square lattice* of N sites, interacting with their nearest neighbors through a Hamiltonian

$$-\beta \mathcal{H} = K \sum_{\langle ij \rangle} \delta_{s_i, s_j}.$$

(a) By comparing the first terms of high and low temperature series, or by any other method, show that the partition function has the property

$$Z(K) = qe^{2NK}\,\Xi\left[e^{-K}\right] = q^{-N}\left[e^K + q - 1\right]^{2N}\Xi\left[\frac{e^K - 1}{e^K + (q-1)}\right],$$

for some function Ξ, and hence locate the critical point $K_c(q)$.

- *As discussed in problem 3 of this chapter, the low temperature series takes the form*

$$Z = qe^{2NK}\left[1 + N(q-1)e^{-4K} + \cdots\right] \equiv qe^{2NK}\,\Xi\left[e^{-K}\right],$$

while at high temperatures

$$Z = \left[\frac{e^K + q - 1}{q}\right]^{2N} q^N \left[1 + N(q-1)\left(\frac{e^K - 1}{e^K + q - 1}\right)^4 + \cdots\right]$$

$$\equiv q^{-N}\left[e^K + q - 1\right]^{2N}\Xi\left[\frac{e^K - 1}{e^K + q - 1}\right].$$

Both of the above series for Ξ are in fact the same, leading to the duality condition

$$e^{-\tilde{K}} = \frac{e^K - 1}{e^K + q - 1},$$

and a critical (self-dual) point of

$$K_c = \tilde{K}_c, \qquad \Longrightarrow \qquad K_c = \ln\left(\sqrt{q} + 1\right).$$

(b) Starting from the duality expression for $Z(K)$, derive a similar relation for the internal energy $U(K) = \langle \beta \mathcal{H} \rangle = -\partial \ln Z / \partial \ln K$. Use this to calculate the exact value of U at the critical point.

- *The duality relation for the partition function gives*

$$\ln Z(K) = \ln q + 2NK + \ln\Xi\left[e^{-K}\right]$$

$$= -N\ln q + 2N\ln\left[e^K + q - 1\right] + \ln\Xi\left[\frac{e^K - 1}{e^K + q - 1}\right].$$

The internal energy $U(K)$ is then obtained from

$$-\frac{U(K)}{K} = \frac{\partial}{\partial K}\ln Z(K) = 2N - e^{-K}\ln\Xi'\left[e^{-K}\right]$$

$$= 2N\frac{e^K}{e^K + q - 1} + \frac{qe^K}{(e^K + q - 1)^2}\ln\Xi'\left[\frac{e^K - 1}{e^K + q - 1}\right].$$

$\ln \Xi'$ *is the derivative of* $\ln \Xi$ *with respect to its argument, whose value is not known in general. However, at the critical point* K_c, *the arguments of* $\ln \Xi'$ *from the high and low temperature forms of the above expression are the same. Substituting* $e^{K_c} = 1 + \sqrt{q}$, *we obtain*

$$2N - \frac{\ln \Xi'_c}{1 + \sqrt{q}} = \frac{2N}{\sqrt{q}} + \frac{\ln \Xi'_c}{1 + \sqrt{q}} \quad \Longrightarrow \quad \ln \Xi'_c = \frac{q-1}{\sqrt{q}} N,$$

and,

$$-\frac{U(K_c)}{K_c} = N \left(2 - \frac{q-1}{\sqrt{q} + q} \right) \quad \Longrightarrow \quad U(K_c) = N K_c \frac{\sqrt{q} + 1}{\sqrt{q}}.$$

6. *Anisotropic random walks:* Consider the ensemble of all random walks on a square lattice starting at the origin (0,0). Each walk has a weight of $t_x{}^{\ell_x} \times t_y{}^{\ell_y}$, where ℓ_x and ℓ_y are the number of steps taken along the x and y directions, respectively.

(a) Calculate the total weight $W(x, y)$ of all walks terminating at (x, y). Show that W is well defined only for $\bar{t} = (t_x + t_y)/2 < t_c = 1/4$.

- *Defining* $\langle 0, 0 | W(\ell) | x, y \rangle$ *to be the weight of all walks of* ℓ *steps terminating at* (x, y), *we can follow the steps in the text. In the anisotropic case, Eq. (7.47) (applied* ℓ *times) is easily recast into*

$$\langle x, y | T^\ell | q_x, q_y \rangle = \sum_{x', y'} \langle x, y | T^\ell | x', y' \rangle \langle x', y' | q_x, q_y \rangle$$

$$= \left(2t_x \cos q_x + 2t_y \cos q_y \right)^\ell \langle x, y | q_x, q_y \rangle,$$

where $\langle x, y | q_x, q_y \rangle = e^{iq_x x + iq_y y}/\sqrt{N}$. *Since* $W(x, y) = \sum_\ell \langle 0, 0 | W(\ell) | x, y \rangle$, *its Fourier transform is calculated as*

$$W(q_x, q_y) = \sum_\ell \sum_{x,y} \langle 0, 0 | T^\ell | x, y \rangle \langle x, y | q_x, q_y \rangle$$

$$= \sum_\ell \left(2t_x \cos q_x + 2t_y \cos q_y \right)^\ell = \frac{1}{1 - \left(2t_x \cos q_x + 2t_y \cos q_y \right)}.$$

Finally, Fourier transforming back gives

$$W(x, y) = \int_{-\pi}^{\pi} \frac{d^2 q}{(2\pi)^2} W(q_x, q_y) e^{-iq_x x - iq_y y}$$

$$= \int_{-\pi}^{\pi} \frac{d^2 q}{(2\pi)^2} \frac{e^{-iq_x x - iq_y y}}{1 - \left(2t_x \cos q_x + 2t_y \cos q_y \right)}.$$

Note that the summation of the series is legitimate (for all q's*) only for* $2t_x + 2t_y < 1$, *i.e. for* $\bar{t} = (t_x + t_y)/2 < t_c = 1/4$.

(b) What is the shape of a curve $W(x, y) = $ constant, for large x and y, and close to the transition?

- *For x and y large, the main contributions to the above integral come from small q's. To second order in q_x and q_y, the denominator of the integrand reads*

$$1 - 2\left(t_x + t_y\right) + t_x q_x^2 + t_y q_y^2.$$

Then, with $q_i' \equiv \sqrt{t_i} q_i$, we have

$$W(x, y) \approx \int_{-\infty}^{\infty} \frac{d^2 q'}{(2\pi)^2 \sqrt{t_x t_y}} \frac{e^{-i q' \cdot \mathbf{v}}}{1 - 2\left(t_x + t_y\right) + \mathbf{q'}^2},$$

where we have extended the limits of integration to infinity, and $\mathbf{v} = \left(\dfrac{x}{\sqrt{t_x}}, \dfrac{y}{\sqrt{t_y}}\right)$. As the denominator is rotationally invariant, the integral depends only on the magnitude of the vector $\mathbf{v}$. In other words, $W(x, y)$ is constant along ellipses

$$\frac{x^2}{t_x} + \frac{y^2}{t_y} = \text{constant}.$$

(c) How does the average number of steps, $\langle \ell \rangle = \langle \ell_x + \ell_y \rangle$, diverge as $\bar{t}$ approaches t_c?

- *The weight of all walks of length ℓ, irrespective of their end point location, is*

$$\sum_{x,y} \langle 0, 0 | W(\ell) | x, y \rangle = \langle 0, 0 | T^\ell | q_x = 0, q_y = 0 \rangle = \left(2 t_x + 2 t_y\right)^\ell = \left(4 \bar{t}\right)^\ell.$$

Therefore,

$$\langle \ell \rangle = \frac{\sum_\ell \ell \left(4\bar{t}\right)^\ell}{\sum_\ell \left(4\bar{t}\right)^\ell} = 4\bar{t} \frac{\partial}{\partial (4\bar{t})} \ln \left[\sum_\ell \left(4\bar{t}\right)^\ell\right] = 4\bar{t} \frac{\partial}{\partial (4\bar{t})} \ln \frac{1}{1 - 4\bar{t}} = \frac{4\bar{t}}{1 - 4\bar{t}},$$

i.e.

$$\langle \ell \rangle = \frac{\bar{t}}{t_c - \bar{t}}$$

diverges linearly close to the singular value of $\bar{t}$.

7. *Anisotropic Ising model:* Consider the anisotropic Ising model on a square lattice with a Hamiltonian

$$-\beta \mathcal{H} = \sum_{x,y} \left(K_x \sigma_{x,y} \sigma_{x+1,y} + K_y \sigma_{x,y} \sigma_{x,y+1}\right);$$

i.e. with bonds of different strengths along the x and y directions.

(a) By following the method presented in the text, calculate the free energy for this model. You do not have to write down every step of the derivation. Just sketch the steps that need to be modified due to anisotropy; and calculate the final answer for $\ln Z/N$.

- *The Hamiltonian*

$$-\beta \mathcal{H} = \sum_{x,y} \left(K_x \sigma_{x,y} \sigma_{x+1,y} + K_y \sigma_{x,y} \sigma_{x,y+1} \right)$$

leads to

$$Z = \sum \left(2 \cosh K_x \cosh K_y \right)^N t_x^{\ell_x} t_y^{\ell_y},$$

where $t_i = \tanh K_i$, and the sum runs over all closed graphs. By extension of the isotropic case,

$$f = \frac{\ln Z}{N} = \ln \left(2 \cosh K_x \cosh K_y \right) + \sum_{\ell_x, \ell_y} \frac{t_x^{\ell_x} t_y^{\ell_y}}{\ell_x + \ell_y} \langle 0| \, W^* \left(\ell_x, \ell_y \right) |0\rangle,$$

where

$$\langle 0| \, W^* \left(\ell_x, \ell_y \right) |0\rangle = \frac{1}{2} {\sum}' (-1)^{\text{number of crossings}},$$

and the primed sum runs over all directed (ℓ_x, ℓ_y)-steps walks from $(0,0)$ to $(0,0)$ with no U-turns. As in the isotropic case, this is evaluated by taking the trace of powers of the $4N \times 4N$ matrix described by Eq. (7.66), which is block diagonalized by Fourier transformation. However, unlike the isotropic case, in which each element is multiplied by t, here they are multiplied by t_x and t_y, respectively, resulting in

$$f = \ln \left(2 \cosh K_x \cosh K_y \right) + \frac{1}{2} \int \frac{d^2 q}{(2\pi)^2} \operatorname{tr} \ln \left[1 - T(\mathbf{q})^* \right],$$

where

$$
\begin{aligned}
\operatorname{tr} \ln \left[1 - T(\mathbf{q})^* \right] &= \ln \det \left[1 - T(\mathbf{q})^* \right] \\
&= \ln \left[\left(1 + t_x^2 \right) \left(1 + t_y^2 \right) - 2 t_x \left(1 - t_y^2 \right) \cos q_x - 2 t_y \left(1 - t_x^2 \right) \cos q_y \right] \\
&= \ln \left[\frac{\cosh 2K_x \cosh 2K_y - \sinh 2K_x \cos q_x - \sinh 2K_y \cos q_y}{\cosh^2 K_x \cosh^2 K_y} \right],
\end{aligned}
$$

resulting in

$$f = \ln 2 + \frac{1}{2} \int \frac{d^2 q}{(2\pi)^2} \ln \left(\cosh 2K_x \cosh 2K_y - \sinh 2K_x \cos q_x - \sinh 2K_y \cos q_y \right).$$

(b) Find the critical boundary in the (K_x, K_y) plane from the singularity of the free energy. Show that it coincides with the condition $K_x = \tilde{K}_y$, where $\tilde{K}$ indicates the standard dual interaction to K.

- *The argument of the logarithm is minimal at $q_x = q_y = 0$, and equal to*

$$\cosh 2K_x \cosh 2K_y - \sinh 2K_x - \sinh 2K_y$$
$$= \tfrac{1}{2} \left(e^{K_x} \sqrt{\cosh 2K_y - 1} - e^{-K_x} \sqrt{\cosh 2K_y + 1} \right)^2.$$

Therefore, the critical line is given by

$$e^{2K_x} = \sqrt{\frac{\cosh 2K_y + 1}{\cosh 2K_y - 1}} = \coth K_y.$$

Note that this condition can be rewritten as

$$\sinh 2K_x = \frac{1}{2} \left(\coth K_y - \tanh K_y \right) = \frac{1}{\sinh 2K_y},$$

i.e. the critical boundary can be described as $K_x = \tilde{K}_y$, where the dual interactions, $\tilde{K}$ and K, are related by $\sinh 2K \sinh 2\tilde{K} = 1$.

(c) Find the singular part of $\ln Z/N$, and comment on how anisotropy affects critical behavior in the exponent and amplitude ratios.

- *The singular part of $\ln Z/N$ for the anisotropic case can be written as*

$$f_S = \frac{1}{2} \int \frac{d^2 q}{(2\pi)^2} \ln \left[\left(e^{K_x} \sqrt{\cosh 2K_y - 1} - e^{-K_x} \sqrt{\cosh 2K_y + 1} \right)^2 + \sum_{i=x,y} \frac{q_i^2}{2} \sinh 2K_i \right].$$

In order to rewrite this expression in a form closer to that of the singular part of the free energy in the isotropic case, let

$$q_i = \sqrt{\frac{2}{\sinh 2K_i}} q_i',$$

and

$$\delta t = e^{K_x} \sqrt{\cosh 2K_y - 1} - e^{-K_x} \sqrt{\cosh 2K_y + 1}$$

(δt goes linearly through zero as $\left(K_x, K_y \right)$ follows a curve which intersects the critical boundary). Then

$$f_S = \frac{1}{\sqrt{\sinh 2K_x \sinh 2K_y}} \int \frac{d^2 q'}{(2\pi)^2} \ln \left(\delta t^2 + q'^2 \right).$$

Thus, upon approaching the critical boundary ($\sinh 2K_x \sinh 2K_y = 1$), the singular part of the anisotropic free energy coincides more and more precisely with the isotropic one, and the exponents and amplitude ratios are unchanged by the anisotropy. (The amplitude itself may vary along the critical line.)

8. Müller–Hartmann and Zittartz estimate of the interfacial energy of the $d = 2$ Ising model on a square lattice.

(a) Consider an interface on the square lattice with periodic boundary conditions in one direction. Ignoring islands and overhangs, the configurations can be labeled by heights h_n for $1 \leq n \leq L$. Show that for an anisotropic Ising model of interactions (K_x, K_y), the energy of an interface along the x direction is

$$-\beta \mathcal{H} = -2K_y L - 2K_x \sum_n |h_{n+1} - h_n| .$$

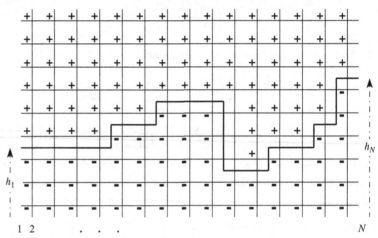

- *For each unsatisfied (+−) bond, the energy is increased by $2K_i$ from the ground state energy, with $i = x$ if the unsatisfied bond is vertical, and $i = y$ if the latter is horizontal. Ignoring islands and overhangs, the number of horizontal bonds of the interface is L, while the number of vertical bonds is $\sum_n |h_{n+1} - h_n|$, yielding*

$$-\beta \mathcal{H} = -2K_y L - 2K_x \sum_{n=1}^{L} |h_{n+1} - h_n| .$$

(b) Write down a column-to-column transfer matrix $\langle h|T|h'\rangle$, and diagonalize it.

- *We can define*

$$\langle h| T |h'\rangle \equiv \exp\left(-2K_y - 2K_x |h' - h|\right),$$

or, in matrix form,

$$T = e^{-2K_y}$$
$$\times \begin{pmatrix} 1 & e^{-2K_x} & e^{-4K_x} & \cdots & e^{-HK_x} & e^{-HK_x} & e^{-2\left(\frac{H}{2}-1\right)K_x} & \cdots & e^{-2K_x} \\ e^{-2K_x} & 1 & e^{-2K_x} & \cdots & e^{-2\left(\frac{H}{2}-1\right)K_x} & e^{-2\left(\frac{H}{2}+1\right)K_x} & e^{-HK_x} & \cdots & e^{-4K_x} \\ & \cdots & & & & & & \end{pmatrix},$$

where H is the vertical size of the lattice. In the $H \to \infty$ limit, T is easily diagonalized since each line can be obtained from the previous line by a single column shift. The eigenvectors of such matrices are composed by the complex roots of unity (this is equivalent to the statement that a translationally invariant system is diagonal in Fourier modes). To the eigenvector

$$\left(e^{i\frac{2\pi}{k}}, e^{i\frac{2\pi}{k}\cdot 2}, e^{i\frac{2\pi}{k}\cdot 3}, \cdots, e^{i\frac{2\pi}{k}\cdot(H+1)} \right)$$

is associated the eigenvalue

$$\lambda_k = e^{-2K_y} \sum_{n=1}^{H+1} T_{1n} e^{i\frac{2\pi}{k}\cdot(n-1)}.$$

Note that there are $H+1$ eigenvectors, corresponding to $k = 1, \cdots, H+1$.

(c) Obtain the interface free energy using the result in (b), or by any other method.

- *One way of obtaining the free energy is to evaluate the largest eigenvalue of T. Since all elements of T are positive, the eigenvector $(1, 1, \cdots, 1)$ has the largest eigenvalue*

$$\lambda_1 = e^{-2K_y} \sum_{n=1}^{H+1} T_{1n} = e^{-2K_y} \left(1 + 2 \sum_{n=1}^{H/2} e^{-2K_x n} \right)$$

$$= e^{-2K_y} \left(2 \sum_{n=0}^{H/2} e^{-2K_x n} - 1 \right) = e^{-2K_y} \coth K_x,$$

in the $H \to \infty$ limit. Then, $F = -Lk_B T \ln \lambda_1$.
Alternatively, we can directly sum the partition function, as

$$Z = e^{-2K_y L} \sum_{\{h_n\}} \exp\left(-2K_x \sum_{n=1}^{L} |h_{n+1} - h_n| \right) = e^{-2K_y L} \left[\sum_d \exp\left(-2K_x |d| \right) \right]^L$$

$$= \left[e^{-2K_y} \left(2 \sum_{d \geq 0} e^{-2K_x d} - 1 \right) \right]^L = \left(e^{-2K_y} \coth K_x \right)^L,$$

yielding

$$F = -Lk_B T \left[\ln(\coth K_x) - 2K_y \right].$$

(d) Find the condition between K_x and K_y for which the interfacial free energy vanishes. Does this correspond to the critical boundary of the original 2d Ising model as computed in the previous problem?

- *The interfacial free energy vanishes for*

$$\coth K_x = e^{2K_y},$$

which coincides with the condition found in problem 7. This illustrates that long wavelength fluctuations, such as interfaces, are responsible for destroying order at criticality.

9. *Anisotropic Landau theory:* Consider an n-component magnetization field $\vec{m}(\mathbf{x})$ in d dimensions.

(a) Using the previous problems on anisotropy as a guide, generalize the standard Landau–Ginzburg Hamiltonian to include the effects of spacial anisotropy.

- *Requiring different coupling constants in the different spatial directions, along with rotational invariance in spin space, leads to the following leading terms of the Hamiltonian,*

$$-\beta \mathcal{H} = \int \mathrm{d}^d x \left[\frac{t}{2} \vec{m}(\mathbf{x})^2 + \sum_{i=1}^d \frac{K_i}{2} \frac{\partial \vec{m}}{\partial x_i} \cdot \frac{\partial \vec{m}}{\partial x_i} + u \vec{m}(\mathbf{x})^4 \right].$$

(b) Are such anisotropies "relevant"?

- *Clearly, the apparent anisotropy can be eliminated by the rescaling*

$$x_i' = \sqrt{\frac{K}{K_i}} x_i.$$

In terms of the primed space variables, the Hamiltonian is isotropic. In particular, the universal features are identical in the anisotropic and isotropic cases, and the anisotropy is thus "irrelevant" (provided all K_i are positive).

(c) In La_2CuO_4, the Cu atoms are arranged on the sites of a square lattice in planes, and the planes are then stacked together. Each Cu atom carries a spin which we assume to be classical, and can point along any direction in space. There is a very strong antiferromagnetic interaction in each plane. There is also a very weak interplane interaction that prefers to align successive layers. Sketch the low-temperature magnetic phase, and indicate to what universality class the order–disorder transition belongs.

- *For classical spins, this combination of antiferromagnetic and ferromagnetic couplings is equivalent to a purely ferromagnetic (anisotropic) system, since we can redefine (e.g. in the partition function) all the spins on one of the two sublattices with an opposite sign. Therefore, the critical behavior belongs to the $d = 3$, $n = 3$ universality class.*

    Nevertheless, there is a range of temperatures for which the in-plane correlation length is large compared to the lattice spacing, while the interplane correlation length is of the order of the lattice spacing. The behavior of the system is then well described by a $d = 2$, $n = 3$ theory.

Solutions to selected problems from chapter 8

1. *Anisotropic nonlinear σ model:* Consider unit n-component spins, $\vec{s}(\mathbf{x}) = (s_1, \cdots, s_n)$ with $\sum_\alpha s_\alpha^2 = 1$, subject to a Hamiltonian

$$\beta \mathcal{H} = \int d^d\mathbf{x} \left[\frac{1}{2T} (\nabla \vec{s})^2 + g s_1^2 \right].$$

For $g = 0$, renormalization group equations are obtained through rescaling distances by a factor $b = e^\ell$, and spins by a factor $\zeta = b^{y_s}$ with $y_s = -\frac{(n-1)}{4\pi} T$, and lead to the flow equation

$$\frac{dT}{d\ell} = -\epsilon T + \frac{(n-2)}{2\pi} T^2 + \mathcal{O}(T^3),$$

where $\epsilon = d - 2$.

(a) Find the fixed point, and the thermal eigenvalue y_T.

- *Setting $dT/d\ell$ to zero, the fixed point is obtained as*

$$T^* = \frac{2\pi\epsilon}{n-2} + \mathcal{O}(\epsilon^2).$$

Linearizing the recursion relation gives

$$y_T = -\epsilon + \frac{(n-2)}{\pi} T^* = +\epsilon + \mathcal{O}(\epsilon^2).$$

(b) Write the renormalization group equation for g in the vicinity of the above fixed point, and obtain the corresponding eigenvalue y_g.

- *Rescalings $x \to b\mathbf{x}'$ and $\vec{s} \to \zeta \vec{s}'$ lead to $g \to g' = b^d \zeta^2 g$, and hence*

$$y_g = d + 2y_s = d - \frac{n-1}{2\pi} T^* = 2 + \epsilon - \frac{n-1}{n-2}\epsilon = 2 - \frac{1}{n-2}\epsilon + \mathcal{O}(\epsilon^2).$$

(c) Sketch the phase diagram as a function of T and g, indicating the phases, and paying careful attention to the shape of the phase boundary as $g \to 0$.

- *The term proportional to g removes full rotational symmetry and leads to a bicritical phase diagram. The phase for $g < 0$ has order along direction 1, while $g > 0$ favors ordering along any one of the $(n-1)$ directions orthogonal to 1. The phase boundaries as $g \to 0$ behave as $g \propto (\delta T)^\phi$, with $\phi = y_g/y_t \approx 2/\epsilon + \mathcal{O}(1)$.*

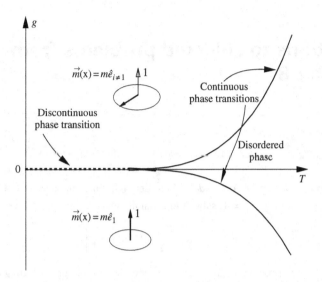

2. *Matrix models:* In some situations, the order parameter is a matrix rather than a vector. For example, in triangular (Heisenberg) antiferromagnets each triplet of spins aligns at 120°, locally defining a plane. The variations of this plane across the system are described by a 3×3 rotation matrix. We can construct a nonlinear σ model to describe a generalization of this problem as follows. Consider the Hamiltonian

$$\beta \mathcal{H} = \frac{K}{4} \int d^d \mathbf{x} \, \text{tr} \left[\nabla M(\mathbf{x}) \cdot \nabla M^T(\mathbf{x}) \right],$$

where M is a *real, $N \times N$ orthogonal matrix*, and "tr" denotes the trace operation. The condition of orthogonality is that $MM^T = M^T M = I$, where I is the $N \times N$ identity matrix, and M^T is the transposed matrix, $M_{ij}^T = M_{ji}$. The partition function is obtained by summing over all matrix functionals, as

$$Z = \int \mathcal{D} M(\mathbf{x}) \delta \left(M(\mathbf{x}) M^T(\mathbf{x}) - I \right) e^{-\beta \mathcal{H}[M(\mathbf{x})]}.$$

(a) Rewrite the Hamiltonian and the orthogonality constraint in terms of the matrix elements M_{ij} ($i, j = 1, \cdots, N$). Describe the ground state of the system.

• *In terms of the matrix elements, the Hamiltonian reads*

$$\beta \mathcal{H} = \frac{K}{4} \int d^d x \sum_{i,j} \nabla M_{ij} \cdot \nabla M_{ij},$$

and the orthogonality condition becomes

$$\sum_k M_{ik} M_{jk} = \delta_{ij}.$$

Since $\nabla M_{ij} \cdot \nabla M_{ij} \geq 0$, any constant (spatially uniform) orthogonal matrix realizes a ground state.

(b) Define the symmetric and anti-symmetric matrices

$$\begin{cases} \sigma = \dfrac{1}{2}\left(M + M^T\right) = \sigma^T \\ \pi = \dfrac{1}{2}\left(M - M^T\right) = -\pi^T. \end{cases}$$

Express $\beta\mathcal{H}$ and the orthogonality constraint in terms of the matrices σ and π.

- As $M = \sigma + \pi$ and $M^T = \sigma - \pi$,

$$\beta\mathcal{H} = \frac{K}{4}\int d^d x\, \mathrm{tr}\left[\nabla\left(\sigma + \pi\right)\cdot\nabla\left(\sigma - \pi\right)\right] = \frac{K}{4}\int d^d x\, \mathrm{tr}\left[\left(\nabla\sigma\right)^2 - \left(\nabla\pi\right)^2\right],$$

where we have used the (easily checked) fact that the trace of the commutator of matrices $\nabla\sigma$ and $\nabla\pi$ is zero. Similarly, the orthogonality condition is written as

$$\sigma^2 - \pi^2 = I,$$

where I is the unit matrix.

(c) Consider *small fluctuations* about the ordered state $M(\mathbf{x}) = I$. Show that σ can be expanded in powers of π as

$$\sigma = I - \frac{1}{2}\pi\pi^T + \cdots.$$

Use the orthogonality constraint to integrate out σ, and obtain an expression for βH to fourth order in π. Note that there are two distinct types of fourth order terms. *Do not include* terms generated by the argument of the delta function. As shown for the nonlinear σ model in the text, these terms do not affect the results at lowest order.

- *Taking the square root of*

$$\sigma^2 = I + \pi^2 = I - \pi\pi^T$$

results in

$$\sigma = I - \frac{1}{2}\pi\pi^T + \mathcal{O}\left(\pi^4\right)$$

(as can easily be checked by calculating the square of $I - \pi\pi^T/2$). We now integrate out σ, to obtain

$$Z = \int \mathcal{D}\pi(\mathbf{x})\exp\left\{-\frac{K}{4}\int d^d x\, \mathrm{tr}\left[\left(\nabla\pi\right)^2 - \frac{1}{4}\left(\nabla\left(\pi\pi^T\right)\right)^2\right]\right\},$$

where $\mathcal{D}\pi(\mathbf{x}) = \prod_{j>i}\mathcal{D}\pi_{ij}(\mathbf{x})$, and π is a matrix with zeros along the diagonal, and elements below the diagonal given by $\pi_{ij} = -\pi_{ji}$. Note that we have *not included the terms generated by the argument of the delta function. Such terms, which ensure that the measure of integration over π is symmetric, do not contribute to the renormalization of K at the lowest order. Note also that the fourth order terms are of two distinct types, due to the non-commutativity of π and $\nabla\pi$. Indeed,*

$$\left[\nabla\left(\pi\pi^T\right)\right]^2 = \left[\nabla\left(\pi^2\right)\right]^2 = \left[(\nabla\pi)\,\pi + \pi\nabla\pi\right]^2$$

$$= (\nabla\pi)\,\pi\cdot(\nabla\pi)\,\pi + (\nabla\pi)\,\pi^2\cdot\nabla\pi + \pi\,(\nabla\pi)^2\,\pi + \pi\,(\nabla\pi)\,\pi\cdot\nabla\pi,$$

and, since the trace is unchanged by cyclic permutations,

$$\mathrm{tr}\left[\nabla\left(\pi\pi^T\right)\right]^2 = 2\,\mathrm{tr}\left[(\pi\nabla\pi)^2 + \pi^2\,(\nabla\pi)^2\right].$$

(d) For an N-vector order parameter there are $N-1$ Goldstone modes. Show that an orthogonal $N\times N$ order parameter leads to $N\,(N-1)/2$ such modes.

- *The anti-symmetry of π imposes $N\,(N+1)/2$ conditions on the $N\times N$ matrix elements, and thus there are $N^2 - N\,(N+1)/2 = N\,(N-1)/2$ independent components (Goldstone modes) for the matrix. Alternatively, the orthogonality of M similarly imposes $N\,(N+1)/2$ constraints, leading to $N\,(N-1)/2$ degrees of freedom. [Note that in the analogous calculation for the $\mathcal{O}\,(n)$ model, there is one condition constraining the magnitude of the spins to unity; and the remaining $n-1$ angular components are Goldstone modes.]*

(e) Consider the quadratic piece of $\beta\mathcal{H}$. Show that the two point correlation function in Fourier space is

$$\langle\pi_{ij}(\mathbf{q})\pi_{kl}(\mathbf{q}')\rangle = \frac{(2\pi)^d\delta^d(\mathbf{q}+\mathbf{q}')}{Kq^2}\left[\delta_{ik}\delta_{jl} - \delta_{il}\delta_{jk}\right].$$

- *In terms of the Fourier components $\pi_{ij}(\mathbf{q})$, the quadratic part of the Hamiltonian in (c) has the form*

$$\beta\mathcal{H}_0 = \frac{K}{2}\sum_{i<j}\int\frac{d^d\mathbf{q}}{(2\pi)^d}q^2|\pi_{ij}(\mathbf{q})|^2,$$

leading to the bare expectation values

$$\langle\pi_{ij}(\mathbf{q})\,\pi_{ij}(\mathbf{q}')\rangle_0 = \frac{(2\pi)^d\,\delta^d(\mathbf{q}+\mathbf{q}')}{Kq^2},$$

and

$$\langle\pi_{ij}(\mathbf{q})\,\pi_{kl}(\mathbf{q}')\rangle_0 = 0,\ \text{if the pairs } (ij) \text{ and } (kl) \text{ are different.}$$

Furthermore, since π is anti-symmetric,

$$\langle\pi_{ij}(\mathbf{q})\,\pi_{ji}(\mathbf{q}')\rangle_0 = -\langle\pi_{ij}(\mathbf{q})\,\pi_{ij}(\mathbf{q}')\rangle_0,$$

and in particular $\langle \pi_{ii}(\mathbf{q}) \pi_{jj}(\mathbf{q}') \rangle_0 = 0$. *These results can be summarized by*

$$\langle \pi_{ij}(\mathbf{q}) \pi_{kl}(\mathbf{q}') \rangle_0 = \frac{(2\pi)^d \, \delta^d \, (\mathbf{q}+\mathbf{q}')}{K q^2} \left(\delta_{ik}\delta_{jl} - \delta_{il}\delta_{jk} \right).$$

We shall now construct a renormalization group by removing Fourier modes $M^>(\mathbf{q})$, *with* $\mathbf{q}$ *in the shell* $\Lambda/b < |\mathbf{q}| < \Lambda$.

(f) Calculate the coarse-grained expectation value for $\langle \text{tr}(\sigma) \rangle_0^>$ at low temperatures after removing these modes. Identify the scaling factor, $M'(\mathbf{x}') = M^<(\mathbf{x})/\zeta$, that restores $\text{tr}(M') = \text{tr}(\sigma') = N$.

- *As a result of fluctuations of short wavelength modes,* $\text{tr}\,\sigma$ *is reduced to*

$$\langle \text{tr}\,\sigma \rangle_0^> = \left\langle \text{tr}\left(I + \frac{\pi^2}{2} + \cdots \right) \right\rangle_0^> \approx N + \frac{1}{2} \langle \text{tr}\,\pi^2 \rangle_0^>$$

$$= N + \frac{1}{2} \left\langle \sum_{i \neq j} \pi_{ij}\pi_{ji} \right\rangle_0^> = N - \frac{1}{2} \left\langle \sum_{i \neq j} \pi_{ij}^2 \right\rangle_0^> = N - \frac{1}{2} \left(N^2 - N \right) \langle \pi_{ij}^2 \rangle_0^>$$

$$= N \left(1 - \frac{N-1}{2} \int_{\Lambda/b}^{\Lambda} \frac{d^d q}{(2\pi)^d} \frac{1}{K q^2} \right) = N \left[1 - \frac{N-1}{2K} I_d(b) \right].$$

To restore $\text{tr}\,M' = \text{tr}\,\sigma' = N$, *we rescale all components of the matrix by*

$$\zeta = 1 - \frac{N-1}{2K} I_d(b).$$

Note: *An orthogonal matrix* M *is invertible* $(M^{-1} = M^T)$, *and therefore diagonalizable. In diagonal form, the transposed matrix is equal to the matrix itself, and so its square is the identity, implying that each eigenvalue is either* $+1$ *or* -1. *Thus, if* M *is chosen to be very close to the identity, all eigenvalues are* $+1$, *and* $\text{tr}\,M = N$ *(as the trace is independent of the coordinate basis).*

(g) Use perturbation theory to calculate the coarse grained coupling constant $\tilde{K}$. Evaluate only the two diagrams that directly renormalize the $(\nabla \pi_{ij})^2$ term in $\beta \mathcal{H}$, and show that

$$\tilde{K} = K + \frac{N}{2} \int_{\Lambda/b}^{\Lambda} \frac{d^d \mathbf{q}}{(2\pi)^d} \frac{1}{q^2} \,.$$

- *Distinguishing between the modes with* $|\mathbf{q}|$ *greater or lesser than* Λ/b, *we write the partition function as*

$$Z = \int \mathcal{D}\pi^< \, \mathcal{D}\pi^> \, e^{-\beta \mathcal{H}_0^< - \beta \mathcal{H}_0^> + U[\pi^<, \pi^>]} = \int \mathcal{D}\pi^< \, e^{-\delta f_b^0 - \beta \mathcal{H}_0^<} \, \langle e^U \rangle_0^> ,$$

where $\mathcal{H}_0$ denotes the quadratic part, and

$$U = -\frac{K}{8} \sum_{i,j,k,l} \int d^d x \left[(\nabla \pi_{ij}) \, \pi_{jk} \cdot (\nabla \pi_{kl}) \, \pi_{li} + \pi_{ij} \left(\nabla \pi_{jk} \right) \cdot (\nabla \pi_{kl}) \, \pi_{li} \right]$$

$$= \frac{K}{8} \sum_{i,j,k,l} \int \frac{d^d q_1 d^d q_2 d^d q_3}{(2\pi)^{3d}} \left[(\mathbf{q}_1 \cdot \mathbf{q}_3 + \mathbf{q}_2 \cdot \mathbf{q}_3) \cdot \right.$$

$$\left. \cdot \pi_{ij} (\mathbf{q}_1) \, \pi_{jk} (\mathbf{q}_2) \, \pi_{kl} (\mathbf{q}_3) \, \pi_{li} (-\mathbf{q}_1 - \mathbf{q}_2 - \mathbf{q}_3) \right].$$

To first order in U, the following two averages contribute to the renormalization of K:

$$(i) \; \frac{K}{8} \sum_{i,j,k,l} \int \frac{d^d q_1 d^d q_2 d^d q_3}{(2\pi)^{3d}} \left\langle \pi_{jk}^> (\mathbf{q}_2) \, \pi_{li}^> (-\mathbf{q}_1 - \mathbf{q}_2 - \mathbf{q}_3) \right\rangle_0^> (\mathbf{q}_1 \cdot \mathbf{q}_3) \, \pi_{ij}^< (\mathbf{q}_1) \, \pi_{kl}^< (\mathbf{q}_3)$$

$$= \frac{K}{8} \left(\int_{\Lambda/b}^{\Lambda} \frac{d^d q'}{(2\pi)^d} \frac{1}{Kq'^2} \right) \left(\int_0^{\Lambda/b} \frac{d^d q}{(2\pi)^d} q^2 \sum_{i,j} \pi_{ij}^< (\mathbf{q}) \, \pi_{ji}^< (-\mathbf{q}) \right),$$

and

$$(ii) \; \frac{K}{8} \sum_{j,k,l} \int \frac{d^d q_1 d^d q_2 d^d q_3}{(2\pi)^{3d}} \left\langle \sum_{i \neq j,l} \pi_{ij}^> (\mathbf{q}_2) \, \pi_{li}^> (-\mathbf{q}_1 - \mathbf{q}_2 - \mathbf{q}_3) \right\rangle_0^>$$

$$(\mathbf{q}_2 \cdot \mathbf{q}_3) \, \pi_{jk}^< (\mathbf{q}_2) \, \pi_{kl}^< (\mathbf{q}_3)$$

$$= \frac{K}{8} \left[(N-1) \int_{\Lambda/b}^{\Lambda} \frac{d^d q'}{(2\pi)^d} \frac{1}{Kq'^2} \right] \left(\int_0^{\Lambda/b} \frac{d^d q}{(2\pi)^d} q^2 \sum_{j,k} \pi_{jk}^< (\mathbf{q}) \, \pi_{kj}^< (-\mathbf{q}) \right).$$

Adding up the two contributions results in an effective coupling

$$\frac{\tilde{K}}{4} = \frac{K}{4} + \frac{K}{8} N \int_{\Lambda/b}^{\Lambda} \frac{d^d q}{(2\pi)^d} \frac{1}{Kq^2}, \quad i.e. \quad \tilde{K} = K + \frac{N}{2} I_d (b).$$

(h) Using the result from part (f), show that after matrix rescaling, the RG equation for K' is given by:

$$K' = b^{d-2} \left[K - \frac{N-2}{2} \int_{\Lambda/b}^{\Lambda} \frac{d^d \mathbf{q}}{(2\pi)^d} \frac{1}{q^2} \right].$$

• *After coarse-graining, renormalizing the fields, and rescaling,*

$$K' = b^{d-2} \zeta^2 \tilde{K} = b^{d-2} \left[1 - \frac{N-1}{K} I_d (b) \right] K \left[1 + \frac{N}{2K} I_d (b) \right]$$

$$= b^{d-2} \left[K - \frac{N-2}{2} I_d (b) + \mathcal{O} (1/K) \right],$$

i.e. to lowest non-trivial order,

$$K' = b^{d-2} \left[K - \frac{N-2}{2} \int_{\Lambda/b}^{\Lambda} \frac{d^d q}{(2\pi)^d} \frac{1}{q^2} \right].$$

(i) Obtain the *differential* RG equation for $T = 1/K$, by considering $b = 1 + \delta\ell$. Sketch the flows for $d < 2$ and $d = 2$. For $d = 2 + \epsilon$, compute T_c and the critical exponent ν.

- *Differential recursion relations are obtained for infinitesimal $b = 1 + \delta\ell$, as*

$$K' = K + \frac{dK}{d\ell}\delta\ell = [1 + (d-2)\,\delta\ell]\left[K - \frac{N-2}{2}K_d\Lambda^{d-2}\delta\ell\right],$$

leading to

$$\frac{dK}{d\ell} = (d-2)\,K - \frac{N-2}{2}K_d\Lambda^{d-2}.$$

To obtain the corresponding equation for $T = 1/K$, we divide the above relation by $-K^2$, to get

$$\frac{dT}{d\ell} = (2-d)\,T + \frac{N-2}{2}K_d\Lambda^{d-2}T^2.$$

For $d < 2$, we have the two usual trivial fixed points: 0 (unstable) and ∞ (stable). The system is mapped onto higher temperatures by coarse graining. The same applies for the case $d = 2$ and $N > 2$.

For $d > 2$, both 0 and ∞ are stable, and a non-trivial unstable fixed point appears at a finite temperature given by $dT/d\ell = 0$, i.e.

$$T^* = \frac{2(d-2)}{(N-2)\,K_d\Lambda^{d-2}} = \frac{4\pi\epsilon}{N-2} + \mathcal{O}\left(\epsilon^2\right).$$

In the vicinity of the fixed point, the flows are described by

$$\delta T' = \left[1 + \frac{d}{dT}\left(\frac{dT}{d\ell}\right)\Bigg|_{T^*}\delta\ell\right]\delta T$$
$$= \left\{1 + \left[(2-d) + (N-2)\,K_d\Lambda^{d-2}T^*\right]\delta\ell\right\}\delta T = (1 + \epsilon\delta\ell)\,\delta T.$$

Thus, from

$$\delta T' = b^{y_T}\delta T = (1 + y_T\delta\ell)\,\delta T,$$

we get $y_T = \epsilon$, and

$$\nu = \frac{1}{\epsilon}.$$

(j) Consider a small symmetry breaking term $-\int d^d\mathbf{x}\ tr\,(hM)$, added to the Hamiltonian. Find the renormalization of h, and identify the corresponding exponent y_h.

- *As usual, h renormalizes according to*

$$h' = b^d \zeta h = (1 + d\delta\ell) \left(K - \frac{N-1}{2K} K_d \Lambda^{d-2} \delta\ell \right) h$$

$$= \left[1 + \left(d - \frac{N-1}{2K} K_d \Lambda^{d-2} \right) \delta\ell + \mathcal{O}\left(\delta\ell^2\right) \right] h.$$

From $h' = b^{y_h} h = (1 + y_h \delta\ell) h$, we obtain

$$y_h = d - \frac{N-1}{2K^*} K_d \Lambda^{d-2} = d - \frac{N-1}{N-2}(d-2) = 2 - \frac{\epsilon}{N-2} + \mathcal{O}\left(\epsilon^2\right).$$

*Combining RG and symmetry arguments, it can be shown that the 3×3 matrix model is perturbatively equivalent to the $N = 4$ vector model at all orders. This would suggest that stacked triangular antiferromagnets provide a realization of the $\mathcal{O}(4)$ universality class; see P. Azaria, B. Delamotte, and T. Jolicoeur, J. Appl. Phys. **69**, 6170 (1991). However, non-perturbative (topological) aspects appear to remove this equivalence as discussed in S.V. Isakov, T. Senthil, and Y.B. Kim, Phys. Rev. B **72**, 174417 (2005).*

3. *The roughening transition:* In an earlier problem we examined a continuum interface model which in $d = 3$ is described by the Hamiltonian

$$\beta \mathcal{H}_0 = \frac{K}{2} \int d^2\mathbf{x} \, (\nabla h)^2,$$

where $h(\mathbf{x})$ is the interface height at location $\mathbf{x}$. For a crystalline facet, the allowed values of h are multiples of the lattice spacing. In the continuum, this tendency for integer h can be mimicked by adding a term

$$-\beta U = y_0 \int d^2\mathbf{x} \cos(2\pi h),$$

to the Hamiltonian. Treat $-\beta U$ as a perturbation, and proceed to construct a renormalization group as follows:

(a) Show that

$$\left\langle \exp\left[i \sum_\alpha q_\alpha h(\mathbf{x}_\alpha) \right] \right\rangle_0 = \exp\left[\frac{1}{K} \sum_{\alpha < \beta} q_\alpha q_\beta C(\mathbf{x}_\alpha - \mathbf{x}_\beta) \right]$$

for $\sum_\alpha q_\alpha = 0$, and zero otherwise. ($C(\mathbf{x}) = \ln|\mathbf{x}|/2\pi$ is the Coulomb interaction in two dimensions.)

- *The translational invariance of the Hamiltonian constrains $\langle \exp[i \sum_\alpha q_\alpha h(\mathbf{x}_\alpha)] \rangle_0$ to vanish unless $\sum_\alpha q_\alpha = 0$, as implied by the following relation*

$$\exp\left(i\delta \sum_\alpha q_\alpha \right) \left\langle \exp\left[i \sum_\alpha q_\alpha h(\mathbf{x}_\alpha) \right] \right\rangle_0 = \left\langle \exp\left\{ i \sum_\alpha q_\alpha [h(\mathbf{x}_\alpha) + \delta] \right\} \right\rangle_0$$

$$= \left\langle \exp\left[i \sum_\alpha q_\alpha h(\mathbf{x}_\alpha) \right] \right\rangle_0.$$

The last equality follows from the symmetry $\mathcal{H}[h(\mathbf{x}) + \delta] = \mathcal{H}[h(\mathbf{x})]$. Using general properties of Gaussian averages, we can set

$$\left\langle \exp\left[i\sum_\alpha q_\alpha h(\mathbf{x}_\alpha) \right] \right\rangle_0 = \exp\left[-\frac{1}{2} \sum_{\alpha\beta} q_\alpha q_\beta \langle h(\mathbf{x}_\alpha) h(\mathbf{x}_\beta) \rangle_0 \right]$$

$$= \exp\left[\frac{1}{4} \sum_{\alpha\beta} q_\alpha q_\beta \left\langle \left(h(\mathbf{x}_\alpha) - h(\mathbf{x}_\beta) \right)^2 \right\rangle_0 \right].$$

Note that the quantity $\langle h(\mathbf{x}_\alpha) h(\mathbf{x}_\beta) \rangle_0$ is ambiguous because of the symmetry $h(\mathbf{x}) \to h(\mathbf{x}) + \delta$. When $\sum_\alpha q_\alpha = 0$, we can replace this quantity in the above sum with the height difference $\left\langle \left(h(\mathbf{x}_\alpha) - h(\mathbf{x}_\beta) \right)^2 \right\rangle_0$ which is independent of this symmetry. (The ambiguity, or symmetry, results from the kernel of the quadratic form having a zero eigenvalue, which means that inverting it requires care.) We can now proceed as usual, and

$$\left\langle \exp\left[i\sum_\alpha q_\alpha h(\mathbf{x}_\alpha) \right] \right\rangle_0 = \exp\left[\sum_{\alpha,\beta} \frac{q_\alpha q_\beta}{4} \int \frac{d^2 q}{(2\pi)^2} \frac{\left(e^{i\mathbf{q}\cdot\mathbf{x}_\alpha} - e^{i\mathbf{q}\cdot\mathbf{x}_\beta} \right)\left(e^{-i\mathbf{q}\cdot\mathbf{x}_\alpha} - e^{-i\mathbf{q}\cdot\mathbf{x}_\beta} \right)}{Kq^2} \right]$$

$$= \exp\left[\sum_{\alpha<\beta} q_\alpha q_\beta \int \frac{d^2 q}{(2\pi)^2} \frac{1 - \cos\left(\mathbf{q}\cdot(\mathbf{x}_\alpha - \mathbf{x}_\beta) \right)}{Kq^2} \right]$$

$$= \exp\left[\frac{1}{K} \sum_{\alpha<\beta} q_\alpha q_\beta C(\mathbf{x}_\alpha - \mathbf{x}_\beta) \right],$$

where

$$C(\mathbf{x}) = \int \frac{d^2 q}{(2\pi)^2} \frac{1 - \cos(\mathbf{q}\cdot\mathbf{x})}{q^2} = \frac{1}{2\pi} \ln\frac{|\mathbf{x}|}{a}$$

is the Coulomb interaction in two dimensions, with a short distance cutoff a.

(b) Prove that

$$\left\langle |h(\mathbf{x}) - h(\mathbf{y})|^2 \right\rangle = -\left. \frac{d^2}{dk^2} G_k(\mathbf{x} - \mathbf{y}) \right|_{k=0},$$

where $G_k(\mathbf{x} - \mathbf{y}) = \left\langle \exp\left[ik\left(h(\mathbf{x}) - h(\mathbf{y}) \right) \right] \right\rangle$.

• *From the definition of $G_k(\mathbf{x} - \mathbf{y})$,*

$$\frac{d^2}{dk^2} G_k(\mathbf{x} - \mathbf{y}) = -\left\langle \left[h(\mathbf{x}) - h(\mathbf{y}) \right]^2 \exp\left[ik(h(\mathbf{x}) - h(\mathbf{y})) \right] \right\rangle.$$

Setting k to zero results in the identity

$$\left\langle \left[h(\mathbf{x}) - h(\mathbf{y}) \right]^2 \right\rangle = -\left. \frac{d^2}{dk^2} G_k(\mathbf{x} - \mathbf{y}) \right|_{k=0}.$$

(c) Use the results in (a) to calculate $G_k(\mathbf{x} - \mathbf{y})$ in perturbation theory to order of y_0^2. (Hint: Set $\cos(2\pi h) = \left(e^{2i\pi h} + e^{-2i\pi h} \right)/2$. The first order terms vanish according to the result in (a), while the second order contribution is identical in structure to that of the Coulomb gas described in this chapter.)

- *Following the hint, we write the perturbation as*

$$-U = y_0 \int d^2x \cos(2\pi h) = \frac{y_0}{2} \int d^2x \left[e^{2i\pi h} + e^{-2i\pi h} \right].$$

The perturbation expansion for $G_k(\mathbf{x}-\mathbf{y}) = \langle \exp[ik(h(\mathbf{x})-h(\mathbf{y}))] \rangle \equiv \langle \mathcal{G}_k(\mathbf{x}-\mathbf{y}) \rangle$ *is calculated as*

$$\langle \mathcal{G}_k \rangle = \langle \mathcal{G}_k \rangle_0 - (\langle \mathcal{G}_k U \rangle_0 - \langle \mathcal{G}_k \rangle_0 \langle U \rangle_0)$$
$$+ \frac{1}{2} \left(\langle \mathcal{G}_k U^2 \rangle_0 - 2 \langle \mathcal{G}_k U \rangle_0 \langle U \rangle_0 + 2 \langle \mathcal{G}_k \rangle_0 \langle U \rangle_0^2 - \langle \mathcal{G}_k \rangle_0 \langle U^2 \rangle_0 \right) + \mathcal{O}(U^3).$$

From part (a),

$$\langle U \rangle_0 = \langle \mathcal{G}_k U \rangle_0 = 0,$$

and

$$\langle \mathcal{G}_k \rangle_0 = \exp\left[-\frac{k^2}{K} C(\mathbf{x}-\mathbf{y}) \right] = \left(\frac{|\mathbf{x}-\mathbf{y}|}{a} \right)^{-\frac{k^2}{2\pi K}}.$$

Furthermore,

$$\langle U^2 \rangle_0 = \frac{y_0^2}{2} \int d^2x' d^2x'' \langle \exp[2i\pi(h(\mathbf{x}')-h(\mathbf{x}''))] \rangle_0$$
$$= \frac{y_0^2}{2} \int d^2x' d^2x'' \langle \mathcal{G}_{2\pi}(\mathbf{x}'-\mathbf{x}'') \rangle_0$$
$$= \frac{y_0^2}{2} \int d^2x' d^2x'' \exp\left[-\frac{(2\pi)^2}{K} C(\mathbf{x}'-\mathbf{x}'') \right],$$

and similarly,

$$\langle \exp[ik(h(\mathbf{x})-h(\mathbf{y}))] U^2 \rangle_0$$
$$= \frac{y_0^2}{2} \int d^2x' d^2x'' \exp\left\{ -\frac{k^2}{K} C(\mathbf{x}-\mathbf{y}) - \frac{(2\pi)^2}{K} C(\mathbf{x}'-\mathbf{x}'') \right.$$
$$\left. + \frac{2\pi k}{K} [C(\mathbf{x}-\mathbf{x}')+C(\mathbf{y}-\mathbf{x}'')] - \frac{2\pi k}{K} [C(\mathbf{x}-\mathbf{x}'')+C(\mathbf{y}-\mathbf{x}')] \right\}.$$

Thus, the second order part of $G_k(\mathbf{x}-\mathbf{y})$ *is*

$$\frac{y_0^2}{4} \exp\left[-\frac{k^2}{K} C(\mathbf{x}-\mathbf{y}) \right] \int d^2x' d^2x'' \exp\left[-\frac{4\pi^2}{K} C(\mathbf{x}'-\mathbf{x}'') \right] \cdot$$
$$\cdot \left\{ \exp\left[\frac{2\pi k}{K} (C(\mathbf{x}-\mathbf{x}')+C(\mathbf{y}-\mathbf{x}'')-C(\mathbf{x}-\mathbf{x}'')-C(\mathbf{y}-\mathbf{x}')) \right] - 1 \right\},$$

and

$$G_k(\mathbf{x}-\mathbf{y}) = e^{-\frac{k^2}{K} C(\mathbf{x}-\mathbf{y})} \left\{ 1 + \frac{y_0^2}{4} \int d^2x' d^2x'' e^{-\frac{4\pi^2}{K} C(\mathbf{x}'-\mathbf{x}'')} \left(e^{\frac{2\pi k}{K}\mathcal{D}} - 1 \right) + \mathcal{O}(y_0^4) \right\},$$

where

$$\mathcal{D} = C(\mathbf{x} - \mathbf{x}') + C(\mathbf{y} - \mathbf{x}'') - C(\mathbf{x} - \mathbf{x}'') - C(\mathbf{y} - \mathbf{x}').$$

(d) Write the perturbation result in terms of an effective interaction K, and show that perturbation theory fails for K larger than a critical K_c.

- *The above expression for $G_k(\mathbf{x} - \mathbf{y})$ is very similar to that obtained in dealing with the renormalization of the Coulomb gas of vortices in the XY model. Following the steps in the text, without further calculations, we find*

$$G_k(\mathbf{x} - \mathbf{y}) = e^{-\frac{k^2}{K}C(\mathbf{x} - \mathbf{y})}\left\{1 + \frac{y_0^2}{4} \times \frac{1}{2}\left(\frac{2\pi k}{K}\right)^2 \times C(\mathbf{x} - \mathbf{y}) \times 2\pi \int dr r^3 e^{-\frac{2\pi \ln(r/a)}{K}}\right\}$$

$$= e^{-\frac{k^2}{K}C(\mathbf{x} - \mathbf{y})}\left\{1 + \frac{\pi^3 k^2}{K^2}y_0^2 C(\mathbf{x} - \mathbf{y})\int dr r^3 e^{-\frac{2\pi \ln(r/a)}{K}}\right\}.$$

The second order term can be exponentiated to contribute to an effective coupling constant K_{eff}, according to

$$\frac{1}{K_{\text{eff}}} = \frac{1}{K} - \frac{\pi^3}{K^2}a^{2\pi/K}y_0^2\int_a^\infty dr r^{3 - 2\pi/K}.$$

Clearly, the perturbation theory is inconsistent if the above integral diverges, i.e. if

$$K > \frac{\pi}{2} \equiv K_c.$$

(e) Recast the perturbation result in part (d) into renormalization group equations for K and y_0, by changing the "lattice spacing" from a to ae^ℓ.

- *After dividing the integral into two parts, from a to ab and from ab to ∞, respectively, and rescaling the variable of integration in the second part, in order to retrieve the usual limits of integration, we have*

$$\frac{1}{K_{\text{eff}}} = \frac{1}{K} - \frac{\pi^3}{K^2}a^{2\pi/K}y_0^2\int_a^{ab} dr r^{3 - 2\pi/K} - \frac{\pi^3}{K^2}a^{2\pi/K} \times y_0^2 b^{4 - 2\pi/K} \times \int_a^\infty dr r^{3 - 2\pi/K}.$$

(To order y_0^2, we can indifferently write K or K' (defined below) in the last term.) In other words, the coarse-grained system is described by an interaction identical in form, but parameterized by the renormalized quantities

$$\frac{1}{K'} = \frac{1}{K} - \frac{\pi^3}{K^2}a^{2\pi/K}y_0^2\int_a^{ab} dr r^{3 - 2\pi/K},$$

and

$${y_0'}^2 = b^{4 - 2\pi/K}y_0^2.$$

With $b = e^{\ell} \approx 1 + \ell$, these RG relations are written as the following differential equations, which describe the renormalization group flows

$$\begin{cases} \dfrac{dK}{d\ell} = \pi^3 a^4 y_0^2 + \mathcal{O}\left(y_0^4\right) \\ \dfrac{dy_0}{d\ell} = \left(2 - \dfrac{\pi}{K}\right) y_0 + \mathcal{O}\left(y_0^3\right). \end{cases}$$

(f) Using the recursion relations, discuss the phase diagram and phases of this model.

- *These RG equations are similar to those of the XY model, with K (here) playing the role of T in the Coulomb gas. For non-vanishing y_0, K is relevant, and thus flows to larger and larger values (outside of the perturbative domain) if y_0 is also relevant ($K > \pi/2$), suggesting a smooth phase at low temperatures ($T \sim K^{-1}$). At small values of K, y_0 is irrelevant, and the flows terminate on a fixed line with $y_0 = 0$ and $K \leq \pi/2$, corresponding to a rough phase at high temperatures.*

(g) For large separations $|\mathbf{x} - \mathbf{y}|$, find the magnitude of the discontinuous jump in $\langle |h(\mathbf{x}) - h(\mathbf{y})|^2 \rangle$ at the transition.

- *We want to calculate the long distance correlations in the vicinity of the transition. Equivalently, we can compute the coarse grained correlations. If the system is prepared at $K = \pi/2^-$ and $y_0 \approx 0$, under coarse graining, $K \to \pi/2^-$ and $y_0 \to 0$, resulting in*

$$G_k(\mathbf{x} - \mathbf{y}) \to \langle \mathcal{G}_k \rangle_0 = \exp\left[-\frac{2k^2}{\pi} C(\mathbf{x} - \mathbf{y})\right].$$

From part (b),

$$\left\langle [h(\mathbf{x}) - h(\mathbf{y})]^2 \right\rangle = -\left.\frac{d^2}{dk^2} G_k(\mathbf{x} - \mathbf{y})\right|_{k=0} = \frac{4}{\pi} C(\mathbf{x} - \mathbf{y}) = \frac{2}{\pi^2} \ln|\mathbf{x} - \mathbf{y}|.$$

On the other hand, if the system is prepared at $K = \pi/2^+$, then $K \to \infty$ under the RG (assuming that the relevance of K holds also away from the perturbative regime), and

$$\left\langle [h(\mathbf{x}) - h(\mathbf{y})]^2 \right\rangle \to 0.$$

Thus, the magnitude of the jump in $\left\langle [h(\mathbf{x}) - h(\mathbf{y})]^2 \right\rangle$ at the transition is

$$\frac{2}{\pi^2} \ln|\mathbf{x} - \mathbf{y}|.$$

4. *Roughening and duality:* Consider a discretized version of the Hamiltonian in the previous problem, in which for each site i of a square lattice there is an integer valued height h_i. The Hamiltonian is

$$\beta \mathcal{H} = \frac{K}{2} \sum_{\langle i,j \rangle} |h_i - h_j|^\infty,$$

where the "∞" power means that there is no energy cost for $\Delta h = 0$; an energy cost of $K/2$ for $\Delta h = \pm 1$ and $\Delta h = \pm 2$ or higher *are not allowed* for neighboring sites. (This is known as the restricted solid on solid (RSOS) model.)

(a) Construct the dual model either diagrammatically, or by following these steps:

(i) Change from the N site variables h_i, to the $2N$ bond variables $n_{ij} = h_i - h_j$. Show that the sum of n_{ij} around any plaquette is constrained to be zero.

(ii) Impose the constraints by using the identity $\int_0^{2\pi} d\theta e^{i\theta n}/2\pi = \delta_{n,0}$, for integer n.

(iii) After imposing the constraints, you can sum freely over the bond variables n_{ij} to obtain a dual interaction $\tilde{v}(\theta_i - \theta_j)$ between dual variables θ_i on neighboring plaquettes.

• *(i) In terms of bond variables $n_{ij} = h_i - h_j$, the Hamiltonian is written as*

$$-\beta \mathcal{H} = -\frac{K}{2} \sum_{\langle ij \rangle} |n_{ij}|^{\infty}.$$

Clearly,

$$\sum_{\substack{\text{any closed loop}}} n_{ij} = h_{i_1} - h_{i_2} + h_{i_2} - h_{i_3} + \cdots + h_{i_{n-1}} - h_{i_n} = 0,$$

since $h_{i_1} = h_{i_n}$ for a closed path.

(ii) This constraint, applied to the N plaquettes, reduces the number of degrees of freedom from an apparent $2N$ (bonds), to the correct figure N, and the partition function becomes

$$Z = \sum_{\{n_{ij}\}} e^{-\beta \mathcal{H}} \prod_{\alpha} \delta_{\sum_{\langle ij \rangle} n_{ij}^{\alpha}, 0},$$

where the index α labels the N plaquettes, and n_{ij}^{α} is non-zero and equal to n_{ij} only if the bond $\langle ij \rangle$ belongs to plaquette α. Expressing the Kronecker delta in its exponential representation, we get

$$Z = \sum_{\{n_{ij}\}} e^{-\frac{K}{2} \sum_{\langle ij \rangle} |n_{ij}|^{\infty}} \prod_{\alpha} \left(\int_0^{2\pi} \frac{d\theta_{\alpha}}{2\pi} e^{i\theta_{\alpha} \sum_{\langle ij \rangle} n_{ij}^{\alpha}} \right).$$

(iii) As each bond belongs to two neighboring plaquettes, we can label the bonds by $\alpha\beta$ rather than ij, leading to

$$Z = \left(\prod_{\gamma} \int_0^{2\pi} \frac{d\theta_{\gamma}}{2\pi} \right) \sum_{\{n_{\alpha\beta}\}} \exp \left(\sum_{\langle \alpha\beta \rangle} \left\{ -\frac{K}{2} |n_{\alpha\beta}|^{\infty} + i (\theta_{\alpha} - \theta_{\beta}) n_{\alpha\beta} \right\} \right)$$

$$= \left(\prod_{\gamma} \int_0^{2\pi} \frac{d\theta_{\gamma}}{2\pi} \right) \prod_{\langle \alpha\beta \rangle} \sum_{n_{\alpha\beta}} \exp \left(\left\{ -\frac{K}{2} |n_{\alpha\beta}|^{\infty} + i (\theta_{\alpha} - \theta_{\beta}) n_{\alpha\beta} \right\} \right).$$

•

Note that if all plaquettes are traversed in the same sense, the variable $n_{\alpha\beta}$ occurs in opposite senses (with opposite signs) for the constraint variables θ_α and θ_β on neighboring plaquettes. We can now sum freely over the bond variables, and since

$$\sum_{n=0,+1,-1} \exp\left(-\frac{K}{2}|n| + i\left(\theta_\alpha - \theta_\beta\right)n\right) = 1 + 2e^{-\frac{K}{2}}\cos\left(\theta_\alpha - \theta_\beta\right),$$

we obtain

$$Z = \left(\prod_\gamma \int_0^{2\pi} \frac{d\theta_\gamma}{2\pi}\right) \exp\left(\sum_{\langle\alpha\beta\rangle} \ln\left[1 + 2e^{-\frac{K}{2}}\cos\left(\theta_\alpha - \theta_\beta\right)\right]\right).$$

(b) Show that for large K, the dual problem is just the XY model. Is this conclusion consistent with the renormalization group results of the previous problem? (Also note the connection with the loop model considered in the problems of the previous chapter.)

• *This is the loop gas model introduced earlier. For K large,*

$$\ln\left[1 + 2e^{-\frac{K}{2}}\cos\left(\theta_\alpha - \theta_\beta\right)\right] \approx 2e^{-\frac{K}{2}}\cos\left(\theta_\alpha - \theta_\beta\right),$$

and

$$Z = \left(\prod_\gamma \int_0^{2\pi} \frac{d\theta_\gamma}{2\pi}\right) e^{\sum_{\langle\alpha\beta\rangle} 2e^{-\frac{K}{2}}\cos\left(\theta_\alpha - \theta_\beta\right)}.$$

This is none other than the partition function for the XY model, if we identify

$$K_{XY} = 4e^{-\frac{K}{2}},$$

consistent with the results of the previous problem in which we found that the low temperature behavior in the roughening problem corresponds to the high temperature phase in the XY model, and vice versa.

(c) Does the one dimensional version of this Hamiltonian, i.e. a 2d interface with

$$-\beta\mathcal{H} = -\frac{K}{2}\sum_i |h_i - h_{i+1}|^\infty,$$

have a roughening transition?

• *In one dimension, we can directly sum the partition function, as*

$$Z = \sum_{\{h_i\}} \exp\left(-\frac{K}{2}\sum_i |h_i - h_{i+1}|^\infty\right) = \sum_{\{n_i\}} \exp\left(-\frac{K}{2}\sum_i |n_i|^\infty\right)$$

$$= \prod_i \sum_{n_i} \exp\left(-\frac{K}{2}|n_i|^\infty\right) = \prod_i \left(1 + 2e^{-K/2}\right) = \left(1 + 2e^{-K/2}\right)^N,$$

($n_i = h_i - h_{i+1}$). The expression thus obtained is an analytic function of K (for $0 < K < \infty$), in the $N \to \infty$ limit, and there is therefore no phase transition at a finite non-zero temperature.

Index

Printed in the United States
By Bookmasters